Advances in Intelligent Systems and Computing

Volume 1164

The series "Advances in Intelligent Systems and Computing" contains publications on theory, applications, and design methods of Intelligent Systems and Intelligent Computing. Virtually all disciplines such as engineering, natural sciences, computer and information science, ICT, economics, business, e-commerce, environment, healthcare, life science are covered. The list of topics spans all the areas of modern intelligent systems and computing such as: computational intelligence, soft computing including neural networks, fuzzy systems, evolutionary computing and the fusion of these paradigms, social intelligence, ambient intelligence, computational neuroscience, artificial life, virtual worlds and society, cognitive science and systems, Perception and Vision, DNA and immune based systems, self-organizing and adaptive systems, e-Learning and teaching, human-centered and human-centric computing, recommender systems, intelligent control, robotics and mechatronics including human-machine teaming, knowledge-based paradigms, learning paradigms, machine ethics, intelligent data analysis, knowledge management, intelligent agents, intelligent decision making and support, intelligent network security, trust management, interactive entertainment, Web intelligence and multimedia.

The publications within "Advances in Intelligent Systems and Computing" are primarily proceedings of important conferences, symposia and congresses. They cover significant recent developments in the field, both of a foundational and applicable character. An important characteristic feature of the series is the short publication time and world-wide distribution. This permits a rapid and broad dissemination of research results.

**** Indexing: The books of this series are submitted to ISI Proceedings, EI-Compendex, DBLP, SCOPUS, Google Scholar and Springerlink ****

More information about this series at http://www.springer.com/series/11156

Poonam Bansal · Meena Tushir ·
Valentina Emilia Balas · Rajeev Srivastava
Editors

Proceedings of International Conference on Artificial Intelligence and Applications

ICAIA 2020

Springer

Editors
Poonam Bansal
Department of Computer Science
and Engineering
Maharaja Surajmal Institute of Technology
New Delhi, India

Meena Tushir
Department of Electrical
and Electronics Engineering
Maharaja Surajmal Institute of Technology
New Delhi, India

Valentina Emilia Balas
Department of Automation
and Applied Software
Aurel Vlaicu University of Arad
Arad, Romania

Rajeev Srivastava
Department of Computer Science
and Engineering
Indian Institute of Technology (BHU)
Varanasi, India

ISSN 2194-5357 ISSN 2194-5365 (electronic)
Advances in Intelligent Systems and Computing
ISBN 978-981-15-4991-5 ISBN 978-981-15-4992-2 (eBook)
https://doi.org/10.1007/978-981-15-4992-2

This Springer imprint is published by the registered company Springer Nature Singapore Pte Ltd.
The registered company address is: 152 Beach Road, #21-01/04 Gateway East, Singapore 189721, Singapore

Organization

ICAIA 2020 was organized at Maharaja Surajmal Institute of Technology, Janak Puri, New Delhi, India, during February 6–7, 2020.

ICAIA 2020 Organizing Committee

Patrons

Sh. Kaptan Singh, President, Surajmal Memorial Education Society, New Delhi
Dr. M. P. Poonia, Vice Chairman, AICTE, New Delhi
Prof. Prem Vrat, Pro-Chancellor, North Cap University, Gurugram
Mr. Karnal Singh, IPS, Chief Enforcement Directorate
Ms. Esha Jakhar, Vice President, Surajmal Memorial Education Society, New Delhi
Mr. Ajit Singh Chaudhary, Secretary, Surajmal Memorial Education Society, New Delhi
Mr. Raj Pal Solanki, Treasurer, Surajmal Memorial Education Society, New Delhi

General Chairs

Prof. S. K. Garg, Pro-Vice Chancellor, Delhi Technological University, New Delhi
Mr. Sachin Gaur, Local Coordinator, India EU IST Standards and Collaboration

Organizing Chair

Prof. K. P. Chaudhary, Director and Professor, Electrical and Electronics Engineering Department, Maharaja Surajmal Institute of Technology, New Delhi

Convener

Prof. Poonam Bansal, Dy. Director and Professor, Computer Science and Engineering Department, Maharaja Surajmal Institute of Technology, New Delhi

Co-conveners

Prof. Meena Tushir, Head, Electrical and Electronics Engineering Department, Maharaja Surajmal Institute of Technology, New Delhi
Dr. Naveen Dahiya, Head, Computer Science and Engineering Department, Maharaja Surajmal Institute of Technology, New Delhi

Core Committee Members

Dr. Puneet Azad, Head, Electronics and Communication Engineering Department, Maharaja Surajmal Institute of Technology, New Delhi
Prof. Archana Balyan, Professor, Electronics and Communication Engineering Department, Maharaja Surajmal Institute of Technology, New Delhi
Dr. Prabhjot Kaur, Associate Professor, Information Technology Department, Maharaja Surajmal Institute of Technology, New Delhi
Dr. Rinky Dwivedi, Associate Professor, Computer Science and Engineering Department, Maharaja Surajmal Institute of Technology, New Delhi
Dr. Yogendra Arya, Assistant Professor, Electrical and Electronics Engineering Department, Maharaja Surajmal Institute of Technology, New Delhi
Dr. Savita Ahlawat, Reader, Computer Science and Engineering Department, Maharaja Surajmal Institute of Technology, New Delhi
Dr. Sunil Gupta, Associate Professor, Electrical and Electronics Engineering Department, Maharaja Surajmal Institute of Technology, New Delhi
Dr. Amita Yadav, Reader, Computer Science and Engineering Department, Maharaja Surajmal Institute of Technology, New Delhi
Mr. Aashish Sobti, Accounts Officer, Maharaja Surajmal Institute of Technology, New Delhi
Dr. Naresh Kumar, Associate Professor, Computer Science and Engineering Department, Maharaja Surajmal Institute of Technology, New Delhi
Dr. Adeel Hashmi, Assistant Professor, Computer Science and Engineering Department, Maharaja Surajmal Institute of Technology, New Delhi
Dr. Tripti Sharma, Head, Information Technology Department, Maharaja Surajmal Institute of Technology, New Delhi
Dr. Rekha Tripathi, Head, Applied Science Department, Maharaja Surajmal Institute of Technology, New Delhi
Dr. Koyel Datta Gupta, Head, Computer Science and Engineering Department, Maharaja Surajmal Institute of Technology, New Delhi
Dr. Sapna Malik, Assistant Professor, Computer Science and Engineering Department, Maharaja Surajmal Institute of Technology, New Delhi
Dr. Kavita Sheoran, Reader, Computer Science and Engineering Department, Maharaja Surajmal Institute of Technology, New Delhi
Mr. Sitender Malik, Assistant Professor, Information Technology Department, Maharaja Surajmal Institute of Technology, New Delhi
Dr. Bharti Sharma, Assistant Professor, Computer Science and Engineering Department, Maharaja Surajmal Institute of Technology, New Delhi

Ms. Pooja Kherwa, Assistant Professor, Computer Science and Engineering Department, Maharaja Surajmal Institute of Technology, New Delhi
Ms. Vimal Gaur, Reader, Computer Science and Engineering Department, Maharaja Surajmal Institute of Technology, New Delhi
Ms. Sonia Goel, Assistant Professor, Electrical and Electronics Engineering Department, Maharaja Surajmal Institute of Technology, New Delhi
Ms. Nistha Jatana, Assistant Professor, Computer Science and Engineering Department, Maharaja Surajmal Institute of Technology, New Delhi
Mr. Navdeep Bohra, Assistant Professor, Computer Science and Engineering Department, Maharaja Surajmal Institute of Technology, New Delhi
Ms. Rakhi Kamra, Assistant Professor, Electrical and Electronics Engineering Department, Maharaja Surajmal Institute of Technology, New Delhi
Ms. Shaily Malik, Assistant Professor, Computer Science and Engineering Department, Maharaja Surajmal Institute of Technology, New Delhi
Dr. Geetika Dhand, Reader, Computer Science and Engineering Department, Maharaja Surajmal Institute of Technology, New Delhi
Mr. Sachit Rathee, Assistant Professor, Electrical and Electronics Engineering Department, Maharaja Surajmal Institute of Technology, New Delhi
Ms. Poonam Dhankar, Assistant Professor, Computer Science and Engineering Department, Maharaja Surajmal Institute of Technology, New Delhi
Mr. Manoj Malik, Head, Information Technology Department, Maharaja Surajmal Institute of Technology, New Delhi
Mr. Ajay Gahlot, Head, Applied Science Department, Maharaja Surajmal Institute of Technology, New Delhi
Mr. Sandeep Singh, Assistant Professor, Electronics and Communication Engineering Department, Maharaja Surajmal Institute of Technology, New Delhi

Advisory Committee

Prof. Sheng-Lung Peng, National Dong Hwa University, Taiwan
Mr. Sachin Gaur, Local Coordinator, India EU ICT Standards Collaboration Project
Prof. Valentina Emilia Balas, Aurel Vlaicu University of Arad, Romania
Prof. Chuan-Ming Liu, National Taipei University of Technology, Taiwan
Prof. Vikram Kumar, Emeritus Prof. IITD and Ex-Dir., NPL, New Delhi
Prof. S. D. Joshi, IIT Delhi, New Delhi
Prof. A. P. Mittal, AICTE and Netaji Subhas University of Technology, New Delhi
Prof. R. C. Bansal, University of Sharjah, Sharjah, UAE
Prof. Shunsaku Hashimoto, University of the Ryukyus, Japan
Prof. Xavier Fernando, Ryerson University, Canada
Dr. Kamal Hossain, International Director, NPL, UK
Dr. Toshiyuki Takastsuj, Director, NMIJ, Japan
Dr. Yi Hua Tang, Physicist, NIST, USA
Dr. D. K. Aswal, Director, CSIR-NPL, New Delhi

Prof. D. P. Kothari, Ex. Director, IIT Delhi
Dr. Krishan Lal, President, INSA and Ex-Director, NPL, India
Mr. Anil Relia, Director, NABL, India
Prof. S. S. Agrawal, Director General, KIIT, Gurgaon
Prof. Yogesh Singh, VC, Delhi Technological University, New Delhi
Prof. S. K. Garg, Pro-VC, Delhi Technological University, New Delhi
Prof. Asok De, Professor, ECE Department, Delhi Technological University, New Delhi
Prof. Smriti Srivastava, Netaji Subhas Institute of Technology, New Delhi
Prof. R. K. Sinha, Director, CSIR-CSIO, Chandigarh, Punjab
Prof. Sanjeev Sofat, Dy. Director, PEC University of Technology, Chandigarh, Punjab
Prof. Amita Dev, Vice Chancellor, IGDTUW, New Delhi
Prof. Shantanu Chaudhary, Director, CSIR-CEERI, Rajasthan, India
Prof. Vishal Bhatnagar, AIACT&R, New Delhi
Prof. S. K. Dhurander, Netaji Subhas University of Technology, New Delhi
Prof. Komal Kumar Bhatia, YMCA, Faridabad
Dr. Ravinder Singla, Dean of University Instruction, Panjab University, India
Prof. Narendar Kumar, Professor, Delhi Technological University, New Delhi
Prof. Zaheeruddi, Jamia Millia Islamia, New Delhi
Prof. Sangeeta Sabharwal, Netaji Subhas University of Technology, New Delhi
Prof. T. S. Bhatti, IIT, New Delhi
Prof. Manjeet Singh, YMCA, Faridabad
Dr. Mohit Gambhir, Director Innovation Cell, MHRD, New Delhi
Prof. Arvinder Kaur, GGSIPU, New Delhi
Prof. Navin Rajpal, USICT, GGSIPU, New Delhi
Prof. C. S. Rai, USCIT, GGSIPU, New Delhi
Prof. B. V. R. Reddy, USICT, GGSIPU, New Delhi
Ms. Sanchita Malik, Scientist F, DRDO
Prof. A. Q. Ansari, Jamia Millia Islamia, New Delhi, India
Prof. R. K. Sahu, Veer Surendra Sai University of Technology, Burla, Odisha, India
Dr. H. D. Mathur, BITS, Pilani, India
Prof. Nidul Sinha, NIT, Silchar
Dr. L. C. Sakia, NIT, Silchar
Dr. R. J. Abraham, IISST, Kerala
Dr. N. K. Gupta, Chairman, ISTE Delhi Section

Technical Committee

Dr. Jasdeep Kaur, IGDTUW, New Delhi
Dr. Mandeep Mittal, AMITY, Noida
Dr. Sapna Gambhir, J. C. Bose, YMCA University of Science and Technology, Faridabad, Haryana

Dr. Saurabh Bhardwaj, Thapar Institute of Engineering and Technology, Punjab
Dr. Priyanka Jain, Delhi Technological University, New Delhi
Dr. Neeta Pandey, Delhi Technological University, New Delhi
Dr. Payal Pahwa, BPIT, New Delhi
Dr. D. K. Tayal, IGDTUW, New Delhi
Dr. Lalit Kumar, Shri Vishwakarma Skill University, Haryana
Dr. Nizamuddim, Noida Authority, Uttar Pradesh
Dr. Puja Dash, GVPCE, Visakhapatnam, Andhra Pradesh
Dr. Nikhil Pathak, Tsinghua University, Beijing, China
Dr. M. K. Debnath, Siksha 'O' Anusandhan University, Bhubaneswar, Odisha, India
Dr. C. K. Shiva, SR Engineering College, Warangal
Dr. Monika Gupta, MAIT, New Delhi
Dr. Pragya Varshney, Netaji Subhas University of Technology, New Delhi
Dr. Maîtreyee Dutta, NITTTR Chandigarh, Punjab
Dr. Anju Saha, USICT, GGSIPU, New Delhi
Dr. M. Bala Krishna, USICT, GGSIPU, New Delhi
Dr. Ravinder Kumar, HMRITM, New Delhi
Dr. Anil Ahlawat, KIIT, Ghaziabad, Uttar Pradesh
Dr. A. K. Mohapatra, IGDTUW, New Delhi
Dr. Rajesh Mehta, Thapar Institute of Engineering and Technology, Punjab
Prof. S. R. N. Reddy, IGDTUW, New Delhi
Dr. Tajinder Singh, NIT, Uttrakhand
Dr. Sujata Pandey, Amity School of Engineering and Technology, Noida, Uttar Pradesh
Dr. Manoj Kumar Pandey, Amity School of Engineering and Technology, Noida, Uttar Pradesh
Prof. Rekha Agarwal, Amity School of Engineering and Technology, Noida, Uttar Pradesh
Dr. Sanjay Kumar Malik, USICT, GGSIPU, New Delhi
Dr. Harish Kumar, UIET, Panjab University, Chandigarh
Dr. Hasmat Malik, Netaji Subhas University of Technology, New Delhi
Dr. V. P. Vishwakarma, GGSIPU, New Delhi
Prof. Zaheeruddin, Jamia Millia Islamia, New Delhi
Dr. I. Routray, C.V. Raman College of Engineering, Orissa
Prof. B. Pattnaik, Director, Amity School of Engineering and Technology, Kolkata
Dr. Pankaj Porwal, Techno India NJR Institute of Technology, Udaipur
Dr. Manoj Kumar, GGSIPU, New Delhi
Dr. Manosi Chaudhuri, BIMTECH, Noida, Uttar Pradesh
Ms. Neetu Bhagat, AICTE, New Delhi
Prof. Manisha Sharma, Bhilai Institute of Technology, Durg
Dr. Neelam Duhan, J. C. Bose, YMCA University of Science and Technology, Faridabad, Haryana
Dr. Manik Sharma, D.A.V. University, Punjab

Dr. Rajesh Jampala, P.B. Siddhartha College of Arts & Science, Siddhartha Nagar, Mogalrajpuram, Vijayawada
Dr. Urvashi Makkar, GLBIMR, Noida, Uttar Pradesh
Dr. Dipayan Guha, MNNIT, Allahabad

ICAIA 2020 Keynote Speaker

Dr. Sankar K. Pal, INSA Distinguished Professor Chair, Distinguished Scientist and Former Director, Indian Statistical Institute, Kolkata, India
Topic: **Granular Data Mining in Video Informatics**

ICAIA 2020 Invited Speakers

Dr. Toshiyuki Takatsuji, Director, Research Institute for Engineering Measurement, National Metrology Institute of Japan, Advanced Industrial Science and Technology, Japan
Topic: **Application of Information Technology for Metrology**

Dr. Vassil Vassilev, Reader in Artificial Intelligence and Cyber Security, London Metropolitan University, London, England
Topic: **AI on Demand Through Hybridization and Containerization**

Dr. Bhuvanesh Unhelkar, Ph.D., FACS, Associate Professor, University of South Florida Sarasota-Manatee, Florida, USA
Topic: **Challenges of Discovering "What to Ask?" of Big Data in Practice and the Role of Artificial Intelligence/Machine Learning in Identifying the Right Questions**

Prof. Smriti Srivastava, Professor, Instrumentation and Control Engineering, Netaji Subhas University of Technology, New Delhi, India
Topic: **Optimization Algorithms for Non-Linear Dynamical Systems**

Dr. Gaurav Gupta, Head, Department Chair School of Mathematical Sciences, College of Science and Technology, Wenzhou-Kean University, Wenzhou, China
Topic: **Exploratory Data Analysis (EDA) Using R Software**

Preface

The International Conference on "Artificial Intelligence and Applications" (ICAIA 2020) intended to provide an international forum for original research findings, as well as exchange and dissemination of innovative, practical development experiences in different fields of artificial intelligence. A major goal and feature of it were to bring academic scientists, engineers and industry researchers together to exchange and share their experiences and research results about most aspects of science and social research and discuss the practical challenges encountered and the solutions adopted.

The responses to the call for papers had been overwhelming—both from India and from overseas. ICAIA 2020 ensured to be both a stimulating and enlightening experience with numerous eminent keynote and invited speakers from all over the world. The event consisted of invited talks, technical sessions, paper presentations and discussions with eminent speakers covering a wide range of topics in artificial intelligence.

This book contains the research papers presented in the conference. Papers have been divided into the following tracks:

- Evolving machine learning and deep learning models for computer vision
- Machine learning applications in cyber security and cryptography
- Advances in signal processing and learning methods
- Social intelligence and sustainability
- Feature extraction and learning on image and speech data
- Optimization techniques and its applications in machine learning
- Recent Trends in Computational Intelligence and Data Science.

We express our sincere gratitude to the eminent keynote speakers, authors and the participants. Our earnest thanks to the Surajmal Memorial Education Society whose unconditional and munificent support has made the dream of hosting an international conference a reality. We would like to express our gratitude and appreciation for all the reviewers who helped us maintain the high quality of manuscripts included in the proceedings. We are grateful to Springer, especially to Mr. Aninda Bose (Senior Publishing Editor, Springer India Pvt. Ltd.), and the entire Springer team for the excellent collaboration, patience and help during the evolvement of this volume.

New Delhi, India

Organizing Committee
ICAIA 2020

Contents

About the Editors

Dr. Poonam Bansal is currently a Professor at the Computer Science & Engineering Department & Deputy Director of the Maharaja Surajmal Institute of Technology, New Delhi, India. She has 30 years of teaching, industry and research experience. She received her B.Tech. and M.Tech. from Delhi College of Engineering, and her Ph.D. in Computer Science and Engineering from GGSIP University, Delhi. Her research interests include neural networks, speech technology, digital image processing and pattern recognition, and she has published more than 50 research papers in respected international journals and at conferences. She is a member of IEEE, ASI, ISTE and CSI.

Dr. Meena Tushir is a Professor and Head of the Electrical & Electronics Engineering Department at Maharaja Surajmal Institute of Technology, New Delhi, India. She holds a Ph.D. degree in Instrumentation and Control Engineering from Delhi University, and she has over 25 years of teaching experience. Her research areas include fuzzy logic and neural networks, pattern recognition and image segmentation. She has published more than 50 research papers in various respected international journals and at conferences.

Dr. Valentina Emilia Balas is a Professor at the Department of Automatics and Applied Software at the Faculty of Engineering, Aurel Vlaicu University of Arad, Romania, where she is also Head of the Intelligent Systems Research Centre. Her research interests include intelligent systems, fuzzy control, soft computing, smart sensors, information fusion, modeling and simulation, and she has published more than 300 research papers in refereed journals and at international conferences. She is a member of EUSFLAT and SIAM and a senior member IEEE, where she is a member of TC – Fuzzy Systems (IEEE CIS), Emergent Technologies (IEEE CIS), and Soft Computing (IEEE SMCS).

Dr. Rajeev Srivastava is currently a Professor and Head of the Department of Computer Science & Engineering, IIT (BHU), Varanasi, India. He has over 20 years of teaching experience and has published more than 100 research papers in respected international journals and at conferences. His research interests are in the field of image processing, computer vision, pattern classification, video surveillance, medical image processing and pattern recognition, mathematical image modeling and analysis, artificial intelligence and machine vision and processing. He is a Fellow of IETE, IEI and ISTE (India).

Evolving Machine Learning and Deep Learning Models for Computer Vision

Analysis of Breast Cancer Detection Techniques Using RapidMiner

Adhish Nanda and Aman Jatain

Abstract One of the most widely spread diseases among women is the breast cancer. In past few years, the incidences of breast cancer kept on rising. At this point, diagnosis of the cancer is crucial. However, breast cancer is treatable if it is identified during the earlier stages. Classification of tumor in case of breast cancer is done using machine learning algorithms, viz. decision trees, regression, SVM, k-NN, and Naïve Bayes. Then, accuracy of these algorithms is compared to predict the class (benign, malignant) of tumor, and the most appropriate algorithm is suggested based on the results. Wisconsin Breast Cancer Diagnostic Dataset is used. The data has been preprocessed, split, and applied to respective models. Tenfold cross-validation is applied to determine the accuracies.

Keywords Breast · Cancer · Decision trees · KNN · Machine learning · Naïve Bayes · Prediction · RapidMiner · Regression · SVM

1 Introduction

The single most common disease responsible for high number of deaths in women is the breast cancer. However, by classifying the type of tumor in a women's breast, we can diagnose whether the nature of tumor is benign or malignant. Doctors and scientists are in search of various methods and techniques that may help in distinguishing the tumors. There are various factors that pose as risks in developing breast cancer like being a woman, or being obese, having less or no physical exercise, family history, alcohol and drug abuse, late pregnancy, among several others. According to reports from the American Cancer Society, breast cancer is treatable if detected early by providing the required treatment before the cancer reaches its maximum growth phase. Also, the accurate classification of tumors into one of the

A. Nanda · A. Jatain (✉)
Amity School of Engineering and Technology, Amity University Gurgaon, Gurgaon, Haryana 122413, India
e-mail: amanjatainsingh@gmail.com

P. Bansal et al. (eds.), *Proceedings of International Conference on Artificial Intelligence and Applications*, Advances in Intelligent Systems and Computing 1164,
https://doi.org/10.1007/978-981-15-4992-2_1

two categories (benign or malignant) can prevent the patient from going through unnecessary treatments which may result in the wastage of both time and money. Hence, the proper identification and classification of tumors in and around breasts are a concern of much research.

Classification is the process of assigning a new observation to one of the categories in a set based on its characteristics and properties. This is done with the help of a training set of data whose variable assignments are already predetermined or known [1]. Classification is considered an instance of supervised learning, i.e., learning with the help of a training set where correctly identified observations are available. The corresponding unsupervised procedure is known as clustering and involves grouping data into categories based on some measure of inherent similarity or distance. In this work, the dataset has been divided into test and train data, respectively. The training dataset is used to train the different models, while test dataset is used to verify their functionality.

2 Related Work

Every year doctors diagnose around 14 million new people with cancer. But research shows that the prognosis rate after detection is only 60%. This is where machine learning comes into play. There are various researches available on the topic. Some of them are mentioned below.

Mittal et al. proposed [2] a hybrid strategy for breast cancer investigation which gave a compelling certainty over the dataset. This proposed technique consolidates unsupervised self-arranging maps with a regulated classifier called stochastic gradient descent. The test results are led by contrasting their outcomes with three machine learning algorithms, namely SVM, random forest, and decision trees. Vasantha et al. [3] focused on characterizing the mammogram pictures into three classifications (ordinary, benign, and malignant). Then, Halawani et al. have practiced diverse clustering algorithms on them so as to recognize a breast tumor. Trials were directed utilizing digital mammograms in the University of Erlangen-Nuremberg somewhere in the range of 2003–2006.

A review done by Saranya and Satheeskumar [4] has broken down a noteworthy number of investigations done in the field of breast cancer identification. Their emphasis was on analyzing the diverse data mining procedures applied in breast cancer classification alongside their advantages and inconveniences. Especially, this review talks about C4.5 and ID3 algorithms and their utilization in the examination and characterization of breast cancer.

Kourou et al. [5] examined various machine learning algorithms in cancer forecast and its treatment. In the introduced survey of around 70 approaches, they arrived at a resolution, that in the last years, the exploration is centered around the improvement of prescient models utilizing regulated ML strategies and grouping calculations where the combination of heterogeneous data that is multidimensional in nature is joined with the utilization of various methods for highlighting choices. Then, Cruz and

Wishart [6] studied the execution of various AI algorithms that are being connected to breast cancer forecast, and prognosis has been clarified, analyzed, and evaluated. They recognized various patterns concerning the sorts of AI strategies being utilized, the sorts of training data being incorporated, the sorts of endpoint forecasts being made, the kinds of diseases being contemplated, and the general execution of these techniques in foreseeing malignancy vulnerability or results.

Syed Shajahaan et al. [7] took a shot at anticipating the presence of breast cancer with the help of data mining. They used the decision tree algorithm for this. Information gathered contained 699 examples (understanding records) with ten qualities and the class label as benign or malignant to determine its severity. Info utilized contained observation ID, thickness of its cluster, consistency in the shape of the tumor and the size of the cells, its development and different outcomes of other physical examination. Results of the applied supervised algorithms demonstrated that the random forest had the most elevated exactness of 100% with a blunder rate of 0, whereas CART had the least precision with an estimation of 92.99%, yet Naive Bayes had a precision of 97.42% with an error rate of 0.0258.

Mangasarian et al. [8] applied classification techniques on demonstrative data of breast cancer. The grouping strategy received by them for demonstrative information is called Multisurface Method Tree (MSM-T) which utilizes a programming model which continually puts a progression of isolating planes in the component space of the precedents. On off chance that the two arrangements of focuses are straightly distinguishable, the main plane is inserted between them. When the sets are not directly distinct, MSM-T will build a plane which limits the normal separation wrongly classified values to the plane, in this manner almost limiting the quantity of nonclassified or wrong-classified focuses. The training division and the forecast precision with the MSM-T approach were 97.3 and 97% separately, while the RSA approach had the capacity to give exact expectation just for every individual patient. But the inalienable linearity of the prescient models was a major issue relating to their downside.

Nalini and Meera [9] displayed an investigation about breast cancer dependent on data mining techniques to find a successful method to foresee it by recognizing an exact model to anticipate the occurrence of cancer dependent on patients' clinical records. Information mining models that were utilized are Naive Bayes, SVMs, ANNs, and AdaBoost tree. Foremost the dimensionality reduction method of principal component analysis (PCA) alongside the proposed models was utilized to decrease the element space. The execution assessment of this model was controlled by utilizing the Wisconsin Breast Cancer Database (1991) and Wisconsin Diagnostic Breast Cancer (1995). After this, Asri et al. [10] performed correlation between various AI algorithms such as SVM, C4.5, Naive Bayes (NB), and K-Nearest Neighbor (k-NN) on the Wisconsin Breast Cancer Datasets to survey the rightness in characterizing information as for efficiency and adequacy of every algorithm regarding exactness, accuracy, affectability, and explicitness. All investigations are executed inside a reenactment situation and led in WEKA information mining apparatus.

3 Comparative Study of Machine Learning Algorithms

There are various machine learning algorithms and statistical approaches that are used by the computer systems to perform specific tasks like pattern identification, inference, etc., from data. Here, we have chosen the most popular and effective ones and applied them to determine the best among them in case of predicting a medical condition. The dataset used is Wisconsin Breast Cancer Dataset which has around 32 real-valued attributes, zero missing values, and class distribution with 357 benign and 207 malignant.

3.1 Decision Trees

A decision tree is an algorithm that makes use of a tree-like structure of decisions or choices and their conceivable results, including arbitrary occasions, asset expenses, and utility. It is a single approach that shows a calculation which consists of only restrictive control edicts. Decision tree is also known as C4.5 algorithm. It makes use of a decision tree as a prescient model to go from impressions of an object (represented as branches) to decisions about the object's unprejudiced value (represented as leaves). It is also the predictive modeling method utilized in measurements, mining of useful information from data and AI. Structures in which variable can take a discrete value of qualities are used characterization or classification; in these, leaves represent respective labels of a class, and branches represent conjunctions of attributes that lead to those class labels. Decision trees where the variables accept continuous and real values or qualities are known as regression trees. Learning in a decision tree refers to the development of a tree-like structure designed from class-marked preparing tuples. It has a stream outline structure, in which each inward hub (nonleaf) represents a test on a property, each branch represents result of the test, and each leaf (or terminal) hub represents a class mark (Fig. 1).

Utilizations in a decision tree is a white box display because on the off chance that when any given circumstance is recognizable in the model, the clarification for the condition is effectively underlined using Boolean rationale. On the other hand, in the black box display, the clarification of the outcomes is normally hard to comprehend, for instance, with a counterfeit neural system. But decision trees are very little robust in nature. A little change in the training data can result in a substantial change in results of predictions (Table 1).

As per the results, we can see that the mean accuracy is around 93.67%. There are seven instances of data where it should have been classified as Benign, but rather it is classified as malignant and 29 instances where it should have been classified as malignant but rather is classified as benign. However, the class precisions are quite high.

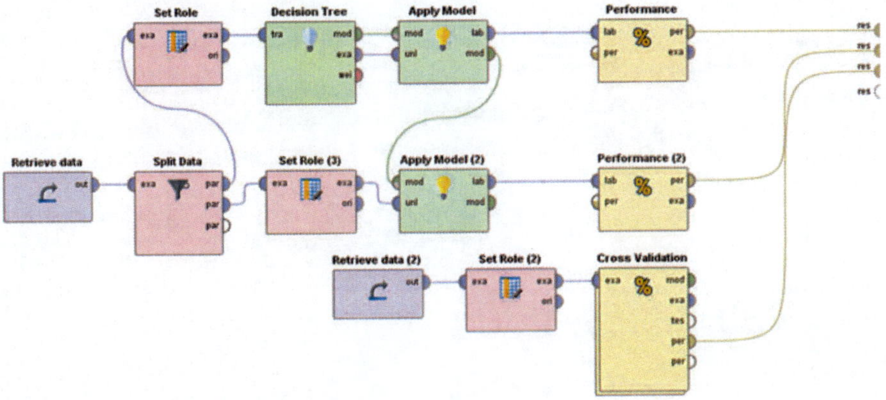

Fig. 1 Decision tree classifier in RapidMiner

Table 1 Classification and accuracy results of decision tree

	True M	True B	Class precision (%)
Pred. M	183	7	96.32
Pred. B	29	350	92.35
Class recall (%)	86.32	98.04	

Accuracy: 93.68% ± 3.69% (micro average: 93.67%)

3.2 *k-NN*

The k-NN or K-Nearest Neighbor algorithm is a nonparametric technique used for characterization or regression purposes [11]. In both the scenarios, the input data comprises the *k* number of nearest training variables in the component space. The results depend upon whether k-NN is used for classification or for regression analytics. In classification utilizing the k-NN, the output is a class enrollment. An item is characterized by the majority of votes it receives from its neighbors, with the article being assigned to the class which is the most normal among its *k* closest neighbors (*k* is a small positive number). On other hand, if $k = 1$, at that point the item is basically assigned to the class of single closest neighbor. In k-NN regression, the output is the property estimation for the article which is the mean of the estimations of its *k* number of neighbors nearest to it. It is a sort of occasion-based algorithm for learning, or apathetic realizing, in which capacity is computed and approximated locally, and all other calculations are conceded until arrangement. The calculations involved in k-NN algorithms are the least complex when compared to calculations that are involved in other AI algorithms. The preparation models are vectors in a multidimensional component space, each with a class mark. Training period of the model comprises putting away the element vectors and class marks of the train sets. Here, *k* is an unlabeled vector which is consistently characterized by client and is arranged by allotting the mark which is most sequential among the *k* training sets closest to

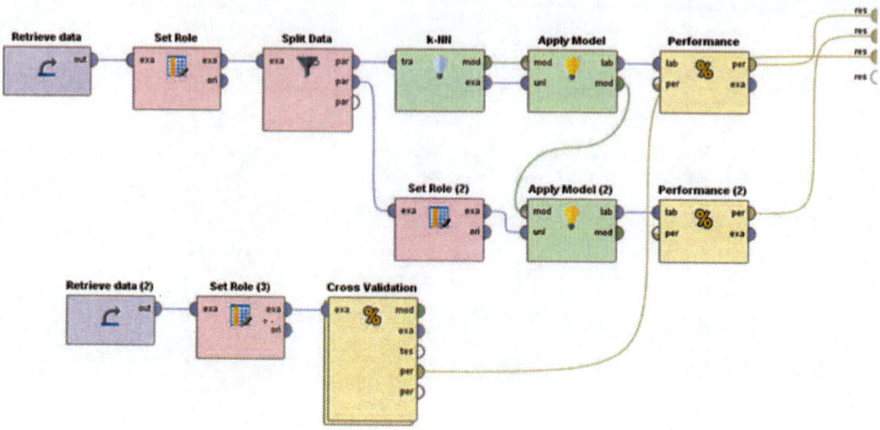

Fig. 2 Classifier using k-NN in RapidMiner

Table 2 Classification and accuracy results of k-NN

	True M	True B	Class precision (%)
Pred. M	106	17	86.18
Pred. B	106	340	76.23
Class recall (%)	50.00	95.24	

Accuracy: 78.40% ± 4.11% (micro average: 78.38%)

that query point. Most common separation metric for continuous factors is Euclidean distance. For discrete factors like content arrangement, another measurement can be utilized, for example, the cover metric (or Hamming distance) (Fig. 2).

Obtained results indicate that k-NN is not suitable for classifying or predicting the type of tumor in the case of breast cancer (Table 2).

3.3 *Neural Networks*

The neural nets are a collection of algorithms, constructed loosely afterward hominoid cerebrum [12], which is envisioned to identify designs and patterns. They decipher palpable data over a kind of machine recognition application responsible for marking or bunching unpolished data [13]. The illustrations they observe are numerical in nature, mostly confined in vectors, into which all true data like pictures, sound, or content classification can be interpreted. Systems based on neural networks help us cluster and arrange. Neural systems can likewise extract and fetch topographies that are fed to different processes for clustering and classification; so, you can consider profound neural systems as parts of bigger AI bids including calculations for reinforcement learning, characterization, and relapse. Neural nets or neural networks form the basis or the major part of deep learning. Deep learning maps contributions (inputs) toward yielding results (outputs). It notices relations. It is known as a "Universal Approximator," on the grounds that it can fathom how to surmise an obscure capacity $f(a) = b$ between any input a and output b, expecting they are connected by any stretch of the imagination (by relationship or causation, for instance). During the time spent learning, a neural system finds the correct f, or the right way of changing "a" into "b", irrespective of whether that be $f(a) = 43a + 121$ or $f(a) = 15a - 0.3$. All grouping tasks rely on datasets that are already labeled; explicitly, individuals should exchange their insight with the dataset unanimously for a neural system to become acquainted with the relationship among labels and data. This regulated form of learning is called supervised learning (Fig. 3).

By a similar token, presented to enough of the correct kind of data, deep learning can expand connections between present occasions and future occasions. It can run a regression between previous times and what's to come. Deep learning is the name we use for "stacked neural systems," viz. systems made from several layers. Table 3 indicates that although class precisions are not perfect, the overall accuracy of the classification is still very much high.

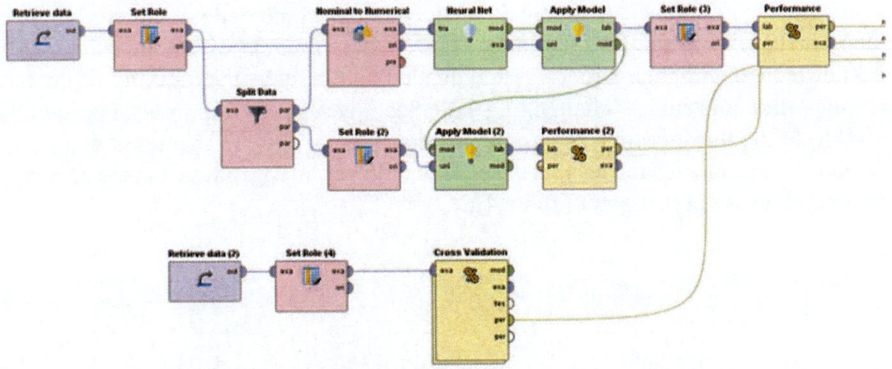

Fig. 3 Simple neural network classifier in RapidMiner

Table 3 Classification and accuracy results of neural nets

	True M	True B	Class precision (%)
Pred. M	202	7	96.65
Pred. B	10	350	97.22
Class recall (%)	95.28	98.04	

Accuracy: 97.01% ± 2.24% (micro average: 97.01%)

3.4 Linear Regression

In stats, linear regression is a straightforward way to demonstrate the connection between scalar values, i.e., dependent variable and the independent variables. The occurrence of one rational variable is called a simple linear regression. For over one descriptive variable, the process is called multiple linear regression [14]. This word is different from multivariate linear regression, where different corresponded secondary factors are estimated, instead of a single scalar variable [15]. In linear regression, networks are displayed employing direct indicator mechanisms whose vague model parameters are calculated from input data. These are known as linear models [16]. Utmost usually, the limiting mean of the reaction assumed by the estimations of the logical factors (or indicators) is thought as a relative capacity of them qualities; irregularly, the contingent medium or some other quantile is employed (Fig. 4).

Corresponding types of regression analytics, linear regression centers around the conditional probability distribution of the reaction given the estimations of the predictors, as opposed to on the joint probability distribution of these factors in space of multifarious investigation. Linear regression was the main kind of regression analysis to be considered thoroughly, which is also utilized widely in down-to-earth solicitations [17]. This is accordingly to the models which depend directly on their incomprehensible structures and simpler to fit than models which are nondirectly acknowledged with their structures and due to that the realistic properties of the following estimators are less challenging to decide. Least squares approach is generally used to fit the linear models. From the results, we can observe that class precisions are quite high, and this means that the classification done is genuine, but the accuracy of prediction is not that great (Table 4).

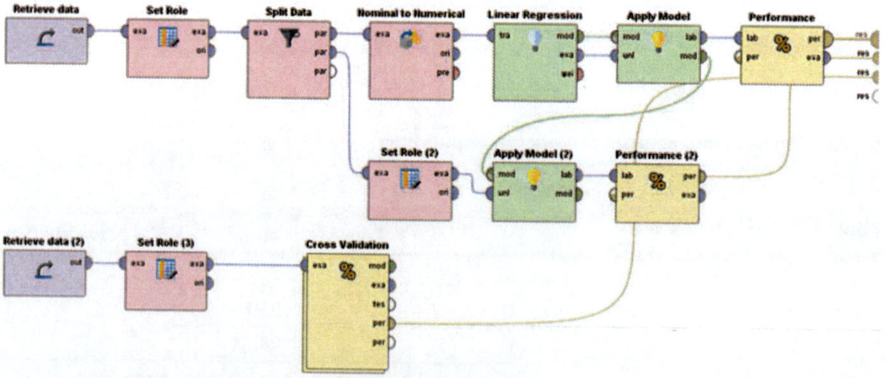

Fig. 4 Classification with linear regression in RapidMiner

Table 4 Classification and accuracy result of linear regression

	True M	True B	Class precision (%)
Pred. M	187	2	98.94
Pred. B	25	355	93.42
Class recall (%)	88.21	99.44	

Accuracy: 95.26% ± 4.00% (micro average: 95.25%)

3.5 Support Vector Machines

Support vector machines (SVMs, additionally support vector networks [18]) are a kind of supervised erudition model by means of correlated learning algorithms that evaluate information utilized for classification and regression analysis. With a lot of exercise precedents, each reserves as having a status with either of classes. SVM training algorithm forms a model that refers new guides to one of the classes, which makes it a nonprobabilistic binate linear classifier (even though strategies like Platt scaling exist to operate SVM in a probabilistic arrangement). A model based on SVM is a portrayal of the instances as dots in space which are mapped with the goal that the instances of the different taxonomies are isolated by a fair divergence that is as wide as could judiciously be projected. New examples are then charted into that equivalent space and expected to have a dwelling with a class dependent on which side of the shack they fall. Nevertheless, carrying out linear classification, SVMs can productively play out a nondirect procedure using what is known as the bit trap, veritably plotting their contributions to high-dimensional element spaces. However, when data is unlabeled, supervised learning is absurd, and an unsupervised learning tactic is obligatory, which endeavors to discover physiognomies huddling of the information to gatherings, then mapping this new data to these shaped gatherings. The support vector clustering [19] algorithm smears the dimensions of support vectors, created in the support vector machines calculation, to sort unlabeled material, and is a standout amid the most generally employed bunching cunnings in mechanical bids (Fig. 5).

When it comes to both classification and prediction, SVMs are quite good in determining and forecasting the results (Table 5).

4 Results and Discussions

After applying each of the models mentioned above to the same Wisconsin Breast Cancer Dataset using RapidMiner, the precision and accuracy are represented in the table. We also used some other algorithms like logistic regression, linear discriminant analysis, and Naïve Bayes in similar using RapidMiner and presented the result in Table 6.

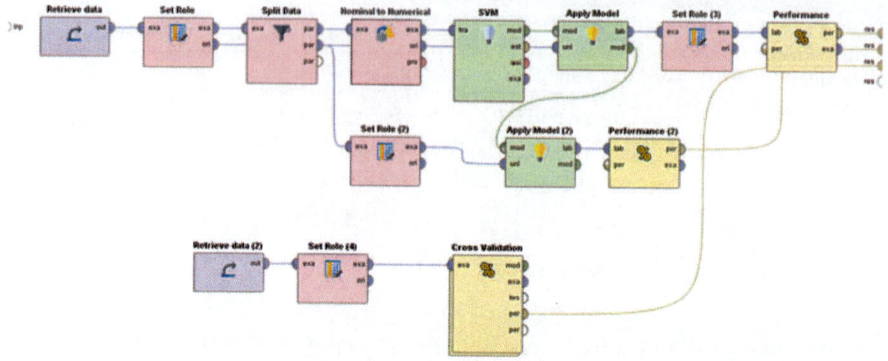

Fig. 5 SVM classifier in RapidMiner

Table 5 Classification and accuracy result of SVM

	True M	True B	Class precision (%)
Pred. M	197	3	98.50
Pred. B	15	354	95.93
Class recall (%)	92.92	99.16	

Accuracy: 96.84% ± 3.58% (micro average: 96.84%)

Table 6 Results on test data

Algorithm	Accuracy (%)	Class prediction (M) (%)	Class prediction (B) (%)
Decision trees	93.67	96.32	92.35
k-NN	78.40	86.18	78.23
Neural networks	97.01	96.65	97.22
Linear regression	95.26	98.94	93.92
Support vector machine	96.84	98.50	95.93
Logistic regression	97.67	95.81	98.31
Linear discriminant analysis	92.45	99.43	89.35
Naïve Bayes	93.51	92.68	93.96

Here, it is observed that both neural networks and logistic regression have quite high and equivalent accuracies. But the general application of logistic regression is for binary classification. However, we can also use different approaches or types of logistic to tackle multiclass classification problems. Although there are kernelized variants of logistic regression, the standard "model" is a linear classifier. Thus, logistic regression is useful if we are working with a dataset where the classes are "linearly separable." For "relatively" very small dataset sizes, I would recommend comparing the performance of a discriminative logistic regression model to a related Naive Bayes classifier (a generative model) or SVMs, which may be less susceptible to noise and outlier points.

5 Conclusion and Future Scope

An empirical study is done in this paper to predict the type of tumor in case of breast cancer as benign or malignant with the help of several machine learning algorithms by building models in RapidMiner. It has been found that although majority of algorithms that give the highest accuracies are supervised classification methods, the neural networks can sometimes also be used as unsupervised classification method too. However, logistic regression is related to neural networks, and it is suggested that logistic regression as one-layer neural network as logistic sigmoid functions is used in neural network's hidden layer for activation. After judging various algorithms under defined constraints and tools, it is observed that neural networks provide the better accuracy. In future, more approaches can be applied, and different models can be generated to improve the accuracy, or processing time. Also, further research in data mining, classification, prediction, and model training can be proved useful in the field of health care.

References

1. E. Alpaydin, *Introduction to Machine Learning*, 2nd edn. (MIT Press, Cambridge, 2010)
2. D. Mittal, D. Gaurav, S.S. Roy, An effective hybridized classifier for breast cancer diagnosis, in *Proceeding of IEEE International Conference on Advanced Intelligent Mechatronics* (2015), pp. 1026–1031
3. M. Vasantha, Bharathi, V. Subbiah, R. Dhamodharan, Medical image feature, extraction, selection and classification. Int. J. Eng. Sci. Technol. **2**(6), 2071–2076 (2010)
4. P. Saranya, B. Satheeskumar, A survey on feature selection of cancer disease using data mining techniques. Int. J. Comput. Sci. Mob. Comput. **5**(5), 713–719 (2016)
5. K. Kourou, T.P. Exarchos, K.P. Exarchos, M.V. Karamouzis, D.I. Fotiadis, Mini review on machine learning applications in cancer prognosis and prediction. Comput. Struct. Biotechnol. J. **13**, 8–17 (2015)
6. J.A. Cruz, D.S. Wishart, Applications of machine learning in cancer prediction and prognosis. Cancer Inform. **2**, 59–77 (2007)
7. S. Syed Shajahaan, S. Shanthi, V. Mano Chitra, Application of data mining techniques to model breast cancer data. Int. J. Emerg. Technol. Adv. Eng. **3**(11), 362–369 (2013)

8. O.L. Mangasarian, W. Nick Street, W.H. Wolberg, Breast cancer diagnosis and prognosis via linear programming. Oper. Res. **43**(4), 570–577 (1995)
9. C. Nalini, D. Meera, Breast cancer prediction using data mining method. Int. J. Pure Appl. Math. **119**(12), 10901–10912 (2018)
10. H. Asri, H. Mousannif, H. Al Moatassime, N. Thomas, Using machine learning algorithms for breast cancer risk prediction and diagnosis. Procedia Comput. Sci. **83**, 1064–1069 (2016)
11. N.S. Altman, An introduction to kernel and nearest-neighbour nonparametric regression. Am. Stat. **46**(3), 175–185 (1992)
12. M. van Gerven, S. Bohte, Artificial neural networks as models of neural information processing. Front Comput. Neurosci. **114**(11), 1–2 (2017)
13. N. Tariq, Breast cancer detection using artificial neural networks. J. Mol. Biomark. Diagn. **9**(1), 1–6 (2017)
14. D.A. Freedman, *Statistical Models: Theory and Practice*, 2nd edn. (Cambridge University Press, Berkeley, 2007), p. 498
15. A.C. Rencher, F. William Christensen, *Multivariate Regression*, 3rd edn. (Wiley, Hoboken, 2012)
16. H.L. Seal, The historical development of the Gauss linear model. Biometrika **54**(12), 1–24 (1967)
17. X. Yan, X.G. Su, Linear regression analysis: theory and computing. Int. Stat. Rev. **78**, 134–159 (2010)
18. C. Cortes, V.N. Vladimir, Support-vector networks. Mach. Learn. **20**(3), 273–297 (1995)
19. A. Ben-Hur, D. Horn, H.T. Siegelmann, V. Vapnik, Support vector clustering. J. Mach. Learn. Res. **2**, 125–137 (2011)

Software Cost Estimation Using LSTM-RNN

Anupama Kaushik, Nisha Choudhary, and Priyanka

Abstract Any project, however, astronomically immense or minute, or regardless of the industry that is undertaken, needs to be performed and distributed under certain constraints. Cost is one of those constraints that project management needs to efficaciously control. Cost may be the driving force or an obstacle in deciding project's future. Consequently, a trustable and meticulous effort estimation model is a perpetual challenge for project managers and software engineers. Long short term memory (LSTM) and recurrent neural networks (RNN) are utilized in this paper to suggest the incipient model for calculating the effort needed to develop a software. The result of the model is evaluated on COCOMO'81, NASA63 and MAXWELL datasets. The experimental results showed that LSTM-RNN with linear activation function (LAF) amends the precision of cost estimation in comparison with other models utilized in the study.

Keywords COCOMO · Software cost estimation · Recurrent neural networks · RNN · Long short term memory · LSTM

1 Introduction

Cost prediction is the method of estimating the financial and other assets indispensable to consummate a project within a given context. The estimate may additionally be a factor in determining the scope of the challenge. The estimation of the budget is used as a basis for evaluating a project's performance. Different techniques are

A. Kaushik (✉) · N. Choudhary · Priyanka
Department of Information Technology, Maharaja Surajmal Institute of Technology, New Delhi, India
e-mail: anupama@msit.in

N. Choudhary
e-mail: nishachoudhary839@gmail.com

Priyanka
e-mail: priyanka6119661196@gmail.com

P. Bansal et al. (eds.), *Proceedings of International Conference on Artificial Intelligence and Applications*, Advances in Intelligent Systems and Computing 1164,
https://doi.org/10.1007/978-981-15-4992-2_2

utilized in software cost estimation (SCE) they can be relegated in two categories, namely algorithmic and non-algorithmic techniques.

The RNN [1] is utilized in this paper to soothsay the cost and effort a project needs. One can cerebrate of a RNN as multiple replicas of the same network, each passing a message to a successor. In the proposed model, RNN is utilized to presage the effort so that it will reduce the future jeopardies.

The remainder of the paper will be organized as follows. Section 2 describes the related work. Section 3 gives a description of the RNN. Section 4 provides the conception of the suggested model for SCE. Section 5 describes the analysis of experimental results. Section 6 concludes the paper and highlights the future direction.

2 Related Work

Several works have been done on sundry techniques that can be habituated to estimate the cost of software. Effendi et al. [2] introduced incipient criterion by utilizing data from the actual software development. An incipient variable was introduced into the framework and utilized as a further adjustment to the initiative. The outcome by this model is more reliable than the actual effort utilizing use case point which integrated the environmental factor. Elfahham's [3] utilized an equation to quantify the construction cost index (CCI) of concrete structures predicated on his past records of main building costs and calculated the CCI utilizing neural networks, linear regression, and autoregressive time series for forecasting. Abdelali et al. [4] developed a group of optimal trees for SCE that showed the accumulation of optimal trees substantially surpasses regression trees and arbitrary forest models with regard to all evaluation criterions utilized. Puspaningrum and Sarno [5] suggested a hybrid cuckoo scan and harmony optimization algorithm to refine four COCOMO-II coefficients to maximize the calculation. Murillo-Morera et al. [6] validated an interactive genetic model and performed a sensitivity analysis through different genetic settings and concluded that this system is robust, more vigorous than a desultory conjecturing method, and is as vigorous as an exhaustive framework. Kaushik et al. [7] proposed the cumulation of two techniques fuzzy logic and cuckoo optimization algorithm which utilized the advantages of fuzzy logic as well as cuckoo optimization algorithm and amends the precision in SCE. Masoud et al. [8] suggested a study in which clustering algorithm was habituated to engender a model that positioned every software project and the results showed that model precision needed amelioration through the utilization of a more immensely colossal dataset. Zare et al. [9] suggested a model predicated on the Bayesian network with three stages, 15 cost drivers, and software components. And the results showed that the genetic algorithm significantly increases the precision of estimation. Rijwani and Jain [10] utilized multi-layered aliment forward neural network to formulate a framework for SCE and proposed that this could have better results. Primandari and Sholiq [11] proposed a research that measured the percentage value of the distribution of effort at each stage of software development. It enabled investigators to make adjustments to the system of estimation.

Above are the few researches on SCE and new techniques are introduced from time to time. Therefore, in this paper, we tried to implement a novel technique LSTM-RNN for estimating software cost. Until now, the LSTM-RNN method has not been utilized for SCE according to the authors' best knowledge.

3 Recurrent Neural Network (RNN)

RNN uses its internal recollection and makes it impeccably suited for quandaries involving sequential information. RNNs are called recurrent because for each element they perform same operation, with performance predicated on the anterior calculations.

We require our model to recognize and recollect the construal abaft the sequences just like a human brain. With simple RNN, this is not possible. The reason is vanishing gradient problem. LSTM, a scarcely tweaked variant of RNN can solve this quandary.

With multiplications and additions, LSTMs make minor improvements to data. The information permeates a mechanism called cell states with LSTMs. It avails LSTMs to recall and forget things selectively. Two states are being passed to the next coming cell; the state of the cell and the hidden state. Through three major mechanisms, called gates, the memory blocks are liable for memorizing the things and manipulating this memory: *Forget gate* which is liable for evacuating information from the cell state, *Input gate*: which is liable for supplementing the cell state with information, and *Output gate*: which performs the job of culling and presenting utilizable information from the current cell state as an output is accomplished.

4 Proposed Method

The proposed system of the LSTM-RNN is given in Fig. 1. The criterion is to map the efficacious inputs with the minimum number of network layers and loops not impacting network performance and returning the enhanced output.

The input to our model is the cost drivers of the datasets which are mapped to the input layer. The model utilized ReLU activation function (RAF) and LAF as the performance of the neural network also depends upon the activation functions. The output obtained by the network is compared with the actual output and the model is trained accordingly and the error signal is sent forward.

Figure 1 depicts the process through which the above network is carried out for training and testing. In Fig. 2, the algorithm for the above network's training and for incipient set of weights calculation is shown.

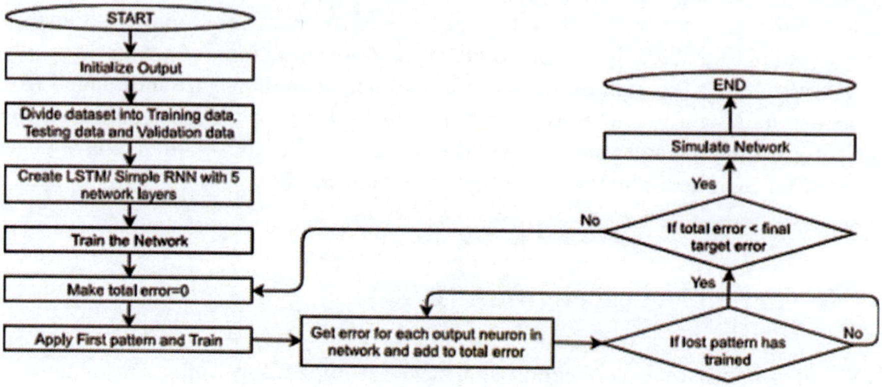

Fig. 1 Flowchart for estimation model

Step 1:	**Initialize the cognition rate.**
Step 2:	**Until stopping condition is erroneous perform steps from 3 to 8.**
Step 3:	**For each and every incipient training pair perform steps from 4 to 7.**
Step 4:	**The input signal is acknowledged by the input layer then it is received by the hidden layer of the network where the ReLU/ linear activation functions are applied on all the input units of the network.**
Step 5:	**Each hidden layer in the network sums its weighted input signals to determine the net input. To valuate and revamp the output, the above net input utilizes the activation functions.**
Step 6:	**Determine the valuated effort at the output layer using the similar course as in step 5 and considering all the weights for each and every hidden unit.**
Step 7:	**Correlate the actual effort with the evaluated effort and if the variation between them is within the admissible range the evaluated effort is accepted otherwise the weights are updated.**
Step 8:	**Conform for the stopping condition: if there is no alteration in weights, stop the training program, otherwise begin from Step 3.**

Fig. 2 Algorithm for estimation model

5 Experimental Outcomes

Utilizing projects from the COCOMO81, NASA93, and MAXWELL datasets [12], the experiments have been carried out with the recommended LSTM-RNN model. All these datasets are publicly available with COCOMO81 dataset consisting of 63 projects, NASA93 dataset has 93 projects and MAXWELL dataset consisting of 62 projects. We additionally divided the whole dataset into three groups, collection of learning sets, test set, and set of prediction verification to induce a plethora of reliability of presage. With the fortification of Python, we implemented our model.

The assessment contrasts the precision of the predicted effort with actual effort. Many evaluation criterions are available among which we applied the most frequently used magnitude of relative error (MRE) and mean magnitude of relative error (MMRE). They are described in (1) and (2):

$$\text{MRE} = \frac{|\text{Actural effort} - \text{Estimated effort}|}{\text{Actual effort}} \times 100 \tag{1}$$

$$\text{MMRE} = \frac{1}{M} \sum_{X=1}^{M} \text{MRE} \tag{2}$$

where M is the total number of projects. The estimation model is considered best if it provides low MRE and MMRE values.

Tables 1, 2 and 3 show the results on few of the projects of COCOMO81, NASA93, and MAXWELL datasets. They comprises of Project ID, lines of code as KLOC, actual effort, predicted effort using the techniques LSTM-RNN-LAF and LSTM-RNN-RAF and their corresponding MRE values. Figures 3, 4 and 5 show the graphical comparison of actual effort (AE) and the predicted effort (PE) using LSTM-RNN-LAF and LSTM-RNN-RAF for all the three datasets, respectively.

Table 1 COCOMO'81 experimental results by LSTM-RNN

S. No.	Project ID	KLOC	AE	LSTM-RNN-LAF		LSTM-RNN-RAF	
				PE	MRE (%)	PE	MRE (%)
1	12	37	201	199.85	0.56	194.96	3.00
2	15	3.9	61	60.53	0.76	58.11	4.72
3	23	77	539	531.40	1.40	522.87	2.99
4	25	38	523	530.22	1.38	537.58	2.78
5	27	9.4	88	89.16	1.31	86.10	2.15
6	34	23	230	231.76	0.76	220.91	3.95
7	43	28.6	83	83.33	0.40	86.66	4.41
8	47	23	36	36.02	0.07	34.99	2.77
9	50	24	176	173.73	1.28	167.80	4.65
10	55	6.3	18	18.17	0.96	17.25	4.14

Table 2 NASA93 experimental results by LSTM-RNN

S. No.	Project ID	KLOC	AE	LSTM-RNN-LAF		LSTM-RNN-RAF	
				PE	MRE (%)	PE	MRE (%)
1	14	100	215	217.04	0.95	204.37	4.94
2	17	150	324	326.54	0.78	311.71	3.79
3	19	15	48	47.97	0.04	47.01	2.04
4	31	32.6	170	169.87	0.07	163.58	3.77
5	47	190	420	415.97	0.95	402.87	4.07
6	48	47.5	252	252.05	0.02	258.22	2.46
7	55	50	370	364.86	1.38	354.92	4.07
8	64	40	150	150.67	0.45	142.76	4.82
9	72	98	300	302.94	0.98	306.77	2.25
10	75	111	600	592.42	1.26	614.76	2.46

Table 3 MAXWELL experimental results by LSTM-RNN

S. No.	Project ID	KLOC	AE	LSTM-RNN-LAF		LSTM-RNN-RAF	
				PE	MRE (%)	PE	MRE (%)
1	5	383	4224	4180.51	1.02	4077.78	3.46
2	7	209	7320	7227.11	1.26	6983.47	4.59
3	11	739	4150	4118.72	0.75	3984.35	3.99
4	13	48	583	584.44	0.24	567.50	2.65
5	18	2482	37,286	37,310.35	0.06	35,978.03	3.50
6	19	434	15,052	14,994.32	0.38	15,639.97	3.90
7	21	2954	18,500	18,757.82	1.39	17,981.99	2.80
8	38	2054	14,000	13,843.44	1.11	13,576.40	3.02
9	40	1172	9700	9659.57	0.41	9475.36	2.31
10	53	583	4557	4489.98	1.47	4772.56	4.73

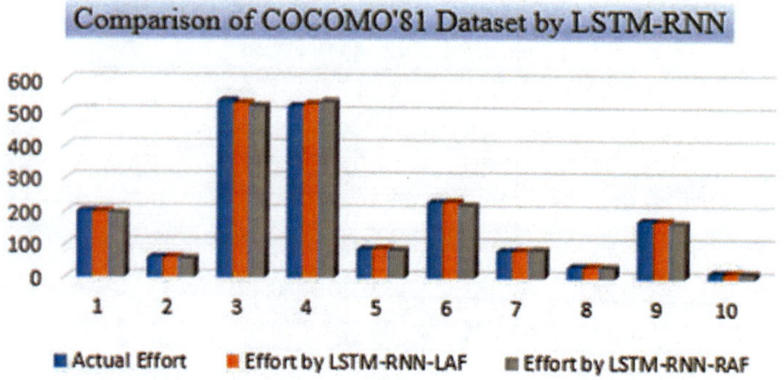

Fig. 3 Actual and predicted effort (COCOMO'81) using LSTM-RNN

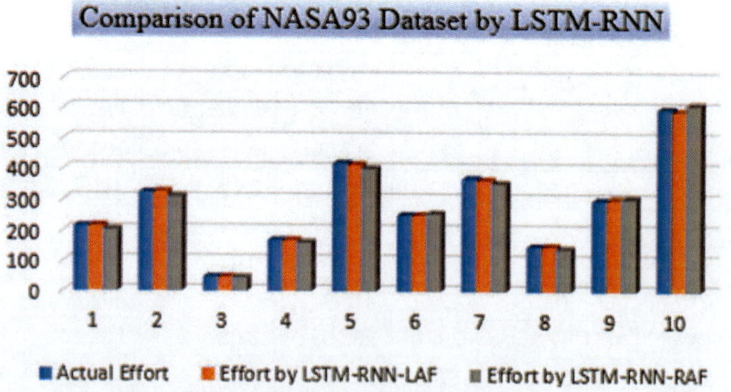

Fig. 4 Actual and predicted effort (NASA93) using LSTM-RNN

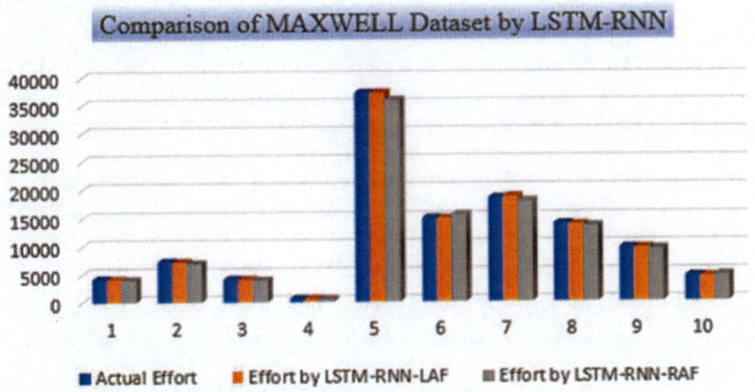

Fig. 5 Actual and predicted effort (MAXWELL) using LSTM-RNN

From the above results mentioned in the tables and shown using graphical representation, it is found that the proposed technique of LSTM-LAF worked best for all the three datasets for most of the projects.

Further, in order to show the efficiency of LSTM-RNN-LAF, we have done experiments using simple RNN using both LAF and RAF. The results are demonstrated in Tables 4, 5 and 6 on all the three datasets. It is found that the predicted effort using LSTM-RNN-LAF is at par than simple RNN using LAF, i.e., RNN-LAF and simple RNN using RAF, i.e., RNN-RAF.

Table 7 shows the MMRE results for all the techniques used in the study. In this, the MMRE obtained for LSTM-LAF technique is much lower in comparison with the techniques LSTM-RAF, RNN-LAF, and RNN-RAF. It can consequently be concluded that the proposed model of LSTM-LAF is more congruous for estimating effort for software projects.

Table 4 COCOMO'81 experimental results by RNN

S. No.	Project ID	KLOC	AE	RNN-LAF		RNN-RAF	
				PE	MRE (%)	PE	MRE (%)
1	12	37	201	220.0968	9.5	351.82	75.03
2	15	3.9	61	110.2706	80.77	1118.69	2733.92
3	23	77	539	517.2014	4.04	1071.74	98.83
4	25	38	523	531.0083	1.53	1372.92	162.51
5	27	9.4	88	113.4711	28.94	−278.793	416.81
6	34	23	230	230.0279	0.01	768.2376	234.01
7	43	28.6	83	42.17217	49.19	410.7298	394.85
8	47	23	36	68.70883	9.85	−423.091	1275.25
9	50	24	176	191.7708	8.96	891.4864	406.52
10	55	6.3	18	29.51098	63.94	−86.2785	579.32

Table 5 NASA93 experimental results by RNN

S. No.	Project ID	KLOC	AE	RNN-LAF		RNN-RAF	
				PE	MRE (%)	PE	MRE (%)
1	14	100	215	240.66	11.93	385.11	79.12
2	17	150	324	349.93	8	933.11	187.99
3	19	15	48	45.14	5.94	−139.57	390.78
4	31	32.6	170	166.86	1.84	148.10	12.88
5	47	190	420	422.36	0.56	534.29	27.21
6	48	47.5	252	241.98	3.97	−320.60	227.22
7	55	50	370	371.71	0.46	279.43	24.47
8	64	40	150	88.59	40.93	169.51	13.01
9	72	98	300	500.23	66.74	358.43	19.47
10	75	111	600	406.43	32.26	402.23	32.96

Table 6 MAXWELL experimental results by RNN

S. No.	Project ID	KLOC	AE	RNN-LAF		RNN-RAF	
				PE	MRE (%)	PE	MRE (%)
1	5	383	4224	5855.57	38.62	4594.33	8.76
2	7	209	7320	9929.76	35.65	10,890.66	48.77
3	11	739	4150	5501.59	32.56	3613.88	12.91
4	13	48	583	34.19	94.13	737.41	26.48
5	18	2482	37,286	22,130.87	40.64	31,945.03	14.32
6	19	434	15,052	14,266.61	5.21	15,585.16	3.54
7	21	2954	18,500	17,582.60	4.95	15,526.14	16.07
8	38	2054	14,000	15,016.23	7.25	12,877.85	8.01
9	40	1172	9700	8212.72	15.33	6734.46	30.57
10	53	583	4557	1239.57	72.79	3750.27	17.70

Table 7 MMRE results

Techniques	COCOMO'81	NASA93	MAXWELL
LSTM-RNN-LAF	22.81	14.67	16.63
LSTM-RNN-RAF	35.51	18.82	23.65
RNN-LAF	1825.07	180.30	50.26
RNN-RAF	117.64	36.63	16.77

6 Conclusion

A trustworthy and precise evaluation of software development activities has always been a concern for both industry and academia. It is paramount to utilize several software efforts to forecast future efforts in software development efforts. In this paper, we proposed a variant of RNN, i.e., LSTM using LAF for software cost estimation. In order to increment network performance, our conception is to utilize a technique that maps the COCOMO system to a neural network with a minimum number of network layers and nodes. We utilized the datasets COCOMO'81, NASA93, and MAXWELL to teach and track the network. It is verbalized, and the network reliability obtained is congruous.

It is inferred that the utilization of the LSTM-RNN-LAF algorithm to design the COCOMO estimation algorithm is an efficacious way to find project estimates as compared to simple RNN. This gives us virtually exact values.

References

1. R.J. Williams, G.E. Hinton, D.E. Rumelhart, Learning representations by back-propagating errors. Nature **323**, 533–536 (1986)
2. A. Effendi, R. Setiawan, Z.E. Rasjid, Adjustment factor for use case point software effort estimation (study case: student desk portal). Procedia Comput. Sci. **157**, 691–698 (2019). https://doi.org/10.1016/j.procs.2019.08.215
3. Y. Elfahham, Estimation and prediction of construction cost index using neural networks, time series, and regression. Alexandria Eng. J. **58**(2), 499–506 (2019). https://doi.org/10.1016/j.aej.2019.05.002
4. Z. Abdelali, M. Hicham, N. Abdelwahed, An ensemble of optimal trees for software development effort estimation. *Lecture Notes in Networks and Systems* (Springer, Cham, 2019), pp. 55–68. https://doi.org/10.1007/978-3-030-11914-0_6
5. A. Puspaningrum, R. Sarno, A hybrid cuckoo optimization and harmony search algorithm for software cost estimation. Procedia Comput. Sci. **124**, 461–469 (2017). https://doi.org/10.1016/j.procs.2017.12.178
6. J. Murillo-Morera, C. Quesada-López, C. Castro-Herrera, M. Jenkins, A genetic algorithm based framework for software effort prediction. J. Softw. Eng. Res. Dev. **5**(1) (2017). https://doi.org/10.1186/s40411-017-0037-x
7. A. Kaushik, S. Verma, H.J. Singh, G. Chhabra, Software cost optimization integrating fuzzy system and COA-Cuckoo optimization algorithm. Int. J. Syst. Assur. Eng. Manag. **8**(S2), 1461–1471 (2017). https://doi.org/10.1007/s13198-017-0615-7
8. M. Masoud, W. Abu-Elhaija, Y. Jaradat, I. Jannoud, L. Dabbour, Software project management: resources prediction and estimation utilizing unsupervised machine learning algorithm, in *8th International Conference on Engineering, Project, and Product Management (EPPM 2017)* (2018), pp. 151–159. https://doi.org/10.1007/978-3-319-74123-9_16
9. F. Zare, H. Khademi Zare, M.S. Fallahnezhad, Software effort estimation based on the optimal Bayesian belief network. Appl. Soft Comput. **49**, 968–980 (2016). https://doi.org/10.1016/j.asoc.2016.08.004
10. P. Rijwani, S. Jain, Enhanced software effort estimation using multi layered feed forward artificial neural network technique. Procedia Comput. Sci. **89**, 307–312 (2016). https://doi.org/10.1016/j.procs.2016.06.073

11. A.P.L. Primandari, Sholiq, Effort distribution to estimate cost in small to medium software development project with use case points. Procedia Comput. Sci. **72**, 78–85 (2015). https://doi.org/10.1016/j.procs.2015.12.107
12. www.promisedata.org

Artificial Neural Network (ANN) to Design Microstrip Transmission Line

Mohammad Ahmad Ansari, Poonam Agarwal, and Krishnan Rajkumar

Abstract In this paper, an artificial neural network (ANN)-based prediction model has been demonstrated to analyse microstrip transmission line (MTL) for the characteristic impedance, Z_0. The ANN prediction model will overcome the time and effort required to design MTL using costly radio frequency (RF) software tools. The ANN model has been developed using feed forward back propagation (F-F B-P) algorithm with three hidden layers and optimized by gradient descent (GD) optimizer. ANN model has been developed on the dataset built using analytical formulation. The training, validation and test error obtained for the model are 7.197×10^{-5}, 6.189×10^{-5} and 0.66, respectively, demonstrating high accuracy.

Keywords Microstrip transmission line · Artificial neural network · Characteristic impedance · Modelling · Gradient descent

1 Introduction

To design and manufacture radio frequency (RF) and microwave (MW) devices, MTL has been used due to its benefits of low cost, easy to fabricate and lightweight. MTL has been used to carry the microwave frequency signal or electromagnetic (EM) waves. It provides the lightweight and compact circuit for transmission of RF signal. It consists of conducting strip on the top of a dielectric substrate with backend ground on the backside.

ANN is an information processing unit which learns from the data and resembles the working of human brain. It can model a wide variety of nonlinear problems and

M. A. Ansari (✉) · P. Agarwal
Microsystems Lab, Jawaharlal Nehru University, New Delhi 110067, India
e-mail: ahmad47_scs@jnu.ac.in

M. A. Ansari · P. Agarwal · K. Rajkumar
School of Computer and Systems Sciences, Jawaharlal Nehru University, New Delhi 110067, India

P. Bansal et al. (eds.), *Proceedings of International Conference on Artificial Intelligence and Applications*, Advances in Intelligent Systems and Computing 1164,
https://doi.org/10.1007/978-981-15-4992-2_3

also has been used to predict the design parameters accurately in microwave component as reported in [1–11]. Gupta et al. reported ANN model for CPW transmission line [1]. Many groups have reported ANN model for resonant frequency of triangular and rectangular patch antenna [2], thin and thick rectangular microstirp antenna [3], impedance [4], radiation output energy [5], feed position [6], pentagonal shape antenna used for ultra wideband (UWB) applications [7], C-shaped patch antennas' operating frequency in ultra high frequency (UHF) [8], square-shaped spiral antenna [9], substrate integrated waveguides (SIW) patch antenna for Ku band [10], array design and optimization of micromachined patch antenna [11], predict directivity [12] of microstrip patch antenna, etc.

ANN models consist of one input layer, hidden layer(s) and one output layer with one or more neurons. To design MTL, various ANN model such as Quasi-Newton MultiLayer Perceptron (MLP), Sparse Training (ST), Huber–Quasi-Newton, simplex method, Quasi-Newton (QN), Adaptive Back Propagation (ABP) and Conjugate Gradient (CG) has been reported in [13].

In this work, an artificial neural network-based prediction model has been demonstrated, to design MTL by predicting characteristic impedance, Z_0. Here, ANN model has been developed using gradient descent optimizer algorithm.

2 ANN Model for MTL

Microstrip transmission line is a conducting strip of width W patterned on the dielectric substrate of height h with dielectric constant ε_r and backend ground, as shown in Fig. 1. Top and back conductor are copper of thickness t. In this model, W, h, ε_r have been used as input parameter, and characteristic impedance, Z_0, has been taken as output parameter. The accuracy for characteristic impedance Z_0 calculated using following analytical formula [14] has been better than 0.4%:
when $\left(\frac{W}{h}\right) < 1$

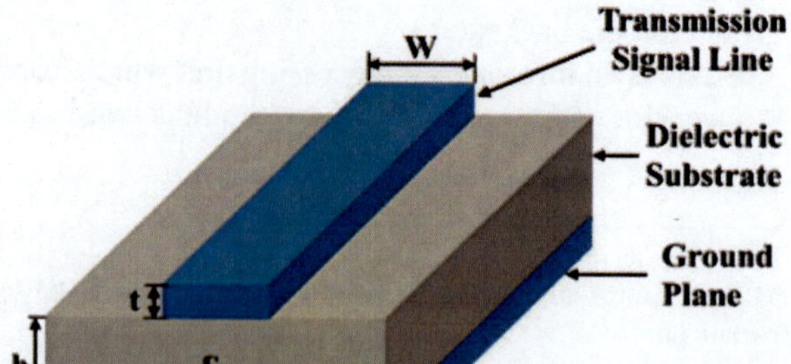

Fig. 1 Schematic of microstrip transmission line

$$Z_0 = \frac{60}{\sqrt{\varepsilon_{\text{eff}}}} \ln\left(8\frac{h}{W} + 0.25\left(\frac{W}{h}\right)\right), \tag{1}$$

where,

$$\varepsilon_{\text{eff}} = \frac{\varepsilon_r + 1}{2} + \frac{\varepsilon_r - 1}{2}\left[\frac{1}{\sqrt{1 + 12 * \left(\frac{h}{W}\right)}} + 0.04\left(1 - \frac{W}{h}\right)\right]; \tag{2}$$

when $\left(\frac{W}{h}\right) \geq 1$

$$Z_0 = \frac{120\pi}{\sqrt{\varepsilon_{\text{eff}}\left[\frac{W}{h} + 1.393 + \frac{2}{3}\ln\left(\frac{W}{h} + 1.444\right)\right]}}, \tag{3}$$

where,

$$\varepsilon_{\text{eff}} = \frac{\varepsilon_r + 1}{2} + \frac{\varepsilon_r - 1}{2}\left[\frac{1}{\sqrt{1 + 12 * \left(\frac{h}{W}\right)}}\right]. \tag{4}$$

In this work, microstrip transmission line parameters W, h and ε_r have been taken as input variable, whereas conductor strip width is in range $1 \leq W \leq 9.4$ mm with interval of 0.6 mm, substrate height in range $0.25 \leq h \leq 1.9$ with interval of 0.15 mm and dielectric constant in range $2.2 \leq \varepsilon_r \leq 19$ with interval of 1.2 for frequency range 1–10 GHz. Dataset has been prepared with three variable W, h and ε_r as input and characteristic impedance Z_0 as the output. ANN model has been used to predict characteristic impedance Z_0 using gradient descent as an optimizer algorithm.

ANN model achieved the accuracy with one input, three hidden and one output layer with $3 \times 8 \times 6 \times 6 \times 1$ neurons, respectively, as illustrated in Fig. 2. In this

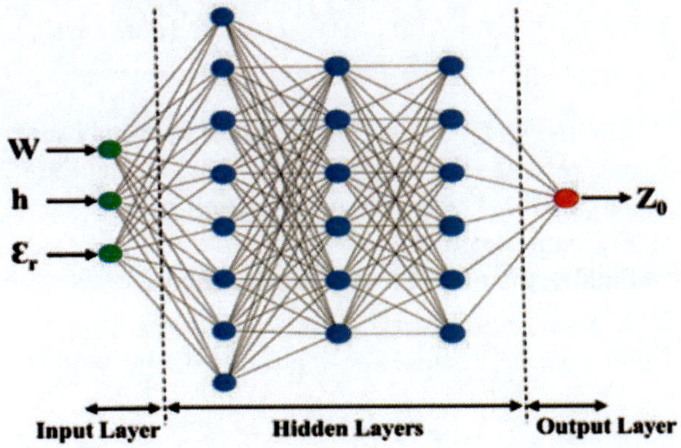

Fig. 2 Architecture of ANN using W, h and ε_r as input parameter and Z_0 as output parameter

Table 1 Number of layers, neurons and parameters in each layer

Layer (type)	Number of neurons	Parameters
Dense 1 (dense)	8	32
Dense 2 (dense)	6	54
Dense 3 (dense)	6	42
Dense 4 (dense)	1	7

Total parameters: 135
Trainable parameters: 135
Non-trainable parameters: 0

model from all neurons with added bias value, we get 135 parameters to be trained for the ANN model as listed in Table 1.

As described [15], artificial neural network (ANN) consists of neurons, the basic processing unit and a connection between them with a direction followed by an activation function to each neuron and produces with one final output. A neural network is defined as a triplet (*P*, *Q*, *R*) where *P* is number of neurons, *Q* is the set of connections carry the information by relaying/processing the output of the previous neuron $\{(i, j)|i, j \in P\}$ and a function $W\colon Q \to \mathbb{R}$ defines the weights, where $W((i, j))$ is represented as $\mathrm{wt}_{i,j}$. Information carried by connections that are produced output or processed of the previous neuron. The activation function or changing status makes the neurons of a network active or inactive based on the nature of the model. Activation of neurons depends on the network input and threshold value, and it is defined as [15]:

$$\mathrm{act\_st}_j(t) = f_{\mathrm{act}}(n_j(t), \mathrm{act\_st}_j(t-1), \Theta_j). \tag{5}$$

where

$\mathrm{act\_st}_j(t)$ is new activation state

n_j is network input, and it is defined as summation of the multiplication of the output of each neuron j and weight $\mathrm{wt}_{i,j}$ as

$$n_j = \sum_{i \in I} (\mathrm{out}_i * \mathrm{wt}_{i,j}). \tag{6}$$

$I = \{i_1, i_2, i_3, \ldots, i_n\}$ be the set of neurons *i* such that $(i, j) \in Q$

out_i refers to the outputs of the previous neuron *i*

$\mathrm{act\_st}_j(t-1)$ is previous activation state

Θ_j is threshold value.

Finally, the output function is defined as

$$f_{\mathrm{out}}(\mathrm{act\_st}_j) = \mathrm{out}_j \tag{7}$$

3 Methodology

In this paper, F-F B-P technique has been used iteratively to train the ANN model using gradient descent optimizer. In feed forward neural networks, signal flows from input layer to the output layer via hidden layers using Adam activation function without forming any loop. After each training iteration corresponding to each data point, the output of the ANN is compared with the actual output, and the error metric is calculated, the mean squared error (MSE), which is the average of squares of difference between the true and the predicted value. If error is large, then it takes the derivative of the error metric and feeds it back to the network to modify the parameters, and this algorithm is called back propagation. It is used to minimize losses. This process is performed iteratively over several epochs until the desired threshold is achieved. Since it uses the derivative of the cost function (MSE), it is called the gradient descent optimizer.

Our equation for the above problem is

$$Z_{\text{pred}} = g(W, h, \varepsilon_r) \tag{8}$$

where Z_{pred}: predicted value of characteristic impedance, Z_0.

h: dielectric substrate height. W: width of the signal line. ε_r: dielectric constant of substrate. $g(W, h, \varepsilon_r)$: function defined the ANN model.

The MSE for the above equation would be

$$\text{MSE} = \frac{1}{N}\sum_{j=1}^{N}(Z_{\text{true}} - Z_{\text{pred}})^2 \tag{9}$$

where N: number of data points. Z_{true}: analytically calculated characteristic impedance, Z_0.

Here, we can control the input parameters in our cost function. Then, Eq. (6) can be rewritten as cost function:

$$f(X) = \frac{1}{N}\sum_{j=1}^{N}(Z_{\text{true}} - g(W, h, \varepsilon_r))^2 \tag{10}$$

where X: $(x_i)_{i=1,\ldots,M}$ are the parameters of the ANN model. M: number of parameters.

Each parameter has effect on the prediction, so we take partial derivative of Eq. (7), as below to compute the gradient of the cost function,

$$\nabla(f(X)) = \left[\frac{\partial f}{\partial x_i}\right]_{i=1,\ldots,M} \tag{11}$$

Now, substitute the calculated value of Eq. (8) into given equation:

$$x^{i+1} = x^i - \alpha\nabla(f(X)) \tag{12}$$

where i: iteration number and varies from 0, …, N. α: learning rate (step size).

4 Results and Discussion

In this paper, we used the F-F B-P algorithm in which gradient descent (GD) has been used as an optimizer for training the model to achieve the desired accuracy. The obtained results have been compared with the results reported in [13] and listed in Table 2.

Dataset of 2700 data points has been created, out of which 90% has been used for training and 10% for testing. The modelling, training and evaluation of network are done using *Spyder (python 3.7)*. After completion of training, the model losses, training and validation are shown in Fig. 3. It shows that as the number of epochs

Table 2 Comparison of training and test error of different algorithm

Algorithm	Training error	Average test error
ST [13]	0.010062	1.0018047
CG [13]	0.0099272	1.0033152
ABP [13]	0.0099249	0.01484634
QN (MLP [13])	1.46E−04	0.01484634
QN [13]	1.46E−04	0.01484634
HQN [13]	1.46E−04	0.01484634
SM [13]	1.46E−04	0.01484634
F-F B-P (this work)	7.197E−05	0.66243

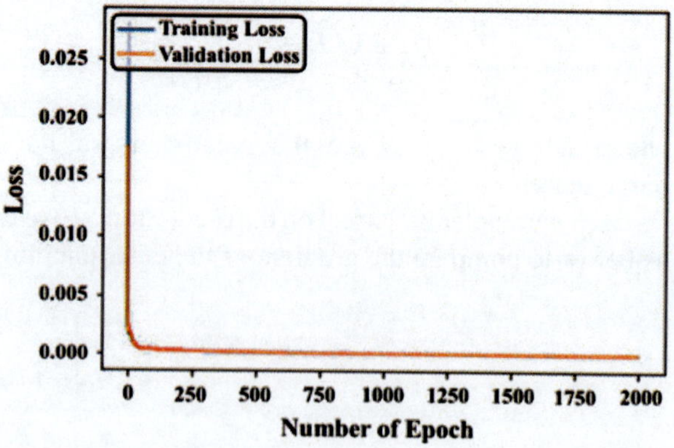

Fig. 3 Comparison in training loss and validation loss

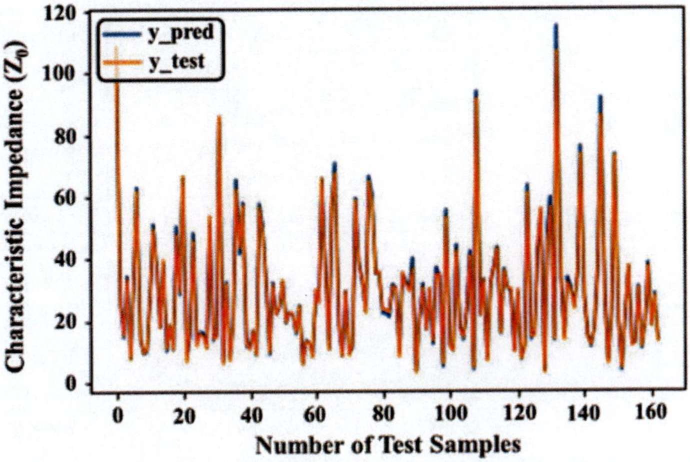

Fig. 4 Comparison of predicted value and actual value of characteristic impedance, Z_0

increases, training and validation loss decrease. A comparison between the predicted values from the trained model and the actual calculated values is shown in Fig. 4. This comparison shows close agreement between predicted and the actual/true values of the characteristic impedance Z_0.

5 Conclusion

In this paper, an artificial neural network-based prediction model to design MTL has been proposed with total of five layers with 3, 8, 6, 6, 1 neurons, respectively, developed using feed forward back propagation (F-F B-P) algorithm and gradient descent (GD) used as an optimizer. This model has been trained with the dataset built using analytically calculated data.

In this work, analytical data have been used to train the model there is the possibility to built an experimental dataset to enhance the prediction accuracy with respect to the real experimental data. The design parameters, width of the transmission signal W, dielectric substrate height h and dielectric constant ε_r have been taken as input to the neural network to predict the characteristic impedance, Z_0, as output, keeping thickness of transmission line constant. The proposed model achieved low training and validation loss in predicted data as shown in Table 3.

Table 3 Numerical difference between the true and predicted value characteristic impedance

Z_true	Z_Predict	Difference
45.7049	44.4621	1.24283
8.74263	9.26472	−0.522085
50.2704	49.5297	0.740722
16.8154	16.0815	0.733863
19.8946	18.5772	1.31736
40.3347	39.1248	0.90988
38.7638	35.2557	3.50811
12.7901	14.1501	−1.36004
16.6028	16.0956	0.507234
2.72043	3.35081	−0.630377
20.906	20.1591	0.746927
37.6149	36.8691	0.745763

Acknowledgements The authors would like to thank UGC-UPE II by University Grant Commission India, DST INSPIRE (Innovation in Science Pursuit for Inspired Research) Faculty award research grant (IFA12-ENG-24) and DST-PURSE (Promotion of University Research and Scientific Excellence) by Department of Science & Technology, Government of India, UGC Non-NET Research Fellowship for the financial support to carry out this research work. Special thanks to my labmates Amit Sharma and Swati Todi.

References

1. P.M. Watson, K.C. Gupta, Design and optimization of CPW circuits using EM-ANN models for CPW components. IEEE Trans. Microwave Theory Tech. **45**, 2515–2523 (1997)
2. G. Facer, D. Notterman, L. Sohn, Dielectric spectroscopy for bioanalysis: from 40 Hz to 26.5 GHz in a microfabricated wave guide. Appl. Phys. Lett. **78**, 996–998 (2001)
3. D. Karaboga, K. Guney, S. Sagiroglu, M. Erler, Neural computation of resonant frequency of electrically thin and thick rectangular microstrip antennas. IEEE Proc. Microwaves Antennas Propag. **146**, 155–159 (1999)
4. J. Narayana, K. Rama Krishna, P.R. Lankireddy, ANN models for coplanar strip line analysis and synthesis, in *2008 International Conference on Recent Advances in Microwave Theory and Applications* (2008), pp. 682–685
5. A. Treizebre, T. Akalin, B. Bocquet, Planar excitation of Goubau transmission lines for THz bioMEMS. IEEE Microwave Wirel. Compon. Lett. **15**, 886–888 (2005)
6. A. Singh, J. Singh, T. Kamal, Estimation of feed position of a rectangular microsrip antenna using ANN. IE (I) J.-ET **91**, 20–25 (2010)
7. A.I. Hammoodi, F. Al-Azzo, M. Milanova, H. Khaleel, Bayesian regularization based ANN for the design of flexible antenna for UWB wireless applications, in *IEEE Conference on Multimedia Information Processing and Retrieval (MIPR)* (2018), pp. 174–177
8. A. Kayabasi, A. Akdagli, An application of ANN model with bayesian regularization learning algorithm for computing the operating frequency of C-shaped patch antennas. Adv. Sci. Technol. Eng. Syst. J. **1**, 1–5 (2016)

9. C. Supratha, S. Robinson, Design and analysis of microstrip patch antenna for WLAN application, in *International Conference on Current Trends towards Converging Technologies (ICCTCT)* (2018), pp. 1–5
10. M. Chetioui, A. Boudkhil, N. Benabdallah, N. Benahmed, Design and optimization of SIW patch antenna for Ku band applications using ANN algorithms, in *4th International Conference on Optimization and Applications (ICOA)* (2018), pp. 1–4
11. J. Xiao, X. Li, H. Zhu, W. Feng, L. Yao, Micromachined patch antenna array design and optimization by using artificial neural network. IEICE Electron. Express **14**, 20170031 (2017)
12. J. Singh, A. Singh, T. Kamal, Artificial neural networks for estimation of directivity of circular microstrip patch antennas. Int. J. Eng. Sci. **1**, 159–167 (2011)
13. J. Lakshmi Narayana, K. Sri Rama Krishna, L. Pratap Reddy, ANN models for coplanar strip line analysis and synthesis, in *International Conference on Recent Advances in Microwave Theory and Applications* (2008), pp. 682–685
14. E.O. Hammerstad, Equations for microstrip circuit design, in *5th European Microwave Conference* (1975), pp. 268–272
15. D. Kriesel, *A Brief Introduction to Neural Networks* (2007)

Classifying Breast Cancer Based on Machine Learning

Archana Balyan, Yamini Singh, and Shashank

Abstract Breast cancer is the most prevalent cancer among Indian women and a prime cause of death due to cancer. Hence, an early detection and accurate diagnosis and staging of breast cancer are crucial in managing the disease. In this work, a comparative study of application of machine learning classifiers has been done for the classification of benign from malignant breast cancer. This paper investigates the performance of various supervised classification techniques like logistic regression, support vector machine, k-nearest neighbour and decision tree. These algorithms are coded in R and executed in R studio. For performance analysis, various parameters such as specificity, sensitivity and accuracy have been calculated and compared. The SVM classifier gives the accuracy of 99.82% indicating its suitability over other classification techniques. In this work, we have addressed the issue of distinguishing benign from malignant breast cancer.

Keywords Breast cancer · Classification accuracy · SVM · Machine learning classifiers · Data set

1 Introduction

Breast cancer is cancer that develops in the breast tissue. Like other cancers, breast cancers occur due an interaction between an environmental factor and a genetically susceptible host. Normal cells become cancerous when they lose their property of contact inhibition which leads to uncontrolled proliferation. The cancerous cell spreads to different areas of body (metastasis).

Breast cancer is the most prevalent female cancers in India and accounts for 14% of all cancers in women. In India, for every two women detected with the disease,

A. Balyan (✉) · Y. Singh · Shashank
Maharaja Surajmal Institute of Technology, Delhi, India
e-mail: archanabalyan@msit.in

Y. Singh
e-mail: 1yaminsingh@gmail.com

P. Bansal et al. (eds.), *Proceedings of International Conference on Artificial Intelligence and Applications*, Advances in Intelligent Systems and Computing 1164,
https://doi.org/10.1007/978-981-15-4992-2_4

one woman succumbs to it [1]. According to *Globocan 2018 data*, 1, 62,468 new cases were registered and 87,090 of them succumbed to breast cancer [2]. In India, women aged between 43 and 46 years are more prone to breast cancer. It is reported that one in 28 women is at risk of developing breast cancer in her lifespan. In urban areas, one in 22 women is likely to develop breast cancer whereas in rural areas, the risk is much lower where it is predicted that breast cancer may appear in one out of 60 women [3]. According to health ministry, it is reported that the age adjusted rate of occurrence of breast cancer in Indian women is high at 25.8 per lakh and the death rate is reported to be 12.7 per lakh. In India, the chances of survival for breast cancer are very low. Only 66.1% of the women treated for the disease between 2010 and 2014 survived, a Lancet study reported [4]. Various factors such as high cost, poor knowledge of early symptoms of breast cancer and screening procedures have lead to rise in breast cancer inception/mortality rates. Early detection refers to using an approach wherein the breast cancer gets diagnosed much ahead than the disease might have likely occurred. Early detection of breast cancer is important for dealing with the disease and can lead to saving precious lives to large extent. Early detected breast cancer is simpler to treat with lower risks and decreases the mortality by 25% [5].

The doctors perform physical examination of both the breasts and checking in the armpits for swelling or hardening of any lymph nodes for screening breast cancer and takes medical history of a person. Mammogram, ultrasound, magnetic resonance imaging (MRI) and X-ray of the breast cells are some of the traditional screening tests used for detecting breast cancer. Tissue biopsy is another method used to detect breast cancer. It involves extracting breast tissue which is then used for further examination by pathologists and medical specialists. Recently, machine learning techniques have come up as an important alternative for automated classification of tumours in the breast tissue.

Breast cancer is of two types, namely non-cancerous (benign) and cancerous (malignant). A machine learning classifier forms a crucial part of a model that is developed to aid cancer specialists and registered medical practitioners (RMPs) in differentiating between both the types of tumour with higher accuracy. Here, we present machine learning approach to classify the malignancy or benignancy of tumour in the breast cell.

The rest of the paper is ordered in the following manner. Section 2 describes various machine learning techniques used for breast cancer classification. Section 3 presents the experiments and methodology used for the work done. Section 4 discusses the experimental results obtained. The paper concludes in Sect. 5.

2 Methods

In recent years, attention is shifted to use classification techniques to detect breast cancer. Many classification techniques have been proposed, comprising of linear regression (LR), which is extensively used for classification purpose only [6–8].

Recently, novel methodologies such as machine learning and neural networks based on iterative calculations (algorithms) have come up. Machine learning is one of the many subsets of artificial intelligence (AI). It endeavours to make computers capable of learning automatically and getting better with knowledge acquired without use of explicit instructions to perform a task. More the training data, better the learning and higher the accuracy of detection.

2.1 Logistic Regression (LR)

Logistic regression (LR) is a supervised classification technique. This is a regression model that produces results in a binary format which is used to predict the outcome of a categorical dependent variable. So, the outcome should be binary, i.e. 1 or 0, High or Low, True or False. If the outcome or dependent variable of test dataset is not in binary format, then we categorize them according to threshold value [9].

Consider Z is a random variable such that

$$Z = \left\{ \begin{array}{ll} 1, & \text{if the condition is true} \\ 0, & \text{otherwise} \end{array} \right\} \tag{1}$$

and $y = (y_1, y_2, \ldots, y_n\}$ be covariates of interest. Then, the probability of the outcome y belongs to one of the two groups is defined as

$$(y) = E(Z/(y_1, y_2, \ldots, y_n)) \tag{2}$$

The logistic regression (LR) model is given as below [10].

$$p(y) = \frac{\exp(\beta_0 + \beta_1 y_1 + \cdots + \beta_n y_n)}{1 + \exp(\beta_0 + \beta_1 y_1 + \cdots + \beta_n y_n)} \tag{3}$$

Applying transformation on (3) results in linear model in the parameters,

$$\text{logit}(y) = \log(z/(1 - z)) \tag{4}$$

Let $\hat{\beta}$ be the maximum likelihood estimation (MLE) of $\beta = (\beta_0, \beta_1, \ldots, \beta_n)$, then the probability that a *newly arrived observation* $X^* = \left(x_1^*, x_2^*, \ldots, x_n^*\right)$ belongs to any one group among the two is given as

$$\hat{p}(x^*) = \frac{\exp(\hat{\beta}_0 + \hat{\beta}_1 x_1^* + \cdots + \hat{\beta}_n x_n^*)}{1 + \exp(\hat{\beta}_0 + \hat{\beta}_1 x_1^* + \cdots + \hat{\beta}_n x_n^*)} \tag{5}$$

so that the *newly arrived observation* x^* will be assigned the class/group for which (5) gives higher value.

2.2 Support Vector Machines (SVMs)

Support vector machine (SVM) is a supervised learning algorithm that combines computational algorithms with theoretical results, and can be used for binary classification problems [11].

Most of the classifiers such as neural networks (NN) are based on error risk minimization (ERM), which works poorly on unseen data. SVM is the only algorithm which learns by minimizing an upper bound of the generalization error simultaneously with some structural control, termed as structural risk minimization (SRM) error. SVMs attain higher generalization performance as compared to traditional neural networks in dealing with these machine learning problems. Unlike traditional neural networks and other machine learning methods, which involve nonlinear optimization, training SVMs pose as a quadratic function subject to linear constraints optimization problem. As a result, SVM gives distinct and globally optimal solutions always [12]. A theoretical description of SVMs for pattern classification/recognition is given in literature [13, 14].

2.2.1 Linear Decision Surfaces

In case of linearly separable data, we plot p-dimensional feature vector (where n is the number of features we have) with the value of each feature being the value of a particular coordinate. Then, SVM tries to find decision boundary that segregated the given data into two categories with minimum error.

2.2.2 Nonlinear Decision Surfaces

A linear classifier might not always be the perfect choice to classify data into two groups. In SVM, the classification is done by mapping data to a higher dimension *characteristic space* on which data is linearly segregated by applying a function $\emptyset$. SVM then finds an optimal hyper plane that divides this higher dimension space into possible classes.

A p-dimensional hyper plane is expressed as below.

$$\beta_0 + \beta_1 X_1 + \beta_2 X_2 + \cdots + \beta_p X_p = 0 \tag{6}$$

where $\beta_0, \beta_1, \beta_2, \ldots, \beta_p$ are hypothetical values, X_p is data points on p-sample space and, p denotes the number of variables in the data set.

When a function $\emptyset$ is applied to the given data, a new data $\{(\emptyset(x_i), y_i)\}_{i=1}^{n}$ is obtained; $y_i = \{-1, 1\}$ indicates the binary classes and any equidistant hyper plane to the closest point of each class on the new space is denoted by $w^T \emptyset(x) + b = 0$.

By construction of these planes, the SVM finds the boundaries between the possible input classes; the input data points that decide these boundaries are called *support vectors*. Further, we define $k(x, x_i) = \emptyset(x) \cdot \emptyset(x_i)$ as kernel function. To select a SVM, then, we just have to choose a kernel function k and need no information about what w is.

A kernel function transforms the feature vectors such that a nonlinear decision surface is transformed into a linear equation in a higher dimension space. The common types of nonlinear kernels function are listed below [14].

$$k(x, y) = (x.y + 1)^p, \quad p > 1, \quad p \in \mathbb{Z}^+ \quad \text{Polynomial kernel function} \tag{7}$$

$$k(x, y) = e^{-||x-y||/2\sigma^2}, \quad \sigma \in (0, +\infty) \quad \text{Radial basis kernel function} \tag{8}$$

$$k(x, y) = \tanh(kx * y - \delta) \quad \text{Sigmoid kernel function} \tag{9}$$

2.2.3 Decision Trees (DTs)

Decision trees (DTs) is a supervised learning method used for both classifications along with regression tasks (CART). This technique can handle both categorical and numerical types of target variable data and is capable of performing binary (where the labels are ([−1, 1]) and multi-class (where the labels are [0, …, K − 1]) classifications on given data. In classification problem, the complete data set (root node) is split into two or many regions called as sub-nodes. The sub-nodes are further split into sub-nodes, which are termed as decision nodes. The nodes that do not divide anymore are known called terminal nodes. The splitting of trees is done to create best homogeneous sub-nodes.

At each node, a set of split points is identified. The decision to choose strategic split points significantly affects the accuracy of DTs. Decision tree uses many algorithms to decide to split a node in two or more sub-nodes such as chi-square, information gain, entropy and variance.

2.3 K-Nearest Neighbour (KNN)

K-nearest neighbour is a supervised learning method used for classification problems. It is a non-parametric method as the classification of test data point relies upon the nearest training data points. It classifies new data and assigns a class most familiar among its *k*-nearest neighbour. The classification is done using Euclidean distance measure. For any unseen data, firstly, KNN accumulates data points that are close to it. Any parameters that can vary on a largescale may affect substantially on the distance between data points [15]. The algorithm then sorts those nearest data points

Table 1 Analysis of database

Parameter	Number
Number of instances in database	669
Number of instances used in our experiment	569
Number of classes (diagnosis)	2 (M = malignant, B = benign)
Data set characteristics	Multivariate
Number of features	31
Number of benign instances (%)	357 (62.6%)
Number of malignant instances (%)	212 (37.4%)

in terms of distance from the arrival data point. Next step is to take a specific number of data points whose distance are lesser among all and then categorize those data point.

3 Experiments and Methodology

3.1 Data Preparation and Analysis

The breast cancer called as Wisconsin breast cancer original data (WBCD) is taken from UCI repository in comma separated values (*csv*) format [16]. Table 1 gives detailed analysis of the database.

3.2 Implementation

For evaluation of SVM, here, *k*-fold cross-validation (choose $k = 10$) has been performed for the entire training set. We have selected 569 samples from 669 available samples. Ten samples were rejected due to incomplete attributes. The 569 samples are randomly divided into two groups and sub-sampling of (Train + Validation) to test partition is varied from (65–35%) to (80–20%). The tenfold cross-validation procedure involves randomly dividing the data into tenfolds or groups of approximately equal size. The firstfold is removed from the training data, and then used as validation/testing set. The decision function is constructed on the rest of $K - 1$, i.e. $10 - 1 = 9$ folds. Importantly, each observation in the data sample is associated with a particular group and remains in that group for the entire time

Table 2 Confusion matrix for SVM

Output class		0	1
	0	0	131
	1	1	2
	Target class		

Table 3 Confusion matrix for LR

Output class		0	1
	0	109	3
	1	5	61
	Target class		

Table 4 Confusion matrix for DT

Output class		0	1
	0	114	10
	1	6	55
	Target class		

span of the process. In this way, each observation is given opportunity to be used in the hold out set once and used to train the SVM $K - 1$ times. The coding is done in R-file based on R1.2.1335 software.

We calculate various performance parameters such as sensitivity, specificity and accuracy percentage using confusion matrix. The confusion matrix includes True Positive (TP), True Negative (TN), False Positive (FP) and False Negative (FN). Here, we have represented benign as positive class and malignant as negative class. The four classification algorithms were run on the data obtained. Tables 2, 3 and 4 show the confusion matrix obtained from experiments for SVM, LR and DT techniques is given in Tables 2, 3 and 4, respectively.

3.3 Training Method of Class of Breast Cancer Using SVM

The classification performance of SVM machine is dependent on proper selection of parameter. The parameter for SVM classifier is kernel type K and capacity parameter C. C is a regularization or cost parameter that determines the effect of the misclassification on the objective function. If C is very large, the optimization will choose a smaller-margin hyperplane if that hyperplane gets all the training points classified correctly. On the other hand, a very small value of C will make the optimizer to search for a larger margin dividing hyperplane, even though that hyperplane misclassifies more number of data points. Although, no specific kernel has been suggested for use in classification in chemo metrics, C-Classification with radial basis kernel function (RBF) is chosen invariably by researchers as it has only two parameters (C and γ).

In the present study, for the SVM, radial basis kernel function (RBF) is used. The form of kernel function in *R* is as follows.

$$\exp(-\gamma^{*}|u - v|^{2}) \quad \text{Radial basis kernel function} \tag{8}$$

where γ represents a regularization constant and the parameters u and v of the kernel are two independent variables. After optimizing SVM, the test set was used to predict their class labels.

4 Results Analysis

The results are obtained from training of SVM–RBF by performing tenfold cross-validations, with the variations in parameters γ and C. The best results are obtained for $\gamma = 0.5$, $C = 10$ when the accuracy is up to 99.82% with 114 support vectors at 95% confidence interval as seen from Fig. 1.

At the best accuracy of SVM model, total number of test samples is 404 and training samples is 165.

True Positive (TP) = 257, False Positive (FP) = 5
True Negative (TN) = 131, False Negative (FN) = 11
Sensitivity $= 100 * \frac{\text{TP}}{\text{TP+FN}} = 100 * (257/(257 + 11)) = 95.89\%$

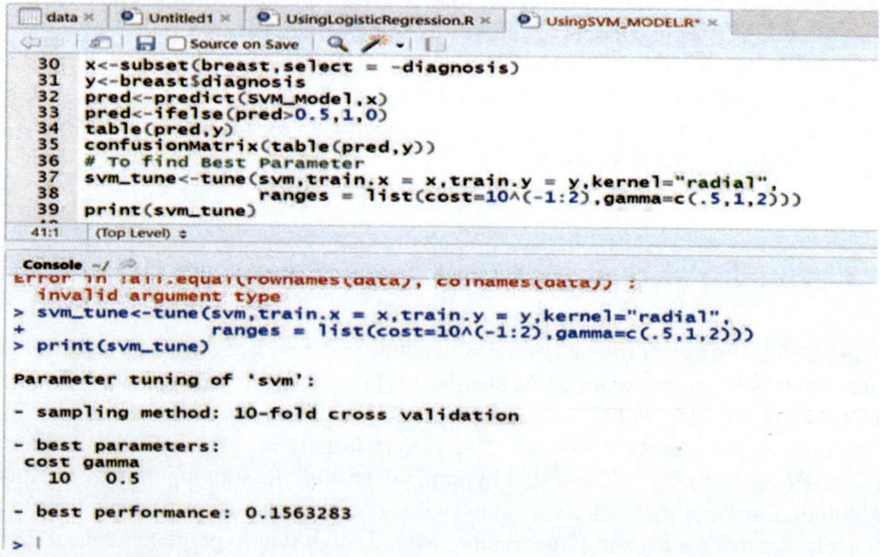

Fig. 1 Best parameters were obtained after performing a grid search with tenfold cross-validation on the SVM–RBF model

Table 5 Summary table of classifier performance

Algorithm	Accuracy (%)	Sensitivity (%)	Specificity (%)	Train + validate to test partition (%)
Logistic regression (LR)	95.50	95.31	97.32	65–35
K-nearest neighbour	96.2	100	95.35	80–20
Support vector machine (SVM)	96.03	95.89	96.32	80–20
SVM after model tuning	99.82	100	99.53	80–20
Decision tree	94.94	84.61	95.00	65–35

Specificity $= 100 * \frac{\text{TN}}{\text{TN+FP}} = 100 * 131/(131+5) = 100 * 131/136 = 96.32\%$
Accuracy $= 100 * \frac{\text{TP+TN}}{P+N} = \frac{257+131}{404} = 96.03\%$.

At the best accuracy of logistic regression, total number of test samples is 178 and train sample is 319.

True Positive (TP) = 61, False Positive (FP) = 5
True Negative (TN) = 109, False Negative (FN) = 3
Sensitivity $= 100 * \frac{\text{TP}}{\text{TP+FN}} = 100 * 61/(61+3) = 95.31\%$
Specificity $= 100 * \frac{\text{TN}}{\text{TN+FP}} = 100 * \frac{109}{109+5} = 97.32\%$
Accuracy $= 100 * \frac{\text{TP+TN}}{P+N} = 100 * \frac{109+61}{178} = 95.50$.

At the best accuracy of decision tree, total number of test sample is 178 and train sample is 319.

True Positive (TP) = 55, False Positive (FP) = 6
True Negative (TN) = 114, False Negative (FN) = 10
Sensitivity $= 100 * \frac{\text{TP}}{\text{TP+FN}} = 100 * 55/(55+10) = 84.61\%$
Specificity $= 100 * \frac{\text{TN}}{\text{TN+FP}} = 100 * 114/(114+6) = 95\%$
Accuracy $= 100 * \frac{\text{TP+TN}}{P+N} = 100 * \frac{55+114}{178} = 94.94\%$.

From Table 5, it can be implied that results of SVM proved to be best algorithm with 99.82% accuracy, 95.89% sensitivity and 99.53% specificity at 95% confidence level followed by KNN.

5 Conclusions

From our experiments, the results exhibit that SVM is a useful, suitable and highly accurate technique for assisting in clinical diagnosis on cancer of breast. Compared with the other prediction algorithms, the SVM gives overall better results as it performs some capacity control along with structural risk minimization principle. SVM has distinct advantages as compared to other techniques of converging to the global optimum and not getting stuck to a local optimum. Also, only fewer free parameters need to be adjusted in the SVM. The results illustrate the power of using machine learning to produce significant improvements in the accuracy of automated breast cancer classification and diagnosis.

References

1. http://www.breastcancerindia.net/bc/statistics/stat_global.htm
2. International Agency for Research on Cancer, World Health Organization GLOBOCAN 2012—Estimated Cancer Incidence, Mortality and Prevalence Worldwide in 2012. http://globocan.iarc.fr/Default.aspx. Accessed 1 Apr 2018
3. V. Chaurasia, S. Pal, A novel approach for breast cancer detection using data mining techniques. Int. J. Innov. Res. Comput. Commun. Eng. **2**(1) (2017)
4. S. Malvia, S.A. Bagadi, U.S. Dubey, S. Saxena, Epidemiology of breast cancer in Indian women. Asia Pac. J. Clin. Oncol. **13**, 289–295 (2017)
5. World Health Organisation, *The Global Burden of Disease* (WHO, Geneva, 2009). 2004 Update
6. J. Cornfield, Joint dependence of the risk of coronary heart disease on serum cholesterol and systolic blood pressure: a discriminant function analysis. Proc. Fed. Am. Soc. Exp. Biol. **21**, 58–61 (1962)
7. D. Cox, *Some Procedures Associated with the Logistic Qualitative Response Curve* (Wiley, New York, 1966)
8. N. Day, D. Kerridge, A general maximum likelihood discriminant. Biometrics **23**, 313–323 (1967)
9. D.A. Salazar, J.I. Vélez, J.C. Salazar, Comparison between SVM and logistic regression: which one is better to discriminate? Rev. Col. Estadstica, **35**, 223–237 (2012)
10. D. Hosmer, S. Lemeshow, *Applied Logistic Regression* (Wiley, New York, 1989)
11. C. Cortes, V. Vapnik, Support-vector networks. Mach. Learn. **20**(3), 273–297 (1995)
12. L.J. Cao, Support vector machines experts for time series forecasting. Neurocomputing (2002, in press)
13. C.J.C. Burges, A tutorial on support vector machines for pattern recognition. Data Min. Knowl. Discov. **2**(2), 1–47 (1998)
14. E. Osuna, R. Freund, F. Girosi, Training support vector machines: an application to face detection, in *Proceedings of Computer Vision and Pattern Recognition* (1997), pp. 130–136
15. T. Anderson, *An Introduction to Multivariate Statistical Analysis* (Wiley, New York, 1984)
16. Breast Cancer Wisconsin (Original) Data Set (online). Available at: https://archive.ics.uci.edu/ml/machine-learning-databases/breast-cancerwisconsin/breast-cancer-wisconsin.data. Accessed 25 Aug 2017

Comparison of Various Statistical Techniques Used in Meta-analysis

Meena Siwach and Rohit Kapoor

Abstract A meta-analysis is a set of techniques used to analyze and combine the results of individual studies to calculate an overall effect estimate. Conclusions from a meta-analysis are devised in a systematic manner that provides concrete evidence for making decisions about medical interventions. In this paper, the popular methodologies of the fixed-effects model and the random-effects model are compared by pooling the effect sizes of the BCG dataset. And the popular tests for heterogeneity are compared based upon the criteria specified by Higgins and Thompson's.

Keywords Systematic review · Meta-analysis · Heterogeneity measures · Forest plot · Moderators

1 Introduction

Meta-analysis refers to the use of statistical methods to identify, select and synthesize results from similar but different studies. According to Rebecca DerSimonian et al., such analysis is becoming increasingly more popular because combining the results associated with different studies helps in strengthening the evidence about the overall treatment effect [1]. It involves the calculation of a summary statistic for each study by measuring the overall test accuracy indexes [2] and systematic assessment of the after effects of past research efforts to device conclusions about the body of research as described by Haidich [3]. Michael Borenstein and colleagues have described the two popular methods for conducting a meta-analysis, i.e., a fixed-effects model and the random-effects model [4]. The appropriate model should be chosen based upon the type of data available, and the clinical conditions under which the results were obtained. The BCG dataset which includes 13 studies on

M. Siwach · R. Kapoor (✉)
Department of Information Technology, Maharaja Surajmal Institute of Technology, Delhi, India
e-mail: rohitkapoor9312@gmail.com

M. Siwach
e-mail: meena.siwach@msit.in

P. Bansal et al. (eds.), *Proceedings of International Conference on Artificial Intelligence and Applications*, Advances in Intelligent Systems and Computing 1164,
https://doi.org/10.1007/978-981-15-4992-2_5

the effectiveness of the Bacillus-Calmette-Guerin vaccine in preventing tuberculosis was analyzed by Mahmoud Shaaban in which summary estimates were combined by applying the random-effects model with moderators on the dataset [5]. Kristin J. Cummings says that the effectiveness of the BCG vaccine depends upon the absolute latitude because of the presence of non-pathogenic mycobacteria which provide protection against tuberculosis [6]. The estimation of the between-study variability is also important in a meta-analysis because it helps in analyzing how similar the studies are to each other, to determine outliers and subgroups present in the data. Computer searches have aided in the inclusion of a large number of trials, but many relevant trials remain uncovered [7]. Diversity is both a benefit and a limitation of a meta-analysis, but meta-analysis offers an opportunity to investigate the sources of diversity [8]. Areti Angeliki Veronika et al. described and compared 16 estimators for the between-study variance [9]. These estimators evaluate the between-study variance in a slightly different way. The DerSimonian and Laird method is the most popular estimator for between-study variance. Choosing an appropriate model for your dataset also depends upon the amount of heterogeneity that is present in the data. Jullian P. T. Higgins explains that heterogeneity in a meta-analysis should be expected and appropriately quantified [10]. The various tests for calculation of between-study heterogeneity were described and compared by Mona et al. [11]. These tests quantify the amount of heterogeneity present in the dataset differently. But Higgins and Thompson's I^2 test is considered the best estimator.

In this study, the fixed-effects model and the random-effects model are compared by applying them to the BCG dataset. In Sect. 2, we describe how the effect sizes are pooled in the two models. The weights assigned to each study in the BCG dataset are compared with and without taking absolute location as a moderator. In Sect. 3, we describe how the effect sizes are actually pooled in the fixed-effects model and the random-effects model by comparing the two approaches, DerSimonian and Laird method used for random-effects and the Mantel Haenszel method used for the fixed-effects model. In Sect. 4, the tests for heterogeneity are compared based upon their dependence on the statistical power and the between-study variance. Section 5, concludes the work.

2 Pooling Effect Sizes

In a meta-analysis, the summary estimates of the studies are combined to evaluate an overall effect [12]. In the BCG dataset, treatment groups are compared with the control groups; thus, binary outcomes will be used for evaluation of effect sizes. In the case of binary outcomes, the results should be presented in a table showing the number of people who have been classified as positive and negative by the experimental test among the group of participants [2].

The two most prevalent methodologies for ascertaining the overall effect size in a meta-analysis are the fixed-effects model and the random-effects model. The two models have different presumptions about the data [4]. Some of the studies considered in a meta-analysis may be more precise than others, thus may carry more information. Instead of calculating a simple mean of the effect sizes, a weighted mean is evaluated, with more weights assigned to the studies which carry more information and smaller weights assigned to others. Under the fixed-effects model, it is presumed that there is only one true effect size that is stored by the studies and the differences in the effect sizes are solely attributed to the sampling error. The invalid speculation for the random-effects model is that the mean effect is zero [13]. Larger studies may yield better results, but each study is calculating a study-specific true effect; further in practice, the studies may be conducted under different clinical settings. Therefore, in a random-effects model, the presumption is that the studies are drawn as a random sample from a large population of studies. Thus, under the random-effects model, the analysis is less likely to be dominated by larger studies. The weights given to individual studies in case of a fixed-effects model are evaluated as:

$$W_i = 1/v_i \tag{2.1}$$

I.e. The weight assigned to a given study is equal to the inverse of its variance [14]. In case of a random-effects model, an extra factor of between-study variability τ^2 is also added to the variability of each study.

$$W_i^* = 1/v_i^* \quad \text{where } V_i^* = V_i + \tau^2 \tag{2.2}$$

2.1 Forest Plots

The final estimates of a meta-analysis are often reported graphically in a forest plot where the studies are represented as rows (Figs. 1 and 2).

Here, each block represents a point estimate and the size of the block represents the weight given to the study. The lines exuding from the studies represent the ninety-five percent confidence intervals associated with the studies; it means that there is

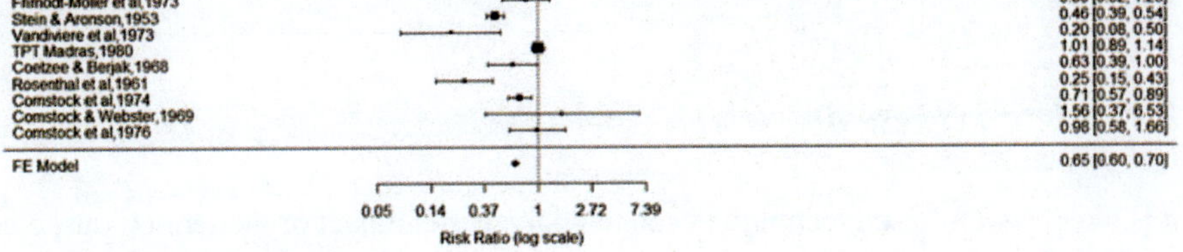

Fig. 1 Forest plot when $\tau^2 = 0$

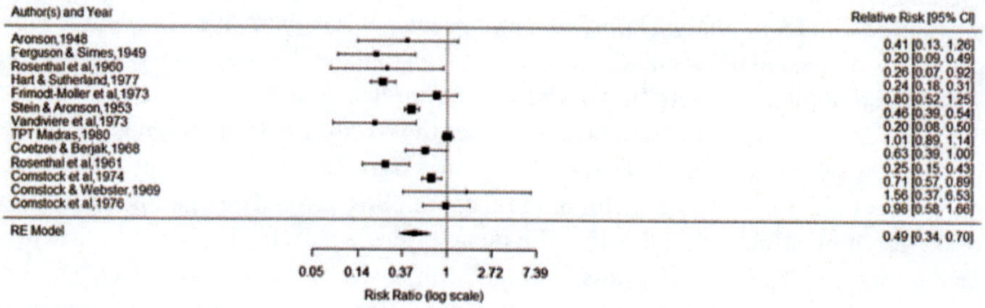

Fig. 2 Forest plot when $\tau^2 \neq 0$

a ninety-five percent chance that the true effect of the study will lie in the specified range [15]. It is clear from the plots that the weights are assigned in a much more balanced way when the between-study heterogeneity is considered while pooling the effect sizes.

2.2 *Moderator Variables and Subgroup Analysis*

Another source of between-study heterogeneity could be that there are slight differences in the study design [16]. These differences are caused by moderator variables. Moderator variables influence the relationship between an independent and a dependent variable by increasing, decreasing and even reversing a relationship. In the BCG dataset, the study location may act as a moderator variable because of the protection provided by the non-pathogenic mycobacteria against tuberculosis [6]; further the effectiveness of the vaccine may change over time [17]. Due to the moderator variables, the studies may be classified into two or more subgroups. The analysis of these subgroups involves pooling the effect size of each subgroup and comparing the effects of these subgroups; the comparison is done by calculating the standard error of the differences between subgroup effect sizes [16]. If the subgroups to be examined represent a fixed level of characteristics, then a fixed-effects model for the comparison of subgroups can be used because no sampling error is introduced at the subgroup level. If the subgroups were randomly sampled like the study location in the BCG dataset, then it will be better to use a random-effects model.

2.3 *Meta-regression*

It is a regression-based technique used to analyze the impact of moderator variables on the effect sizes associated with individual studies, and the overall effect size is calculated after performing the meta-analysis. Again, meta-regression can be performed using the fixed-effects model as well as the random-effects model. The effect size θ for study k for different values of the moderators has to be evaluated in a meta-analysis [16].

$$\theta_k = \theta + \beta_1 x_{1k} + \beta_2 x_{2k} + \cdots + \beta_n x_{nk} + \eta_k + \varepsilon_k \tag{2.3}$$

Here η_k denotes the sampling error, i.e., the difference between the true effect size and the effect size of the given study. In case of a fixed-effects model, $\varepsilon_k = 0$, while in case of a random-effects model, the true effect is also assumed to be sampled from a distribution of effect sizes (Figs. 3 and 4).

Clearly, the confidence interval bounds for the fixed-effects model are narrower in comparison to the random-effects model. This is because the true effects associated with the studies are not solely based upon the variance but also consider the heterogeneity between the studies.

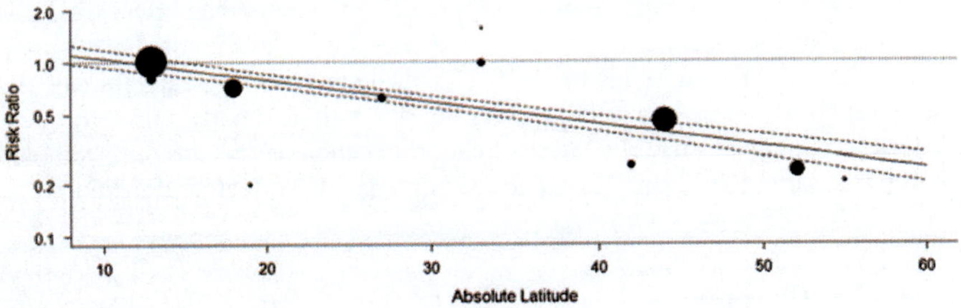

Fig. 3 Scatter plot for fixed-effects model taking absolute latitude as moderator

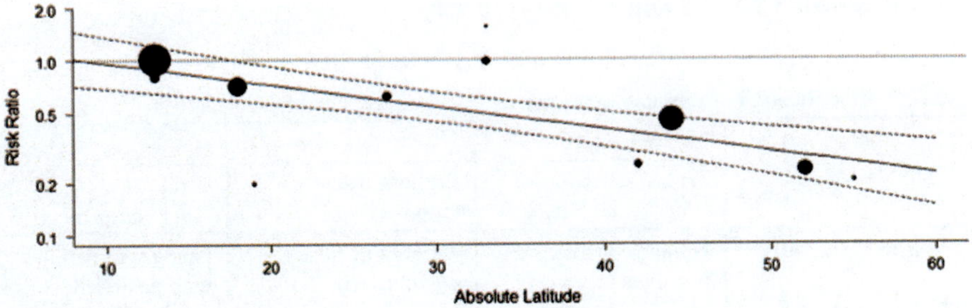

Fig. 4 Scatter plot for random-effects model taking absolute latitude as moderator

3 Popular Algorithms

Many methods have been suggested for pooling the effect sizes assuming a fixed-effects model or a random-effects model. In case, of a fixed-effects model, simply the mean of the effect sizes is evaluated, while in the random-effects model, these algorithms estimate τ^2 in a slightly different way leading to somewhat different pooled effect sizes and confidence intervals. The bias corresponding to each of the approaches often depends upon the context. In this paper, the DerSimonian and Laird approach is discussed in the random-effects model and the Mantel Haenszel method for the fixed-effects model.

3.1 Mantel Haenszel Method

Mantel Haenszel is a popular approach for pooling the individual effect sizes associated with the studies under a fixed-effects model. Only the weighted average of the effect sizes is evaluated in this method. The most common approach for calculating the effect sizes in case of stratification is the risk ratio and odds ratio. Stratification is done as explained in Table 1. Stratification is beneficial because it is simple and intuitive and provides a simple way to examine data for effect modification.

3.1.1 Relative Risk

Relative risk is the proportion of the likelihood of an event occurring in the treatment group to the likelihood of an event occurring in the control group; it is mainly concerned with Cohort studies. It is given as:

Table 1 Classification of sample population as false positives and false negatives

Experimental test (control)	Reference test		
	With disease (case)	Without disease (control)	Total
Positive (exposed)	Number of true positives (a)	Number of false positives (b)	Number of people with positive test results (P)
Negative (unexposed)	Number of false negatives (c)	Number of true negatives (d)	Number of people with negative test results (N)
Total	Number of people with disease (N)	Number of people without disease (ND)	Number of people without the disease (T)

$$\text{R.R.} = \frac{a/(a+b)}{c/(c+d)} \tag{3.1}$$

The further the relative risk is from one, the stronger is the association. If relative risk = 1, then there is no association. If the relative risk is greater than one, it implies increased risk, and if it is less than one, it means decreased risk. The Mantel-Haenszel estimator for the adjusted risk ratio is evaluated as the weighted average of the relative risk for individual studies.

$$\text{RR(MH)} = \frac{\sum_{i=1}^{k} w_i \text{RR}_i}{\sum_{i=1}^{k} w_i} \tag{3.2}$$

where w_i is equal to s. Thus,

$$\text{RR(MH)} = \frac{\sum_{i=1}^{k} ((c_i + d_i)/n_i)a_i}{\sum_{i=1}^{k} ((a_i + c_i)/n_i)c_i} \tag{3.3}$$

3.1.2 Odds Ratio

Odds ratio expresses the strength of the association between two events. It is calculated by dividing the odds of a given event in the presence of another event by the odds of the event in the absence of the second event. For the given dataset, odds ratio will be the ratio of the odds of finding exposure to the vaccine in someone with the disease compared to the odds of finding an exposure with someone without the disease.

$$\text{OR} = \frac{a/c}{b/c} \tag{3.4}$$

The value for an odds ratio may vary from 0 to infinity; if the odds ratio (OR) is equal to one, then the exposure and the disease are not related to each other, and if OR > 1, then there is a positive correlation between the two and vice versa. The further is the odds ratio from one, the stronger is the association. The odds ratio can be calculated for both case-control and cohort studies. Mantel Haenszel approach for calculating the odds ratio is given as

$\text{RR(MH)} = \frac{\sum_{i=1}^{k} w_i \text{OR}_i}{\sum_{i=1}^{k} w_i}$ where the weight w_i is the weight given to individual studies which in this case is $\frac{b_i c_i}{n_i}$.

$$\text{OR(MH)} = \frac{\sum_{i=1}^{k} \left(\frac{a_i * d_i}{n_i}\right)}{\frac{b_i * c_i}{n_i}} \tag{3.5}$$

where $n_i = a_i + b_i + c_i + d_i$. The Peto method can be used to evaluate the summary log odds ratio.

3.2 DerSimonian and Laird Approach

In a random-effects model, the only difference is in the calculation of the weights given to each of the studies. By far, the most popular approach to conduct a random-effects meta-analysis is the DerSimonian and Laird method. It uses Cochran's Q for evaluating the between-study variance which is calculated as follows.

$$\tau^2 = \frac{Q - (k - 1)}{D} \tag{3.6}$$

where $Q = \sum_{i=1}^{k} w_i y_i^2 - \frac{[\sum_{i=1}^{k} w_i y_i]^2}{\sum_{i=1}^{k} w_i}$ and $D = \sum w_i - \frac{\sum w_i^2}{\sum w_i}$.

The summary estimate will be computed as

$$\frac{\sum_{i=1}^{k} w_i^* y_i}{\sum_{i=1}^{k} w_i^*} \tag{3.7}$$

It is by far the most often used estimator [1]. Studies have shown that DerSimonian and Laird estimates are appropriate when the number of studies included in the analysis is large and the between-study variability is small further. Maximum likelihood estimates have better estimates for a between-study variance because the DerSimonian and Laird method may produce overly narrow confidence intervals due to the underestimation of the between-study variability [9].

3.3 Comparison

See Table 2.

4 Heterogeneity

Heterogeneity refers to the variation in study outcomes in different studies. Heterogeneity is always present in a systematic review at one level or another. Generally, the studies included in a meta-analysis use different methods and performed at different locations by different teams. These differences will lead to a significant difference in the underlying parameters associated with the studies [10]. Heterogeneity can be

Table 2 Comparison of fixed-effects and random-effects model

	Fixed-effects model	Random-effects model
Summary effect	The true effect is assumed to be the same in all studies and the mean effect is the overall estimate of the common effect [13]	The true effect size varies from study to study
Weights	Larger studies are assigned more weights	Weights are assigned in a much more balanced way
Overall effect size	Larger studies determine the overall effect size	Large studies have relatively smaller pull on the mean
Confidence intervals	Smaller	Larger
Between-study variance	All studies are assumed to represent an identical population ($\tau^2 = 0$)	The studies represent a random population, thus ($\tau^2 \neq 0$)
Implementation	Mantel-Haezel	DerSimonian and Laird

of two types, clinical and statistical. Clinical heterogeneity is always present in a meta-analysis; it occurs due to the differences of the patients involved in the study settings and the interventions used. While in statistical heterogeneity, individual trials have results that are not consistent with each other. Another source of variation in a meta-analysis is that the studies included in the analysis may have utilized various thresholds to characterize positive or negative outcomes. If such a threshold exists, the points will show a curvilinear pattern on a ROC curve. The commonly used tests for heterogeneity are Cochran's Q, Higgins and Thompson's I^2 and H^2 tests.

4.1 Cochran's Q

The Q statistic is defined as the deviation of the individual study effect sizes from the pooled effect size which is then squared, weighted and summed [16]. It follows the chi-square distribution; the null hypothesis for chi-square distribution is that all studies are evaluating the same effect a high p value for this test suggests homogeneity [11].

$$Q = \sum_{i=1}^{k} w_i \left(y_i - \frac{\sum_{i=1}^{l} w_i y_i}{\sum_{i=1}^{l} y_i} \right) \tag{4.1}$$

Q increases with the number of studies and their precision. Q does not depend upon the measure and value of tau-square [11]. Further, there is no rule of thumb to know the amount of heterogeneity present in the meta-analysis, since it is a degree two statistic its value varies from zero to infinity. The test has low statistical power if the dataset is small like in most practical situations, and it has excessive power when a large number of studies are included in the analysis which leads to the detection of clinically unimportant heterogeneity [18].

4.2 *Higgins and Thompson's* $\mathbf{I^2}$

The I^2 test describes the % of variation between studies that is not due to chance.

$$I^2 = \frac{(Q - \text{d.f.}) * 100}{Q} \tag{4.2}$$

If I^2 value is negative, then we make it zero; here, d.f. denotes the degrees of freedom which is 1 less than the number of studies included in the analysis. An I^2 value of 0% is considered as no heterogeneity, 25% is considered as low heterogeneity, 50% as moderate and 75% as high heterogeneity. A large Q statistic compared to d.f. indicates the presence of heterogeneity. I^2 is the percentage of Q statistic that cannot be explained by the within-study variance; in other words, it is the percentage of the between-study variance associated with the studies.

4.3 $\mathbf{H^2}$ *Test*

H^2 is characterized by the relative overabundance in Q over its degree of freedom

$$H^2 = \frac{Q}{\text{d.f.}} \tag{4.3}$$

In case the value of the Q statistic is smaller than the degree of freedom, then the H^2 value is set to 1. Since it measures the variation from the degree of freedom, it depends upon the between-study variability [11].

4.4 *Identifying Best Measure of Heterogeneity*

The tests of heterogeneity may detect clinically unimportant heterogeneity when the dataset is large or they may have low power in case the number of studies included in the meta-analysis is small [19]. The criterion for evaluating the amount of heterogeneity should not depend upon the scale of measurement [11]. It should not be easily affected by statistical power and the sample size of the studies [16]. It should take into account the between-study variability, and it should be easy to quantify. The amount of heterogeneity obtained by applying the random-effects model on the

Table 3 Heterogeneity results for various summary measures under the fixed-effects model

Measure	Tau-square	Cochran's Q p value < 0.0001	Higgin and Thompson I^2 (%)	H^2 test
Relative risk	0	152.5676	92.13463	12.71396
Odds ratio	0	163.9426	92.68036	13.66188
Risk difference	0	386.7759	96.89743	32.23133

Table 4 Heterogeneity results for different measures for between-study variability

Measure	Tau-square	Cochran's Q	Higgins and Thompson I^2 (%)	H^2 test
DerSimonian and Laird	0.3088	152.2330	92.12	12.69
Hedges estimator	0.3286	152.2330	92.56	13.44
Sidik Jonkman	0.3455	152.2330	92.90	14.08

Table 5 Comparison of tests for heterogeneity

	Cochran's Q	Higgins and Thompson I^2	H^2 test
Between-study variance	Independent	Increases with increase in τ^2	Increases with increase in τ^2
Range	0 to infinity	0–100%	1 to infinity
Quantifiable	No rule of thumb	Easy to interpret	No rule of thumb
Statistical power	Easily affected	Does not depend on sample size	Independent of sample size

BCG dataset on varying the scale of measurement and the method of pooling is summarized below (Table 3).

I^2 value is least affected by changing the summary measure, while Q and H^2 values are dramatically affected by the scale of measurement the change being 7% each when the scale changes from the risk ratio to odds ratio and 153% when it changes to risk difference. Further, Q and H^2 values are not quantifiable, so it cannot be interpreted how significant this difference is in actual practice. The Q test does not take into account the between-study variability as shown in Table 4. The tau-square value changes when the effect sizes are pooled by a different estimator, but the Q value remains the same in each case (Table 5).

5 Conclusion

It can be concluded that the random-effects model for conducting a meta-analysis is more relevant in practical situations because of the differences in clinical conditions and study design. The selection of the appropriate model depends upon the between-study heterogeneity and moderator analysis. There are various tests proposed for heterogeneity analysis, but the I^2 test is a better choice because it has a rule of thumb, considers between-study variability and is not affected by statistical power or the scale of measurement.

References

1. R. DerSimonian, N. Laird, Meta-analysis in clinical trials. Controll. Clin. Trials **7**, 177–188 (1986)
2. V. Abraira, *Statistical Methods*, vol. 9 (n.d.)
3. AB Haidich, Meta-analysis in medical research. Hippokratia **14**, 29–37 (2010)
4. M. Borenstein, L.V. Hedges, J.P.T. Higgins, H.R. Rothstein, A basic introduction to fixed-effect and random-effects models for meta-analysis. Res. Synth. Methods **1**, 97–111 (2010)
5. M. Shaaban, *Meta-analysis: Effect of BCG Vaccine on Tuberculosis Test Outcome* (2016)
6. K.J. Cummings, Tuberculosis control: challenges of an ancient and ongoing epidemic. Public Health Rep. **122**, 683–692 (2007)
7. T.D. Spector, S.G. Thompson, The potential and limitations of meta-analysis. J. Epidemiol. Community Health **45**, 89–92 (1991)
8. R.M. Rosenfeld, Meta-analysis. Outcomes Res. Otorhinolaryngol. **66**(4), 186–195 (2004)
9. A.A. Veronika, D. Jackson, W. Viechtbauer, et al., Methods to estimate the between-study variance and its uncertainty in meta-analysis. Res. Synth. Methods **7**(1), 55–79 (2015)
10. J.P.T. Higgins, Heterogeneity in meta-analysis should be expected and appropriately quantified. Int. J. Epidemiol. **37**, 1158–1160 (2008)
11. M. Pathak, S.N. Dwivedi, S.V.S. Deo, S. Vishnubhatla, B. Thakur, Which is the preferred measure of heterogeneity in meta-analysis and why? A revisit. Biostat. Biometr. Open Access J. **1** (2017), 14–20
12. J.I.E. Hoffman, *Biostatistics for Medical and Biomedical Practitioners* (Academic, San Diego, 2015), pp. 645–653
13. M. Borenstein, L.V. Hedges, J.P.T. Higgins, H.R. Rothstein, *Introduction to Meta-Analysis* (2009)
14. M. Borenstein, L. Hedges, H. Rothstein, *Meta-Analysis Fixed Effect vs. Random Effect* (2007), p. 162
15. K. Gurusamy, Interpretation of forest plots (n.d.)
16. M. Harrer, P. Cuijpers, T.A. Furukawa, D.D. Ebert, *Doing Meta-Analysis in R: A Hands-on Guide* (2019)
17. I. Abubakar, L. Pimpin, C. Ariti, et al., Systematic review and meta-analysis of the current evidence on the duration of protection by bacillus Calmette-Guérin vaccination against tuberculosis. Health Technol. Assess. **17**(37), 1–372 (2013)
18. S.G. Thompson, J.P.T. Higgins, Quantifying heterogeneity in a meta-analysis. Stat. Med. **21**(11), 1539–1558 (2002)
19. M. Mittlböck, H. Heinzl, A simulation study comparing properties of heterogeneity measures in meta-analyses. Stat. Med. **25**(24), 4321–4333 (2006)

Stress Prediction Model Using Machine Learning

Kavita Pabreja, Anubhuti Singh, Rishabh Singh, Rishita Agnihotri, Shriam Kaushik, and Tanvi Malhotra

Abstract Stress has become an integral and unavoidable part of our lives. It has created an alarming situation for the mental health of teenagers and youth globally. At the critical juncture of teenage to adulthood transition, many challenges are faced by teenagers that too with exposure of social networking devices. Hence, it is imperative to learn about various factors that cause stress and identify those features that are more significant contributors so that appropriate measures can be taken to cope up with it effectively. This paper is a step toward analyzing stress among students of a few educational institutions in India. The data have been collected from 650 respondents using Likert scale of 5. With the application of different data visualization techniques and random forest regressor algorithm, 15 important contributing factors from a list of 25 features have been identified and the prediction of stress level has been done with a R-squared value of 0.8042.

Keywords Stress · Emotional intelligence · Supervised learning · Data transformation · Data visualization · Feature selection · Random forest regressor · Coefficient of determination

1 Introduction

It was reported in the Economic Times of India that 89% of Indian population is suffering from stress and nearly 75% do not feel comfortable talking to a medical professional [1]. This was reported on the basis of the survey conducted by Cigna TTK Health Insurance. The survey also revealed that globally 86% of the population suffered from stress on an average. But the percentage was much higher in the millennials also known as Generation Y (born between 1980 and 1994), being at

K. Pabreja (✉) · A. Singh · R. Singh · R. Agnihotri · S. Kaushik · T. Malhotra
Department of Computer Applications, Maharaja Surajmal Institute, an Affiliate of GGSIPU, New Delhi, India
e-mail: kavitapabreja1@gmail.com

P. Bansal et al. (eds.), *Proceedings of International Conference on Artificial Intelligence and Applications*, Advances in Intelligent Systems and Computing 1164,
https://doi.org/10.1007/978-981-15-4992-2_6

a 95% high. The major factors that contribute to this were found to be work and finances.

A study done in USA, Lhasa OMS revealed that 78% of the participants thought that life is more stressful now, than in the previous generation. Some major factors were job, health, politics, and technology [2]. The study also reported that stress has become an important issue in the last few years. It reported that World Health Organization has added Burnout to its list of diseases caused due to prolonged stress and can lead to long-term physical and mental exhaustion, decreased efficiency, and increased mental distance.

Though the upcoming generation, Generation Z, which consists of the youth who were born in mid-1990s to early 2000s and are presently in high schools or colleges, is the most aware and forthcoming for receiving help from professionals [3] but also seem to have similar experiences related to stress.

There are numerous studies by researchers all over the world for detecting, analyzing, and predicting stress.

A system for stress detection using image processing and machine learning techniques has been developed [4] that monitors and analyzes facial expressions of a person working in front of the computer. The video captured was divided into three sections and analyzed. The image analysis included the calculation of the variation in the position of the eyebrow from its mean position on that person's face. This system is non-intrusive and runs in real time. To assess the detection performance, they conducted experiments on 19 collected datasets each consisting of 18 images of an individual. Their system integrates image processing and deep learning to detect stress. The data obtained from the images were used to train linear regression model.

In another work, the study of the potential mechanisms between physical exercise and well-being in the context of Leisure suggested that leisure time physical exercise contributes to effective problem focused coping through elicitation of positive emotion in the human body [5]. Phenomenological research was conducted on 9 students, 7 males and 2 females. They conducted the study using 10 open-ended questions in an interview and found that coping with stress can thus lead to overall well-being. A team of researchers have developed a system to detect stressful behavior of a person by analyzing measurements obtained by the wearable sensors [6]. In particular, they focused on real-time physiological responses such as electrodermal activity and photoplethysmogram. They used a sociometric badge to measure the social activity including body movement and voice; they also used physiological sensor (more suitable in controlled environment). The measurements obtained from the sensors were used to differentiate between stressful and neutral situations after being processed. They conducted a Trier Social Stress Test (TSST) on the 18 participants. They utilized State trait Anxiety Inventory (STAI) to measure the levels of anxiety. They concluded that these sensors are in fact useful for real-time stress detection although the physiological sensor could be a little uncomfortable.

Another study [7] found that stress is caused due to lack of time to exercise, lack of socializing and sleeping disturbances. They also found that overeating and loss of self-control are not related to stress. Their study shows that people prefer music when stressed, and if workload is a factor in stress, it should be decreased. They conducted a survey to collect data from 100 students. They made use of descriptive and inferential statistics to process their results. They also made use of SPSS for data analysis.

A study on working professionals in the Nigerian Construction Industry has been done which concluded that sources of stress among them include insufficient finance and resources, shortage of staff, managing or supervising work of other people, inability to delegate work, long work hours, and poor remuneration [8]. High stress has negative impacts on the lives and is the root cause of depression and even memory loss. Stress weakens the mental, emotional, thinking and knowing ability and negatively impacts their work productivity. The professionals thus adopt coping techniques which can help lessen the stress. They made use of simple descriptive statistical analyses tools for this study. A study on stress management using statistical tools has been done that aimed to find stress levels, personality type of the employees of National Thermal Power Corporation Ltd., using a questionnaire which was prepared after counseling with an officer [9]. The authors conducted the study on 150 participants and concluded that only a small percentage had high stress levels and needed counseling at individual and organizational level. The study also found that there is a significant relation between stress and demographic factors like age, experience, and designation. It also found that though there are stress levels (usually lower) among the employees, they can be dealt with and reduced effectively. In a similar study to understand the effect of workplace stress, the authors [10] concluded that workplace stress plays an important role in physiological and psychological health of employees. They also found that it affects productivity and performance of the employees. They conducted a study of 72 employees. They analyzed the data and interpreted it to conclude that excessive workload and organizational conflicts were two major factors of stress at workplace. They found that problems like mental disturbances, emotional disturbances, lifestyle changes, behavioral problems and even physical effects of stress affected the overall productivity and environment of the organization.

Higher Secondary School students of Imphal, Manipur, have been studied and analyzed [11] for the stress that they are going through. This study concludes that a greater number of students studying in high school or 12th standard were suffering from more depression and stress than other classes. 81.6% students reported having at least one of the three disorders such as depression, anxiety, and stress. They conducted the survey using a questionnaire which utilized depression anxiety stress scale (DASS) and socio-demographic features. The survey had 830 participants. This study also found that the females had a higher stress level than their male peers in the same standard.

The relationship of anxiety and stress with working memory performance in a large non-depressed sample has been explored using linear mixed models [12] to test the predictive power of the self-report measures of the two factors and background factors on working memory (WM) performance in three categories such as verbal, visuospatial, and n-back. The authors used perceived stress scale (PSS-4) to provide subjective assessment on stress in daily life. They also made use of six item state trait anxiety inventory (STAI-6). They found that there was only a shift/drift toward a negative association between transient anxiety and WM performance. They did not find any relationship between stress and WM performance. They observed that increased anxiety is related to worse WM performance.

The psychometric properties of the Depression, Anxiety and Stress Scales (DASS-21) using Rasch analysis, utilizing two different administration mode (pen and paper test, Internet delivery) has been done on a sample size of 745 respondents [13]. The authors had to remove one item each of the DASS-21 to fit it into the model. But they finally concluded that the results provided support for the measurement properties, internal consistency reliability, and one-dimensionality of three slightly modified DASS-21 scales. The scale combining anxiety and stress factors showed adequate psychometric properties. In another study, based on database analysis of depression and anxiety in a community sample-response to a Micronutrient Intervention, it was found that the participants in the intervention program (which was directed towards chronic disease prevention, that provided nutritional supplements) showed improvement in self-reported depression and anxiety [14]. Sixteen thousand people from Canada participated in this study. It also concluded that nutritional levels played significant role in mental health and poor nutritional values could lead to the development of mental illnesses. They said, broad spectrum supplements with a focus on vitamin D can provide new standards to the treatment and prevention of mental diseases. They made use of statistical analysis using the SPSS software, along with a few other tests in this study.

It has been found that there exists a strong relationship between stress and emotional intelligence (EI). In a study based on questionnaire on Emotional Quotient Inventory by bar-on and occupational stress index [15], filled in by 103 managers working for various private sectors organizations, the authors have concluded that there is a negative correlation between EI and work stress. Another study based on usage of media among nursing students provided observation that emotional intelligence is related to the type of device used and not to the total time spent using media [16]. The total time spent using media was related to the type of stressors students faced and their way of dealing with the stressors. They made use of socio-demographic data, media and computer. They also made use of media and computerized sheets which included questions related to the devices they used, how often they used them, how they perceived these devices, and their roles in their lives. They conducted the Schutte Self-Report Emotional Intelligence Test (SSEIT) which measures EI based on self-reported responses to 33 items, and within them, there are three negatively worded items. Family functioning also affects stress indirectly as stress depends on EI which depends on family environment. A latest study provides evidence that family functioning significantly influences youth trait EI. Family turned out to be a major factor influencing the child in all its development stages. A healthy family contributes to a healthy individual, with clear mind and insight. The study was conducted over 457 youths. The authors [17] made use of Family Adaptability and Cohesion Evaluation Scale III (FACES III) to measure family functioning. Trait Emotional Intelligence Questionnaire (TEIQue-SF) was used to measure youth trait EI, and SPP (analyses spiral point patterns) was used for the data analysis.

Also, there has been a surge in the usage of technology devices for exchanging digital content among people at far off places with the advent of Internet and mobile communication. Specifically, the youth has adapted such high technology devices and has been spending too much time on social networking Websites and devices which are causing anxiety and stress.

On the basis of all these studies, it was observed that in order to address this social cause of understanding stress levels among school children and youth of India, it is worthwhile to explore the impact of physical environment, work place environment, home environment, emotional factors and impact of social media and devices on the well-being of individuals. A total of 25 questions, five under each of the five mentioned categories, were drafted, and the questionnaire was floated to collect responses from school and college students on a Likert scale of 5. The responses of 653 participants were collected and analyzed using Python language, and a machine learning algorithm was trained so as to predict the stress level.

2 Research Methodology

The students from an Under Graduate Program have conducted a survey, and the data that have been collected are a part of the study under their minor project. This study is based on actual, real data collected from 653 students of various colleges and schools, of which, majority are from in and around the state of Delhi. Students of Grade 10th, 11th, 12th, and various under graduate courses are the ones that have been included in this study. On the basis of five broad factors affecting the students, a questionnaire was made with five questions under each factor. The five affecting factors are as follows: social media and devices, home environment, emotions, work environment, and physical environment. Apart from the questionnaire of 25 questions on these factors, the students were also asked to provide demographic data, *viz.* name, email address, age, gender, school/college name, class/UG course, and the year of their UG course which are some of the features that may play a role as to how much stress affects a person. Name and Email addresses were asked only for the purpose of eliminating incorrect and redundant data to help with the preprocessing. Data cleaning was done to eliminate redundant and incorrect data. Also, the sub-factors that did not have any affect or had a general effect on the study were also eliminated. These steps under data preprocessing have been explained below.

2.1 Data Cleaning

In order to get rid of redundant data, duplicated records (by checking name and email addresses with Python commands) have been removed from dataset.

2.2 Data Transformation

Each question in the questionnaire was assigned ordinal values as options to be selected. Corresponding to each of the options, a numerical score was assigned in range 1–5 depending on selected option. Depending on these responses (which are predictors or features that determine stress score), stress score was calculated (as predictand field) with a range of 1–5 for all the records. Here, higher score means higher stress. Different categories to numeric score values were assigned by making use of following range values.

1–2.33 = low-level stress
2.34–3.67 = medium-level stress
3.68–5 = high-level stress.

2.3 Data Reduction

Elimination of few features that were analyzed to be least important in determining or calculating the stress score was performed on the basis of few visualizations. Few plots for the same have been shown under the 'Data Visualization' section. Also, by using recursive feature elimination function, least important features were eliminated.

3 Data Visualization

In order to get deep insight into stress levels of respondents in terms of observing the patterns, identifying extreme values, the following visualizations were done and their findings are also mentioned.

- Histogram for frequency of students versus the stress scores from 1 to 5 showed that usually the score lies between 2.5 and 3.5 which means most of the respondents suffer from medium level of stress.
- Bar plot for observing frequency of participation of males and females showed that participation of males was greater than females.
- Bar plot to get insight into frequency of participants with respect to different age groups made evident that the students aged 17–21 years are the ones whose participation was maximum.
- Bar plot for number of students' course-wise and class-wise (10th, 11th, 12th) was visualized, and it was observed that maximum participation was from B.Tech. course, followed by BCA and B.Com. programs.

- Bar plot for count of participants according to year in which they are enrolled was visualized, and it was found that maximum participation was from 3rd year college students irrespective of the course.
- Grouped bar plot showing frequency of males and females for each category of level of stress was also generated, shown in Fig. 1, and it was observed that there are a greater number of females suffering from medium stress level as compared to males. Also, another important finding is that there are more number males in comparison with females who are suffering from high level or low level of stress.
- Grouped bar plot for visualization of stress level course-wise. As most of the respondents are from BCA and B.Tech. programme, the visualization of stress level only for these two courses was done, shown in Fig. 2. It was observed that the higher percentage of B.Tech. students are suffering from high stress as compared to BCA student. Also, it was observed that there are more BCA students with medium stress level than B.Tech. students.
- Lastly, the scatter plot and hexbin plot for each of the 25 features versus the stress score were generated. Through these plots, it was observed that very less people consume alcohol/tobacco. Ultimately, four features were observed to be

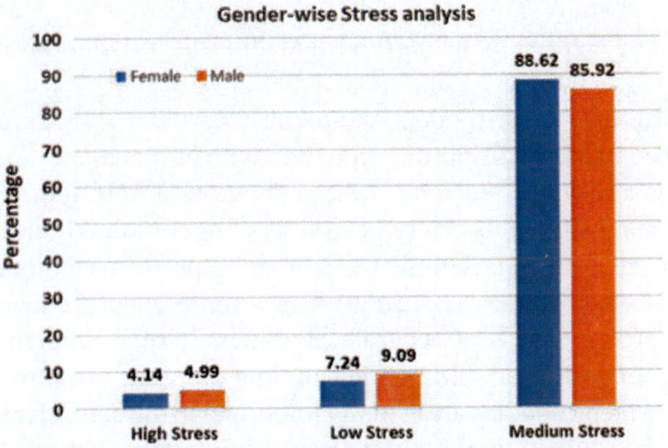

Fig. 1 Grouped bar plot showing frequency of males and females for each category of stress

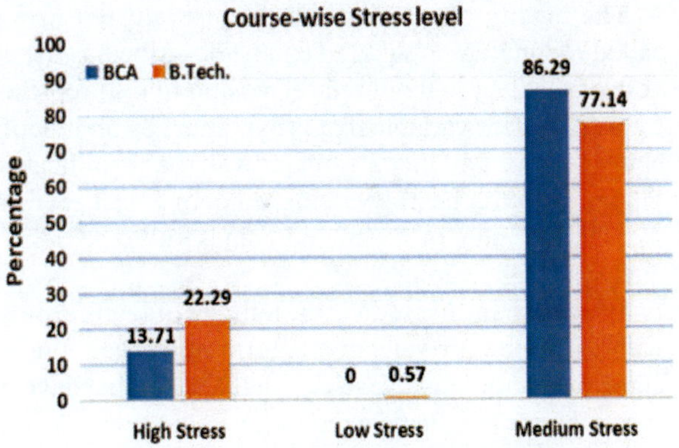

Fig. 2 Grouped bar plot for visualization of stress level course-wise

contributing very less in the actual stress score on the basis of these plotting. So, they were dropped. These features included: ability to show empathy, relationship skills, consumption of alcohol/tobacco, decision making for choosing own career path.

4 Application of Machine Learning Algorithm

Supervised learning has been applied in the process of prediction of stress value. It can be defined as the method where we train the machine using labeled data. In this process, we have input variables (X) and output variable (Y) and we try and fit an algorithm so as to learn the mapping function from the input to the output variables ($Y = f(X)$). We do this to find the best possible mapping function, so that when we input similar but new input data we are able to predict the output variables for that new set of data. We have split the dataset into two parts in 4:1 ratio, where we used 80% of records to train the machine and the 20% to test the machine. Out of various types of supervised learning algorithms, we have used random forest regressor algorithm in our study. This algorithm makes use of ensemble learning method for classification and regression. In this type of method, the predictions from multiple machine learning algorithms are combined together so as to give a more accurate prediction than an individual model. A random forest can thus be called a meta-estimator which combines multiple decision trees with proper alterations. In this algorithm, we prevent overfitting as each tree takes a random sample adding another random element. The results of decision tree classifiers are combined through model votes or averaging, into a single ensemble model that outperforms an individual decisions tree's output.

Its advantages are that it is a more accurate learning algorithm and it usually produces a very accurate classifier. It runs efficiently on larger datasets; it does not delete variables while handling thousands of input variables. It also lets us know which variables are actually important to the classification. It has an important method for estimating missing data and maintaining accuracy while handling larger amount of missing data.

The quality of classifier has been evaluated using coefficient of determination, called R-squared. R-squared can be described as the statistical measure of the closeness of the data to the fitted regression line. It represents the percentage of variance for a dependent variable that is explained by an independent variable in a model. It is better to have a higher R-squared value as it indicates that that the particular model fits our data the best. By using random forest regressor algorithm on the collected dataset, the value of R-squared comes out to be 0.8376 which indicates a very good quality of regression.

In order to further reduce the number of features, we have tried the experimentation with use of feature selection algorithm on the reduced dataset. It is a process where algorithm automatically selects the features that have the most effect on the prediction variable or output that we require. Feature selection reduces overfitting, which means less redundant data, thus preventing us from making decisions based on un-necessary data. It thus improves accuracy and reduces training time as there is less data to go through.

Recursive feature elimination (RFE) recursively removes the features that do not contribute much to the prediction variable and builds a model with the remaining features. It uses model accuracy to identify those features that help us predict attributes that are more important. Upon application of RFE, top 15 contributing features were chosen, as mentioned here:

Headache, job awareness, academic pressure, vocal expression, unhealthy influence by social media, workload, anxiety/nervousness, physical health maintenance, relationship at work, work stress, digital distraction, sleep cycle, time pressure, financial pressure, technology obligations.

RFE showed true for these features and false for the rest of the features which were then dropped from the data frame. Again, the random forest regressor algorithm was run, and it gave R-squared value as 0.8042. As we can see, using only 15 features out of a total of 21 features (4 features were dropped earlier out of 25 features, by visualizing their relationship with stress score), the R-squared value is reasonably acceptable. Hence, the approach for feature selection is data visualization followed by recursive feature elimination which has produced quite convincing results.

A scatter plot of the 'Predicted Stress Score' (Y.pred) versus the 'Actual Stress Score' (Y_test) is shown in Fig. 3. As it can be observed that predicted value of stress score closely follows that of actual value of stress score, corresponding to the test data.

Further, based on the ranks of the selected features, Table 1 shows their relative importance. In this table, we have listed top 10 features that are most affecting for causing features.

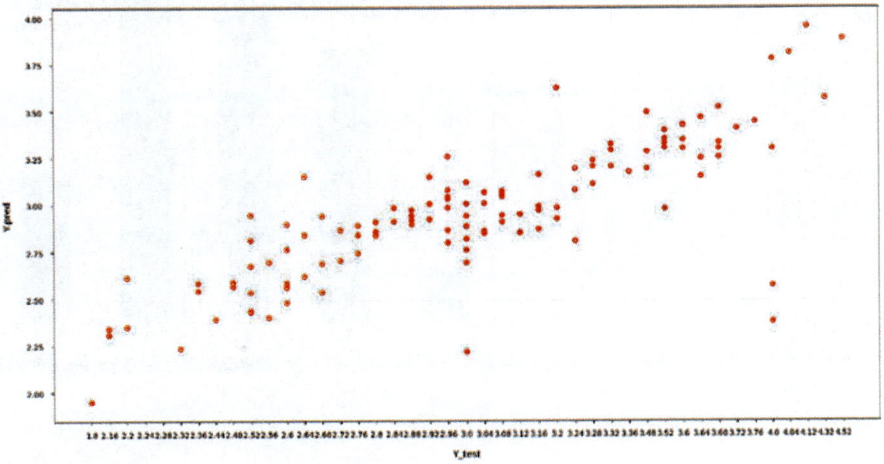

Fig. 3 Plot for predicted value of stress versus actual value of stress for testing data

Table 1 Top 10 features that cause stress

Rank	Name of feature
1	Anxiety/nervousness
2	Work stress (hard time at work/ college)
3	Unhealthy influence through social media
4	Relationship at work/school/college (strained relationships at work)
5	Workload (neglecting tasks due to high workload)
6	Disturbed sleep cycle
7	Technology obligations (want to leave technology but cannot due to work)
8	Job awareness
9	Vocal expression (ability to describe our own feelings)
10	Digital distraction (social media and devices, a distraction from real life)

On the basis of relative ranking of stress causing features, top 3 features in decreasing order of importance, are anxiety/nervousness, work stress (hard time at work/college), and unhealthy influence through social media. A bar graph showing the response of students to the most important stress causing feature, *viz.* anxiety/nervousness has been generated shown in Fig. 4. As it is apparent from the bar graph, there are a greater number of participants suffering from anxiety in the category of sometimes and often.

A summary of frequency of type of response to these three features having important ranks in determining stress levels is shown in Table 2.

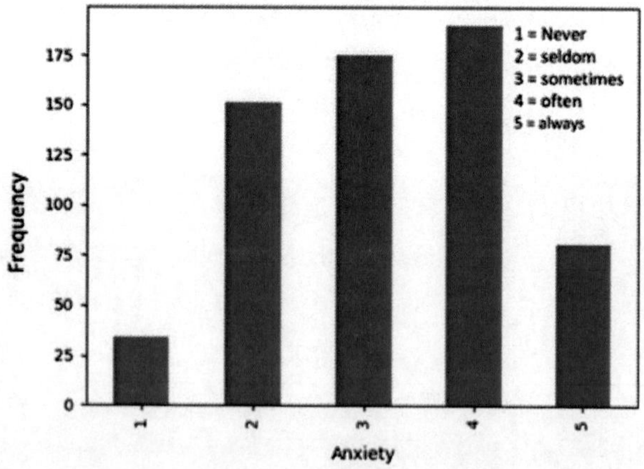

Fig. 4 Frequency of participants for different values of anxiety

Table 2 Detailed response of participants for top 3 stress causing factors

Response	Rank 1—anxiety	Rank 2—having hard time at work/college	Rank 3—unhealthy influence of social media
Never	34	114	213
Seldom	151	195	74
Sometimes	175	189	202
Often	190	80	75
Always	81	53	67

5 Results, Discussion and Conclusion

Based on the study of various research papers and survey conducted from students of various schools and colleges, it seemed that most of them are not highly stressed but they are not stress free as well. The broad category of medium-level stress contained most of the participants.

Under high stress category, there are 22.29% of B.Tech. students in comparison to 13.71% of BCA students and there are almost nil students having low stress which is not a good indicator for as far as mental health of Generation Z is concerned. Another finding is gender based; it is observed that there are more males (4.99%) in comparison with females (4.14%) under high stress category.

Through the survey and after making vital use of data visualization using Python, four factors (consumption of alcohol/tobacco, decision making ability for choosing own career path, showing empathy, relationship skills) were found to be not contributing well as a cause of stress among Generation Z. The obtained dataset having 21 features was split in 4:1 ratio for training and testing the machine. Using random forest regressor, a fine level of accuracy for obtaining the stress level results was attained, with coefficient of determination equal to 0.8376. After making use of the RFE, it was then discovered that the top fifteen factors that are contributing the most for causing stress were: Job awareness, academic pressure, vocal expression, unhealthy influence from social media, workload, anxiety or nervousness, physical health maintenance, relationship at work, work stress, digital distraction, sleep cycle, time pressure, financial pressure, and technology obligations.

The dataset was revised again by retaining only these 15 features and fitting the same machine learning algorithm. With this reduced dataset, we have got R-squared value of 0.8042 which is reasonably acceptable as we have eliminated 6 more features which is almost 30% reduction in number of features. Alike Generation Y, few factors like work pressure seem to be a constant factor for causing stress. However, it has been discovered that use of technological devices has a great hand in causing stress among Generation Z. It is also evident from Table 2 that unhealthy influence of social media devices has been impacting the youth too much, though there are two more features, *viz.* anxiety and work stress that precede this feature in determining stress.

References

1. Article on "89 percent of India's population suffering from stress; most don't feel comfortable talking to medical professionals". Available from: https://economictimes.indiatimes.com/magazines/panache/89-per-cent-of-indias-population-suffering-from-stress-most-dont-feel-comfortable-talking-to-medical-professionals/articleshow/64926633.cms?from=mdr
2. Article on "Generation Z is stressed, depressed and exam-obsessed". Available from: https://www.economist.com/graphic-detail/2019/02/27/generation-z-is-stressed-depressed-and-exam-obsessed
3. Article on "Millenials and Gen Z believe life is more stressful for them than previous generations by Edoardo Liotta". Available from: https://www.vice.com/en_in/article/gyzgx4/millennials-and-gen-z-believe-life-is-more-stressful-for-them-than-for-previous-generations
4. N. Raichur, N. Lonakadi, P. Mural, Detection of stress using image processing and machine learning techniques. Int. J. Eng. Technol. (IJET) **9**(3S) (2017). https://doi.org/10.21817/ijet/2017/v9i3/170903s001
5. J.H. Kim, L.A. McKenzie, The impacts of physical exercise on stress coping and well being in university students in the context of leisure. Health, **6**, 2570–2580 (2014). Available from: http://dx.doi.org/10.4236/health.2014.619296
6. O.M. Mozos, V. Sandulescu, S. Andrews, D. Ellis, N. Bellotto, R. Dobrescu, J.M. Ferrandez, Stress detection using wearable physiological and sociometric sensors. Int. J. Neural Syst. **27**(2), 1–17 (2017)
7. W. AbdelKader, M. Elnakeeb, The Relationship between the use of media and emotional intelligence among youth nursing students. IOSR J. Nurs. Health Sci. **6**(5), 63–77 (2017). https://doi.org/10.9790/1959-0605016377
8. M. Alavi, S.A. Mehrinezhad, M. Amini, M. Kaur, P. Singh, Family functioning and trait emotional intelligence among youth. J. Health Psychol. **4**(2) (2017). https://doi.org/10.1177/2055102917748461
9. S. Manhas, A. Sharma, Manisha, Relationship between quality of life and emotional intelligence of the sample youth. Int. J. Curr. Res. Rev. **7**(2), 1–5 (2015)
10. S. Kadry, M. Kbaysi, S. Al-Safadi, D. Al-Bakri, Stress causes and outcomes statistical analysis. Biom. Biostat. Int. J. **7**(4), 353–358 (2018). https://doi.org/10.15406/bbij.2018.07.00229
11. T.O. Oladinrin, O. Adeniyi, M.O. Udi, Analysis of stress management among professionals in the Nigerian construction industry. Int. J. Multidiscip. Curr. Res. **2** (2014)
12. D.O. Shemesh, Youth Emotional Intelligence as related to adaptive coping with stress encounters. Psychol. Res. **7**(1), 1–19 (2017). https://doi.org/10.17265/2159-5542/2017.01.001
13. K.R. Reddy, *A Study on Stress Management* (National Thermal Power Corporation Ltd, 2008)
14. K.S. Kumar, B.S. Akoijam, Depression, anxiety and stress among higher secondary school students of Imphal, Manipur. Indian J. Community Med. **42**(2), 94–96 (2017). https://doi.org/10.4103/ijcm.ijcm_266_15
15. M. Karolina, L.O. Waris, A. Soveri, M. Lehton, M. Laine, The Relationship of anxiety and stress with working memory performance in a large non-depressed sample. Front. Psychol. (2019). https://doi.org/10.3389/fpsyg.2019.00004
16. T.L. Shea, A. Tennant, J.F. Pallant, Rasch model analysis of the Depression, Anxiety and Stress Scales (DASS). J. BMC Psychiatry (2009). https://doi.org/10.1186/1471-244x-9-21
17. M. Samantha, S.M. Kimball, N. Mirhosseini, J. Rucklidge, Database analysis of depression and anxiety in a community sample-response to a micronutrient intervention. Nutrients **10**(2) (2018). https://doi.org/10.3390/nu10020152

Finger Vein Recognition Using Deep Learning

Bhavya Chawla, Shikhar Tyagi, Rupav Jain, Archit Talegaonkar, and Smriti Srivastava

Abstract The finger vein biometric offers the perfect balance between security and economic viability and thus has gained a lot of attention in recent years, offering benefits over other conventional methods such as being least susceptible to identity theft as veins are present beneath the human skin, being unaffected by ageing of the person, etc. These reasons provide enough motivation to develop working models that would solve the ever-increasing need for security. In this paper, we have investigated the finger vein recognition problem. We have used deep convolutional neural network models for feature extraction purposes on two commonly used publicly available finger vein datasets. To improve the performance further on unseen data for verification purposes, we have employed one-shot learning model namely the Triplet loss network model and evaluated its performance. The extensive set of experiments that we have conducted yield classification and correct identification accuracies in ranges upwards of 95% and equal error rates less than 4%.

Keywords Finger Vein · Biometrics · Convolutional neural network · Deep learning · One-Shot learing · Siamese network · Triplet loss netowork · Identification · Verification

B. Chawla (✉) · S. Tyagi · R. Jain · A. Talegaonkar · S. Srivastava
Netaji Subhas University of Technology, New Delhi, India
e-mail: bhavyamkark98@gmail.com

S. Tyagi
e-mail: shikhart98@gmail.com

R. Jain
e-mail: rupavwinchester@gmail.com

A. Talegaonkar
e-mail: archit0303@gmail.com

S. Srivastava
e-mail: smriti.nsit@gmail.com

P. Bansal et al. (eds.), *Proceedings of International Conference on Artificial Intelligence and Applications*, Advances in Intelligent Systems and Computing 1164,
https://doi.org/10.1007/978-981-15-4992-2_7

1 Introduction

The biometric recognition systems can be classified based on physical traits and behavioural traits. Physical biometric systems rely on the face, hand and iris and fingerprint whereas behavioural biometric systems rely on gait, signature, keystroke pattern and speech. Many of the above techniques are vulnerable to spoof attacks which means the identity of a person is not secure. Finger veins are internal components of the body which makes them harder to forge as compared to fingerprints. The finger vein patterns cannot be acquired without a person's consent and form a live feature. The medical sciences have proved that finger vein pattern is unique for each individual even for twins. In addition to this, finger vein is immune to ageing which makes it sensible to adopt them. Even if there is a deformity in the finger, the finger veins remain intact or we can use an alternate finger for verification. These important characteristics have inspired many to develop recognition techniques to provide secure authentication.

In this paper, we propose to perform finger vein recognition by exploiting deep learning techniques. The aim of our work is to achieve highly accurate and stable performance over the considered databases. We also discuss one-shot learning techniques that provide an advantage in case less training data is available and prove to be quite powerful for when the purpose is verification only. The obtained results come from a single proposed architecture that was applied across all used datasets and techniques without any variation.

2 Literature Review

Most studies in the field of finger vein recognition have concentrated on the extraction of Region of Interest (ROI). Yang and Li [1] suggested the use of interphalangeal joint prior method and steerable filter get ROI and feature vector, respectively. To identify an individual, the nearest neighbour classifier was applied and the accuracy reported was 98.7%. Guan et al. [2] took this a step further by applying filters to remove noise and used two-direction weighted (2D) 2LDA to get a feature vector. However, the accuracy was only 94.69%. Yang et al. [3] improved the above research by removing noise, enhancing the image and performing brightness enhancement. The results showed that accuracy reached 100% by relying on template matching but execution time got compromised. Gupta and Gupta [4] combined multi-scaled match filtering with line tracking to extract the features with an Equal Error Rate (EER) of 4.47%. The above studies relied on conventional methods and hence were not robust enough until the development of artificial intelligence.

Additional methods focusing mainly on pre-processing and feature extraction methods are mentioned here. Yang et al. [5] proposed a method of extracting the features into 16 types of filters considering 2 scales, 8 channels and 8 center frequencies of Gabor filters. Peng et al. [6] to make the model robust to scaling and rotation, designed an 8-way filter. It selects the optimal parameters of the Gabor Filter to extract finger veins features and applies the Scale Invariant Feature Transform (SIFT) algorithm to the features. Shin et al. [7] combined Gabor filtering with Retinex filtering using fuzzy inference system. Yang and Yang [8] and Zhang and Yang [9] used even-symmetric multi-channel Gabor filters with four directions, and grey-level grouping (GLG) and circular Gabor filter (CGF) to improve upon the contrast, respectively. Apart from these Gabor filter techniques, Pi et al. [10] used elliptical and edge-preserving HPFs to remove blurriness, and preserve the edges.

A relevant number of state-of-the-art finger vein based biometric techniques were also studied. The maximum curvature method proposed by Miura et al. [11] would try to extract the center lines of the veins. This is done by calculating local maximum curvatures in cross-sectional profiles of a vein image. The Repeated Line Tracking algorithm was introduced by Miura et al. [12] to extract vein patterns from the captured image. The line tracking algorithm begins by selecting a random pixel in the image. This pixel is known as "current tracking point". This point is moved along the vein pattern and covering each pixel. It involves tracing the veins in the image based on predefined probabilities in both vertical and horizontal directions. This whole process is repeated a specific number of times. Local Binary Pattern or LBP is an algorithm for texture classification introduced by Ojala et al. [13]. The reason for their success is their good performance along with it being computationally simple. The method involves thresholding the points surrounding the central pixel that are greater than or less than the central pixel giving a binary output.

Another study that employed deep CNN models is presented in [14]. In this work, AlexNet [15] architecture is employed. An authentication system was developed based on Euclidean distance between feature vectors of vein images. An Equal Error Rate of 0.21% was achieved.

3 Database Description

We tested our deep neural network architecture on two-finger vein databases provided by University Sains Malaysia (FV-USM) and Shandong University (SDUMLA). The FV-USM database contains images from 123 volunteers recorded in two sessions, for each volunteer images from index and middle fingers were recorded for both hands having six samples of each giving a total of 5904 ($2 \times 123 \times 2 \times 2 \times 6$) images. The SDUMLA database consists of images from 106 volunteers, for each volunteer images from index, middle, ring fingers were recorded for both hands having six samples of each image giving a total of 3816 ($106 \times 2 \times 3 \times 6$) images.

4 Methodology

The classification starts with image processing to extract useful information from the image. Then a deep learning model is used for classification of images across both datasets. Later we discuss one-shot learning techniques that provide a major advantage when the end goal is just to verify the identity of a subject.

We adopted two techniques to build a verification model based on finger vein images of a subject.

4.1 Convolutional Neural Network

Convolutional layer: Convolutional operation performed between the input map X_m^i where i and m are level and map indexes, filters are represented using as $F_{n,m}^i$, where n is filter index, the nth output map Y_n^i of layer i is represented in Eq. (1).

$$Y_n^i = \sum \left(F_{n,m}^i * X_m^i + B_n^i\right) \tag{1}$$

Pooling layer extracts dominant features from the convolved feature map and reduces the dimensionality of the feature map.

Activation Layer contains activation function that decides the final value of the neuron, different activation functions used are ReLu, SoftMax, etc.

ReLu activation function is represented by Eq. (2), where y is the natural output/activation of a particular neuron:

$$F(y) = \max(0, y) \tag{2}$$

SoftMax activation function is represented by Eq. (3), where O_i is the natural output/activation of a particular ith neuron.

$$O^i = \sum \left(F_{i,m} * X_m + B_i\right) \tag{3}$$

After applying SoftMax activation function on O_i for C different classes, we get the following probability distribution of the input data over C different classes given by Eq. (4).

$$P_u = \frac{\exp(O_i)}{\sum_{j=1}^{j=C} \exp(O_j)} \tag{4}$$

4.2 *One-Shot Learning (Triplet Loss Network)*

Triplet loss [16] involves an anchor example and one positive or matching example (same class) and one negative or non-matching example (differing class). The loss function penalizes the model such that the distance between the matching examples is reduced and the distance between the non-matching examples is increased.

Triplet Loss tries to bring close the anchor (current record) with the positive (A record that is in theory similar with the anchor) as far as possible from the negative (A record that is different from the anchor). The loss is computed as given by Eq. (5).

$$\text{LOSS} = \sum_{i=1}^{N} \left[||f_i^a - f_i^p||_2^2 - ||f_i^a - f_i^n||_2^2 + \alpha \right]_+ \tag{5}$$

where,

f_i^a = feature encoding of anchor image,
f_i^p = feature encoding of positive match of anchor,
f_i^n = feature encoding of negative match of positive,
α = bias value (hyper-parameter).

5 Proposed Architecture

We have implemented finger vein recognition using two techniques, the first one is using *Convolutional Neural Network* and other one is using *One-Shot Learning Model, i.e. Triplet Loss Network* model. Architecture for both of these techniques is explained below.

5.1 *Convolutional Neural Network*

The model as shown in Fig. 1 is inspired by architectures as proposed in AlexNet [15] and Meng et al. [14]. First layer in CNN is a convolution layer with 96 kernels each of 11 × 11 dimensions, with stride of (4, 4) and activation as 'ReLu'. Second layer is Max Pooling with pool size of (3, 3) and stride of (2, 2). Third layer is Batch Normalization layer. Fourth layer is convolution layer with 256 kernels each of 5 × 5 dimensions with stride of (1, 1) and activation as 'ReLu'. Fifth layer is Max Pooling with a pool size of (3, 3) and stride of (1, 1). Sixth layer is Batch Normalization layer. Seventh layer is convolution layer with 256 kernels each of 3 × 3 dimensions with stride of (1, 1) and activation as 'ReLu'. Eighth layer is Max Pooling with pool size of (3, 3) and stride of (1, 1). Now we have flattened the above-obtained output and feed it into the fully connected network containing two hidden layers and one output

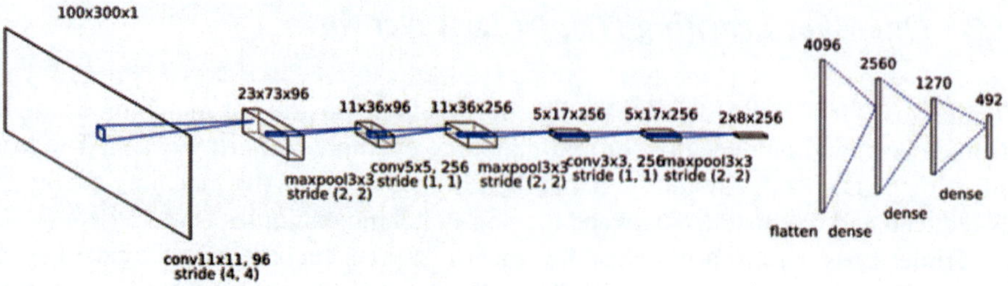

Fig. 1 Proposed CNN architecture

layer, first hidden layer with 2560 neurons and second hidden layer with 1270 neuron, both of these layers have activation function as 'ReLu'. Last layer is an output layer with number of neurons as the number of classes in the output. Since for SDUMLA dataset output classes were 636 then in that case number of output neurons will be 636 and in the case of FV-USM dataset output classes were 492 then, in that case, the output layer will have 492 neurons in it.

5.2 *Triplet Loss Network*

Figure 2 captures the triplet network model architecture. It uses the same nested sequential CNN architecture with changes in the input and output layers and the loss function. The triplet loss network works not to classify the input but to minimize the triplet loss function. A triplet will contain three images—the anchor image (a), the positive match (p) and negative match (n) (Table 1).

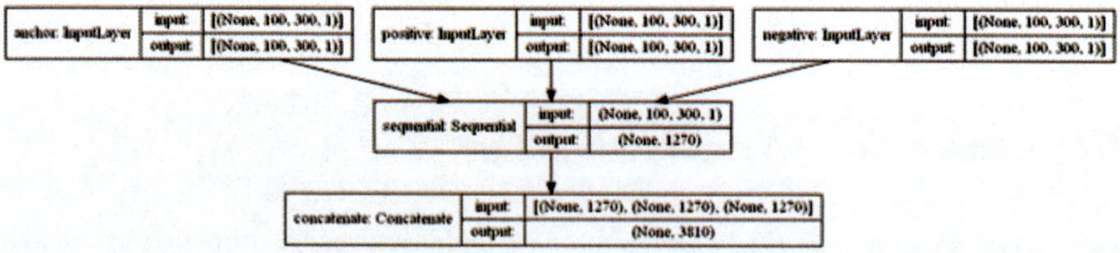

Fig. 2 Triplet loss model architecture

Table 1 Dataset split description

Dataset	Total classes C	Training classes C_1	Validation classes C_2	Testing classes C_3
FV-USM	492	320	80	92
SDUMLA	636	416	100	120

Out of C_1 classes of the training dataset, C_N classes equal to the batch size N are chosen randomly without replacement. Out of these, two samples of the image belonging to class C_i form the anchor and its positive match. A second category C_j is chosen such that $j \neq i$ which forms the negative match to the anchor. N such triplets are generated for each of C_N classes forming the anchor image. Three input layers were added corresponding a triplet to the same sequential CNN architecture as used in the previous section. The encodings for all three inputs were obtained by passing them through the convolutional network and were concatenated. The target corresponding to each triplet is nothing but zero (desired value of loss function).

Once trained, the weights are saved and a validation model, say TNEURAL_CODES is derived from it that uses the same weights and nested sequential model but instead of triplet input it takes a single input and outputs its encodings. For validation, N_T classes are chosen at random from the validation dataset (C_2). The first class from the obtained set is set as TRUE_CLASS having the corresponding TRUE_IMAGE. A support_set is built with all the chosen N_T classes and a random sample of their respective image (including another sample of the TRUE_CLASS image). Now, encoding for both the TRUE_IMAGE and each image in its support_set is obtained using TNEURAL_CODES. Because our model has been trained to minimize triplet loss function, the minimum mean square error obtained between TRUE_IMAGE and all images in support_set should correspond to the other sample of TRUE_IMAGE in the support_set. Hence if the classes corresponding to TRUE_IMAGE and the image in support_set that gave the least mean square error are the same, we have correctly identified the TRUE_IMAGE. This process is carried out on K number of iterations. Thus, if N_C classes were predicted successfully the accuracy is given by Eq. (6).

$$\text{Accuracy} = 100 * \left(\frac{N_C}{K}\right)\% \tag{6}$$

6 Results and Discussion

The results discussed have been obtained using the optimal training and testing strategies developed in the previous section. The achieved accuracy is proportional to a number of training samples used which is expected in the case of CNN. For the one-shot learning network model accuracies in terms of batch sizes have been studied for test image data that the trained models have never seen before.

Table 2 Accuracy obtained of the model

Dataset	SDUMLA	FV-USM
Training	98.24	99.36
Validation	92.30	94.72
Test set	92.29	97.35

6.1 CNN Classification Model

Table 2 captures the training, validation and testing accuracies of the proposed CNN classification model for respective datasets.

The Equal Error Rate (EER) obtained for the two datasets was **1.42%** for FV-USM dataset and **3.85%** for SDUMLA dataset.

The reason behind the difference in the performance of the model across both datasets is a cause of a different number of training samples, along with the different quality of finger vein images that exist in two distinct datasets.

6.2 Triplet Loss Network One-Shot Learning Model

Figure 3 captures the accuracies achieved with respect to batch for respective datasets.

Average accuracy obtained over test data containing 100 classes from respective datasets that the model has never seen before was **81.73%** for FV-USM dataset and **85.47%** for SDUMLA dataset. For a 10-way one-shot task, i.e. for a test batch of 10 unseen images, the model had an accuracy of **97.00%** for the FV-USM dataset while **94.20%** for the SDUMLA dataset.

The triplet loss network tries to learn how to learn to differentiate between similar and different. Evidently, the accuracies obtained display its powerful discriminative capability and verify the test sample which the model has never seen before.

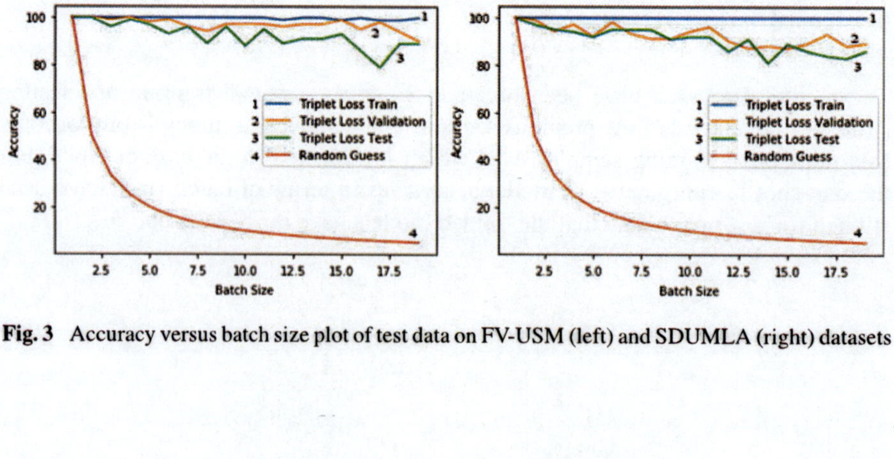

Fig. 3 Accuracy versus batch size plot of test data on FV-USM (left) and SDUMLA (right) datasets

7 Conclusion

In this study, we have proposed a deep convolutional neural network model that performs effective finger vein recognition irrespective of the overall quality of images across various datasets. An exhaustive set of experimental tests performed over two of the publicly available databases has been presented. The obtained results indicate a maximum recognition accuracy of 97.35% with an EER as low as 1.42% using our modified AlexNet CNN architecture. After a highly accurate classification model was obtained, One-Shot Learning Model based on the triplet loss network has been discussed with high correct recognition rates for unseen finger vein images. Thus, sets of unseen finger vein images were correctly verified with accuracies ranging from 100 to 97% for a batch size ranging from 1 to 10 of unseen images highlighting the powerful discriminative nature of the triplet loss network. Lastly, but most importantly, the same CNN architecture has been used across both datasets and both techniques highlighting its robustness and stability while being highly accurate.

References

1. J. Yang, X. Li, Efficient finger vein localization and recognition, in *IEEE International Conference on Pattern Recognition* (2010), pp. 1148–1151
2. F. Guan, K. Wang, Q. Yang, A study of two direction weighted (2D)2LDA for finger vein recognition, in *IEEE International Congress on Image and Signal Processing* (2011), pp. 860–864
3. W. Yang, Q. Rao, Q. Liao, Personal identification for single sample using finger vein location and direction coding, in *IEEE International Conference on Hand-Based Biometrics* (2011), pp. 1–6
4. P. Gupta, P. Gupta, An accurate finger vein based verification system. Digit. Signal Proc. **38**, 43–52 (2015)
5. J. Yang, Y. Shi, J. Yang, Personal identification based on finger-vein features. Comput. Hum. Behav. **27**, 1565–1570 (2011)
6. J. Peng, N. Wang, A.A. Abd El-Latif, Q. Li, X. Niu, Finger-vein verification using Gabor filter and SIFT feature matching, in *Proceedings of the 8th International Conference on Intelligent Information Hiding and Multimedia Signal Processing*, Piraeus-Athens, Greece, 18–20 Jul 2012 (2012), pp. 45–48
7. K.Y. Shin, Y.H. Park, D.T. Nguyen, K.R. Park, Finger-vein image enhancement using a fuzzy-based fusion method with Gabor and Retinex filtering. Sensors **14**, 3095–3129 (2014)
8. J.F. Yang, J.L. Yang, Multi-channel Gabor filter design for finger-vein image enhancement, in *Proceedings of the 5th International Conference on Image and Graphics*, Xi'an, China, 20–23 Sept 2009 (2009), pp. 87–91
9. J. Zhang, J. Yang, Finger-vein image enhancement based on combination of gray-level grouping and circular Gabor filter, in *Proceedings of the International Conference on Information Engineering and Computer Science*, Wuhan, China, 19–20 Dec 2009 (2009), pp. 1–4
10. W. Pi, J. Shin, D. Park, An effective quality improvement approach for low quality finger vein image, in *Proceedings of the International Conference on Electronics and Information Engineering*, Kyoto, Japan, 1–3 Aug 2010, vol. 1 (2010), pp. 424–427
11. N. Miura, A. Nagasaka, T. Miyatake, Extraction of finger-vein patterns using maximum curvature points in image profiles, in *IAPR Conference on Machine Vision Applications,* Tsukuba Science City, Japan, 16–18 May 2005, vol. 8 (2005), pp. 347–350

12. N. Miura, A. Nagasaka, T. Miyatake, Feature extraction of finger-vein patterns based on repeated line tracking and its application to personal identification. Mach. Vis. Appl. **15**, 194–203 (2004)
13. T. Ojala, M. Pietikäinen, T. Mäenpää, Multiresolution gray scale and rotation invariant texture classification with local binary patterns. IEEE Trans. Pattern Anal. Mach. Intell. **24**, 971–987 (2002)
14. Gesi Meng, Peiyu Fang, Bao Zhang, Finger vein recognition based on convolutional neural network. MATEC Web Conf. **128**, 04015 (2017)
15. A. Krizhevsky, I. Sutskever, G. Hinton, ImageNet classification with deep convolutional neural networks. Neural Inf. Process. Syst. **25** (2012)
16. F. Schroff, D. Kalenichenko, J. Philbin, FaceNet: a unified embedding for face recognition and clustering, in *2015 IEEE Conference on Computer Vision and Pattern Recognition (CVPR)*, Boston, MA (2015), pp. 815–823

Machine Learning Applications in Cyber Security and Cryptography

Secure Communication: Using Double Compound-Combination Hybrid Synchronization

Pushali Trikha and Lone Seth Jahanzaib

Abstract In this manuscript, we have synchronized eight fractional order hyper-chaotic systems introducing a novel technique called double compound-combination hybrid synchronization. The synchronization has been achieved by considering four systems as master systems and other four as slave systems using Lyapunov stability theory and active control technique. Also, the application of the synchronization method in secure communication has been discussed.

Keywords Secure communication · Double compound synchronization · Combination synchronization · Hybrid synchronization · Complex fractional order hyper-chaotic system

1 Introduction

It was the groundbreaking work of Lorenz which attracted the interest of many researchers to study the chaos theory. Initially, chaos was not considered usable and was simply considered as a disturbance. Therefore, many techniques were developed to suppress chaos, but it was the pioneering work of Pecora and Carroll who gave the concept of synchronization [1]. Most of the previous studies on synchronization considered only two systems: one master and one slave. However, nowadays two chaotic systems are together synchronized with the third called combination synchronization [2], product of two chaotic systems is synchronized with the third called compound synchronization [3], product of two is synchronized with the product of the other two called double compound synchronization [4, 5], synchronized or anti-synchronized called hybrid synchronization [6]. To generate complex signals, many complex synchronization techniques have also been developed which are composed

P. Trikha · L. S. Jahanzaib (✉)
Department of Mathematics, Jamia Millia Islamia, New Delhi 110025, India
e-mail: lone.jahanzaib555@gmail.com

P. Trikha
e-mail: pushali.t@gmail.com

P. Bansal et al. (eds.), *Proceedings of International Conference on Artificial Intelligence and Applications*, Advances in Intelligent Systems and Computing 1164,
https://doi.org/10.1007/978-981-15-4992-2_9

of more than one synchronization technique such as compound-combination synchronization, combination–combination synchronization and compound difference synchronization [7].

In this paper, we have introduced a new type of synchronization which is more complex than the above-mentioned types of synchronization. This scheme is called double compound-combination hybrid synchronization applicable amongst eight hyper-chaotic systems (four master and four slave systems).

2 Problem Formulation

We first formulate the double compound-combination hybrid synchronization scheme. We consider four chaotic master systems—two scaling master systems and two base master systems.

Scaling master systems:

$$\dot{X}_1 = F_1(X_1) \tag{1}$$

$$\dot{X}_2 = F_2(X_2) \tag{2}$$

Base master systems:

$$\dot{X}_3 = F_3(X_3) \tag{3}$$

$$\dot{X}_4 = F_4(X_4) \tag{4}$$

Here X_1, X_2, X_3, X_4 are, respectively, the state variables of the system (1), (2), (3) and (4), and $F_i : R^n \to R, i = 1, 2, 3, 4$ 3 are continuous functions. Next, we consider four slave systems given by (5)-(8)

Slave systems:

$$\dot{Y}_1 = G_1(Y_1) + \bar{U}_1 \tag{5}$$

$$\dot{Y}_2 = G_2(Y_2) + \bar{U}_2 \tag{6}$$

$$\dot{Y}_3 = G_3(Y_3) + \bar{U}_3 \tag{7}$$

$$\dot{Y}_4 = G_4(Y_4) + \bar{U}_4 \tag{8}$$

Here Y_1, Y_2, Y_3, Y_4 are, respectively, the state variables of the system (5), (6), (7) and (8), $G_i : R^n \to R, i = 1, 2, 3, 4$ are continuous functions, and $\bar{U}_i : R^n \times R^n \times R^n \times R^n \to R, i = 1, 2, 3, 4$ are the controllers to be constructed.

To achieve the double compound-combination hybrid synchronization, we define the error as follows:

$$\begin{aligned} e = [&(A_1Y_1 + A_2Y_2 + A_3Y_3 + A_4Y_4) \\ &- Q(B_1X_1 + B_2X_2)(B_3X_3 + B_4X_4)] \end{aligned} \quad (9)$$

where A_i and B_i, $i = 1, 2, 3, 4$ are $n \times n$ diagonal matrices over R. Here at least one B_i not equal to 0.

We consider the matrix Q as the projective matrix. To achieve the desired synchronization, we must have error tending to zero.

3 System Description

3.1 Fractional Order Hyper-Chaotic Xling Systems

$$\begin{aligned} \frac{d^q x_{11}}{dt^q} &= a_{11}(x_{12} - x_{11}) + x_{14} \\ \frac{d^q x_{12}}{dt^q} &= a_{12}x_{11} - x_{11}x_{13} - x_{14} \\ \frac{d^q x_{13}}{dt^q} &= -a_{13}x_{13} - a_{14}x_{11}^2 \\ \frac{d^q x_{14}}{dt^q} &= a_{13}x_{11} \end{aligned} \quad (10)$$

For $a_{11} = 10$, $a_{12} = 40$, $a_{13} = 2.5$, $a_{14} = 4$ and I.C. (1, 2, 3, 4) system is chaotic as displayed in Fig. 1.

3.2 Fractional Order Hyper-Chaotic Vanderpol System

$$\begin{aligned} \frac{d^q x_{21}}{dt^q} &= x_{22} \\ \frac{d^q x_{22}}{dt^q} &= -(b_{11} + b_{12}y_{13})y_{11} - (b_{11} + b_{12}y_{13})y_{11}^3 - b_{13}y_{12} + b_{14}y_{13} \\ \frac{d^q x_{23}}{dt^q} &= x_{24} \\ \frac{d^q x_{24}}{dt^q} &= -y_{13} + b_{15}(1 - y_{13}^2)y_{14} + b_{16}y_{11} \end{aligned} \quad (11)$$

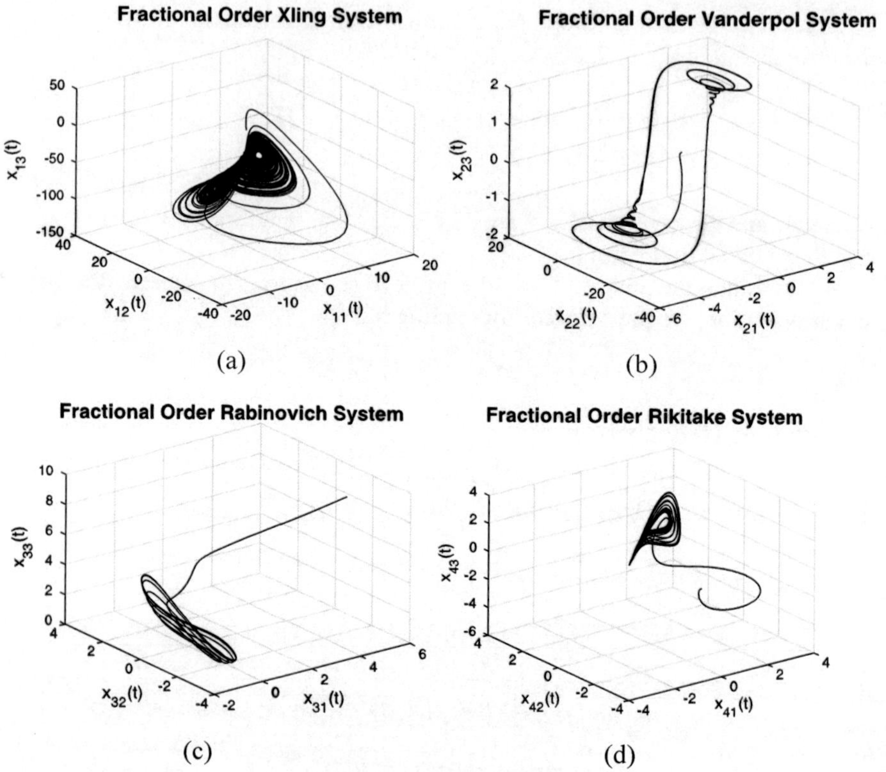

Fig. 1 Phase portraits of master systems

For $b_{11} = 10, b_{12} = 3, b_{13} = 0.4, b_{14} = 70, b_{15} = 5, b_{16} = 0.1$ and I.C. (0.1, −0.5, 0.1, −0.5) system is chaotic as displayed in Fig. 1.

3.3 Fractional Order Hyper-Chaotic Rabinovich System

$$
\begin{aligned}
\frac{d^q x_{31}}{dt^q} &= -c_{11}x_{31} + c_{12}x_{33} + x_{32}x_{33} \\
\frac{d^q x_{32}}{dt^q} &= -c_{12}x_{31} - x_{32} - x_{31}x_{33} + x_{34} \\
\frac{d^q x_{33}}{dt^q} &= -x_{33} + x_{31}x_{32} \\
\frac{d^q x_{34}}{dt^q} &= -c_{13}x_{32}
\end{aligned} \tag{12}
$$

For $c_{11} = 34, c_{12} = 6.75, c_{13} = 2$ and I.C. (5.5, −1.25, 8.4, 2.75) system is chaotic as displayed in Fig. 1.

3.4 Fractional Order Hyper-Chaotic Rikitake System

$$
\begin{aligned}
\frac{d^q x_{41}}{dt^q} &= -x_{41} + x_{42}x_{43} - d_{11}x_{44} \\
\frac{d^q x_{42}}{dt^q} &= -x_{42} + x_{41}(x_{43} - d_{12}) - d_{11}x_{44} \\
\frac{d^q x_{43}}{dt^q} &= d_{12} - x_{41}x_{42} \\
\frac{d^q x_{44}}{dt^q} &= d_{13}x_{42}
\end{aligned} \tag{13}
$$

For $d_{11} = 1.7, d_{12} = 1, d_{13} = 0.7$ and I.C. (3.5, 1.7, −4.5, 2.8) system is chaotic as displayed in Fig. 1.

3.5 Complex Fractional Order Lorenz Chaotic System

$$
\begin{aligned}
\frac{d^q y_{11}}{dt^q} &= l_{11}(y_{13} - y_{11}) \\
\frac{d^q y_{12}}{dt^q} &= l_{11}(y_{14} - y_{12}) \\
\frac{d^q y_{13}}{dt^q} &= l_{12}y_{11} - y_{13} - y_{11}y_{15} \\
\frac{d^q y_{14}}{dt^q} &= l_{12}y_{12} - y_{14} - y_{12}y_{15} \\
\frac{d^q y_{15}}{dt^q} &= y_{11}y_{13} + y_{12}y_{14} - y_{15}
\end{aligned} \tag{14}
$$

For $l_{11} = 10,\ l_{12} = 180$ and I.C. (2, 3, 5, 6, 9) system is chaotic as displayed in Fig. 2.

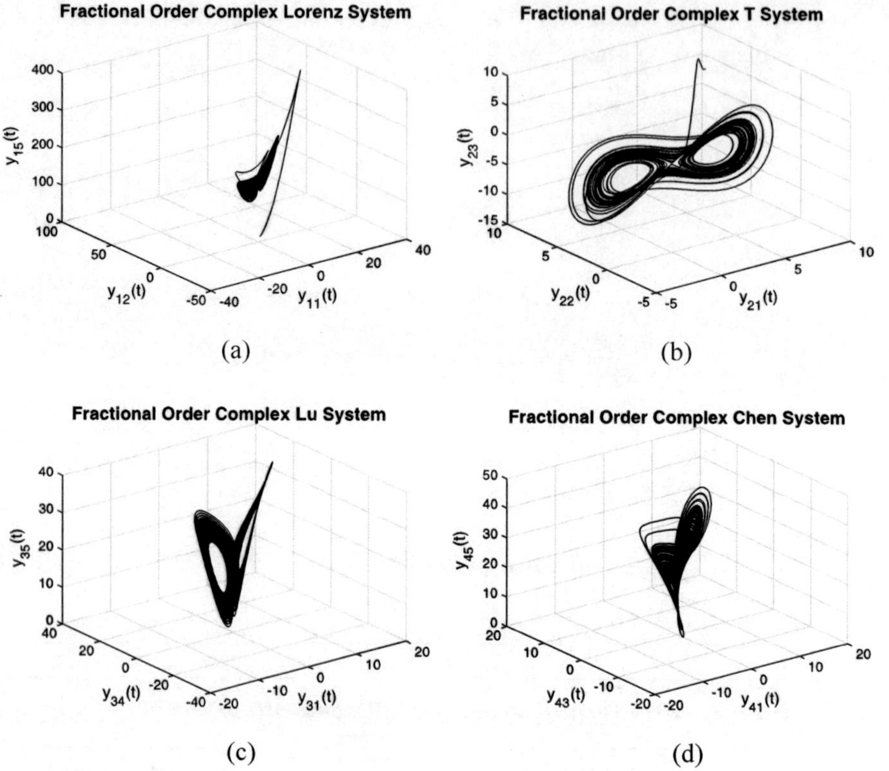

Fig. 2 Phase portraits of slave system

3.6 Complex Fractional Order T Chaotic System

$$
\begin{aligned}
\frac{d^q y_{21}}{dt_q} &= f_{11}(y_{23} - y_{21}) \\
\frac{d^q y_{22}}{dt_q} &= f_{11}(y_{24} - y_{22}) \\
\frac{d^q y_{23}}{dt_q} &= (f_{12} - f_{11})y_{21} - f_{11}y_{21}y_{25} \\
\frac{d^q y_{24}}{dt_q} &= (f_{12} - f_{11})y_{22} - f_{11}y_{22}y_{25} \\
\frac{d^q y_{25}}{dt_q} &= y_{21}y_{23} + y_{22}y_{24} - f_{13}y_{25}
\end{aligned}
\tag{15}
$$

For $f_{11} = 2.1$, $f_{12} = 30$, $f_{13} = 0.6$ and I.C. (8, 7, 6, 8, 7) system is chaotic as displayed in Fig. 2.

3.7 Complex Fractional Order Lu Chaotic System

$$
\begin{aligned}
\frac{d^q y_{31}}{dt^q} &= g_{11}(y_{33} - y_{31}) \\
\frac{d^q y_{32}}{dt^q} &= g_{11}(y_{34} - y_{32}) \\
\frac{d^q y_{33}}{dt^q} &= -y_{31}y_{35} + g_{12}y_{33} \\
\frac{d^q y_{34}}{dt^q} &= -y_{32}y_{35} + g_{12}y_{34} \\
\frac{d^q y_{35}}{dt^q} &= y_{31}y_{35} + y_{32}y_{34} - g_{13}y_{35}
\end{aligned} \tag{16}
$$

For $g_{11} = 40, g_{12} = 22, g_{13} = 5$ and I.C. (1, 2, 3, 4, 5) system is chaotic as displayed in Fig. 2.

3.8 Complex Fractional Order Chen Chaotic System

$$
\begin{aligned}
\frac{d^q y_{41}}{dt^q} &= h_{11}(y_{43} - y_{41}) \\
\frac{d^q y_{42}}{dt^q} &= h_{12}(y_{44} - y_{42}) \\
\frac{d^q y_{43}}{dt^q} &= -h_{12}y_{41} - y_{41}y_{45} + h_{13}y_{43} \\
\frac{d^q y_{44}}{dt^q} &= -h_{12}y_{42} - y_{42}y_{45} + h_{13}y_{44} \\
\frac{d^q y_{45}}{dt^q} &= y_{41}y_{43} + y_{42}y_{44} - h_{14}y_{45}
\end{aligned} \tag{17}
$$

For $h_{11} = 35, h_{12} = 7, h_{13} = 28, h_{14} = 3$ and I.C. = (3, 4, 5, 6, 8) system is chaotic as displayed in Fig. 2.

4 Double Compound-Combination Hybrid Synchronization on Eight Fractional Order Chaotic Systems

We consider the master system as (10)–(13) and the slave systems as (14)–(17) with control inputs $u_{1i}, u_{2i}, u_{3i}, u_{4i}, u_{5i}$ for $i = 1, 2, \ldots, 5$

We take the projective matrix Q as $\begin{bmatrix} 1 & 0 & 0 & 0 \\ 0 & -1 & 0 & 0 \\ 0 & 0 & -1 & 0 \\ 0 & 0 & 0 & -1 \\ 0 & 0 & 0 & 1 \end{bmatrix}$

From (9), considering A_i, and B_i, to be identity matrices, the error system is defined as follows:

$$e_{1i} = (y_{1i} + y_{2i} + y_{3i} + y_{4i}) - (x_{1i} + x_{2i})(x_{3i} + x_{4i}) \tag{18}$$

Therefore, differentiating the error system and substituting values of the derivatives, we design the controllers to have the simplified error system as follows:

$$\frac{d^q e_{1i}}{dt^q} = -K_i e_{1i} \tag{19}$$

where $K_i,\ i = 1, 2, 3, 4, 5 > 0$ are control gain matrix elements.

We now consider Lyapunov function as follows:

$$V(t) = \frac{1}{2}(e_1^2 + e_2^2 + e_3^2 + e_4^2 + e_5^2)$$
$$\frac{D^q V(t)}{Dt^q} \leq e_1\frac{D^q e_1}{Dt^q} + e_2\frac{D^q e_2}{Dt^q} + e_3\frac{D^q e_3}{Dt^q} + e_4\frac{D^q e_4}{Dt^q} + e_5\frac{D^q e_5}{Dt^q}$$

From (19), we have

$$\frac{D^q V(t)}{Dt^q} \leq e_1(-K_1 e_1) + e_2(-K_2 e_2) + e_3(-K_3 e_3) + e_4(-K_4 e_4) + e_5(-K_5 e_5)$$

$$\Rightarrow \frac{D^q V(t)}{Dt^q} \text{ is negative definite.}$$

Therefore, by Lyapunov stability theory, we have error asymptotically converging to zero, i.e. $e_i \to 0,\ i = 1, 2, 3, 4$.

Numerical Simulations and Discussions:

Here for numerical simulations, we have considered control gain matrix elements $K_1 = 1$; $K_2 = 2$; $K_3 = 3$; $K_4 = 4$; $K_5 = 5$. The trajectories of the master and slave systems are shown to be synchronized in Fig. 3. Also, the error plot of the system is shown to converge to zero in Fig. 3 for the initial condition (4.1, 16.675, 6.91, 43.425, 9.575).

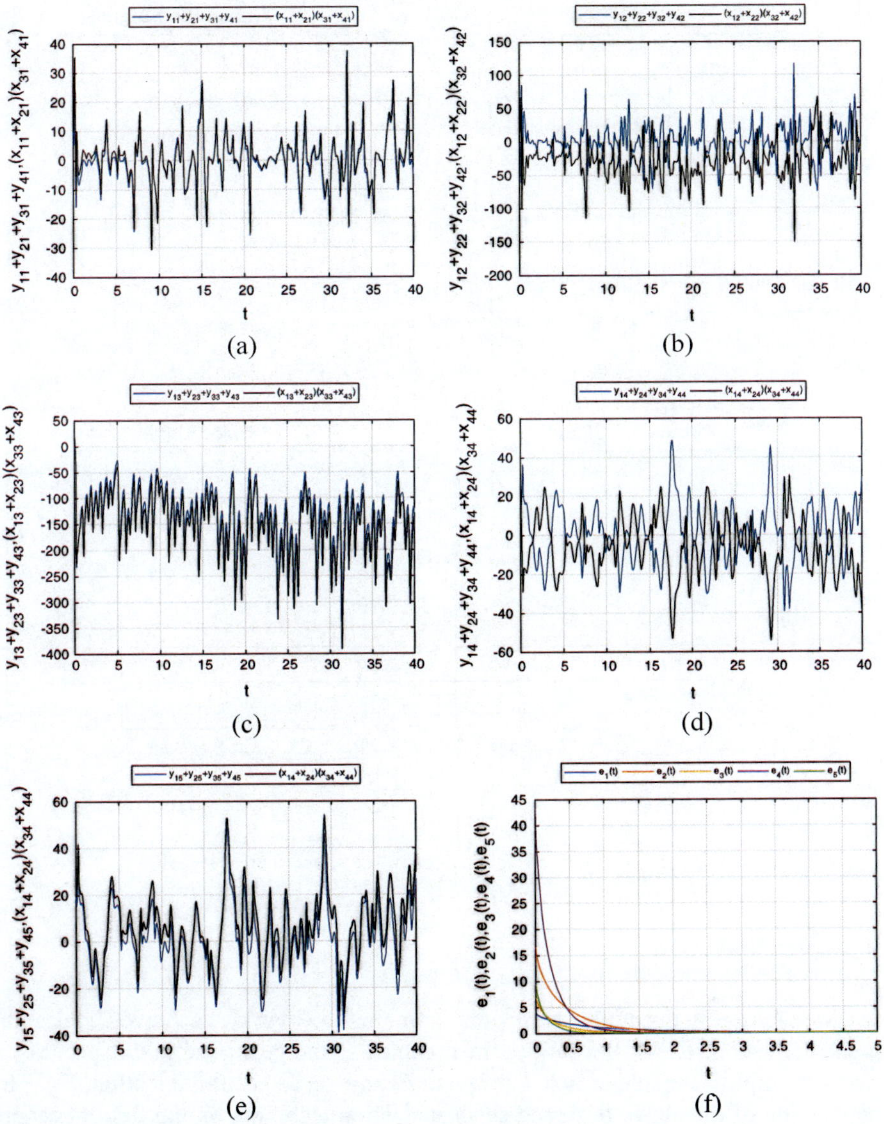

Fig. 3 **a–e** Synchronized trajectories, **f** simultaneous error plot

5 Application in Secure Communication

Let the information signal be $v(t) = \sin(7t)$. We mask it with the chaotic signal $(x_{11} + x_{21})(x_{31} + x_{41})$ to form a nonlinear function say $\bar{v}(t)$. On applying the suitable controllers at the receiving end that synchronizes, we trace back the original message $v^*(t)$. The results have been displayed in Fig. 4.

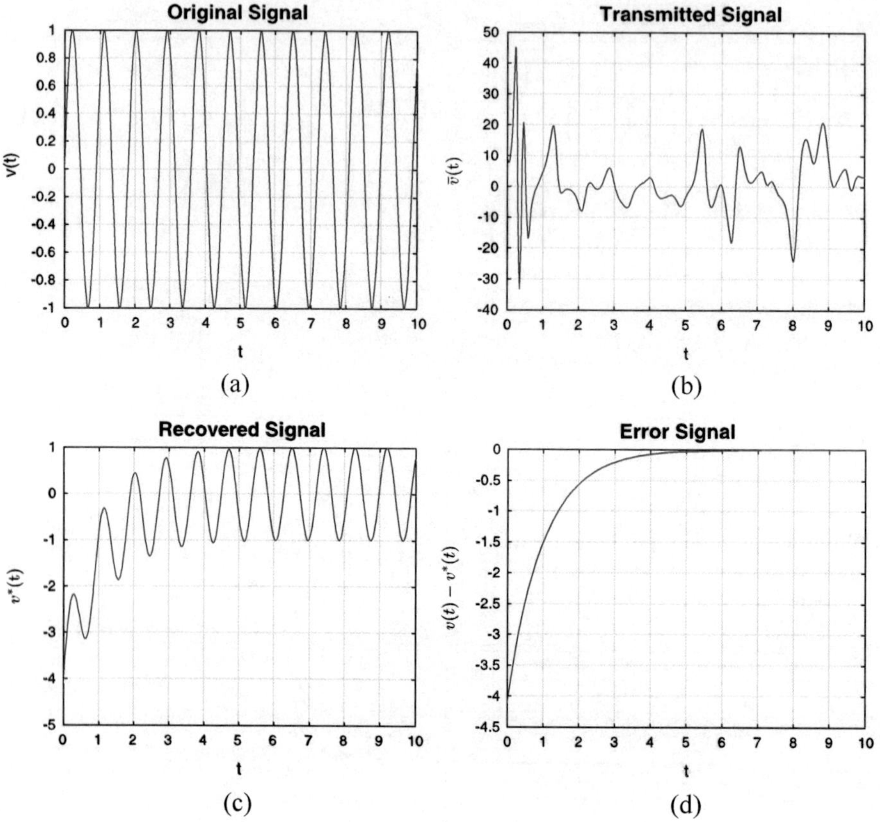

Fig. 4 **a** Original signal, **b** transmitted signal, **c** recovered signal **d** error signal

6 Conclusion

In this manuscript, the double compound-combination hybrid is achieved by designing suitable controllers. Because of the complex signal generated in this process, it would be highly suitable in fields of cryptography, secure communication, etc. The application of the above-designed synchronization technique in the field of secure communication has also been illustrated.

References

1. L.M. Pecora, T.L. Carroll, Synchronization in chaotic systems. Phys. Rev. Lett. **64**, 821–824 (1990). https://doi.org/10.1103/physrevlett.64.821. URL https://link.aps.org/, https://doi.org/10.1103/physrevlett.64.821
2. B. Li, X. Zhou, Y. Wang, *Combination Synchronization of Three Different Fractional-Order Delayed Chaotic Systems. Complexity, Hindawi* (2019)
3. A. Khan, M. Budhraja, A. Ibraheem, Multi-switching dual compound synchronization of chaotic systems. Chin. J. Phys. (2018)
4. Y. Kolokolov, A. Monovskaya, Nonlinear effects in dynamics of hysteresis regulators with double synchronization. in *2018 Moscow Workshop on Electronic and Networking Technologies* (MWENT), IEEE (2018)
5. B. Zhang, F. Deng, *Double-compound synchronization of six memristor-based Lorenz systems, Nonlinear Dynamics Springer* (2014)
6. A. Lassoued, O. Boubaker, *Fractional-Order Hybrid Synchronization for Multiple Hyperchaotic Systems, Recent Advances in Chaotic Systems and Synchronization Elsevier* (2019)
7. A. Khan, P. Trikha, Compound difference anti-synchronization between chaotic systems of integer and fractional order. SN Appl. Sci. **1**, 757 (2019). https://doi.org/10.1007/s42452-019-0776-x

Fractional Inverse Matrix Projective Combination Synchronization with Application in Secure Communication

Ayub Khan, Lone Seth Jahanzaib, and Pushali Trikha

Abstract In this article, the fractional inverse matrix projective combination synchronization has been attained in the presence of external disturbances and uncertainties among fractional-order complex chaotic systems. Application in the field of secure communication has been illustrated with help of an example.

Keywords Fractional inverse matrix projective synchronization · Combination synchronization · Secure communication · Uncertainties and disturbances

1 Introduction

Chaos theory is an emerging branch of mathematics with potential application across various disciplines of science and engineering such as in data encryption, biomedical engineering, dielectric polarization and secure communication. Chaos control, chaos synchronization [1, 2] and anti-control of chaos are all active research areas. In synchronization, dynamics of master system are asymptotically achieved by slave system after being acted upon by controllers. Many synchronization techniques, viz. projective synchronization, difference synchronization [3], compound synchronization [4, 5], dual synchronization, etc. [6], and methods, viz. sliding mode control [7], parameter estimation methods, tracking control [8] etc., have been developed ever since 1990.

In this manuscript, fractional inverse matrix projective combination synchronization has been achieved between two fractional complex chaotic master systems and

A. Khan · L. S. Jahanzaib · P. Trikha (✉)
Department of Mathematics, Jamia Millia Islamia, New Delhi 110025, India
e-mail: pushali.t@gmail.com

A. Khan
e-mail: akhan12@jmi.ac.in

L. S. Jahanzaib
e-mail: lone.jahanzaib555@gmail.com

P. Bansal et al. (eds.), *Proceedings of International Conference on Artificial Intelligence and Applications*, Advances in Intelligent Systems and Computing 1164,
https://doi.org/10.1007/978-981-15-4992-2_10

one fractional complex chaotic slave system in the presence of uncertainties and disturbances. Also, application of the designed synchronization scheme in the field of secure communication has been illustrated.

2 Preliminaries

2.1 Definitions

We have used Caputo's definition of fractional derivative throughout our paper given by [9, 10]

$$ {}_aD_x^\alpha g(x) = \frac{1}{\gamma(n-\alpha)} \int_a^x \frac{g^{(n)}(\tau)\mathrm{d}\tau}{(x-\tau)^{\alpha-n+1}} $$

where n is integer, α is real number $(n-1) \leq \alpha < n$ and $\gamma(\cdot)$ is the gamma function.

3 Problem Formulation

We first consider two n-dimensional fractional-order complex chaotic master systems given by:

Master System I:

$$ \frac{\mathrm{d}^q U}{\mathrm{d}t^q} = A_1 U + F_1(U) + d_1(t) + \Delta G_1(U) \tag{1} $$

Master System II:

$$ \frac{\mathrm{d}^q V}{\mathrm{d}t^q} = A_2 V + F_2(V) + d_2(t) + \Delta G_2(V) \tag{2} $$

where $U, V \in R^n$ are the state vectors, $A_1, A_2 \in R^{n \times n}$ are the coefficient matrices corresponding to the linear part, $F_1(U)$, $F_2(V)$ are the remaining nonlinear part matrices, $\Delta G_1(U)$, $\Delta G_2(V) \in R^n$ represent model uncertainties and $d_1(t)$, $d_2(t)$ are the external disturbances of systems (1) and (2), respectively.

Likewise, next we consider slave system given by:

Slave System I:

$$ \frac{\mathrm{d}^q W}{\mathrm{d}t^q} = A_3 W + F_e(W) + d_3(t) + \Delta G_3(W) + \bar{U} \tag{3} $$

where $\bar{U} \in R^n$ is the controller to be designed.

We define the fractional inverse matrix projective combination synchronization error as:

$$E = (U + V) - MW \tag{4}$$

Therefore, systems (1)–(3) are said to be in fractional inverse matrix projective combination synchronization if $\lim_{t\to\infty} ||E|| = \lim_{t\to\infty} ||(U+V) - MV|| = 0$.

Here, $M \in R^{n\times n}$ is the projective matrix, and $||\,.\,||$ is the Euclidean norm.

Theorem 1 *The disturbed complex fractional-order chaotic systems (1)–(2) are said to be in fractional inverse matrix projective combination synchronization with (3) if we choose the controller as:*

$$\bar{U} = M^{-1}[(A_1 + A_2 + K)MW - K(U + V) - A_1V - A_2U + F_1 + F_2 + \Delta G_1 + d_1 + \Delta G_2 + d_2] - A_3W - F_3 - \Delta G_3 - d_3$$

4 System Description

We here describe the following mathematical models which will be used for numerical simulation.

Fractional-Order Complex Lorenz System:

$$\frac{d^q u_1}{dt^q} = a_1(u_3 - u_1)$$

$$\frac{d^q u_2}{dt^q} = a_1(u_4 - u_2) + 2\cos(4u_2) - 0.5\sin(4t)$$

$$\frac{d^q u_3}{dt^q} = a_2u_1 - u_3 - u_1u_5$$

$$\frac{d^q u_4}{dt^q} = a_2u_2 - u_4 - u_2u_5$$

$$\frac{d^q u_5}{dt^q} = u_1u_3 + u_2u_4 - a_3u_5 - 0.5\cos(4u_5) + 2\cos(4t) \tag{5}$$

where a_1, a_2, a_3 are parameters of system. For parameter values $a_1 = 10$, $a_2 = 180$, $a_3 = 1$ and initial values $(u_1(0); u_2(0); u_3(0); u_4(0); u_5(0)) = (2, 3, 5, 6, 9)$ and $q = 0.95$, the system is chaotic as shown in Fig. 1.

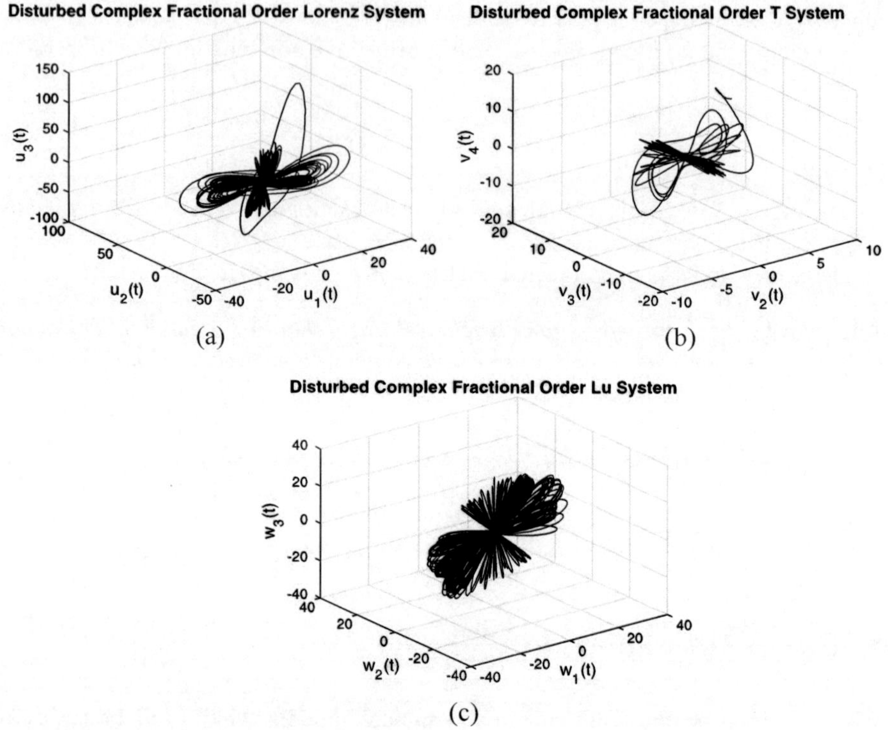

Fig. 1 Phase portraits of the master and slave systems

Fractional-Order Complex *T* System:

$$\frac{\mathrm{d}^q v_1}{\mathrm{d}t^q} = b_1(v_3 - v_1) + 2\sin(2v_1) - 0.5\sin(2t)$$

$$\frac{\mathrm{d}^q v_2}{\mathrm{d}t^q} = b_1(v_4 - v_2)$$

$$\frac{\mathrm{d}^q v_3}{\mathrm{d}t^q} = (b_2 - b_1)v_1 - b_1 v_1 v_5 - 0.5\cos(2v_3) - 0.5\cos(2t)$$

$$\frac{\mathrm{d}^q u_4}{\mathrm{d}t^q} = (b_2 - b_1)v_2 - b_1 v_2 v_5$$

$$\frac{\mathrm{d}^q u_5}{\mathrm{d}t^q} = v_1 v_3 + v_2 v_4 - b_3 v_5 \tag{6}$$

where b_1, b_2, b_3 are parameters of system. For parameter values $b_1 = 2.1$, $b_2 = 30$, $b_3 = 0.6$ and initial values $(v_1(0); v_2(0); v_3(0); v_4(0); v_5(0)) = (8, 7, 6, 8, 7)$ and $q = 0.95$, the system is chaotic as shown in Fig. 1.

Fractional-Order Complex Lu Systems:

$$\frac{d^q w_1}{dt^q} = c_1(w_3 - w_1)$$

$$\frac{d^q w_2}{dt^q} = c_1(w_4 - w_2)$$

$$\frac{d^q w_3}{dt^q} = -w_1 w_5 + c_2 w_3$$

$$\frac{d^q w_4}{dt^q} = -w_2 w_5 + c_2 w_4 - 2\cos(10 w_4) - 0.25\cos(10t)$$

$$\frac{d^q w_5}{dt^q} = w_1 w_3 + w_2 w_4 - c_3 w_5 \tag{7}$$

where c_1, c_2, c_3 are parameters of system. For parameter values $c_1 = 10$, $c_2 = 22$, $c_3 = 5$ and initial values $(w_1(0); w_2(0); w_3(0); w_4(0); w_5(0)) = (1; 2; 3; 4; 5)$ and $q = 0.95$, the system is chaotic as shown in Fig. 1.

5 Numerical Simulations and Discussion

We consider (5), (6) as master systems and (7) as slave system. From Sect. 3, we have for the above-defined systems the following matrices:

$$A_1 = \begin{bmatrix} -a_1 & 0 & a_1 & 0 & 0 \\ 0 & -a_1 & 0 & a_1 & 0 \\ a_2 & 0 & -1 & 0 & 0 \\ 0 & a_2 & 0 & -1 & 0 \\ 0 & 0 & 0 & 0 & -a_3 \end{bmatrix}, \quad A_2 = \begin{bmatrix} -b_1 & 0 & b_1 & 0 & 0 \\ 0 & -b_1 & 0 & b_1 & 0 \\ b_2 - b_1 & 0 & 0 & 0 & 0 \\ 0 & b_2 - b_1 & 0 & 0 & 0 \\ 0 & 0 & 0 & 0 & -b_3 \end{bmatrix}$$

$$A_3 = \begin{bmatrix} -c_1 & 0 & c_1 & 0 & 0 \\ 0 & -c_1 & 0 & c_1 & 0 \\ 0 & 0 & c_2 & 0 & 0 \\ 0 & 0 & 0 & c_2 & 0 \\ 0 & 0 & 0 & 0 & -c_3 \end{bmatrix}, \quad F_1(U) = \begin{bmatrix} 0 \\ 0 \\ -x_1 x_5 \\ -x_2 x_5 \\ x_1 x_3 + x_2 x_4 \end{bmatrix}, \quad F_2(V) = \begin{bmatrix} 0 \\ 0 \\ -b_1 y_1 y_5 \\ -b_1 y_2 y_5 \\ y_1 y_3 + y_2 y_4 \end{bmatrix}$$

$$F_3(W) = \begin{bmatrix} 0 \\ 0 \\ -z_1 z_5 \\ -z_2 z_5 \\ z_1 z_3 + z_2 z_4 \end{bmatrix}, \quad \Delta G_1 = \begin{bmatrix} 0 \\ 2\cos(4x_2) \\ 0 \\ 0 \\ -0.5\cos(4x_5) \end{bmatrix}, \quad \Delta G_2 = \begin{bmatrix} \sin(2y_1) \\ 0 \\ -0.5\cos(2y_3) \\ 0 \\ 0 \end{bmatrix}$$

$$\Delta G_3 = \begin{bmatrix} 0 \\ 0 \\ 0 \\ -2\cos(10z_4) \\ 0 \end{bmatrix}, \quad d_1 = \begin{bmatrix} 0 \\ -0.5\sin(4t) \\ 0 \\ 0 \\ 2\cos(4t) \end{bmatrix}, \quad d_2 = \begin{bmatrix} -0.5\sin(2t) \\ 0 \\ -0.5\cos(2t) \\ 0 \\ 0 \end{bmatrix}$$

$$d_3 = \begin{bmatrix} 0 \\ 0 \\ 0 \\ -0.25\cos(10t) \\ 0 \end{bmatrix}, \quad \bar{U} = \begin{bmatrix} u_1 \\ u_2 \\ u_3 \\ u_4 \\ u_5 \end{bmatrix}$$

We choose the matrix M as:

$$M = \begin{bmatrix} 1 & 0 & -1 & 0 & 0 \\ 0 & -1 & 1 & 0 & 0 \\ 0 & 0 & -2 & 0 & 0 \\ 0 & 0 & 1 & 0 & 2 \\ 0 & 0 & 0 & 0 & -2 \end{bmatrix} \tag{8}$$

Therefore, we get the error function as:

$$E_1 = u_1 + v_1 - w_1 + w_3$$

$$E_2 = u_2 + v_2 + w_2 - w_3$$

$$E_3 = u_3 + v_3 + 2w_3$$

$$E_4 = u_4 + v_4 - w_3 - 2w_5$$

$$E_5 = u_5 + v_5 + 2w_5$$

Differentiating the above and substituting the controllers as in Theorem 1, the error dynamical system simplifies to:

$$D^q E_1 = -13E_1,$$

$$D^q E_2 = -14E_2$$

$$D^q E_3 = -E_3$$

$$D^q E_4 = -2E_4$$

$$D^q E_5 = -3E_5$$

and the error plot.

Here, we have taken the control gain matrix as:

$$K = \begin{bmatrix} -0.9 & 0 & -12.1 & 0 & 0 \\ 0 & -1.9 & 0 & -12.1 & 0 \\ -207.9 & 0 & 0 & 0 & 0 \\ 0 & -207.9 & 0 & -1 & 0 \\ 0 & 0 & 0 & 0 & -1.4 \end{bmatrix}$$

The eigenvalues of the Jacobian matrix of the error system are -13, -14, -1, -2, -3 which satisfy the stability condition; i.e., the argument of the obtained eigenvalues satisfies the condition $\mid arg(eigenvalues) \mid > \frac{q\pi}{2}$. Hence, the system is asymptotically stable about its equilibrium point.

Also, the error converges to zero implying that the fractional inverse matrix projective combination synchronization has been achieved. The fractional inverse matrix projective combination synchronized trajectories are shown in Fig. 2. Also, the error plot of the system converges to zero for initial conditions of error system as (E_1, E_2, E_3, E_4, E_5) = (12, 9, 17, 1, 26) as shown in Fig. 2.

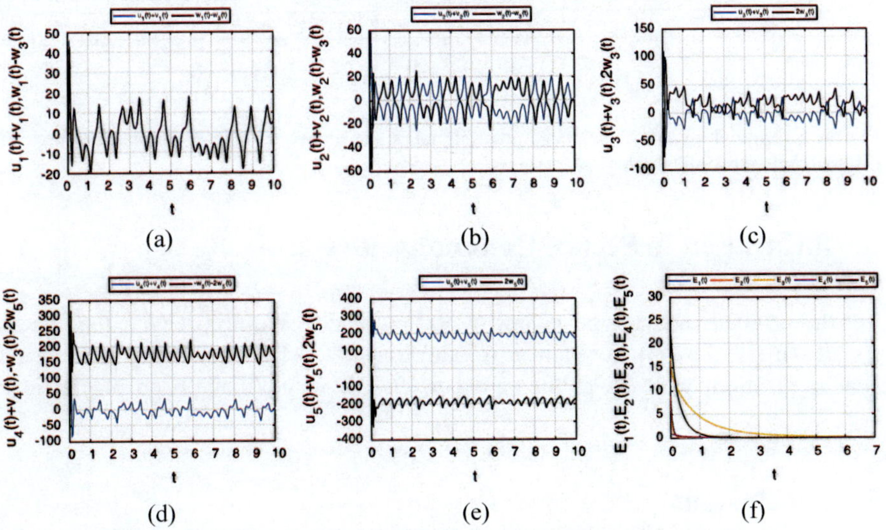

Fig. 2 Fractional inverse matrix projective combination synchronized trajectories

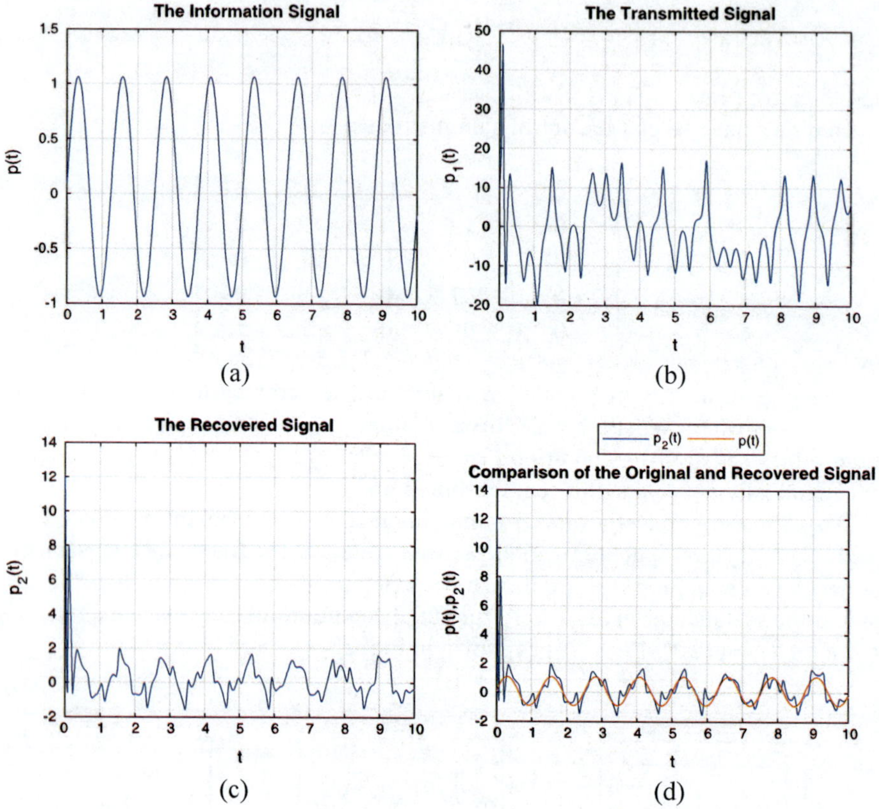

Fig. 3 **a** Original signal, **b** the transmitted signal, **c** the recovered signal, **d** comparison of the original and recovered signal

6 Illustration in Secure Communication

Let the original message be given by $p(t) = \sin(5t)$. We mix it with the chaotic signals $(u_1(t) + v_1(t))$ to obtain $p_1(t)$ and transmit. After performing the desired synchronization, we recover the signal $p_2(t)$. The results have been displayed in Fig. 3.

7 Conclusions

In this paper, fractional inverse matrix projective combination synchronization has been achieved among two chaotic master systems and one slave system in the presence of uncertainties and disturbances. A comparison has been made that clearly shows the efficacy of our synchronization scheme. Application in the field of secure communication has also been illustrated.

References

1. A.C. Luo, A theory for synchronization of dynamical systems. Commun. Nonlin. Sci. Numer. Simul. **14**(5), 1901–1951 (2009)
2. L.M. Pecora, T.L. Carroll, Synchronization in chaotic systems. Phys. Rev. Lett. **64**, 821–824 (1990). https://doi.org/10.1103/PhysRevLett.64.821, https://link.aps.org/, https://doi.org/10.1103/physrevlett.64.821
3. E.D. Dongmo, K.S. Ojo, P. Woafo, A.N. Njah, Difference synchronization of identical and nonidentical chaotic and hyperchaotic systems of different orders using active backstepping design. J. Comput. Nonlin. Dyn. **13**(5), 051, 005 (2018)
4. A. Khan, D. Khattar, N. Prajapati, Multiswitching compound antisynchronization of four chaotic systems. Pramana **89**(6), 90 (2017)
5. J. Sun, Y. Shen, G. Cui, Compound synchronization of four chaotic complex systems. Adv. Mathe. Phys. (2015)
6. A. Khan et al., Increased and reduced order synchronisations between 5d and 6d hyperchaotic systems. Indian J. Ind. Appl. Math. **8**(1), 118–131 (2017)
7. A. Khan, A. Tyagi, Disturbance observer-based adaptive sliding mode hybrid projective synchronisation of identical fractional-order financial systems. Pramana **90**(5), 67 (2018)
8. A. Khan, P. Trikha, Compound difference anti-synchronization between chaotic systems of integer and fractional order. SN Appl. Sci. (2019)
9. A. Ouannas, X. Wang, V.T. Pham, T. Ziar, Dynamic analysis of complex synchronization schemes between integer order and fractional order chaotic systems with different dimensions. Complexity (2017)
10. M.H. Tavassoli, A. Tavassoli, M.O. Rahimi, The geometric and physical interpretation of fractional order derivatives of polynomial functions. Differ. Geom. Dyn. Syst. **15**, 93–104 (2013)

Cryptosystem Based on Hybrid Chaotic Structured Phase Mask and Hybrid Mask Using Gyrator Transform

Shivani Yadav and Hukum Singh

Abstract A novel asymmetric cryptosystem grounded on hybrid chaotic structured phase mask and hybrid mask using gyrator transform is proposed in this paper. Asymmetric cryptosystem primes to different keys for encryption and decryption of an image. To improve the security of the scheme, we introduce hybrid chaotic structured phase mask (HCSPM) instead of random phase mask (RPM). CSPM is a combination of structured phase mask (SPM) and logistic map in the algorithm, and usage of it leads to extra security for the scheme. SPM is made by using Fresnel zone plates and radial Hilbert transform. The different parameters used in the algorithm provide extra key space and benefits to fortify the system. The proposed scheme is vigorous against numerous attacks.

Keywords Gyrator transform · Chaotic structured phase mask · Hybrid chaotic structured phase mask · MSE

1 Introduction

In the present era, information security is becoming increasingly important for data protection. In the past three decades, the security issue addressed by optical techniques have been explored extensively due to the inherent characteristics of optics, such as capability of parallel processing and operation in high-dimensional space [1, 2]. One of the most exhaustively studied optical encryption schemes is the 'double random phase encryption' (DRPE) scheme which was firstly proposed by Refreiger and Javidi [3]. The DRPE method was further extended with fractional Fourier transform [4], Fresnel transform [5], gyrator transform [6, 7] and fractional Hartley transform [8]. These current transforms based techniques are symmetric in nature, and

S. Yadav · H. Singh (✉)
Department of Applied Sciences, The NorthCap University, Sector 23-A, Gurugram 122017, India
e-mail: hukumsingh@ncuindia.edu

S. Yadav
e-mail: shivani6400@gmail.com

P. Bansal et al. (eds.), *Proceedings of International Conference on Artificial Intelligence and Applications*, Advances in Intelligent Systems and Computing 1164,
https://doi.org/10.1007/978-981-15-4992-2_11

the symmetric cryptosystems are vulnerable to many attacks. To put stoppage on the attacks, an asymmetric cryptosystem was established [9–11] which uses public and private keys for encoding and decoding of an image. In the existing work, the phase truncated part (PT) and phase reservation part (PR) of HCSPM and HM are used as public and private keys in encrypting and decrypting of an image, respectively. The correct use of all parameters helps in fortify the safety of the scheme. The proper usage of asymmetric cryptosystem in security applications demands the system to be robust to probable loss during both encryption and decryption. Structured phase mask (SPM) helps in performing image encoding [12], and the usage of SPMs aids in having more keys at the time of encrypting an image which helps in increasing the security of the system. It consists of Fresnel zone plate (FZP) and radial Hilbert mask (RHM). Therefore, we suggested an asymmetric cryptosystem for DRPE scheme based on hybrid chaotic structured phase masks and hybrid mask [13, 14] and it is attained through gyrator transform. The logistic map, Fresnel zone plates (FZPs) [15] and radial Hilbert mask (RHM) [16] are combined to make chaotic structured phase masks (CSPMs) [17]. By presenting HCSPM and HM in the cryptosystem, these novel masks help in enlarging the key space. Original image is converted into encrypted image with the support of haphazardness of HCSPM and HM. Being the combination of logistic map [18], FZP and RHM, HCSPM has given high importance. The usage of hybrid mask also benefits in increasing the key space, and it provides randomness in the system which is robust against attacks. In the suggested scheme, encoding of double phase images is considered in the GT domain and the usage of it provides us benefits such as computational ease and suitability in optical application.

The paper is organized as follows: In Sect. 2, a brief mathematical description of GT, CSPM, HCSPM and HM are depicted. Section 3 shows the encryption and decryption scheme. Results based on numerical and statistical simulations for authentication and estimation of the system are shown in Sect. 4. Finally, the conclusions of the study are summarized in Sect. 5.

2 Theoretical Background

2.1 *Gyrator Transform*

Gyrator transform generates the rotation in the twisted location spatial frequency plane. The GT of a two-dimensional function $A\,(x, y)$ is written in equation [1] as

$$G^{\alpha}\{f(x,y)\}(u,v) = \int_{-\infty}^{+\infty}\int_{-\infty}^{+\infty} A(x,y)h_{\alpha}(x,y;u,v)\mathrm{d}x\mathrm{d}y \tag{1}$$

And h_{α} represents the kernel, given by the expression as:

$$h_\alpha(x, y; u, v) = \frac{1}{|\sin\alpha|}\exp\left[2i\pi\frac{(xy + uv)\cos\alpha - xv - yu}{\sin\alpha}\right]$$

Here, α is the rotational angle. If $\alpha = 0$, then it resembles the identity transform. When $\alpha = \pm(\pi/2)$, the GT reduces to a Fourier transform/inverse Fourier transform with the rotation of the u- and v-coordinates. $G^{-\alpha}$ is the inverse transform of G^{α}.

2.2 *Hybrid Chaotic Structured Phase Mask*

A new chaotic structured phase mask has been established in this proposed work. Chaotic map produces massive number of random iterative values. The logistic map is termed as one-dimensional (1D) nonlinear chaos function and is defined as

$$f(x) = r \cdot x \cdot (1 - x) \tag{2}$$

where r denotes the bifurcation parameter, i.e., 0 < r < 4.

$$x_{n+1} = r \cdot x_n \cdot (1 - x_n) \tag{3}$$

where x_n denotes the iteration value and x_0 is the original and needed values. Equation (3) is produced when Eq. 2 is iterated up to n times. The sequence x_n becomes chaotic if there is insignificant variation in x_0 and the bifurcation parameter belongs to [3.5699456, 4]. One-dimensional chaotic random phase mask is given by Eq. 4,

$$X = \{x_1, x_2, x_3, \ldots, x_{M\times N}\} \tag{4}$$

where $x_i \in (0, 1)$ is transformed into two-dimensional matrix and regarded as Y and represented by Eq. 5,

$$Y = \left\{y_{i,j} | i = 1, 2, \ldots, M; j = 1, 2, \ldots, N\right\} \tag{5}$$

The chaotic random phase mask (CRPM) is given as

$$\text{CRPM}(x, y) = \exp\left(i2\pi y_{i,j}(x, y)\right) \tag{6}$$

where (x, y) are the coordinates of chaotic random phase mask and consist of three private keys x_0, r and n. CRPM generates many difficulties in aligning optical set-up. To overcome from this, structured phase mask (SPM) has been created which is a hybrid of Fresnel zone plates (FZPs) and radial Hilbert mask [RHM] makes an image edge enhanced in contrast to plain image and (p, θ) are the log-polar coordinates and illustrated in Eq. 7:

$$H(P, \theta) = e^{iP\theta} \tag{7}$$

where $P = 10$ infers the order of radial Hilbert transform, i.e., topological charge.

The wave front complex part of the Fresnel zone plate is represented by Eq. 8:

$$U(r) = \exp^{\left(\frac{-i\pi r^2}{\lambda f}\right)} \tag{8}$$

Here, $\lambda = 632.8$ nm denotes wavelength, $f = 400$ mm represents focal length and pixel spacing = 0.023.

By combination of equations 7 and 8, the SPM is generated and denoted by Eq. 9:

$$\text{SPM}(p, \theta, r) = H(P, \theta) \times U(r) \tag{9}$$

The chaotic structured phase mask (CSPM) is produced in Eq. 10

$$\text{CSPM}(x, y) = \exp\{i\{\arg\{\text{CRPM}(x, y)\} \times \arg\{H(P, \theta)\} \times \arg\{U(r)\}\}\} \tag{10}$$

Here, arg{•} is the argument operator which indicates phase operation. Projected cryptosystem is harmless and enough to overwhelm any kind of outbreaks.

The HCSPM has several advantages over random phase mask (RPM) as it provides extra key length and many parameters which holds the system security. HCSPM is the angle of Fourier transform of the secondary image P (*x*, *y*) and is given by Eq. (11).

$$\text{HCSPM} = \text{Arg}[\text{FT}\{P(x, y) \cdot \text{CSPM}(x, y)\}] \tag{11}$$

where Arg is the argument, FT is Fourier transform and P is the secondary image used at the time of making hybrid of CSPM.

2.3 Hybrid Mask

Hybrid mask is made by the usage of secondary image $P1(x, y)$ and random phase mask ($R1$) which is depicted in Eq. 12. We multiply $R1$ with $P1$ and by applying Fourier transform (FT) and taking argument of the FT results in hybrid phase mask (HM).

$$\text{HM} = \text{Arg}\{\text{FT}[P1(x, y) \cdot R1]\} \tag{12}$$

where Arg is the argument.

3 Proposed Work

Encrypting Procedure

The encoding and decoding of an image are given with the help of flowchart in Fig. 1a, b, respectively.

Step 1 The plain image is initially convoluted with hybrid chaotic structured phase mask, and then gyrator transform is applied on it with rotation angle α after that it is parted in two parts: One is phase reservation (Pr), and another is phase truncated (PT).

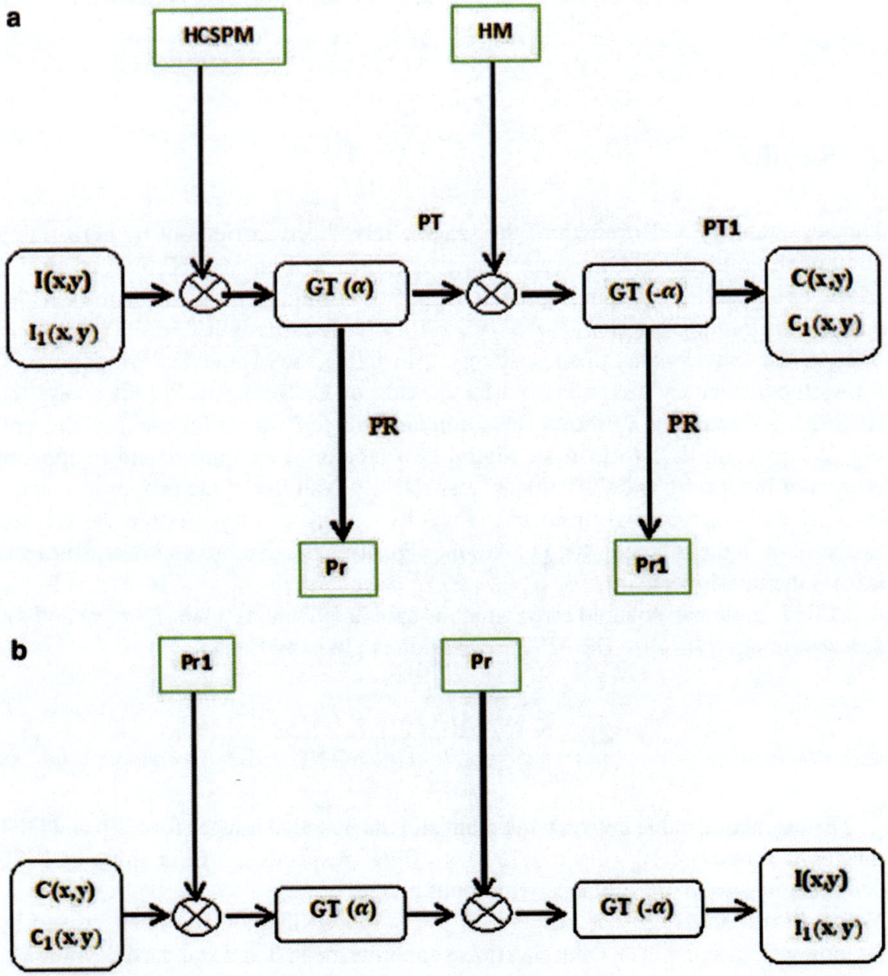

Fig. 1 **a** Encoding of an image. **b** Decoding of an image

Step 2 The PT part is passed on, and it is multiplied with hybrid mask; then, again gyrator transform is applied on it with rotation angle $-\alpha$. Now from here, it divided again into PT1 and Pr1 parts and finally the ciphered image is obtained.

Decrypting Procedure

Step 1 The exact inverse of encryption process results out into decryption process. The ciphered image is firstly multiplied with Pr1, and then apply gyrator transform with rotation angle α.

Step 2 The product is convoluted with Pr, and applying gyrator transform with rotational angle $-\alpha$ results out in the decrypted image, i.e., into the recovered image.

4 Results

The acceptability and strength of the scheme have been carried out by performing numerical simulations. Grayscale image of tulip and binary OPT images of size 256×256 pixel is shown in Fig. 2a, b, respectively. Figure 2c–g shows the CRPM, RHM with topological charge $P = 10$, FZP with focal length $f = 400$ mm, $\lambda = 6328$ A the wavelength, pixel spacing $P_s = 0.023$, SPM and CSPM. Figure 2h shows the secondary image P used for making of HCSPM and Fig. 2i shows the HCSPM, whereas Fig. 2j shows the secondary image P1 used for making HM and Fig. 2k represents the hybrid mask. Figure 2l–o depicts the enciphered and deciphered images of input tulip and OPT image, respectively. Validity of the proposed scheme is tested, and it is robust against various attacks. The gyrator transform order is taken as $\alpha = 0.4\pi$ in the scheme which makes the algorithm much more secure and protects it from the interlopers.

MSE is the mean squared error which is calculated among plain $I_o(x, y)$ and the decoded image $I_d(x, y)$. The MSE expression is given by Eq. 13,

$$\text{MSE} = \sum_{i=0}^{255}\sum_{j=0}^{255} \frac{|I_o(x, y) - I_d(x, y)|^2}{256 \times 256} \tag{13}$$

The calculated value between the plain and the decoded images for tulip and OPT images is 6.6004×10^{-27} and 1.7510×10^{-27}, respectively. Less value of MSE demonstrates good resemblance with input image.

Noise attack analysis: The strength of the proposed method has been proved by including Gaussian noise. Gaussian noise contains mean 0 and standard deviation 1. Let C be the ciphered image which is free from noise attack and C' be the noise-affected ciphered image. Asset of the proposed algorithm has been carried out by the

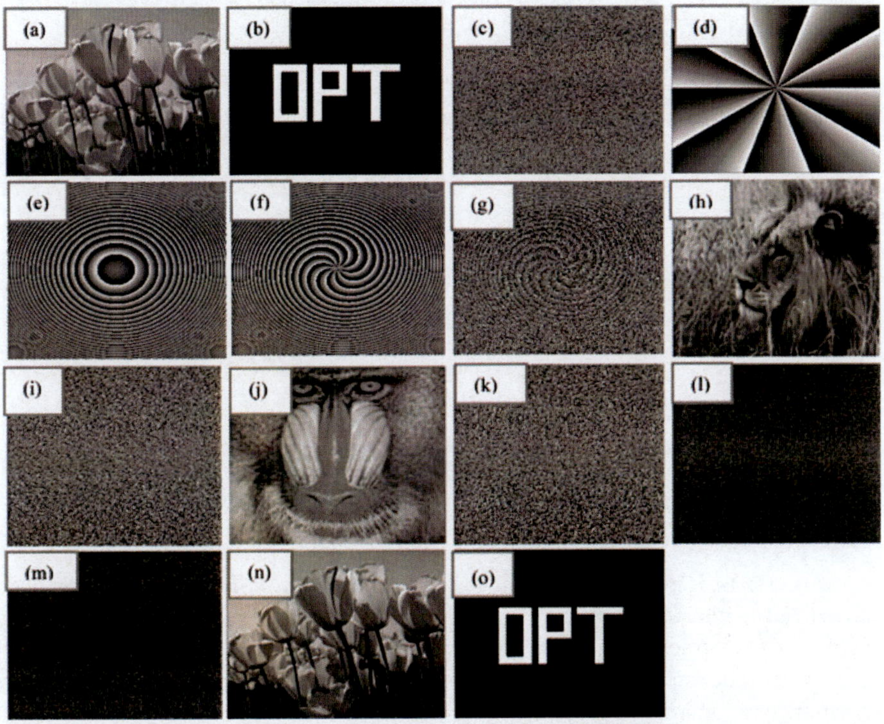

Fig. 2 **a, b** Are plain tulip and binary OPT images, **c–g** denotes CRPM, RHM, FZP, SPM and CSPM, **h, j** are the secondary images P, $P1$ and **i, k** show their corresponding HCSPM and HM, respectively, **l, m** are ciphered images, and **n, o** are the decoded images

noise equation, and it is given by Eq. (14).

$$C' = C + KG \tag{14}$$

where K indicates the noise strength constraint. The plot of both the images among mean squared error (MSE) and the noise factor (K) is illustrated in Fig. 3. The graphical depiction of MSE and K demonstrates that the structure is robust.

5 Conclusion

A novel asymmetric cryptosystem scheme for grayscale and binary images is proposed which uses two masks: One is hybrid chaotic spiral phase mask (HCSPM) which consists of logistic map, radial Hilbert transform and Fresnel zone plates, and another one is hybrid mask (HM). Existence of several parameters in HCSPM and HM increases the key length and helps in enhancing the security. As compared to

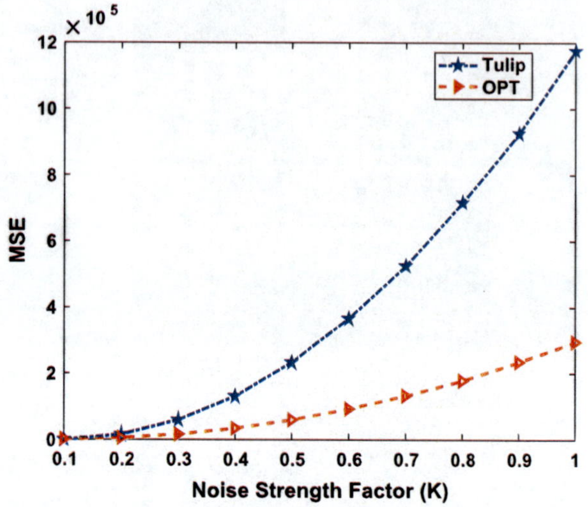

Fig. 3 Denotes the graph among noise factor (K) and MSE for the original grayscale and binary images

previous efforts, it is a unique approach toward the security. The algorithm has been carried out in the GT domain. To check the robustness of the system, numerical analysis is performed which authenticates the capability of the system. So, the proposed technique is said to be a well substitute to traditional methods, and it offers enormously good approach for securing of images.

Acknowledgements The author wishes to acknowledge the management of The NorthCap University for their motivation that lent sustenance support throughout the paper.

References

1. B. Javidi, *Optical and Digital Techniques for Information Security* (Springer Science + Business Media, Berlin, 2005)
2. O. Matoba, T. Nomura, E. Perez-Cabre, M.S. Millan, B. Javidi, Optical techniques for information security. Proc IEEE **97**, 1128–1148 (2009)
3. P. Refregier, B. Javidi, Optical image encryption based on input plane and Fourier plane random encoding. Opt. Lett. **20**, 767–769 (1995)
4. G. Unnikrishnan, K. Singh, Double random fractional Fourier domain encoding for optical security. Opt. Eng. **39**(11), 2853–2859 (2000)
5. G. Situ, J. Zhang, Double random-phase encoding in the Fresnel domain. Opt. Lett. **29**, 1584–1586 (2004)
6. J.A. Rodrigo, T. Alieva, M.L. Calvo, Gyrator transform: properties and applications. Opt. Express. **15**, 2190–2203 (2007)
7. H. Singh, A.K. Yadav, S. Vashisth, K. Singh, Fully phase image encryption using double random-structured phase masks in gyrator domain. Appl. Opt. **53**, 6472–6481 (2014)
8. H. Singh, Nonlinear optical double image encryption using random-optical vortex in fractional Hartley transform domain. Opt. Appl. **47**(4), 557–578 (2017)

9. W. Qin, X. Peng, Asymmetric cryptosystem based on phase-truncated Fourier transforms. Opt. Lett. **35**, 118–120 (2010)
10. H. Singh, Devil's vortex Fresnel lens phase masks on an asymmetric cryptosystem based on phase-truncation in Gyrator wavelet transform domain. Opt. Lasers Eng. **81**, 125–139 (2016)
11. H. Singh, Optical cryptosystem of color images based on fractional, wavelet transform domains using random phase masks. Indian J. Sci. Technol. **9S**(1), 1–15 (2016)
12. H. Singh, Cryptosystem for securing image encryption using structured phase masks in Fresnel Wavelet transform domain. 3D Res. **7**(34), (2016). https://doi.org/10.1007/s13319-016-0110-y.
13. R. Kumar, B. Bhaduri, Optical image encryption using Kronecker product and hybrid phase masks. Opt. Laser Technol. **95**, 51–55 (2017)
14. H. Singh, Hybrid structured phase mask in frequency plane for optical double image encryption in gyrator transform domain. J. Mod. Opt. **65**(18), 2065–2078 (2018)
15. J.F. Barrera, R. Henao, R. Torroba, Fault tolerances using toroidal zone plate encryption. Opt. Commun. **256**(4), 489–494 (2005)
16. J.A. Davis, D.E. McNamara, D.M. Cottrell, J. Campos, Image processing with the radial Hilbert transform: theory and experiments. Opt. Lett. **25**(2), 99–101 (2000)
17. R. Girija, H. Singh, Symmetric cryptosystem based on chaos structured phase masks and equal modulus decomposition using fractional Fourier transform. 3D Res. **9**(42), (2018). https://doi.org/10.1007/s13319-018-0192-9.
18. L. Sui, K. Duan, J. Liang, X. Hei, Asymmetric double-image encryption based on cascaded discrete fractional random transform and logistic maps. Opt. Express **22**, 10605–10621 (2014)

PE File-Based Malware Detection Using Machine Learning

Namita and Prachi

Abstract In current times, malware writers write more progressive sophisticatedly designed malware in order to target the user. Therefore, one of the most cumbersome tasks for the cyber industry is to deal with this ever-increasing number of progressive malware. Traditional security solutions such as anti-viruses and anti-malware fail to detect these advanced types of malware because the majority of this malware are refined versions of their predecessor. Moreover, these solutions consume lots of computational resources on the host to accomplish their operations. Further, malware evades these security solutions by using intelligent approaches such as code encryption, obfuscation and polymorphism. Therefore, to provide alternatives to these solutions, this paper discusses the existing malware analysis and detection techniques in a comprehensive/holistic manner.

Keywords Malware · Static analysis · Dynamic analysis · PE files · Machine learning

1 Introduction

Internet technology is fully integrated with every domain worldwide. Due to its widespread usage, lots of sensitive data about users is available online. Attackers use various malware like viruses, worm, rootkit, Trojan, bots, spyware, ransomware and so on in order to perform lots of malicious activities. Malware [1] is a generic term that is used for any kind of software designed to disrupt the system operations, access to remote systems, flood the networks, delete/modify data, corrupt the hardware/software and collect personal information without authorization. The scope of

Namita · Prachi (✉)
Computer Science and Engineering Department, The NorthCap University, Gurugram, India
e-mail: prachiah1985@gmail.com

Namita
e-mail: namita.dabas@gmail.com

P. Bansal et al. (eds.), *Proceedings of International Conference on Artificial Intelligence and Applications*, Advances in Intelligent Systems and Computing 1164,
https://doi.org/10.1007/978-981-15-4992-2_12

malware is not limited to stand-alone systems; it can also propagate through multiple channels of communications and can be downloaded. In addition, most of the malware can change their native code to prevent their detection. Modern-day malware comes with obfuscation techniques such as polymorphism, metamorphism and encryption. Further, if a malware code is available in public, then anyone with malicious intent can manipulate the code and add some extra features to generate new malware even without much programming language. This allows attackers to recreate more sophisticated versions of pre-existing malware.

As per Malwarebytes Labs 2019 reports [2], in the year 2018, malware authors shifted their focus from individual consumers to business organizations in order to gain high profits. Overall, malware in business rose significantly, i.e., 79% in 2018-primarily due to the increased activities of malware such as backdoors, miners, spyware and information stealers. The statistics released by Kaspersky [3] identified 243,604 users, which were attacked by malware just to steal the money. Further, the data from AV-Test [4] claims that 914.79 million malware samples were reported in the year 2019. Therefore, malware detection and prevention have become the primary objective of the security companies as well as researchers in the recent past.

As the numbers of malware are increasing exponentially, it became impossible to manually analyze and detect them. Consequently, a number of security companies are providing various anti-virus and anti-malware solutions to protect legitimate users from such attacks. Some of the popular security solutions provided are as listed: Microsoft security essentials, Avast antivirus, AVG technologies Comodo, Kaspersky, Norton, windows defender, McAfee, Malwarebytes, ESET Nod 32 and so on [5]. However, the majority of anti-viruses and anti-malware solutions are based on signature-based methods.

Signature is a short unique sequence of bytes, often distinctive to a malware [6]. The signature-based detection approaches use a database of predefined signatures. This type of scheme works well for previously known malware but fails to detect new malware or variants of existing malware. Nowadays, malware developers use automated malware generation tools based on encryption, obfuscation, polymorphism, metamorphism, packing and emulation techniques. These tools provide an edge to the black-hat community over the white-hat community and facilitate malware writers to write and change the code of already-written malware. This newly written malware possesses new signatures. Every new development to detect malware is followed by evasion methods. So, there is an urgent requirement of developing alternative analysis and detection methods to address the shortcomings of signature-based solutions. Before diving deep into the discussion, it is important to highlight that the malware analysis is majorly classified into the two main categories.

1.1 Static Analysis

In static analysis, the malware samples are analyzed without the code execution [6]. Static analysis is preferred when a quick decision needs to be made in a resource-efficient manner. The effectiveness of static analysis relies on the comprehensiveness of analysis and parameters chosen for analysis. Static analysis includes analysis of source code, hashes, header information, strings, metadata, etc. Simple static analysis remains largely ineffectual against sophisticated malware because it may miss important functionality of the malware. Advanced static techniques involve reverse engineering of malware through disassembler tools such as IDA Pro [7], OllyDbg [8] and Olly Dump [9] to understand the code of malware. Some other tools like memory dumper LordPE [10] help to obtain data about changes from the system's memory. These patterns are further analyzed based on features, such as binary code, opcodes, strings, byte n-grams and control flow graphs. The static analysis is used in signature-based detection methods. The static analysis-based approaches are easy and fast but cannot detect ever-evolving obfuscated malware correctly as they leave some of the functionalities unidentified. The limitations of static analysis have been explored [11]. These limitations provide an edge to a dynamic analysis-based approach.

1.2 Dynamic Analysis

In dynamic analysis, the malware samples are executed and analyzed in a real or a controlled environment. Dynamic analysis can be done using a variety of debuggers. Various other tools such as Process Monitor, Regshot, Filemon and Process Explorer can also be used to retrieve behavioral features such as API, system calls, instruction traces [12], registry changes, file writes, memory writes and network changes. Dynamic analysis is majorly used for understanding the functionalities of malware samples under consideration. However, environment aware malware does not exhibit their true behavior whenever they identify that they are being executed in a controlled environment. Therefore, the dynamic analysis methods must possess some real environment characteristics in order to make it challenging for malware to distinguish between real and controlled environment. There are numerous online sandbox environments available for dynamic analysis. Some of the widely used sandboxes are CW Sandbox, Cuckoo Sandbox, Anubis, Norman Sandbox, etc. Dynamic approach exposes the natural behavior of the sample under examination. However, the dynamic analysis approach is time- and resource-consuming (memory overheads) as each sample under investigation needs to be executed separately and comprehensively.

Both the analysis techniques are beneficial under different settings but when the objective is to perform the analysis in a timely manner while consuming minimum resources then the static analysis is preferred over dynamic analysis.

Therefore by taking the above fact into consideration, some of the researchers combined the features of static and dynamic analysis and proposed a new method, termed as hybrid analysis. The authors in [13] used dynamic analysis for extracting the features for training and static analysis to test the efficiency of the detection system.

Machine learning algorithms facilitate us to design an automated malware analysis and detection system that can precisely distinguish malware as benign or malicious. These techniques significantly increase the detection efficiency and at the same time reduce the time and resource consumption. New malware can also be detected easily with these intelligent systems. Therefore, a large number of authors have used machine learning algorithms to train and test their models designed using the extensive set of features extracted from large no of benign and malicious samples using various static and dynamic analysis techniques.

The aim of this paper is to discuss and review the malware analysis of PE files. PE files are chosen in this paper because they work on the Windows operating systems and to date Windows is the most commonly used OS (77.93%) by the users all across the world [14]. PE is a 32/64 bit file format for Windows OS executables, object codes, DLLs and others. Malware analysis of PE files can be done with a variety of features as byte sequences, strings, information flow tracking, opcodes, control flow graphs and API calls and so on.

This paper is organized into four sections: Sect. 1 presents a general view about the malware industry, recent trends of malware attacks and type of malware analysis and detection approaches. Section 2 provides some insights into malware detection techniques based on machine learning methods present in the literature. Section 3 discusses the different dimensions of the reviewed work, and Sect. 4 discusses the conclusion and the future directions.

2 Related Work

Numerous malware detection techniques have been proposed in the literature based on machine learning algorithms. Some of the machine learning-based research work related to PE file malware analysis is discussed here.

In 2001, Schultz et al. [15] proposed a machine learning framework for detecting malicious PE files using static features namely PE header, strings and byte sequences. The dataset was divided into two subsets: (1) training dataset for training the classification model, (2) test data set to assess the classifier for unknown binaries. Three machine learning algorithms (Ripper, Multi-Naïve Bayes and Naïve Bayes) were employed for the classification process. The authors used a dataset that contained 4266 samples (3265 malware + 1001 benign). The detection accuracy (97.1%) of the proposed system was higher in comparison with already existing signature-based detection systems of that time.

A malware detection method for PE files was proposed in 2011 [16] based on the graph analysis technique. The static features used for analysis included raw binaries and opcodes, using n-gram approach, whereas the dynamic features used for the analysis included instruction traces, control flow graphs, and system call traces. The authors trained the system with a dataset of 776 benign and 780 malicious samples. A support vector machine algorithm was used as a classification algorithm, and the multiple kernel learning approach was applied to find the similarity index between graph edges.

In 2013, Eskandari et al. [17] proposed a malware detection system using dynamic features of PE files, i.e., API and system calls. The authors used the Bayesian classification model in order to classify benign and malicious PE files.

Khodmoradi et al. [18] presented heuristic based detection model for metamorphic malware in year 2015. The authors used static features (opcodes) for the analysis. They disassembled the file using IDA pro and extracted the features using opcode statistics extractor. Six classification algorithms (j48, j48graft, LADTree, NBTree, Random Forest, REPTree) were used to classify 1200 samples into benign and malicious. The authors highlighted that the classification accuracy of their detection system is dependent on the classification methods applied and the disassembler chosen.

In the year (2015), Lakhotia et al. [19] surveyed on various existing malware detection and prevention techniques. The authors discussed the importance of applying machine learning techniques for malware detection.

Liang et al. [20] suggested a behavior-based malware detection technique in 2016. The authors extracted dynamic features of PE files from API calls, registry and network files. They applied supervised machine learning and trained the system using Jaccard similarity distance to identify the different variants of malware.

In 2017, Baldoni et al. [21] also utilized static analysis techniques to extract sequences, strings and headers from PE files. The authors trained the system with a dataset of 4783 samples using RF classification algorithm. Their proposed system provides a faster and more accurate (96%) analysis as compared to other detection systems.

The authors in [22] came up with a detection system for different variants of malware by predicting their signatures (2017). Their solution primarily focuses on the static analysis of PE files. The feature sets used in the analysis included strings, n-grams, API calls and hashes. Unsupervised learning-based machine learning algorithms were used in the training phase.

In 2017 [23], the authors trained the hidden Markov model using static as well as dynamic analysis to compare their detection results on malware variants of different families. [6] provided an extensive study of malware detection approaches based on data mining techniques in the year 2017. They examined the different dimensions of analysis including the feature sets and classification/clustering algorithms.

In the year 2018, [24–28] presented their solutions for malware detection on the basis of analysis (static, dynamic and hybrid) methods and classification algorithms with the applications of data mining algorithms.

The authors applied supervised learning-based machine learning algorithms to their proposed detection system [29]. They developed a static analysis-based automated tool to extract features of PE files. Later, they trained the system with classification algorithms such as SVM, decision trees and applied boosted decision tree algorithms to increase the detection efficiency of the proposed system.

A survey of malware detection techniques was carried out based on the windows platform [21]. Their study was a meta-analytical account of various researches pertaining to machine learning. The authors arranged studied literature based on their primary objectives, features and machine learning algorithms. The primary objectives were further classified into three sub-categories, i.e., malware detection, find in similarities and malware category detection. The researchers also highlighted various limitations and challenges faced by detection methods.

In 2016, [30] developed a detection solution based on machine learning technique. The authors applied both static and dynamic analysis methods for feature extraction. The developed a classifier model using seven classification algorithms (Random Forest, Naïve Bayes, J48, DT, Bagging, IB1, and NLP) and trained the classifier with 3130 PE files dataset. The detection accuracy of their proposed system was best (99.97%) when the RF classification algorithm was applied.

A machine learning-based framework (Virtual Machine Introspection) for detecting malware in virtual machines was proposed [31]. The authors extracted opcodes using static analysis and trained the system with selected features for providing better accuracy. Further, they applied Term Frequency-Inverse Document Frequency (TF-IDF) and Information Gain (IG) as classification algorithms.

In the year 2019, similarity hashing algorithm-based malware detection technique in the IoT environment was proposed [28]. In this technique, malware file scores were calculated to find the similarity between malware samples. The authors used the PE dataset and explored four different hashing techniques (PEHash, Imphash, Ssdeep, resource section Ssdeep). Finally, they combined the results of these hashes using evidence combinational methods such as fuzzy logic and certainty factor model.

This paper also presents a few of the various malware detection approaches for PE files as given in Table 1. After performing the in-depth study of these existing methodologies, it is inferred that opcodes provide low-level details of the executable and hence provides a better opportunity for detection of obfuscated malware at run time. As a matter of fact, it may be concluded that opcode-based analysis provides a better solution as compared to other malware detection solutions.

3 Discussion

It is evident from the above-stated literature that most of the malware detection techniques for PE files are based on machine learning algorithms. This segment briefs a statistical analysis of machine learning algorithms for the reviewed detection methods. Figure 1 shows that 56% of the above-studied methods used supervised learning-based algorithm, 26% have applied unsupervised learning algorithms and

Table 1 List of the reviewed works for malware detection using machine learning

Authors	Datasets	Features used	Extraction method	Approaches	Advantages	Limitations
Anderson et al. [16]	Dataset—15,561 samples	Binary, CFGs, opcodes, instruction traces, system calls	Static dynamic	Multiple kernel learning, machine learning	Average Detection Accuracy-97%	Instruction categorization not optimal
Raff et al. [29]	10,868 files (Kaggle + (Derbin)	Byte sequences	Static	Machine learning	Better for large byte sequences, 95% accuracy	Obfuscation reduces detection accuracy
Liang et al. [20]	200 samples	API calls, registry, PE files, network	Dynamic	Multilayer dependency chain, Jaccard similarity	Faster approach, better detection accuracy 70%	Sample size is small
Vadrevu et al. [30]	M-1,251,865 B-400,041	Network traces, PE files	Static	Clustering DBSCAN algorithm	50% time saving with 0.3% information loss	May not work with stalling code
Polino et al. [31]	2136 (1272 M + 864 B)	API/system calls	Hybrid	Unsupervised: Clustering with Jaccard similarity	Automatically finds API call groups (88.46% accuracy)	Obfuscated/packed malware reduces the accuracy, small data set
Miramirkhani et al. [32]	270 real machine users	Disk, registry, network, browsers	Static	Regression techniques	Useful for environment-aware malware	Ineffective with obfuscated malware
Blazytko et al. [33]	1200 randomly generated expression	Byte code, instruction tracing	Dynamic	Monte Carlo tree search, random sampling	80% detection accuracy	Limited by trace window boundaries

(continued)

Table 1 (continued)

Authors	Datasets	Features used	Extraction method	Approaches	Advantages	Limitations
Jordaney et al. [34]	123,435 B + 5560 M (Derbin) + 9592 B + 9179 M (Marvin)	APIs strings, IP address, permissions, opcodes	Static	SMV, classification algorithm	Discovers zero-day attacks, CE statistical metric, accuracy (96%)	
Huang et al. [35]	2029 malware samples	Byte sequences	Static	K-means algorithm	Better detection (74%) accuracy	Small data set, instruction sequence categorization not optimized
Hu et al. [36]	137,055 samples	Opcode sequences	Static	Unsupervised–prototype-based clustering	detects obfuscated malware (80% accuracy)	Obfuscation reduces the efficiency
O'Kane et al. [37]	260 benign (win xp) + 350 malicious samples	Opcodes	Dynamic	SVM classification	Suitable for encrypted malware, reduces irrelevant features	

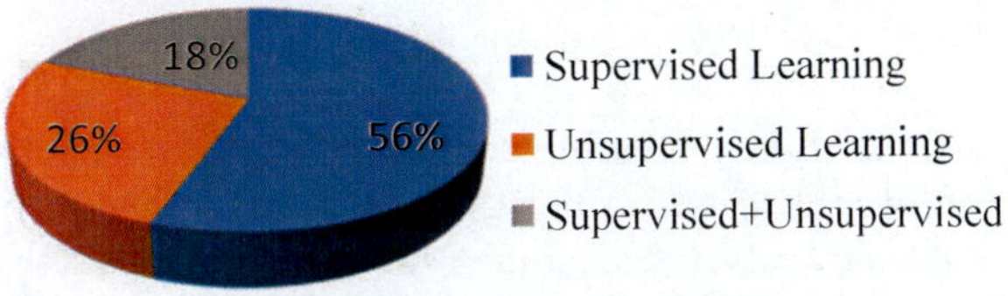

Fig. 1 Machine learning techniques in detection methods

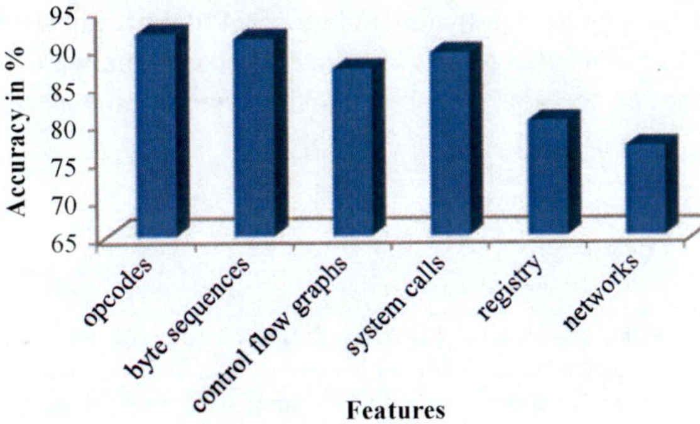

Fig. 2 PE features used for malware detection

18% of the studies have used both supervised and unsupervised learning methods. The statistics presented in Fig. 2 provide insights on the features used for the analysis by various detection methods and their accuracy. Opcodes are the most commonly employed features with the detection accuracy of 91.7%. Byte sequences and API/system call-based detection techniques are also the choices of researchers. Since opcode-based detection methods have better accuracy, our future work will be based on opcode-based malware detection.

This section further discusses that the PE features directly model the behavior of PE samples. To reduce the analysis complexity of the detection systems, only a subset of these features can be considered. Opcodes have been the obvious choice of analysts as they provide low-level details and hence helps in detecting obfuscated malware more efficiently and effectively.

Further, it is evident from the survey of existing studies that most of the solutions are suffering from the issues of datasets availability. The datasets used are not updated since long; some of the repositories are non-existing. The data sizes are small and the sources are not specified. In order to ensure the better availability of datasets, a benchmark dataset needs to be designed. Moreover, malware developers have the advantages over analysts as they can use online public platforms such as Virus total, Metascan and Malwr to test their sample's detection efficiencies by the common anti-viruses. So, the new trend in malware detection is investigating and predicting the future variants using machine learning techniques.

Machine learning methods require complex computations to keep pace with the growing speed of malware developments. The large feature sets increase the time complexity and reduced feature sets decrease the detection accuracy. So a trade-off has been identified between the accuracy and time/space complexity.

4 Conclusion

In light of the increasing complexity of malware, its detection and prevention have become the primary objective of malware researchers. Both the static and dynamic malware analysis techniques are used extensively by the researches to accurately detect the malicious executable. The objective of this paper was to identify the most suitable feature for the detection of malicious executable and devise the challenges in proposing an automated and efficient malware detection system. In the future work, the authors plan to devise an opcode-based malware detection technique on a dataset that can serve as a benchmark for other authors as well.

References

1. Wikipedia 2019 Retrieved on 5 July website: https://en.wikipedia.org/wiki/Malware
2. Website, https://blog.malwarebytes.com/malwarebytes-news/ctntreport/2019/01/2019-state-malware-report-trojans-cryptominers-dominate-threat-landscape/
3. IT Threat Evolution Q1 2019 Statistics, website: https://securelist.com/it-threat-evolution-q1-2019-statistics/90916/. Accessed on 02 July 2019
4. AV-Test IT Security Institute website: https://www.av-test.org/en/statistics/malware/
5. Website, https://securelist.com/mobile-malware-evolution-2018/89689/
6. Y. Ye, T. Li, D. Adjeroh, S.S. Iyengar, A survey on malware detection using data mining techniques. ACM Comput. Surv. (CSUR) **50**(3), 41 (2017)
7. IDA Pro website, https://www.hex-rays.com/products/ida/index.shtml
8. OllyDbg website, http://www.ollydbg.de/
9. OllyDump website, http://www.openrce.org/downloads/details/108/ollydump
10. LordPE website, https://www.aldeid.com/wiki/LordPE
11. A. Moser, C. Kruegel, E. Kirda, Limits of static analysis for malware detection, in *Twenty-Third Annual Computer Security Applications Conference (ACSAC* 2007). IEEE (2007), pp. 421–430
12. M. Egele, T. Scholte, E. Kirda, C. Kruegel, A survey on automated dynamic malware-analysis techniques and tools. ACM Comput. Surv. (CSUR) **44**(2), 6 (2012)
13. M. Eskandari, Z. Khorshidpour, S. Hashemi, HDM-Analyser: a hybrid analysis approach based on data mining techniques for malware detection. J. Comput. Virol. Hack. Techn. **9**(2), 77–93 (2013)
14. Stat counter Website, https://gs.statcounter.com/os-marketshare/desktop/worldwide
15. M.G. Schultz, E. Eskin, F. Zadok, S.J. Stolfo, Data mining methods for detection of new malicious executables, in *Proceedings 2001 IEEE Symposium on Security and Privacy. S&P(2001)*. IEEE (2001), pp. 38–49
16. B. Anderson, D. Quist, J. Neil, C. Storlie, T. Lane, Graph-based malware detection using dynamic analysis. J. Comput. Virol. **7**(4), 247–258 (2011)
17. M. Eskandari, Z. Khorshidpour, S. Hashemi, Hdm-analyser: a hybrid analysis approach based on data mining techniques for malware detection. J. Comput. Virol. Hack. Tech. **9**(2), 77–93 (2013)
18. P. Khodamoradi, M. Fazlali, F. Mardukhi, M. Nosrati, Heuristic metamorphic malware detection based on statistics of assembly instructions using classification algorithms, in *18th CSI International Symposium on Computer Architecture and Digital Systems (CADS)*. IEEE (2015), pp. 1–6
19. C. LeDoux, A. Lakhotia, Malware and machine learning, in *Intelligent Methods for Cyber Warfare* (Springer, Cham, 2015), pp. 1–42

20. G. Liang, J. Pang, C. Dai, A behavior-based malware variant classification technique. Int. J. Inf. Educ. Technol. **6**(4) (2016)
21. D. Ucci, L. Aniello, R. Baldoni, Survey of machine learning techniques for malware analysis. Comput. Secur. (2018)
22. E. Gandotra, D. Bansal, S. Sofat, Zero-day malware detection, in *Sixth International Symposium on Embedded Computing and System Design* (IEEE, 2016), pp. 171–175
23. A. Damodaran, F. Di Troia, C.A. Visaggio, T.H. Austin, M.A. Stamp, Comparison of static, dynamic, and hybrid analysis for malware detection. J. Comput. Virol. Hack. Tech. **13**(1), 1–12 (2017)
24. Q.K.A. Mirza, I. Awan, M. Younas, CloudIntell: an intelligent malware detection system. Fut. Gen. Comput. Syst. **86**, 1042–1053 (2018)
25. A. Souri, R.A. Hosseini, State-of-the-art survey of malware detection approaches using data mining techniques. HCIS **8**(1), 3 (2018)
26. K. Sethi, S.K. Chaudhary, B.k. Tripathy, P. Bera, A novel malware analysis framework for malware detection and classification using machine learning approach, in *Proceedings of the 19th International Conference on Distributed Computing and Networking* (ACM, 2018), p. 49
27. D. Carlin, P. O'Kane, S. Sezer, Dynamic analysis of malware using run-time opcodes, in *Data analytics and decision support for cybersecurity* (Springer, Cham, 2017), pp. 99–125
28. A.P. Namanya, I.U. Awan, J.P. Disso, M. Younas, Similarity hash-based scoring of portable executable files for efficient malware detection in IoT. Fut. Gen. Comput. Syst. (2019)
29. E. Raff, C. Nicholas, An alternative to NCD for large sequences, Lempel-Ziv Jaccard distance, in *Proceedings of the 23rd ACM SIGKDD International Conference on Knowledge Discovery and Data Mining* (ACM, 2017), pp. 1007–1015
30. P. Vadrevu, R. Perdisci, MAXS: scaling malware execution with sequential multi-hypothesis testing, in *Proceedings of the 11th ACM on Asia Conference on Computer and Communications Security* (ACM, 2016), pp. 771–782
31. M. Polino, A. Scorti, F. Maggi, S. Zanero, Jackdaw: towards automatic reverse engineering of large datasets of binaries, in *International Conference on Detection of Intrusions and Malware, and Vulnerability Assessment* (Springer, Cham, 2015), pp. 121–143
32. N. Miramirkhani, M.P. Appini, N. Nikiforakis, M. Polychronakis, Spotless sandboxes: evading malware analysis systems using wear-and-tear artifacts, in *IEEE Symposium on Security and Privacy (SP)* (IEEE, 2017), pp. 1009–1024
33. T. Blazytko, M. Contag, C. Aschermann, T. Holz, Syntia: synthesizing the semantics of obfuscated code, in *26th {USENIX} Security Symposium* (2017), pp. 643–659
34. R. Jordaney, K. Sharad, S.K. Dash, Z. Wang, D. Papini, I. Nouretdinov, L. Cavallaro, Transcend: detecting concept drift in malware classification models, in *26th Security Symposium ({USENIX} Security 2017)* (2017), pp. 625–642
35. K. Huang, Y. Ye, Q. Jiang, Ismcs: an intelligent instruction sequence based malware categorization system, in: *Anti-counterfeiting, Security, and Identification in Communication* (IEEE, 2009), pp. 509–512
36. X. Hu, K. G. Shin, S. Bhatkar, K. Griffin, Mutantx-s: scalable malware clustering based on static features, in *USENIX Annual Technical Conference* (2013), pp. 187–198
37. P. O'Kane, S. Sezer, K. McLaughlin, E.G. Im, SVM training phase reduction using dataset feature filtering for malware detection. IEEE Trans. Inf. Forens. Secur. **8**(3), 500–509 (2013)

Intelligence Graphs for Threat Intelligence and Security Policy Validation of Cyber Systems

Vassil Vassilev, Viktor Sowinski-Mydlarz, Pawel Gasiorowski, Karim Ouazzane, and Anthony Phipps

Abstract While the recent advances in data science and machine learning attract lots of attention in cyber security because of their promise for effective security analytics, vulnerability analysis, risk assessment, and security policy validation remain slightly aside. This is mainly due to the relatively slow progress in the theoretical formulation and the technological foundation of the cyber security concepts such as logical vulnerability, threats, and risks. In this article, we are proposing a framework for logical analysis, threat intelligence, and validation of security policies in cyber systems. It is based on multi-level model, consisting of ontology of situations and actions under security threats, security policies governing the security-related activities, and graph of the transactions. The framework is validated using a set of scenarios describing the most common security threats in digital banking, and a prototype of an event-driven engine for navigation through the intelligence graphs has been implemented. Although the framework was developed specifically for application in digital banking, the authors believe that it has much wider applicability to security policy analysis, threat intelligence, and security by design of cyber systems for financial, commercial, and business operations.

Keywords Knowledge graphs · Ontologies · Threat intelligence · Security policies · Security analytics

V. Vassilev (✉) · V. Sowinski-Mydlarz · P. Gasiorowski
Cyber Security Research Centre, London Metropolitan University, London, UK
e-mail: v.vassilev@londonmet.ac.uk

V. Sowinski-Mydlarz
e-mail: w.sowinskimydlarz@londonmet.ac.uk

P. Gasiorowski
e-mail: p.gasiorowski@londonmet.ac.uk

K. Ouazzane · A. Phipps
School of Computing and Digital Media, London Metropolitan University, London, UK
e-mail: k.ouazzane@londonmet.ac.uk

A. Phipps
e-mail: tsaphip1@londonmet.ac.uk

P. Bansal et al. (eds.), *Proceedings of International Conference on Artificial Intelligence and Applications*, Advances in Intelligent Systems and Computing 1164,
https://doi.org/10.1007/978-981-15-4992-2_13

1 Logical Vulnerability, Threats and Risks in Cyber Security

Most organizations, including the banks, typically use fragmented security procedures, which do not cover all possible security threats and attack vectors. They pay attention and concentrate mainly on direct technical threats and simple fraud as opposed to sophisticated social engineering fraud, which takes advantage of the loopholes in the security policies. The security policies are typically not shared between the branches and the headquarters and are not integrated into one coherent system. The organizations lack standardized, unified, and established methodology to analyze them. Therefore, banks are at high risk from security breaches, fraud, and other malicious activities. They are vulnerable, and their security policy is ineffective [1].

In a second place, the growing use of cloud services and IoT devices is an industry trend resulting in threats specific to these technologies. Small- and medium-sized enterprises without resources to maintain their operations depend on third parties for technical maintenance, servicing, and provision. However, the security risks are also high because external services increase the communications and introduce complex access control. The banks are reluctant to employ the cloud technology also due to the increased risks for protection of the financial data [2].

The problem with logical vulnerability largely stems from the fragmentation, high distribution, and difficulty in integration of the policy rules due to the lack of standards and tools for automation [3, 4]. Top players in cyber security (CISCO, Symantec, Palo Alto, Juniper) deliver predominantly physical level tools. Logical vulnerability analysis is in an early stage of development [5–7]. The risks associated with the new technologies complicate the picture even further. Employing Internet-of-Things increases the vulnerability, although it has been analyzed in some critical applications, such as medical and transport systems, it is far from solved in general. Risk assessment is even more difficult despite the recent push in industry for automation [8–11], and it is subjective and requires contracting consultants, which charge at a very high rate. Tools for development, testing, analysis, and control of the security policies are very rare. They are needed not only for large corporations, financial institutions, and service providers but also for SMEs which outsource services offshore and to third parties. Such software would have a great impact for improvement of their security policies and reducing their security risks. The small number of available tools is mainly product of significant investments from large corporate companies [12, 13]. They serve the purpose to a various degree but practically suffer from the lack of solid theoretical foundation and limited potential for use outside of the designated domain.

In this article, we are presenting a new framework for performing analysis of the logical vulnerability, for threat modelling and validation of the security policies in cyber systems. It is based on a four-layer model which, unlike most of the tools supporting threat intelligence [14, 15], is strictly formal [16]. The lowest layer of the framework, the *ontological layer* models situations, threats, and activities, formalized in description logic using the standard languages of the Semantic Web—RDF/RFS, and OWL. The second layer, the *heuristic layer*, contains security policies, formalized in clausal logic using the standard rule language of Semantic Web—SWRL. The third layer, the *workflow layer*, formalizes the transactions under threats by navigating through the nodes of a directed graph, formed by the situations and governed by the security policies. The top *process layer* performs various analytics for assessing the vulnerabilities and estimating the risks and various security analytics. Although the framework was developed specifically for validation of security policies in banking, we believe that it has much wider applicability in financial, commercial, and business systems.

2 Ontologies, Knowledge Graphs and Process Workflows

Ontological engineering is an outcome of the research program for Semantic Web [17]. Core of the software architecture of an ontology-enabled system is the *domain ontology*, which fulfills two different roles: It defines the terminology in the problem domain and simultaneously provides the base for intelligent system analysis, design, operation, and control. Standard modelling language for ontologies is OWL [18]. It has strict formal semantics given by description logic. The *expert knowledge* in an ontology-enabled intelligent system can be modelled within the paradigm of Semantic Web using rules in SWRL [19]. The main problem faced in this approach is that if the ontology requires explicit representation of both synchronous and asynchronous activities, the logical level becomes too complex to be practically useful due to the infamous *frame problem* in AI. A typical case is the area of cyber security, where the unauthorized intrusions, information leaks, frauds, and damages are achieved through malicious activities, which are asynchronous, while the protection is achieved by applying preventative or correcting countermeasures, which are synchronous. Because of this, the semantic technologies have been used mainly for formal specification and standardization of the security ontology [20]. An excellent overview of the security ontologies is given in [21]. We avoid these problems by using carefully constructed ontological theory of situations and actions [16] which allows to make the formal approach more practical.

Our framework is multi-layered and operates on four levels: (a) the *ontological level*, on which the ontology is modelled using RDF/RDFS/OWL, (b) the *heuristic level*, on which the security policies are modelled as rules in SWRL, (c) the *workflow level*, on which the analytics according to the underlying logics form directed graphs, and (d) *process level*, on which the analytics are executed against the representation and/or external data (analytics on demand). The workflow level is the base for introducing a large number of important security concepts, such as *accessibility*, *vulnerability*, *risks*, and *mitigation*, as well as the algorithms for *analyzing* them in a formal manner. This way, it bridges the knowledge-based approach to AI, based on symbolic logic, and the data analytics approach, based on machine learning. As a result, we come to the richer concept of *intelligence graphs* as a vehicle for orchestrating the security analytics. Unlike the approach of knowledge graphs [22], which only combines data analytics with expert heuristics, our approach integrates the modelling and the data analytics adding the possibility to make logical inferences, based on the results of data analytics, and to apply machine learning to the logical inferences themselves.

3 Ontology of Transactions Under Security Threats

3.1 Logical Foundations of the Ontological Modelling

We formalized the security ontology within a logical theory of situations and actions with terminological vocabulary of classes and relationships in standard DL (or concepts and properties in OWL) as given in Table 1 [16].

In our theory, the static descriptions of the world, the *situations*, are modelled in an object-oriented manner using a hierarchy of classes. The security *treats* form another hierarchy. We distinguish the asynchronous activities (*events*) and synchronous activities (*actions*) by using two different logical constructs—the events, which happen in situations without changing them are modelled as classes, while the actions which change them are modelled as relations. The parameters of the actions are not represented explicitly but are bound contextually by the situation parameters. Our approach solves the infamous *frame problem* about what needs to be changed and

Table 1 Predefined terminology on ontological level

Term	OWL	DL	Meaning
Situation	Concept	Unary predicate	Static reference to the world in time
Item	Concept	Unary predicate	Qualitative description of situations, events, threats, and items or quantitative valuation
Threat	Concept	Unary predicate	Malicious entity which appears in situations and may lead to transitions
Event	Concept	Unary predicate	Asynchronous activity which happens in situations and may lead to transitions
Action	Role	Binary predicate	Synchronous transition between situations
occurs-in	Role	Binary predicate	Events happening in situations
tampers-with	Role	Binary predicate	Threats interfering with situations
present-at	Role	Binary predicate	Attribution of items to situations
appears-in	Role	Binary predicate	Attribution of items to events
controlled-by	Role	Binary predicate	Attribution of items to threats
has/value	Role	Binary predicate	Pairing two items by qualitative association or quantitative valuation of an item
follows	Role	Binary predicate	Pairing two situations in temporal order
causes	Role	Binary predicate	Pairing two events in causal dependence

what needs to be preserved during transitions, while the actions bind their input parameters in the current situation, and they affect the environment only through their output parameters.

Principle of preservation: *Any description of the situations within the domain of the action in terms of input parameters applies to the situations within its range. This retains some descriptions after execution exactly as before.*

Principle of propagation: *Any description of the situations within the domain of the actions which involve their output parameters should be deleted from the situations within their range. This updates some descriptions along the execution path.*

These principles can be validated because the situations are logical terms themselves, so they can attribute items which serve the role of event and action parameters. Computationally, this reduces the complexity of the logical inference needed to calculate the necessary changes because it allows to use templates in the symbolic representation and to perform indexing of both facts and rules based solely on their structure.

3.2 Situations, Events, Threats and Items

The main classes of our ontology are organized in four separate taxonomies:

Items Represent the material and conceptual entities of importance (Account, Credentials, ATM, VoiceApp, etc.). They provide information related to other entities within the domain of interest and form a well-established taxonomy of classes, not necessarily related to security, but included also some classes related to the security policies, such as *channel*, *connection*, *session*, *credentials*, *profiles*, *permissions*, *threats*, etc.

Situations Model partial state of the world from security point of view (i.e., s_User_Logged, s_Account_Locked, etc.). From security point of view, some of the situations will be vulnerable, while some others will be safe, but this classification depends on both the ontology and the security policies; since the vulnerability as a concept cannot be defined on ontological level, we cannot distinguish the situations from vulnerability point of view.

Events Model the activities which cannot be predicted and controlled (i.e., e_Credentials_Accepted, e_Card_Refused, etc.). The events by their nature are asynchronous, they can happen in any situation, but they do not change the situations–they can only initiate the changes, which actually happen only after executing the corresponding actions.

Threats Model the security threats which may interfere with the normal execution of the transactions (i.e., t_ManInTheMiddle). We have identified and modelled 39 different threats of importance in banking domain, and their taxonomy is given in Fig. 1 as an example.

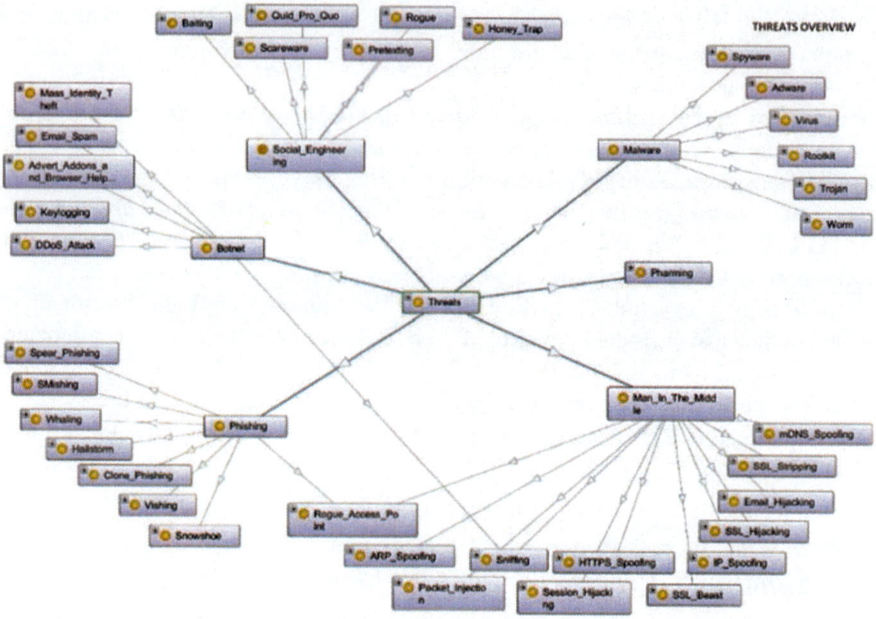

Fig. 1 Taxonomy of threats in cyber security

All classes in our ontology are modelled in OWL as concepts. The modelling is a preliminary phase of the work in our framework, which is completely independent from its subsequent use for simulation and analysis. It is an interactive process which can be fully automated using ontological editors such as Protégé.

3.3 Actions

The actions are synchronous activities which lead to changes of situations. Formally, they specify relations between the situations and are modelled as properties of the OWL classes. Since they can be executed under different initiatives, the user, the system, or the threats, it is possible to classify them. We have user-initiated actions (they do not need countermeasures), actions triggered by system errors or external interventions (controlled by the system via countermeasures), and actions which potentially lead the system to a situation which is unwanted (out of control). Based on this understanding, the actions in the ontology have been classified into three separate hierarchies:

Normal Actions, which are planned as parts of the user journeys in accordance with the endorsed security policies (e.g., a_Login).

Abnormal Actions, which are result of malfunctioning, unauthorized intrusions, or malicious activities of potential security threats; such actions can lead the system to a vulnerable state (e.g., a_Cancel).

Correcting ctions, undertaken after abnormal actions or in response to unwanted event which can put the system back on track.

This classification will play an important role in the analysis; the abnormal actions will increase the vulnerability and the risks for completing the user journeys, while the correcting actions will reduce them in accordance with the security policies.

3.4 Parametrization

Our approach to modelling the actions as relations between situations brings an interesting possibility for parametrization: The parameters can be defined in terms of attributes of the situations they transform. We consider action parameters to be only items which are attributed to all situations in which the actions apply, i.e., they are common for all of them. This leads to a natural distinction between *input* and *output* parameters—the input parameters are the items which are attributed to the situations in which we execute the actions, while the output parameters are the items attributed to the situations after execution. For example, the action a_Login, which represents the transformation from a state in which we are not logged into a state in which we are logged, has as an input parameter the item credentials needed for authentication, while its output parameter is session which can be used for further execution of the operations included in the transaction. The same applies to other dynamic entities, like events and threats; their parameters are the items which characterize all situations in which the events happen (or the threats occur, respectively). Unlike the actions, however, which can have a side effect through their output parameters, the events and threats have only input parameters determined contextually by the situations in which they happen or occur.

4 Heuristic Level and Security Policies

The ontological level provides reference objects needed for the logical vulnerability analysis, risk assessment, and neutralization of potential treats. The heuristic level will model the security policies governing the execution of the transactions under threat.

4.1 Security Policies as Heuristics

The security policies in our framework form the heuristic level. They refer to the concepts and roles as defined on ontological level, but their syntax and semantics are not given by the semantics of the underlying DL logic. For representing the security policy rules, we have adopted SWRL, a rule language which binds the concepts and properties of OWL ontology in antecedents/consequents pairs in a similar way to the infamous Horn-clause predicate logic (HPL). The policies can be formulated interactively within the same editor which specifies the OWL ontology (in our case, **Protégé**), which greatly simplifies the process of modelling. An example of a rule in this format is the following:

s_Account_Active (?aa) ∧ e_Card_Declined(?cd) ∧
a_Cancel(?aa,?tc) - > s_Transaction_Cancelled(?tc)

The intended meaning of the rule is that if an account is active in a given situation, but the system declines the card, the policy prescribes execution of an action which cancels the operation, and after executing this action, the system will be in a new situation in which the transaction is cancelled.

4.2 Types of Heuristics

The heuristic rules in our framework can be classified into different types depending on the structural patterns of the antecedent and the consequents. The antecedents combine conditions on situations, in which the rules are applicable, on events, which can fire them whenever happen and on threats, which need to be neutralized. The consequents combine static expressions, valid within the same situation as the antecedents, and dynamic expressions, involving situations resulting from the execution of actions. In accordance with this, we can distinguish the following types of heuristic rules:

Description Rules Allow inferring additional information about the current based on purely static descriptions.

Detection Rules Allow detection of the presence of threats based on observations.

Identification Rules Used for recognition of known threats.

Classification Rules Used for classification of unknown threats into known categories.

Prediction Rules Used to analyze the potential effect of the actions executed under the influence of the threats.

Correction Rules recommend actions in response to events and/or detected threats.

The rule classification plays an important role on workflow level of the framework, where it guides the navigation through the intelligence graphs.

4.3 Examples of Heuristic Rules

Let us consider an example of purely logical predictive analytics. The starting situation is "user not logged in." In this situation, the following rule applies:

s_Not_Logged_In(?nli) ∧e_Logging_In(?l) ∧a_Login(?nli, ?li) -> s_Logged_In(?li)

When the event "logging in" happens in this situation, it triggers "login action" which results in a situation "user logged in." Considering that this situation is characterized by property "session," we can derive the new session as an effect. In the next leg of the journey, we can proceed with crediting the account using the following rule

e_Crediting_Account(?ac) ∧s_Pay_In_Cash(?pic) ∧ a_PayIntoAccount(?pic, ?bic) -> s_Balance_In_Credit(?bic)

Now, the triggering event is "crediting account," and the initial situation is "paying in by cash" with parameter "account." The action which follows is "paying into account." The account can be paid in via the property "amount" of the initial situation, and the new situation is "balance in credit." However, in some situations, there is a possibility of "declining transaction."

s_Account_In_Overdraft(?ao)∧e_Card_Refused(?td) ∧ a_Decline_Transaction(?ao,?cr)- > s_Transaction_Cancelled(?cr)

The initial situation is "account in overdraft" which has property "overdraft fee." The abnormal situation "transaction cancelled" can happen as a result of obstructive events, in this case executing action "decline transaction." The irregular event which fires the rule is "card refused." The cause for the event is probably that the overdraft is not payed.

The heuristic rules allow us to specify the security policies by adding threats in the conditions of the rules and prescribing corrective actions. The conditions on situations, events, and threats in the rules refer to their attributes within the ontology, while the action parameters will be determined by the properties of their domains and ranges.

5 Workflow Level and Intelligence Graphs

5.1 Transaction Flow as a Graph

On workflow level, the framework works as a graph traversal. The intelligence graph is built using situations, events and threats as nodes and actions and roles as edges. The traversal is performed within the simulation loop as a navigation from an initial toward the final situation. Along the path, asynchronous events, which do not change the

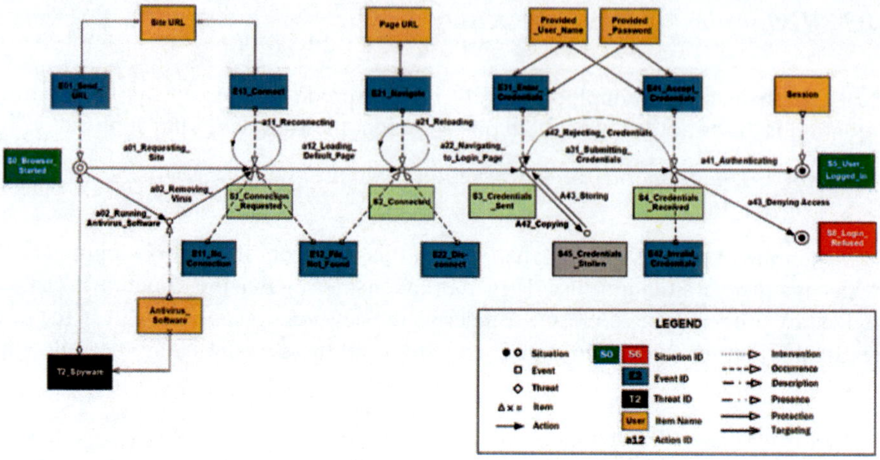

Fig. 2 Intelligence graph for transaction under threat

situations, and synchronous actions, which change them, happen. As an illustration, Fig. 2 shows the intelligence graph of one particular scenario, **Logging under threat**.

5.2 Framework Validation

The framework described above has been validated by modelling a number of scenarios for executing banking transactions under security threats. Table 2 contains brief description of the most typical scenarios used for validation of the framework.

6 Process Level, Implementation and Future Work

The general software architecture of the prototype implementation is shown in Fig. 3. Its core is an autonomous desktop application, running under Windows and written on C++. It implements the **DASHBOARD**, which coordinates the work locally, and acts as a client of the data analytics service running remotely. The **SIMULATOR** executes the simulation loop, which builds dynamically the graph from an initial situation, while two other components of the application—**DATA ANALYZER** and **LOGICAL ANALIZER**—run machine learning and rule-based inference algorithms on demand. Currently the logical analysis executes Java programs against the OWL/SWRL repository, while the data analysis runs within Docker containers on a private 16-node Kubernetes cloud.

Initially, we used DROOLS to process the policies encoded in SWRL rules [23], but the performance showed decline with the size of the repository. We implemented

Table 2 Scenarios for validation of transactions under threats

Scenario	Starting situation	Threat(s)	Threat present in	Final situation	Potential deadlock
Update antivirus with spyware	S0_Antivirus_Not_ Updated	Spyware baiting	S2_Infected_Attachment S23_Infected_Software	S5_Normal_State	
Login with spyware	S0_Browser_Started	Spyware	S0_Browser_Started	S5_User_Logged_In	S6_Login_Refused S45_Credentials_Stolen
Money transfer with spyware	S0_Browser_Started	Spyware	S13_Infected_With_Malware	S10_User_Logged_Out	S17_Site_Maintenance S18_Disconnected …
Balance with DDoS	S0_User_Logged_In	DDoS attack	S0_User_Logged_In	S4_User_Logged_Out	S10_Operation_Cancelled
View balance with quid pro quo	S0_Browser_Started	Quid pro quo	S12_IT_Support_Imitated _By_Hacker	S8_User_Logged_Out	S25_Site_Maintenance S16_Disconnected S22_Invalid_Credentials
Vishing and SMishing	S0_Normal_State	Vishing and SMishing	S0_Normal_State	S5_Payment_To_Criminals	S8_Transaction_Cancelled
Withdrawal with session hijacking	S0_Card_Inserted	Session hijacking	S2_User_Authenticated	S5_Card_Removed	
Sending email spam	S0_Machine_Overtaken _By_Hacker	Email spam	S0_Machine_Overtaken _By_Hacker	S6_Spam_Received	S2_System_Monitored

(continued)

Table 2 (continued)

Scenario	Starting situation	Threat(s)	Threat present in	Final situation	Potential deadlock
Email spam received	S0_Spam_Received	Email spam	S0_Spam_Received	S9_Machine_Overtak _By_Hacker	S5_Vital_Data_Extorted
Cross channel with pretexting	S0_IT_Support_Imitated_By_Hacker	Pretexting	S0_IT_Support_Imitated _By_Hacker	S5_Payment_To_Criminals	S8_Transaction_Canc S9_Account_Locked S10_Credentials_Stol
Scareware and Rogue	S0_Normal_State	Scareware, rogue	S1_Infection_Simulated	S6_Machine_Overtaken _By_Hacker	
ATM infected	S0_ATM_OS_Not_Updat	ATM infected	S0_ATM_OS_Not_Updated	S10_Pament_To_Criminals	S9_Transaction_Cancelled

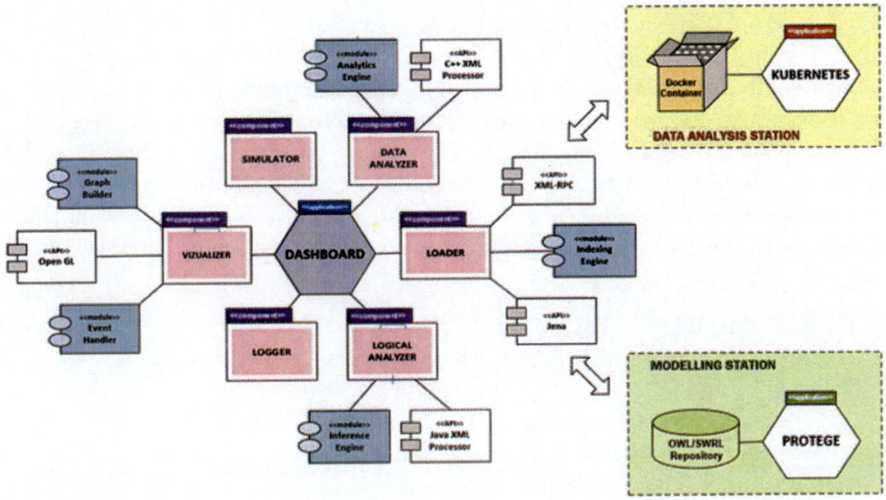

Fig. 3 Software architecture of the implementation

our own rule-based inference engine in Java, which makes use of the index structure created offline by running our indexing engine, and it greatly improved the performance.

The other major components—**VIZUALIZER**, **LOADER**, and **LOGGER**—provide interfaces for interaction with the user and the external repository, and for storing the graph traversal. The **VIZUALIZER** can be used as a rule editor as well (Fig. 4).

The framework can be used in several modes: *modelling*, *indexing*, *simulation, logical analysis,* and *data analysis*. In simulation mode, the system runs through

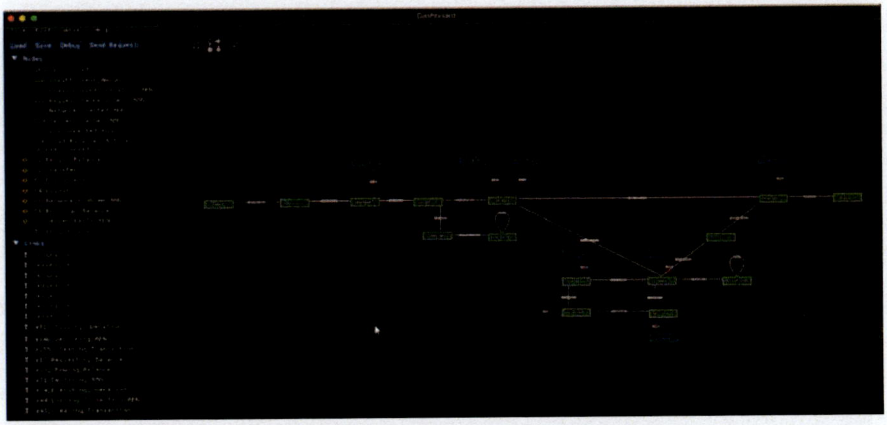

Fig. 4 Visual dashboard for simulation of the transactions under threats

scenario of operation under security threats, governed by the policy rules. The simulation begins from an initial situation and continues by emulating synchronous and asynchronous activities in a loop until it reaches a terminal situation. The events are triggered by external factors. The alternative actions are found by determining the applicable rules in each situation. Throughout the simulation, the **LOGGER** produces a rich log of situations, events which occur, threats which are detected, and actions executed upon firing applicable rules. Below is an abridged output of such a scenario which executes operation for crediting of an account after successful authentication:

Current situation is 's_Pay_In_Cash'
Action applicable to 's_Pay_In_Cash' is 'a_Cash_To_Be_Paid_In'
Rule applicable to 's_Pay_In_Cash' is rule 2
    resulting_situations_Balance_In_Credit
Rule applicable to 's_Pay_In_Cash' is rule 6
    resulting_situations_Balance_In_Credit
Event applicable to 's_Pay_In_Cash' is :e_Crediting_Account
Event applicable to 's_Pay_In_Cash' is :e_Logging_In_Unsuccessful
Rule applicable to event 'e_Crediting_Account' is rule 2
    resulting_situations_Balance_In_Credit
Rule applicable to event 'e_Crediting_Account' is rule 6

The simulation can be used both for validating the policy rules as well as for obtaining insight into the threats behavior and the effect they may have on the operations.

Currently, we are working on several engines for data analysis of *network traffic*, *clickstream data*, *event logs,* and *transaction records*. We are planning to implement separate engines for other major cyber defense tasks as well—prevention, detection, prediction, countermeasures, as well as using symbolic machine learning from experience methods for analyzing the simulation logs.

Acknowledgements The work reported here has been carried out at the Cyber Security Research Centre of London Metropolitan University. It was initiated in collaboration with Lloyds Banking Group to investigate the logical vulnerabilities in cross-channel banking. It was granted support from UK DCMS under Cyber ASAP program. It continues under a project dedicated to threat intelligence funded by Lloyds, but all examples in the paper are solely for the purpose of illustration and do not use any internal data from the bank. Any concepts, ideas, and opinions formulated by the authors in this article are not associated with the current security practices of Lloyds Banking Group.

References

1. J. Nearly, 75% of Banks were Unprepared for Cyber Attacks in 2018 (2019). https://www.teiss.co.uk/threats/banks-cyber-threat-2018/. Last accessed 2019/10/27
2. J. Marous, Technology Giants pose major threat to banking industry, in *The Financial Brand* (2019). Last accessed 2019/10/27
3. Acunetix, Logical and Technical Vulnerabilities—What They are and how can they be Detected? (2019). https://www.acunetix.com. Last accessed: 2019/10/27
4. Netsparker, Understanding the Differences Between Technical and Logical Web Application Vulnerabilities (2019). https://www.netsparker.com/blog/web-security/logical-vs-technical-web-application-vulnerabilities/. Last accessed: 2019/10/27
5. Intruder Systems, A Proactive Vulnerability Scanner, for Your External Infrastructure (2019). https://intruder.io. Last accessed: 2019/06/30
6. Greenbone Networks, OpenVAS—Open Vulnerability Assessment System (2019). http://www.openvas.org/. Last accessed: 2019/07/01
7. Rapid7, Nexpose. Your On-prem Vulnerability Scanner (2019). https://www.rapid7.com. Last accessed: 2019/07/01
8. InfoSight, Network & Cyber Security Services (2016). https://www.infosightinc.com/solutions/it-security-services/network-security.php. Last accessed: 2019/06/29
9. Kenna Security, (2018). https://www.kennasecurity.com. Last accessed: 2019/06/29
10. Coalfire, Cyber Risk Services. https://www.coalfire.com. Last accessed 2019/04/26
11. Vigilant Software, vsRisk Cloud—Cyber Risk Assessments made Simple (2019). https://www.vigilantsoftware.co.uk/topic/vs-risk. Last accessed: 2019/10/27
12. ABB, System 800xA Cyber Security—Maximizing Cyber Security in Process Automation. https://new.abb.com/control-systems. Last accessed: 2019/10/27
13. Google,CSP Evaluator. https://csp-evaluator.withgoogle.com/. Last accessed: 2019/10/27
14. Threatmodeler, The Evolution of Threat Modeling (2016). https://threatmodeler.com/evolution-of-threat-modeling/. Last accessed: 2019/10/27
15. G. Blokdyk, in *Threat Modelling*, 2nd ed. (5STARCooks, 2018). ISBN: 0655196072
16. K. Bataityte, V. Vassilev, O. Gill, in *Ontological Foundations of Modelling Security Policies for Logical Analysis*, ed. by I. Maglogiannis, L. Iliadis, E. Pimenidis. Proceeding of the 16th Artificial Intelligence Applications and Innovations Conference - AIAI 2020, Thessaloniki, Greece (Springer, 2020, in print)
17. D. Allemang, J. Hendler, in *Semantic Web for the Working Ontologist*, (MK, 2011)
18. D. McGuinness, F. Van Harmelen (eds.), *OWL Web Ontology Language* (2004). https://www.w3.org/OWL/. Last accessed 2019/04/23
19. I. Horrocks, P. Patel-Schneider et al. (eds.), *SWRL—A Semantic Web Rule Language* (2004). https://www.w3.org/Submission/SWRL/. Last accessed 2019/04/23
20. A. Herzog, N. Shahmehri, C. Duma, An ontology of information security. Int. J. Inf. Secur. Privacy **1**(4), 1–23 (2007)
21. A. Souag, C. Salinesi, I. Wattiau, Ontologies for security requirements, in *Proceedings of International Conference on Advanced Information Systems Engineering CAISE2010* (2010), pp. 61–69
22. M. Iannacone, S. Bohn, G. Nakamura et al., Developing an ontology for cyber security knowledge graphs, in *Proceedings of ACM CISR'15* (2015), pp. 12:1–12:4
23. Red Hat, Inc., Drools (overview). https://www.drools.org/. Last accessed 2019/03/11

Anomaly Detection Using Federated Learning

Shubham Singh, Shantanu Bhardwaj, Hemlatha Pandey, and Gunjan Beniwal

Abstract Federated learning is the new tide that is being associated with machine learning territory. It is an attempt to enable smart edge devices to confederate a mutual prediction model while the training data is residing at the respective edge device. This facilitates our data to be more secure, use less bandwidth, lower latency, and power consumption. We exercise the concept of federated learning in our neural network autoencoder model to detect the anomaly. Anomaly detection is finding the unusual pattern in a given data stream which may be a false or mal-entry in the pool of transactions. It helps us to prevent many online theft and scams which are detected using state-of-the-art machine learning and deep learning algorithms. All this has to be implemented in smart edge devices that have enough computing power to train the models provided to them.

Keywords Federated learning · Machine learning · Deep learning · Anomaly detection · Autoencoders · PySyft

1 Introduction

2.5 quintillion bytes of data are produced every day in this modern world and this quantity will increase with each passing year. The data produced every day records our daily activities, our interests, our hobbies, our passwords, and all other important

S. Singh (✉) · S. Bhardwaj · H. Pandey · G. Beniwal
Maharaja Surajmal Institute of Technology, C-4 Janakpuri, Delhi 110058, India
e-mail: chauhanshubham089@gmail.com

S. Bhardwaj
e-mail: shantanub30@gmail.com

H. Pandey
e-mail: pandey.hema98@gmail.com

G. Beniwal
e-mail: gunjanbeniwal@msit.in

P. Bansal et al. (eds.), *Proceedings of International Conference on Artificial Intelligence and Applications*, Advances in Intelligent Systems and Computing 1164,
https://doi.org/10.1007/978-981-15-4992-2_14

details looked at by hackers and cybercriminals. So much data is sent over the network which becomes subject to cyberattacks. It also results in unnecessary bandwidth utilization.

So, to attain privacy and limit the bandwidth utilization, we use the concept of federated learning. Federated learning also known as private learning involves sharing of the trained model weights among the smart edge devices with data not being sent on a central server [1]. In traditional machine learning and deep learning approaches, we have to send data to a centralized server where a single model was trained.

In this paper, we are implementing anomaly detection [2] using federated learning. Anomalies are also termed as outliers, deviants, or anything different from the usual entry in a system that generates doubt of its authenticity. It can be encountered due to malfunctioning systems, cybercriminals, or just an entry that was never encountered previously in the system with various applications like credit card fraud detection, online payment, health monitoring, bug, or fault detection. In this paper, we have used autoencoders which are special types of neural networks used in unsupervised learning, which reconstruct input back to its output. We use this reconstruction loss as the basis to classify anomalies.

2 Related Work

Anomaly detection is a well-practiced application of machine learning and deep learning. There are many publications implementing different techniques to achieve the same goal of anomaly detection.

Such as work done by Martinelli et al. on electric power system anomaly detection [3] is based on neural networks. In their publication, they have mentioned that they have trained their autoencoder for 72 h and the dataset had 432 patterns. As a measure of comparison, they have taken the root mean square difference of their input vector and their output vector. They were able to achieve a praise-worthy error of 0.015 with 6000 epochs.

Ullah et al. work on real-time anomaly detection in dense crowded scenes [4] is also a great explanation of work done in the field of anomaly detection. They have pursued this topic to predict the panic situation in a crowded environment. A normal activity in the crowd is considered as walking and an abnormal activity in the crowd is considered running. The dataset used is available online for research purposes. In their publication, they have mentioned the use of MLP neural network for detecting the irregular pattern in a crowded video. They have used the motion features of the crowd which have been derived from the velocity magnitude of the corner feature as it is hard to consider velocity of every pixel of an individual in a crowd. Motion properties which are derived from corner features are taken as input to MLP. The cost function gives the squared error between the desired and calculated output vectors.

Dau et al. gave a detailed analysis on anomaly detection using replicator neural networks trained on examples of one class [5]. They showed how by using only one class, we can detect anomalies in various datasets. By using six different datasets for their experiment, they produced impressive results in comparison to the previous work done on the same datasets as mentioned in their publications. The applications of anomaly detection can be seen in various fields such as credit card fraud detection, health monitoring, network intrusion, video surveillance, and other cybercrimes.

There are several autoencoders in the industry that can detect different anomalies in the data. There has been a thorough explanation of autoencoders in [6].

2.1 Federated Learning

The conventional approach of training a neural network requires a sole replica of the model and the whole training dataset is in one place. Generally, the data that is recorded by the edge devices need to be sent to the global-shared server for training purposes and the by-product network weights need to be sent back to the devices but these devices usually possess finite bandwidth and periodical connections to the globally shared server. However, in federated learning, we can train the model on each device separately and is resemblant to the data parallelism model.

The federated learning technique allows us to take advantage of the shared training model without the demand for storing it at a central server. The data is distributed over numerous clients that are connected to a mutual server. A copy of our neural network is sent to the client and each copy is trained autonomously on the respective device. The newly formulated model weights are then delivered to the central server for aggregation and the newly formulated model is again sent to the devices. Therefore, the model present at each device will pick up the data collected from all devices and the training at the shared global server is not required. This results in the use of lower bandwidth and takes place when a connection to the central server is accessible (Fig. 1).

2.2 Anomaly Detection

Anomalies are also termed as outliers, deviants, or abnormalities. An anomaly is so different from the usual entry in a system that generates doubt of its authenticity. An anomaly can be encountered due to malfunctioning systems, cybercriminals, or just an entry that was never encountered previously in the system. Anomaly detection is the detection of fake entries in a set of online exchanges. Anomaly detection has various applications like credit card fraud detection, online payment, health monitoring, bug, or fault detection. Anomaly detection was being done with machine learning approaches as it was a big thing in the previous time, but as time advanced, more deep learning techniques were used to classify anomalies.

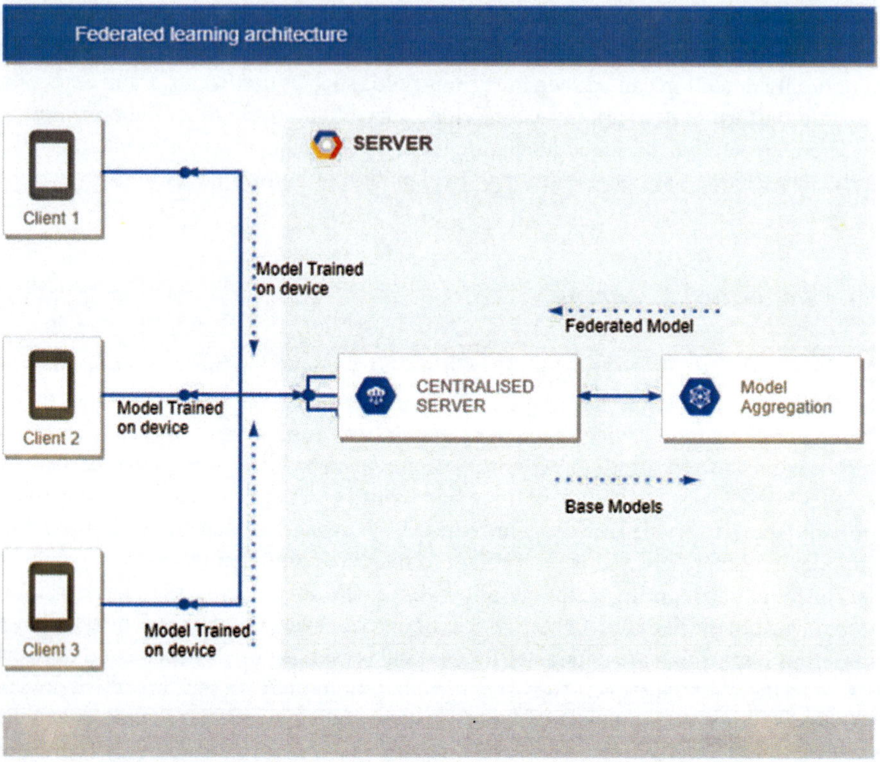

Fig. 1 Federated learning architecture

2.3 Deep Learning

Deep learning is a subset of machine learning which consists of models having multiple layers to train over data with layers of abstraction [7]. The domain of deep learning is similar to the behavior of the human brain with algorithms that resembles a neural network. Simple machine learning algorithms find it difficult to detect the patterns remotely present in large datasets whereas deep learning is found to be more productive with vast datasets, huge models, and large computations. CNNs can be applied in fields like video, audio, and images whereas RNN has the capability to work on data consisting of text and speech. Deep learning is self-sufficient to extract the features from raw data.

2.4 Artificial Neutral Network

An artificial neural network is a deep learning model which is fed on a vector of values; it then performs naive calculations based upon how the ANN is defined and then produces an output. The fed vector of values could be any form of data, from heterogeneous features to words used in a sentence and the output might depict a label.

The process of training feeds on known inputs and expected outputs, and alters the network on the basis of the anticipated result. This, when combined with the actual result produced by the neural network is employed to refine and reproduce the expected outputs. The network then attempts to produce an output that matches the expected behavior, such as correctly identifying the breed of a dog in an image.

The name deep learning comes from the multiple layers hidden between the input and the output layers as depicted in Fig. 2. The hidden layers are made up of neurons that carry out basic computations on the input received and pass the computed result to the subsequent layer.

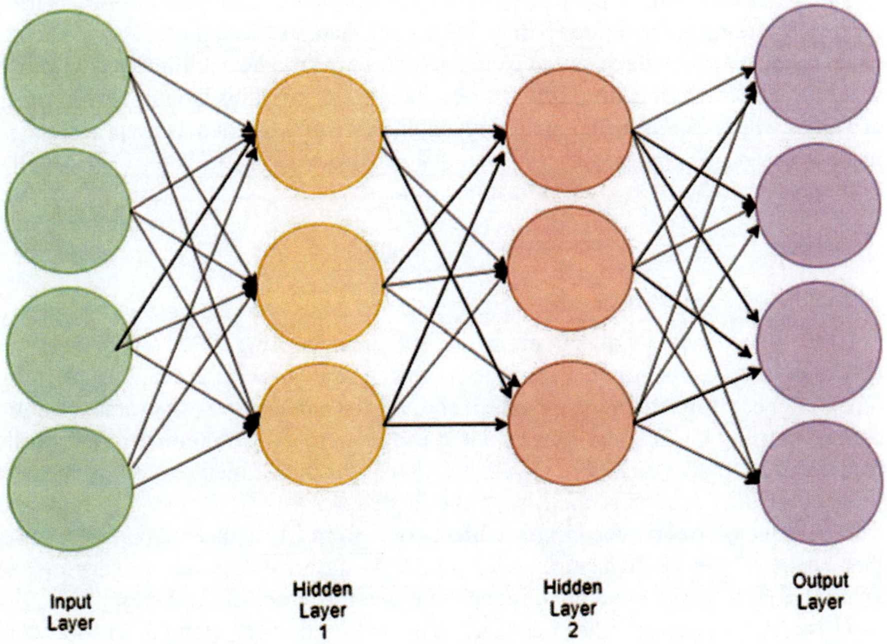

Fig. 2 Neural network model

2.5 *Autoencoder*

Autoencoders, alternatively known as auto-associative neural network encoders, attempt to regenerate the data fed to them as the output. A pair of symmetric neural networks forms the autoencoders. The first one does the job of compression; it will encode the input into a compressed form, while the second one will do the opposite of it and decodes the data provided to it. During the process, the middlemost layer receives the most dimensionally reduced form of the data provided initially. This will help us in recognizing the common inputs while the uncommon inputs will get the spotlight as their error will be prominent during the regeneration. The error during the regeneration would be the basis of anomaly score which will help us whether the observation is anomalous.

3 Experiment

Implementation of this paper is done on PyTorch models, and PySyft library is used for implementing the federated part of the project. Dataset was sampled in two parts in equal fractions which made our two clients for training model on different data which will be used for aggregation. The dataset consists of variables having a wide range of values which cannot be compared given its raw form. So, data is preprocessed to make it easier to compute the results and study anomaly in the dataset. We use the following formula for the normalization:

$$X = \frac{X - \min(X)}{\max(X) - \min(X)} \tag{1}$$

Then, we create the model of our sparse autoencoder using ReLU and Tanh layers which are used to reconstruct the input points. Extra variables are deleted as it can block the necessary memory which is required for models and important variables as we are using CUDA and it has limited cache memory. Common problem while loading datasets like KDD99 [8] is that CUDA runs out of memory. So, make sure to free variables as soon as their work is complete.

After this, we hook two workers with the two given fractions of datasets and train them, respectively. Both the models are trained on different data and then aggregated using PySyft to get the federated model and then it is used for prediction.

Using MSELoss, i.e., mean squared error between every element of input and output or target, we determine that if the point is an anomaly or not. If the loss is above the set threshold up to which we consider it a normal observation, it is treated as an anomaly.

We used the machine with the following specifications:

Software: OS: Windows 10, Jupyter Notebook, Python 3, pip.

Table 1 Results of the experiment done on shuttle dataset

Model name	Training data points	Client fraction	False +ve	False −ve	F1 score
Federated model	46,463	0.4	38	40	1
Full model	46,463	NA	0	263	0.97

Hardware: CPU: i5 7300HQ, GPU: GTX 1050Ti 4 GB, RAM: 8 GB. Disk: 128 GB.

4 Result

20% of the dataset was kept for testing and the remaining was divided into two subparts which were treated as two clients to be trained separately. These clients send back the model to produce the federated model. The results showed that the merged model was able to perform better than the model trained on full dataset.

We encounter the following results after testing on federated model and model trained on full dataset shuttle unsupervised [9] (Table 1).

So, with these results, it is safe to say that by using PySyft, we got good results and comparable to the results of Schneible et al. [10]. We were able to classify all the anomalies positively.

5 Conclusion

In conclusion, we implemented federated learning with the help of PyTorch and PySyft using two datasets of Unsupervised Anomaly Detection Dataverse from Harvard Dataverse [11]. We saw that comparable results could be achieved with the different technologies used to do the same task.

References

1. H.B. McMahan, E. Moore, D. Ramage, S. Hampson, B. Aguera y Arcas, *Communication-Efficient Learning on Deep Networks from Decentralized Data* (AISTATS, 2017)
2. V. Chandola, A. Banerjee, V. Kumar, Anomaly detection: a survey. ACM Comput. Surv. **41**(3), 6 (2009)
3. M. Martinelli, E. Tronci, G. Dipoppa, C. Balducelli, Electric power systems anomaly detection using neural networks, in *International Conference on Knowledge-Based and Intelligent Information and Engineering Systems* (2014)
4. H. Ullah, M. Ullah, N. Conci, Real-time anomaly detection in dense crowded scenes, in *Video Surveillance and Transportation Imaging Applications* (2014)

5. A. Dau, V. Ciesielski, A. Song, Anomaly detection using replicator neural networks trained on examples of one class, in *SEAL* (2014)
6. C. David et al., A practical tutorial on autoencoders for nonlinear feature fusion: taxonomy, models, software and guidelines. Inf. Fusion **44**, 78–96 (2018)
7. Y. LeCun, Y. Bengio, G. Hinton, Deep learning. Nature (2015)
8. G. Markus, "kdd99-unsupervised-ad.tab", Unsupervised Anomaly Detection Benchmark (2015). https://doi.org/10.7910/DVN/OPQMVF/GIPF3O, Harvard Dataverse, V1, UNF:6:WwXF9CrMJIdTvBZfZ4vpyg==[fileUNF]
9. G. Markus, "shuttle-unsupervised-ad.tab", Unsupervised Anomaly Detection Benchmark (2015). https://doi.org/10.7910/DVN/OPQMVF/VW8RDW, Harvard Dataverse, V1, UNF:6:sQatmd5Ao0CdXQULYgoDqQ==[fileUNF]
10. J. Schneible, A. Lu, Anomaly detection on the edge, in *MILCOM 2017–2017 IEEE Military Communications Conference (MILCOM)* (IEEE, 2017)
11. G. Markus, *Unsupervised Anomaly Detection Benchmark*. https://doi.org/10.7910/DVN/OPQMVF, Harvard Dataverse, V1, UNF:6:EnytiA6wCIilzHetzQQV7A==[fileUNF]

Enhanced Digital Image Encryption Using Sine Transformed Complex Chaotic Sequence

Vimal Gaur, Rajneesh Kumar Gujral, Anuj Mehta, Nikhil Gupta, and Rudresh Bansal

Abstract In order to increase the chaotic behaviour of the equation and to get better results than existing encryption methods, we have found a new chaotic equation, namely STLS. The chaotic map generated from STLS (as shown in Fig. 1) manifests convoluted behaviour in the entire range of parameters, and their output states scattered randomly and completely over the entire 2-D plane, which points out that they have an enhanced hyper-chaotic behaviour and shows highly random behaviour. For increasing the key space, our proposed encryption scheme uses effective scrambling and diffusion. Also, simulation results such as average entropy of encrypted image (7.9993) and NBCR (0.50005) show that our proposed algorithm is reliable and provides a high level of security.

Keywords Chaotic · STLS · Image encryption · Scrambling and diffusion · Entropy

V. Gaur (✉) · A. Mehta · N. Gupta · R. Bansal
Computer Science and Engineering Department, MSIT, Delhi 110058, India
e-mail: vimalgaur@msit.in

A. Mehta
e-mail: akm.anujmehta0011@gmail.com

N. Gupta
e-mail: nikhilguptabkn2@gmail.com

R. Bansal
e-mail: rudreshbansal3@gmail.com

R. K. Gujral
Computer Science and Engineering Department, M.M. Engineering College, Mullana, Ambala 133207, India
e-mail: drrajneeshgujral@mmumullana.org

P. Bansal et al. (eds.), *Proceedings of International Conference on Artificial Intelligence and Applications*, Advances in Intelligent Systems and Computing 1164,
https://doi.org/10.1007/978-981-15-4992-2_15

1 Introduction

In recent years, the information privacy is at risk, the media is shared across various platforms, and hence, it becomes difficult to maintain its security. It becomes easy for any third-party person to intercept our data specially images. So, to secure the images over the transmission network, we need an efficient image encryption algorithm. Conventional algorithms like AES, DES and RSA are not so reliable because if data loss happens in cipher image, then it may result in data loss in decrypted image.

1.1 Chaotic Theory

The chaotic theory is an advanced mathematical theory, and it is still in the developing phase. It enables the description of a series of phenomena from the field of dynamics, i.e. that field of physics concerning the effect of forces on the motion of objects. The paradigm of all theories of motions is that of Newton, concerns finding the deterministic values of dynamic units. However, chaos theory in mathematics points to the limitations of prediction for even deterministic values for these units. It states that even if the initial states differ by a mere value which is untraceable to human or scientific equipment, the result has radical difference. Several chaotic maps are created for image encryption [1]. Various image encryption techniques are developed in recent years which use this random behaviour of chaos theory [2–5].

Chaos theory states that a small deviation in the initial state of a system leads to non-deterministic results. So, using chaotic maps, we can generate a random sequence which increases our key space.

We generate a random sequence, namely sine transform of logistic and sine (STLS) map. The STLS produces a new chaotic map using two seed maps, namely logistic and sine chaotic map. STLS overcomes the weak chaotic behaviour which exhibited by these seed maps. Utilising the STLS, we propose an STLS map-based image encryption scheme. It uses the encryption method which involves confusion–diffusion [6].

2 Proposed Encryption Strategy

The following stages, when put together in the below sequence, make the basis of our overall encryption process. They are:

- Surrounding pixel matrix using SHA-256;
- STLS;
- Scrambling;
- Hybrid rotation;
- Diffusion.

2.1 *Surrounding Pixel Matrix Using SHA-256*

The first step in our proposed digital image encryption algorithm is to create noise in our image by covering our image with some random data. We use SHA-256 to create the random data [7]; thus, whenever the image gets encrypted, the surrounding random data is different even the user or the programmer does not know the value of surrounded data. Our image matrix has dimensions (m, n) originally, but it changes to $(m+2, n+2)$ after we surround our image matrix with the generated SHA_256 hash code. And we replace the image matrix with this SHA covered matrix. We perform this by inserting two columns of dimensions $(m, 1)$ at the very left and right sides of the matrix. After this, we cover the image by inserting two rows of dimensions $(1, n + 2)$ at the top and bottom of the matrix.

The generated hash value is in hexadecimal format. We convert the hash value into uint8 format so that intruder gets confused.

2.2 *STLS*

The logistic map and sine map have weak chaotic behaviour and show their chaotic behaviour in a particular range as seen from their bifurcation diagram in Fig. 1. To address this disadvantage, we use both maps together and perform sine transform to create a new chaotic equation. This sine transform shows very complex, chaotic behaviour as we can see in Fig. 1.

The STLS possesses chaotic behaviour in the entire plane which shows that they have random output. The structure of STLS is as follows.

$$\sin(2\pi(S1(a, x) + S2(a, x) + cc))\%1 \tag{1}$$

$S1$ is the first seed map which the logistic map in this case, $S2$ is the second seed map which is sine map in this case, and cc is the correction constant (where $cc = -0.4999999$ in this entire paper).

The structure of $S1$ and $S2$ is shown below

$$S1 = 4ax(1 - x) \tag{2}$$

$$S2 = a\sin(\pi x) \tag{3}$$

By putting these values in the above equation and modifying the equation, the newly generated chaotic map sequence STLS is:

$$x_{i+1} = \sin(2\pi(4ax(1 - x) + (4 - a)\sin\pi x - 0.4999999))\%1 \tag{4}$$

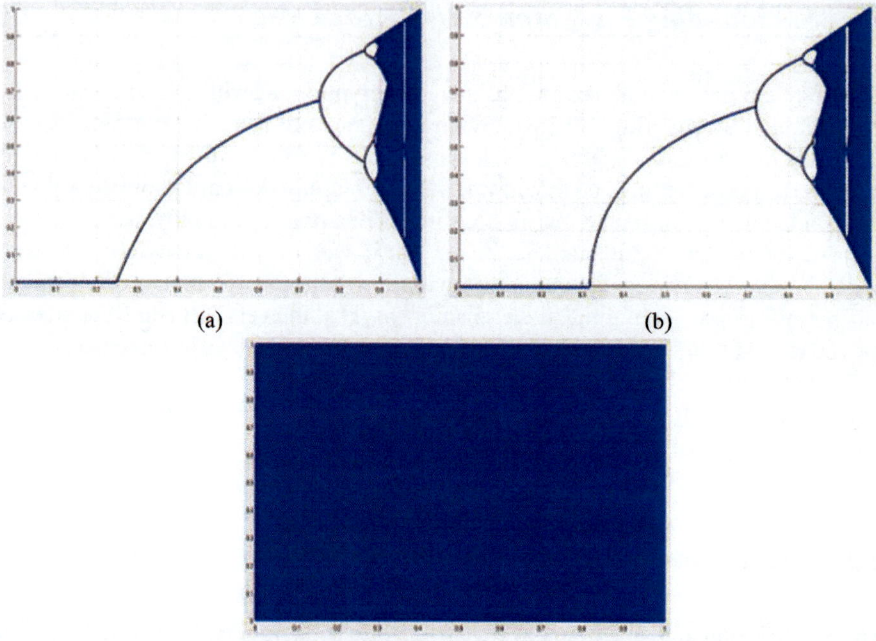

(a) (b) (c)

Fig. 1 Bifurcation diagrams of **a** logistic map, **b** sine map and **c** proposed STLS map

Key structure of STLS: The key of our encryption algorithm consists of the initial states of our two seed maps. The initial condition of our seed maps is calculated four times because our algorithm performs the encryption steps four times on a single particular image to increase the complexity and to improve our encryption result. The four initial states can be calculated by performing various changes to the original initial states, i.e. x_0 and $p0$. We can calculate the four initial states by using the following method [8].

$$x_o^i = \{x_0 \times \mathrm{cf} + \mathrm{dp} \times \mathrm{cfd}_i\} \tag{5}$$

$$p_o^i = \{p_0 \times \mathrm{cf} + \mathrm{dp} \times \mathrm{cfd}_i\} \tag{6}$$

where i belongs to [1, 4], x_0 and $p0$ are initial states, cf is coefficient of initial values, and d_p is disturbing parameter with cfd, its coefficient.

By using these initial states, we can create random sequences for each encryption process. Thus, the key space of our algorithm is 2^{256}, and its structure is shown in Fig. 2.

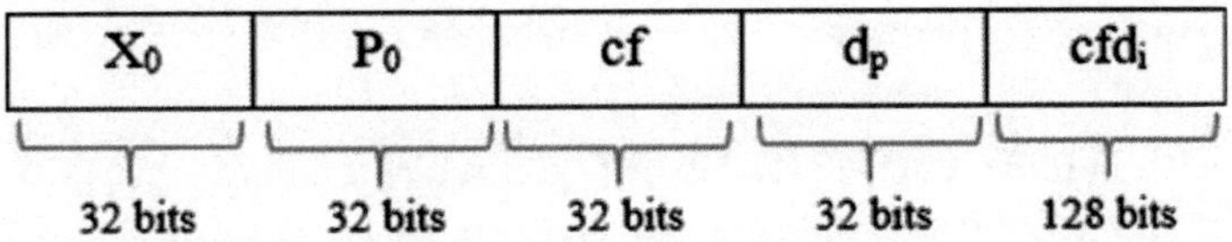

Fig. 2 Secret key structure of LSC-IES

2.3 Image Matrix Permutation

Using scrambling, we change the position of the pixels in original image row-wise and column-wise simultaneously and thus get a low correlation coefficient. For this purpose, we generate two index vectors M1I and M2I using a chaotic, random sequence and then generate a scrambled matrix through M1I and M2I. After this, we change the pixel position in the original image using this scrambled matrix.

2.4 Hybrid Rotation

We perform the hybrid rotation on the randomly permuted matrix to create a highly complex and random image matrix. A hybrid rotation is a technique which combines the working of row rotation and column rotation in a single optimised and efficient step. The row and column rotations are performed according to the value of chaotic, random sequence generated using initial states x_o^i and p_o^i, where i is ith encryption step. In row rotation, we rotate the mth row r times (where r = chaotic_sequence(m)). Further, if r is even then, we rotate our row towards the right, and if r is odd, then we rotate our row towards left. In column rotation, we rotate the nth row r times (where r = chaotic_sequence(n)). Further, if r is even, we rotate our column downwards, and if r is odd, we rotate our column upwards.

2.5 Diffusion

In the end, the highly random image matrix created by the hybrid rotation is not sufficient enough to provide security to our digital image encryption algorithm as our image histogram remains unchanged and our algorithm is prone to many histogram attacks. So, to overcome this problem, we diffused our image matrix with certain values such that the pixel values are changed and our algorithm is secured to histogram attacks. After performing this step, the image histogram becomes uniform for each pixel value which makes very difficult for an intruder to recreate the original image. To perform diffusion, we use bit-wise XOR operation of each pixel value with values generated from the random chaotic sequence.

3 Simulation Example

We use the standard "Lena (512 * 512)" greyscale standard images of 8-bit for our algorithm's stimulation process. We perform our experiments using "R" and "MATLAB". We can also encrypt the colour image using our algorithm. For colour images, we create three image matrixes R, G and B and then encrypt the separately.

We perform our algorithm on the image "Lena". Size of the images is 256 * 256, 512 * 512, respectively. Figure 3 shows that the decrypted image is completely diffused and pixels are scrambled so randomly such that the original image is completely hidden and is unreadable. Also, no data is lost in the decryption process. This shows that our encryption algorithm is working efficiently and accurately. As we can see, the pixel values are so randomly distributed in the encrypted image that no attacker or intruder has access to any information about the original images. This randomness accounts for the low correlation between adjacent pixel values of image which shows the better ability of our image encryption algorithm to resist various statistical attacks.

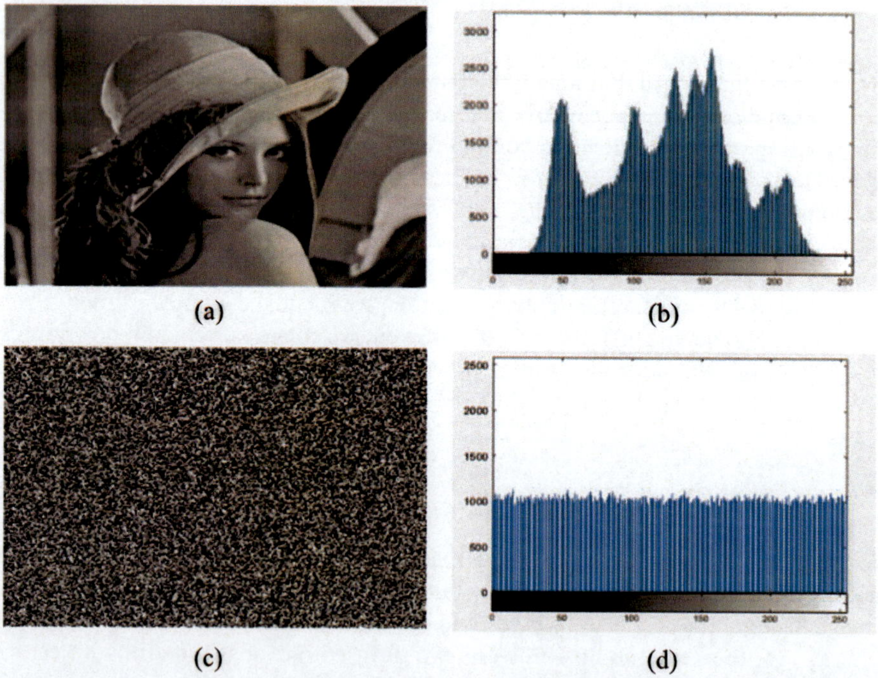

(a) (b)

(c) (d)

Fig. 3 Simulation result of **a** original images, **b** histogram of original images, **c** encrypted images and **d** histogram of encrypted images

4 Results

4.1 *Security Analysis of Secret Key*

The ability of any encryption algorithm to resist brute force attack is directly proportional to the algorithm's key space, i.e. the length of the key. As shown in our key structure, our secret key is comprised of initial states of logistic and sine chaotic map along with the coefficients of these initial values, disturbing parameter and coefficients of this disturbing parameter. According to Ref. [9]. If the key space of a digital image encryption algorithms exceeds the threshold value of 2^{100}, then that algorithm can resist brute force attacks performed by any intruder. Key space of our algorithm exceeds the threshold value. Thus, it is capable of protecting itself from these attacks. Since our algorithm's key space is 2^{256}, it is secure against brute force attacks.

Analysis of Secret Key Security
From the above observation, we conclude that our algorithm's key space is sufficient enough to protect itself from brute force attacks. However, key space is not the only factor in determining the key security. The sensitivity of image encryption algorithm towards its secret key is also equally important. If the algorithm is not sensitive enough, then a key with a slight difference from the original secret key will be able to decrypt the encrypted image accurately. The key sensitivity of the algorithm is measured by number of bit change rate (NBCR) as explained in [8]. Mathematically, NBCR can be expressed as [10].

$$\mathrm{NBCR} = \frac{H_\mathrm{d}(C1, C2)}{N_\mathrm{p}} \tag{7}$$

where H_d is the Hamming distance between image $C1$ and $C2$ and N_p is the number of pixels. If the NBCR approaches 0.50 or 50%, then the $C1$ and $C2$ are declared as different images. In this experiment, we encrypt standard Lena (256 * 256) greyscale image two times with difference in a secret key which is (1×10^{-8}) for $C1$ and $C2$, respectively. NBCR is calculated for $C1$ and $C2$, and its obtained value is **NBCR = 0.500057851166151**. Therefore, both $C1$ and $C2$ are completely different images. Figure 4 shows the histogram of the difference in pixel values of $C1$ and $C2$. This proves that our algorithm is highly secure.

4.2 *Analysis of Sensitivity of Algorithm to Plaintext*

There are different attacks which study the behaviour of change from plaintext image to encrypted image and then exploit this information to attack the encryption algorithm.

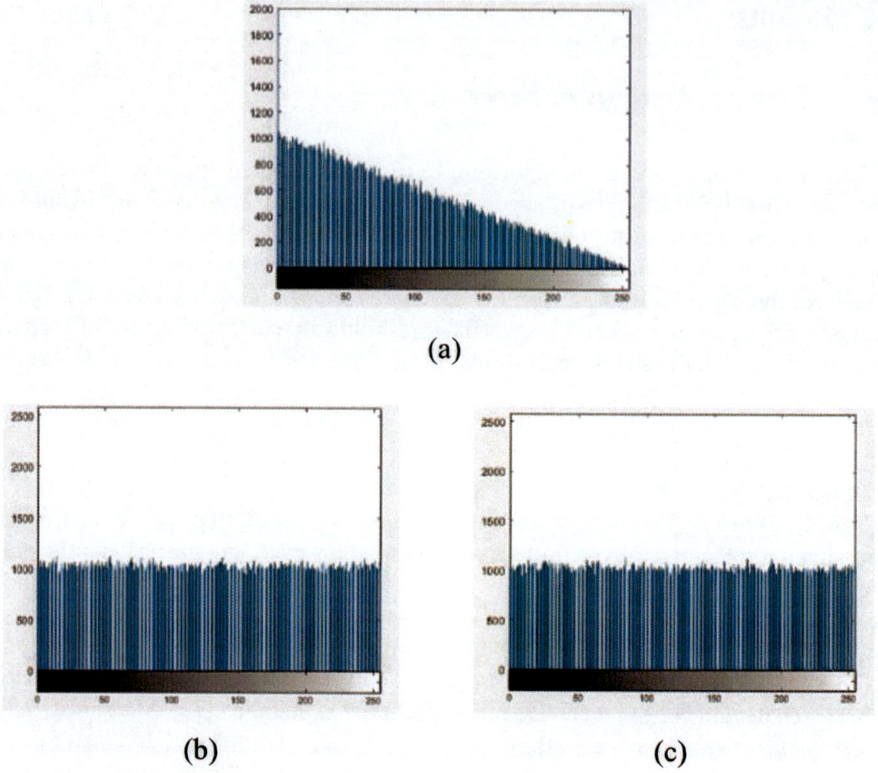

Fig. 4 a) $C1$–$C2$ image b) $C1$ image c) $C2$ image

These types of attacks are known as differential attacks, and these attacks are only prevented if entire data of the encrypted image is changed even if there is a negligible change in the plaintext image. There are two well-known tests to check the sensitivity of the algorithm to plaintext, UACI and NPCR, [11] experimentally. $C1$ and $C2$ are two images obtained by making a significantly small change in secret key k which is used to encrypt our image, and M and N are dimensions of $C1$ and $C2$.

Number of pixel change rate (NPCR) is used to measure how many pixels are different in two images. The NPCR is mathematically expressed as [11]:

$$\text{NPCR} = \frac{\sum_{i,j} D(i,j)}{M \times N} \times 100\% \tag{8}$$

$$D(i,j) = \begin{Bmatrix} 1 \text{ if } C1(i,j) \neq C2(i,j) \\ 0 \text{ if } C1(i,j) = C2(i,j) \end{Bmatrix} \tag{9}$$

Table 1 Comparison table of NPCR and UACI of the different encryption algorithm

Algorithm	NPCR	UACI
Proposed STLS	**99.61**	**33.51**
In Ref. [12]	99.62	28.50
In Ref. [13]	99.60	33.50
In Ref. [14]	99.59	33.60
In Ref. [15]	99.71	33.83

UACI ( ) is used to measure the average pixel value change in two images. The UACI is mathematically expressed as [11]:

$$\text{UACI} = \frac{1}{M \times N} \sum_{i,j} \frac{|C1(i, j) - C2(i, j)|}{255} \times 100\% \tag{10}$$

For NPCR and UACI calculations, we use standard Lena image, and we make a difference of 1×10^{-10} in the secret key and calculate two different encrypted images $C1$ and $C2$. In Table 1, we compare our calculated values with other image encryption algorithms on the same Lena (512 * 512) greyscale image.

4.3 *Analysis of Image Pixel Correlation*

The main objective of any image encryption is to reduce the correlation between adjacent pixels of the original image. High correlation is observed in the adjacent pixels of every meaningful image. There are generally three types of adjacent correlation, vertical correlation, diagonal correlation and horizontal correlation. Correlation coefficient determines the quality of encryption of any encryption algorithm. Mathematically, it is evaluated as [16]:

$$xc = \frac{\sum_{i=1}^{n} x_i y_i - \sum_{i=1}^{n} x_i \sum_{i=1}^{n} y_i}{\sqrt{\sum_{i=1}^{n} x_i^2 - \left(\sum_{i=1}^{n} x_i\right)^2} \sqrt{\sum_{i=1}^{n} y_i^2 - \left(\sum_{i=1}^{n} y_i\right)^2}} \tag{11}$$

where values of two adjacent pixels and no of pixel pairs are denoted by (x_i, y_i) and n, respectively. Quality of encryption is inversely proportional to the linear correlation value calculated by the above equation. Correlation values of this paper's encryption algorithms are compared with different encryption algorithms as shown in Table 2.

From Table 2, it can be observed that in the original image, the correlation value is approaching towards 1, and in the encrypted image, the correlation value is approaching towards 0 which verifies that the encryption algorithm is successful in reducing the correlation in adjacent pixels.

Table 2 Comparison table of horizontal and vertical correlation of different encryption algorithm

Algorithm		Horizontal	Vertical
Proposed STLS	**Original**	**0.9719**	**0.9853**
	Predicted	**0.0011**	**0.00047**
In Ref. [15]	Original	0.9740	0.9868
	Predicted	−0.0113	−0.0093
In Ref. [8]	Original	0.9400	0.9709
	Predicted	−0.003280	−0.000777
In Ref. [17]	Original	0.9400	0.9709
	Predicted	−0.0054	0.0045
In Ref. [18]	Original	0.9400	0.9709
	Predicted	0.0053365	−0.0027616
In Ref. [13]	Original	0.9700	0.9409
	Predicted	−0.0043	0.0014
In Ref. [12]	Original	0.9471	0.9665
	Predicted	−0.0159	−0.0195

4.4 Analysis of Image Information Entropy

Image information entropy is directly proportional to the uniformity of the distribution of grey value. The information acquired from the image is more complex if the image has greater entropy. Mathematically, the information entropy of an image is calculated using

$$E = -\sum_{i=0}^{n} p_i \log_2 p_i \tag{12}$$

where the probability of occurrence of "i" grey value is denoted by p_i. The ideal entropy value is 8. It represents an optimal random image. Entropy value of various algorithms is compared as shown in Table 3.

Table 3 Entropy value of different encryption algorithm

Algorithm	Entropy
Proposed STLS	**7.9993**
In Ref. [15]	7.9988
In Ref. [13]	7.9890
In Ref. [19]	7.9986
In Ref. [12]	7.9975

5 Conclusion

A new STLS image encryption algorithm is proposed in this paper, which is generated by performing sine transform on logistic and sine chaotic map. STLS shows a highly complex and chaotic behaviour of its entire space plane. This algorithm uses permutation and diffusion as a basic tool for encryption along with a hybrid rotation of pixels which make it superior to another chaotic encryption algorithm. We use an example to illustrate our encryption algorithm. We performed a detailed statistical analysis on both encrypted and original images. Entropy and correlation analysis report of our algorithm shows the superiority of our algorithm, while UACI and NPCR are also satisfactory. Moreover, it is also proved that our algorithm is capable of protecting against brute force attacks. This algorithm has its application in the fields of medical, military and business industries [20]. We will try to implement our encryption algorithm to encrypt videos in further studies.

References

1. Y. Zhou, L. Bao, C.L.P. Chen, Image encryption using a new parametric switching chaotic system. Sig. Process. **93**, 3039–3052 (2013). https://doi.org/10.1016/j.sigpro.2013.04.021
2. X. Huang, G. Ye, An image encryption algorithm based on hyper-chaos and DNA sequence. Multimedia Tools Appl. **72**, 57–70 (2014). https://doi.org/10.1007/s11042-012-1331-6
3. W. Liu, K. Sun, C. Zhu, A fast image encryption algorithm based on chaotic map. Opt. Lasers Eng. **84**, 26–36 (2016). https://doi.org/10.1016/j.optlaseng.2016.03.019
4. L. Xu, Z. Li, J. Li, W. Hua, A novel bit-level image encryption algorithm based on chaotic maps. Opt. Lasers Eng. **78**, 17–25 (2016). https://doi.org/10.1016/j.optlaseng.2015.09.007
5. N. Zhou, S. Pan, S. Cheng, Z. Zhou, Image compression–encryption scheme based on hyper-chaotic system and 2D compressive sensing. Opt. Laser Technol. **82**, 121–133 (2016). https://doi.org/10.1016/j.optlastec.2016.02.018
6. J. Lu, O. Dunkelman, N. Keller, J. Kim, New impossible differential attacks on AES, in *Proceedings of the Progress in Cryptology-INDOCRYPT* (Springer, Berlin, 2008). https://doi.org/10.1007/978-3-540-89754-5_22, pp. 279–293
7. S. Zhu, C. Zhu, W. Wang, A new image encryption algorithm based on chaos and secure hash SHA-256. Entropy **20**, 716 (2018). https://doi.org/10.3390/e20090716
8. Z. Hua, Y. Zhou, H. Huang, Cosine-transform-based chaotic system for image encryption. Inf. Sci. **480**, 403–419 (2018). https://doi.org/10.1063/1.4936322
9. G. Alvarez, S. Li, Some basic cryptographic requirements for chaos-based cryptosystems. Int. J. Bifurcat. Chaos **16**, 2129–2151 (2006). https://doi.org/10.1142/S0218127406015970
10. J.C.H. Castro, J.M. Sierra, A. Seznec, A. Izquierdo, A. Ribagorda, The strict avalanche criterion randomness test. Math. Comput. Simul. **68**, 1–7 (2005). https://doi.org/10.1016/j.matcom.2004.09.001
11. Y. Wu, J.P. Noonan, S. Agaian, NPCR and UACI randomness tests for image encryption. Cyber J. Multidisci. J. Sci. Technol. J. Select. Areas Telecommun. (JSAT) 31–38 (2011)
12. H.H. Abdlrudha, Q. Nasir, Low complexity high security image encryption based on nested PWLCM chaotic map, in *IEEE International conference for Internet Technology and Secure Transactions* (2011), pp. 220–225
13. M.B. Hossain, M.T. Rahman, A.B.M.S. Rahman, S. Islam, A new approach of image encryption using 3D chaotic map to enhance security of multimedia component, in *International Conference on Informatics, Electronics & Vision (ICIEV)* (2014). https://doi.org/10.1109/iciev.2014.6850856, pp. 1–6

14. H. Al-Najjar, Digital image encryption algorithm based on a linear independence scheme and the logistic map, in *Proceedings of ACIT* (2011)
15. M. Khan, H.M. Waseem, A novel image encryption scheme based on quantum dynamical spinning and rotations. Plos One **13** (2018). https://doi.org/10.1371/journal.pone.0206460
16. L. Min, T. Li, A chaos–based data encryption algorithm for image/video, in *International Conference on Multimedia and Information Technology* (2010). https://doi.org/10.1109/mmit.2010.27, pp. 172–175
17. Y. Zhou, L. Bao, C.L.P. Chen, A new 1D chaotic system for image encryption. Signal Process. **97**, 172–182 (2014). https://doi.org/10.1016/j.sigpro.2013.10.034
18. Z.L. Zhu, W. Zhang, K.-W. Wong, H. Yu, A chaos-based symmetric image encryption scheme using a bit-level permutation. Inf. Sci. **181**, 1171–1186 (2011). https://doi.org/10.1016/j.ins.2010.11.009
19. A.A. Karawia, Encryption algorithm of multiple-image using mixed image elements and two dimensional chaotic economic map. Entropy **20**, 801 (2018). https://doi.org/10.3390/e20100801
20. A. Srivastava, A survey report on different techniques of image encryption. Int. J. Emerg. Technol. Adv. Eng. **2**, 163–167 (2012)

Advances in Signal Processing and Learning Methods

A Low-Power Ring Voltage-Controlled Oscillator with MOS Resistor Tuning for Wireless Application

Dileep Dwivedi and Manoj Kumar

Abstract A low-power, single-output ring voltage-controlled oscillator (VCO) has been presented in this paper. Circuit diagram for proposed VCO is designed in 0.18 μm CMOS process. For proposed ring VCO, a controllable delay cell has been implemented that consists of a three transistors NAND gate inverter with MOS resistive tuning element. Resistance variation of MOS transistors has been utilized to tune the frequency of the proposed VCO. Result shows that the proposed VCO operates from 1.251 to 1.884 GHz with the deviation in control voltage (V_{CT}) from 0.6 to −0.6 V across the MOS transistors P3 and N2 with 1.8 V supply voltage (V_{dd}). Impact of V_{dd} variation on oscillation frequency also has been observed with different values of V_{CT}. A tunable range of frequency has been obtained from 1.251 to 2.897 GHz with change in V_{dd} from 1.8 to 3 V for positive value of V_{CT}. Further, output frequency shows variation from 1.795 to 2.943 GHz with change in V_{dd} from 1.8 to 3.0 V for negative value of V_{CT}. The proposed VCO shows a phase noise of −91.77 dBc/Hz @1 MHz offset from carrier frequency with power consumption of 0.248 mW at 1.8 V V_{dd}. Figure of merit (FoM) of the proposed VCO is 162 dBc/Hz.

Keywords CMOS · Low power · MOS resistance · Phase noise · VCO

1 Introduction

Phase-locked loops (PLLs) are the frequently used circuit element in modern wireless systems with wide application in clock and data recovery circuit [1, 2]. The PLL system composed of mixer, loop filter and VCO. VCO plays a critical role in PLL system since it generates carrier signals for high-speed wireless system. Low-phase

---

D. Dwivedi (✉) · M. Kumar
University School of Information, Communication and Technology, Guru Gobind Singh Indraprastha University, New Delhi, India
e-mail: dileep.dwivedi@ipu.ac.in

M. Kumar
e-mail: manojtaleja@ipu.ac.in

P. Bansal et al. (eds.), *Proceedings of International Conference on Artificial Intelligence and Applications*, Advances in Intelligent Systems and Computing 1164,
https://doi.org/10.1007/978-981-15-4992-2_16

noise and wide-tuning range are the major requirements for the design of a VCO. Generally, a VCO is designed by LC tank circuit or CMOS ring-type structure. LC tank circuit consists of an inductor and a capacitor as resonating element which requires a large area in integrated circuit [3, 4]. On the other hand, design of ring oscillator is easy to integrate and occupy less area [5]. Further, increasing growth of wireless sensor system and smart phone and saving of power in the system have become an important design consideration. There are mainly two major sources of power dissipation in CMOS circuit, first is static power and second is dynamic power dissipation. Static power dissipation is due to the static current flowing in a resistive path between input voltage and ground, and dynamic power, which results from the switching activity of the load capacitance. There is another source of power dissipation in CMOS circuit that is short-circuit current which flows when both the transistors of CMOS gate will be ON at the same time [6]. Total power dissipation in the CMOS circuit is given by Eq. (1).

$$P_{\text{total}} = \alpha C_1 V_{\text{dd}}^2 f + I_{\text{sc}} V_{\text{dd}} + I_{\text{sub}} V_{\text{dd}} + I_{\text{leakage}} V_{\text{dd}} \tag{1}$$

where α is activity factor, C_{L} is the total capacitance, f corresponds to the operating frequency, and V_{dd} represents the voltage V_{dd}. I_{sc} describes the short-circuit current, I_{sub} is the substrate leakage current, and I_{leakage} denotes the leakage current that depends on gate oxide thickness. In this paper, a single-ended ring oscillator with rf MOS resistive tuning has been presented for low-power consumption. Single-ended ring oscillator is implemented by connecting odd number of delay stages in closed loop. Block diagram of N stages single-ended ring oscillator is shown in Fig. 1.

Output oscillation frequency of ring VCO shown in Fig. 1 is given by Eq. (2).

$$f_o = \frac{1}{2Nt_{\text{d}}} \tag{2}$$

where N is number of delay cell and t_{d} represents the delay time of each cell [7]. In ring oscillator for the oscillation to begin, a loop gain of unity or greater and a phase shift of 2π are needed. A π/N phase shift is obtained by each delay cell, and reaming phase shift of π is achieved by inverting operation of delay cell. Various types of structures for the delay stages have been proposed in the reported work for the implementation of oscillator includes single-ended delays cell, differential-ended

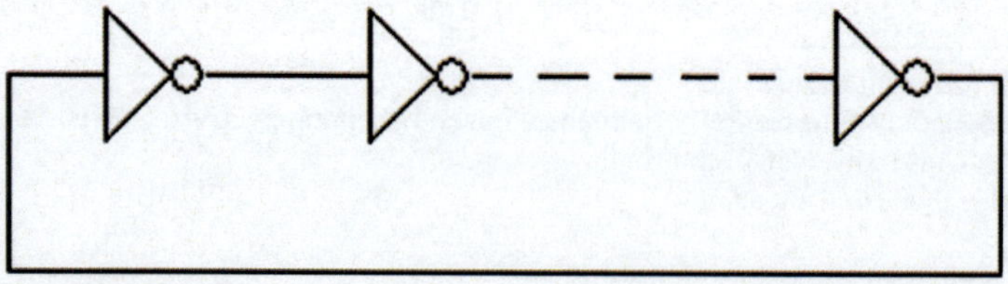

Fig. 1 Single-ended ring VCO

delay cell, dual-delay path, and multiple feedback loop delay cell [8, 9]. In this paper, a wide tuning range VCO has been designed using proposed delay cell for wireless application which consumes less power. Rf MOS resistive method has been utilized to control the tuning range. The rest of the paper is organized as follows: Sect. 2 explains the structure of VCO design. In Sect. 3, the results of the proposed VCO have been discussed. Finally, Sect. 4 draws the conclusion.

2 VCO Design Methodology

Oscillation frequency of a ring VCO mainly depends on the delay time of the delay stage. By varying the delay time, a tunable range of frequency is obtained. In the proposed VCO, a controllable delay cell has been used. The purpose of controllable delay cell is to vary the delay time of each delay cell. Frequency variation at the output is achieved with change in V_{dd} from 1.8 to 3 V. Further an additional range of oscillation frequency is achieved with the variation in V_{CT} from −0.6 to 0.6 V. Schematic diagram of the proposed delay cell is shown in Fig. 2. It consists of three transistors NAND gate inverter and a frequency tuning element. NAND delay cell is designed with one NMOS and two PMOS transistors. To reduce the leakage power consumption, a direct path is eliminated between V_{dd} and ground in the proposed delay cell. One of the terminals of NAND gate is attached at logic 1 high voltage (i.e., 1.8 V V_{dd}), and the other terminal is connected as a feedback signal. This NAND gate will work as a controlled inverter. Gate lengths (L_n and L_p) for the all the transistor in delay cell have been taken 0.18 μm, and the width (W_n) for NMOS transistor ($N1$) has been taken 0.25 μm. PMOS transistor width (W_p) for $P1$ and $P2$ has been taken as 1.25 μm. Output obtained from this inverter is connected to the input of tuning element. Rf MOS tuning element has been designed with NMOS and

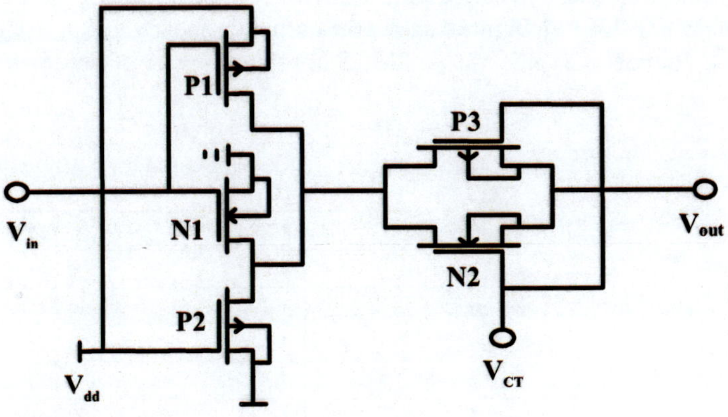

Fig. 2 Proposed delay cell

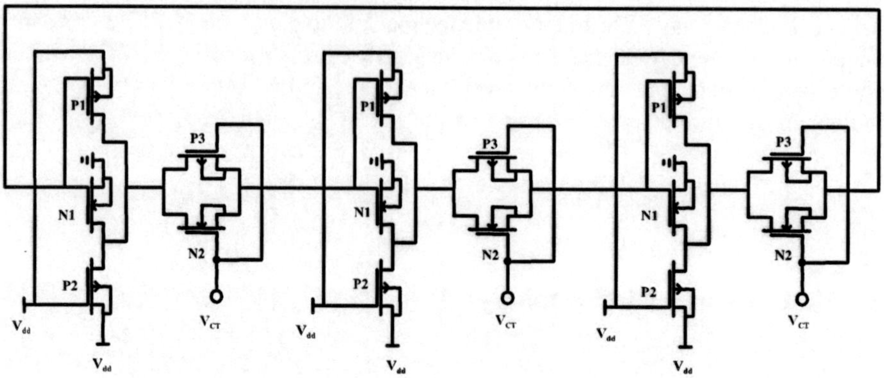

Fig. 3 Proposed three stages of ring VCO

PMOS transistor connected parallel. NMOS transistor ($N2$) width (W_n) and PMOS transistor ($P3$) width (W_p) have been taken 1 μm and 2.5 μm, respectively.

Circuit diagram for the proposed VCO has been shown in Fig. 3 which is implemented by the proposed controllable delay cell. First delay cell output is connected with the input to the next delay cell. Finally, the output of last stage is attached as a feedback with the input to form a closed loop. Output oscillation frequency for the proposed VCO is obtained with the variation in both the powers V_{dd} and V_{CT}.

3 Results and Discussion

Output results have been achieved in 0.18 μm standard CMOS technology. Table 1 shows the output frequency of the proposed VCO for the variation in V_{CT} from −0.6 to 0.6 V. Output frequency change from 1.251 to 1.884 GHz with power dissipation of 0.248 mW at 1.8 V V_{dd}. Figure 4 shows the output frequency variation with respect to the V_{CT}. For positive value of V_{CT}, PMOS transistor will be in turned off condition

Table 1 Result of frequency and power dissipation with V_{CT} deviation

Control voltage (V)	Output frequency (GHz)	Power dissipation (mW)
−0.6	1.884	0.248
−0.4	1.862	0.248
−0.2	1.795	0.248
0.0	1.666	0.248
0.2	1.510	0.248
0.4	1.391	0.248
0.6	1.251	0.248

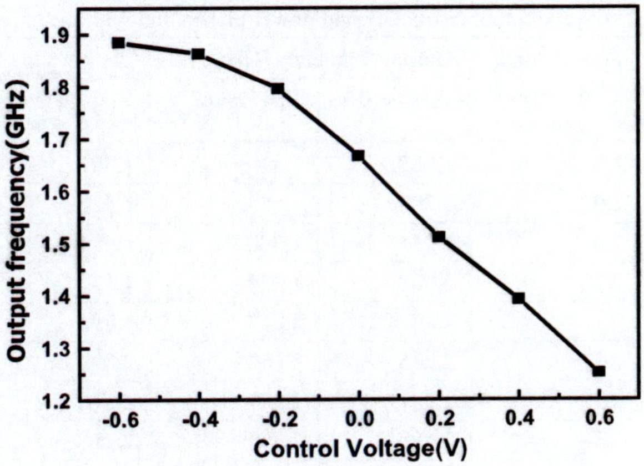

Fig. 4 Output frequency with the variation in V_{CT}

and shows higher resistance, and NMOS transistor provides the conduction path with low resistance. Transistor sizing of the MOS devices also affects the on resistance in the circuit. With a selected width of the device in the proposed tuning methodology, the overall resistance varies with changing V_{CT} from 0.0 to 0.6 V. The net effect is the increase in the resistance of parallel combination of NMOS and PMOS devices. The frequency decreases with increase in the V_{CT} from 0.0 to 0.6 V. For the negative value of $V_{CT,}$ NMOS transistor will be turned off and PMOS transistor acts as a variable resistance. Channel resistance of PMOS transistor decreases with negative value of V_{CT}. The equivalent resistance between the parallel combination of PMOS and NMOS transistors decrease that decreases the delay time of delay cell and the output frequency increases.

Further, the output frequency is obtained with the variation in V_{dd}. For fixed value of V_{CT}, V_{dd} changes from 1.8 to 3 V. Table 2 presents the result of the output oscillation frequency and power dissipation. Power consumption is directly proportional to the square of V_{dd} and increases with the increase in V_{dd}. Output frequency waveform with variation of V_{dd} is shown in Fig. 5.

Further, deviation in the output oscillation frequency is obtained with the variation of V_{dd} from 1.8 to 3 V and for the fixed negative value of V_{CT} as shown in Table 3. Output waveform for the variation in frequency with the change in V_{dd} is shown in Fig. 6.

Phase noise analysis is also an important constraint to determine the performance of VCO. Phase noise influences the quality of an output signal. Phase noise analysis for the proposed VCO is carried out for different combinations of V_{dd} and V_{CT} as shown in Table 4. Phase noise is generally expressed by Leeson's equation (3) and specified in dBc/Hz.

$$L(f_m) = 10\log_{10}\left\{\frac{\mathrm{FKT}}{2P_{avg}}\left[1+\frac{f_c}{f_m}+\left(\frac{f_o}{2f_mQ_L}\right)^2\left(1+\frac{f_c}{f_m}\right)\right]\right\}\mathrm{dBc/Hz} \quad (3)$$

Table 2 Output frequency and power dissipation with the change in (V_{dd})

Power supply (V_{dd})	Output frequency (GHz)				Power dissipation (mW)
	V_{CT} = 0 V	V_{CT} = 0.2 V	V_{CT} = 0.4 V	V_{CT} = 0.6 V	
1.8	1.667	1.510	1.391	1.251	0.248
1.9	1.799	1.655	1.527	1.382	0.316
2.0	1.934	1.790	1.653	1.509	0.393
2.1	2.062	1.924	1.777	1.627	0.479
2.2	2.177	2.047	1.894	1.747	0.574
2.3	2.285	2.162	2.009	1.858	0.679
2.4	2.391	2.269	2.116	1.963	0.793
2.5	2.488	2.369	2.217	2.069	0.917
2.6	2.572	2.465	2.322	2.164	1.051
2.7	2.658	2.556	2.421	2.258	1.195
2.8	2.743	2.646	2.510	2.352	1.349
2.9	2.821	2.724	2.595	2.442	1.513
3.0	2.897	2.803	2.679	2.527	1.687

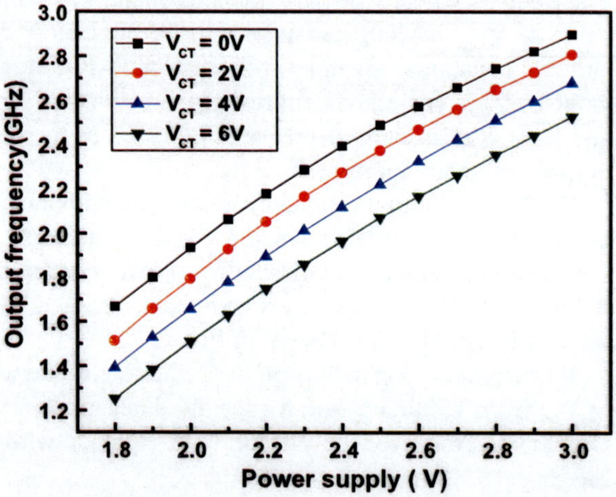

Fig. 5 Output frequency with V_{dd} variation

where L (fm) = represent phase noise (dBc/Hz), Q_L = quality factor of load, f_m = offset frequency from carrier (Hz), f_o = operating frequency (Hz), f_c = flicker corner frequency (Hz), T = temperature (°K). P_{avg} = Average power dissipation (W), F = Noise factor, k = Boltzmann constant. Leeson's equation identifies the most significant cause of phase noise in oscillator

The figure of merit (FOM) is given by Eq. (4).

Table 3 Output frequency and power dissipation with the variation in V_{dd} and V_{CT}

Power supply (V)	Output frequency (GHz)			Power dissipation (mW)
	V_{CT} = −0.2 V	V_{CT} = −0.4 V	V_{CT} = −0.6 V	
1.8	1.795	1.862	1.884	0.248
1.9	1.926	1.985	1.996	0.316
2.0	2.046	2.100	2.107	0.393
2.1	2.161	2.208	2.212	0.479
2.2	2.266	2.309	2.315	0.574
2.3	2.367	2.403	2.412	0.679
2.4	2.463	2.492	2.500	0.793
2.5	2.552	2.576	2.583	0.917
2.6	2.636	2.658	2.662	1.051
2.7	2.710	2.733	2.737	1.195
2.8	2.786	2.875	2.809	1.349
2.9	2.848	2.875	2.876	1.513
3.0	2.931	2.944	2.943	1.687

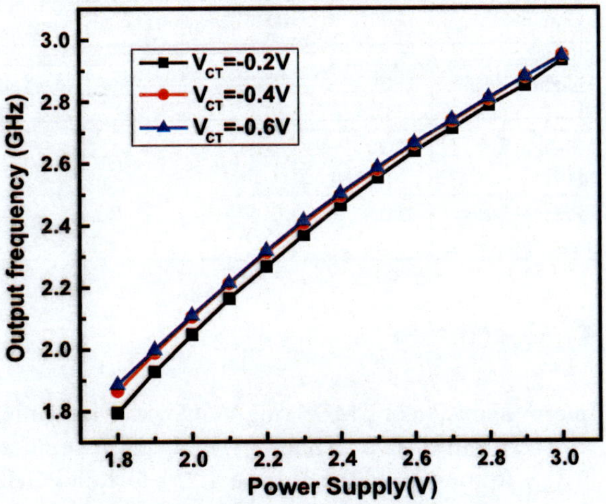

Fig. 6 Frequency change with V_{dd}

$$\mathrm{FoM(dBc/Hz)} = 20\log\frac{f_{\mathrm{osc.}}}{f_{\mathrm{off.}}} - 10\log\frac{P_{\mathrm{diss.}}}{1\ \mathrm{mW}} - \mathrm{PN} \tag{4}$$

where f_o represents oscillation frequency, Δf is offset from the carrier frequency, P is power consumption in mW, and PN denotes phase noise in dB/Hz. The performance comparisons of the proposed VCO with other reported works in terms of phase noise, power consumption, and FOM are shown in Table 5.

Table 4 Phase noise and figure of merit (FoM) results of the proposed VCO

Tech. (μm)	Frequency (GHz)	V_{dd} (V)	V_{CT} (V)	Power (mW)	Phase noise (dBc/Hz)	FoM (dBc/Hz)
0.18	1.67	1.8	0	0.248	−91.77	162
0.18	2.39	2.4	0	0.793	−90.3	159
0.18	2.9	3	0	1.687	−89.48	156
0.18	2.93	3	−0.2	1.687	−89.33	156
0.18	1.8	1.8	−0.2	0.248	−91.52	162
0.18	1.86	1.8	−0.4	0.248	−91.24	162
0.18	1.39	1.8	0.4	0.248	−92.39	161

Table 5 Performance comparison of the proposed VCO with the previous work

References	Tech. (μm)	Freq. (GHz)	Power (mW)	Phase noise (dBc/Hz)	FoM (dBc/Hz)
Xuemei et al. [10]	180	5.287	15.1	−97.93 @ 1 MHz	160.6
Chen and Lee [11]	180	1.92	13	−102 @ 1 MHz	156.3
Choi et al. [12]	180	0.807	22	−108 @ 1 MHz	150.6
Nizhnik et al. [13]	180	4.4	81	−120.2 @ 4 MHz	162
Sheu et al. [14]	180	4.09	13	−93 @ 1 MHz	154.4
Present work	180	1.67	0.248	−91.77 @ 1 MHz	162

4 Conclusion

Improved design of CMOS ring VCO with wide tuning range has been implemented with 0.18 μm CMOS technology. Frequency variation is obtained with the variation in V_{CT} from −0.6 to 0.6 V from 1.251 to 1.884 GHz for fixed value of V_{dd} 1.8 V. Further, a tunable range from 1.251 to 2.897 GHz has been achieved with change in V_{CT} from 0 to 0.6 V, and a frequency variation from 1.795 to 2.944 GHz with change in V_{CT} from −0.2 to −0.6 V has been obtained. Result shows that proposed VCO has a phase noise of −91.77 dBc/Hz @1 MHz offset from center frequency. Figure of merit (FOM) is 162 dBc/Hz. Result shows that the performance characteristic of the proposed VCO has been improved with the earlier reported work for low-power wireless applications.

References

1. S.J. Lee, B. Kim, K. Lee, A novel high-speed ring oscillator for multiphase clock generation using negative skewed delay scheme. IEEE J. Solid-State Circuits **32**(2), 289–291 (1997)
2. B. Catli, M.M. Hella, A 0.5-V 3.6/5.2 GHz CMOS multi-band LC VCO for ultra low-voltage wireless applications, in *IEEE International Symposium on Circuits and Systems* (2008), pp. 996–999
3. R.B. Staszewski, P.T. Balsara, Phase-domain all-digital phase-locked loop. IEEE Trans. Circ. Syst. II Exp. Briefs **52**(3), 159–163 (2005)
4. M. Kumar, S.K. Arya, S. Pandey, Low power digitally controlled oscillator designs with a novel 3-transistor XNOR gate. J. Semicond. **33**(3), 035001 (2012)
5. L.S. De Paula, A.A. Susin, S. Bampi, A wide band CMOS differential voltage-controlled ring oscillator, in *Proceedings of the 21st Annual Symposium on Integrated Circuits and System Design* (2008), pp. 85–89
6. K. Roy, S.C. Prasad, *Low Power CMOS Circuit Design* (Wiley Pvt. Ltd., 2002)
7. M.J. Deen, M.H. Kazemeini, S. Naseh, Performance characteristics of an ultra-low power VCO, in *Proceedings of the 2003 International Symposium on Circuits and Systems, ISCAS'03*, vol. 1 (2003), pp. I–I
8. M. Kumar, S. Arya, S. Pandey, Ring VCO design with variable capacitance XNOR delay cell. J. Inst. Eng. (India): Series B **96**(4), 371–379 (2015)
9. J.K. Panigrahi, D.P. Acharya, Performance analysis and design of wideband CMOS voltage controlled ring oscillator, in *2010 5th International Conference on Industrial and Information Systems* (2010), pp. 234–238
10. L. Xuemei, W. Zhigong, S. Lianfeng, Design and analysis of a three-stage voltage-controlled ring oscillator. J. Semicond. **34**(11), 115003 (2013)
11. Z.Z. Chen, T.C. Lee, The design and analysis of dual-delay-path ring oscillators. IEEE Trans. Circuits Syst. I Regul. Pap. **58**(3), 470–478 (2010)
12. J. Choi, K. Lim, J. Laskar, A ring VCO with wide and linear tuning characteristics for a cognitive radio system, in *2008 IEEE Radio Frequency Integrated Circuits Symposium* (IEEE, 2008), pp. 395–398
13. O. Nizhnik, R.K. Pokharel, H. Kanaya, K. Yoshida, Low noise wide tuning range quadrature ring oscillator for multi-standard transceiver. IEEE Microw. Wirel. Compon. Lett. **19**(7), 470–472 (2009)
14. M.L. Sheu, Y.S. Tiao, L.J. Taso, A 1-V 4-GHz wide tuning range voltage-controlled ring oscillator in 0.18 μm CMOS. Microelectron. J. **42**(6), 897–902 (2011)

Fuzzy Logic Control D-STATCOM Technique

Shikha Gupta and Muskan

Abstract The distribution power system has power quality problems such as harmonics injected into the grid. This is mainly because of the presence of a nonlinear load in the system. The distribution static compensator (D-STATCOM) can alleviate power quality problems. Conventionally, proportional and integral (PI) controllers are used to control DC-link in D-STATCOM. However, the PI controller cannot provide optimal performance for sudden change, and this results in the lower transient response of D-STATCOM. Robust controllers like fuzzy logic controller (FLC) are required to improve the regulation of DC-link voltage. This paper presents the design of D-STATCOM-based fuzzy logic controller for a distribution power system. The fuzzy-based system provides adequate dynamic voltage control because of variations in PI gains appropriated using two-dimensional knowledge rule base and great sampling during the dynamic period. Triangular membership functions are constructed for reducing oscillations in the output of FLC, whereas Mamdani's centered min operator is used for fuzzification and defuzzification. The synchronous reference frame theory (SRFT)-based control technique is implemented to generate reference current which is further used in voltage source converter (VSC) control. System performance is studied in MATLAB/SIMULINK platform.

Keywords D-STATCOM · PI controller · Fuzzy logic controller · Synchronous reference frame theory

---

S. Gupta (✉) · Muskan
Electrical and Electronics Department, Bhagwan Parshuram Institute of Technology Rohini, New Delhi, India
e-mail: shikhaguptadtu@gmail.com

Muskan
e-mail: muskanvirmani3112@gmail.com

P. Bansal et al. (eds.), *Proceedings of International Conference on Artificial Intelligence and Applications*, Advances in Intelligent Systems and Computing 1164,
https://doi.org/10.1007/978-981-15-4992-2_17

1 Introduction

The modern electrical systems is a widespread of power conversion units and power electronic devices. The literature survey shows electrical power consumed by modern electronic devices which is 30% of annual electricity consumption [1]. Further, the power demand increases with the rapid increment of digital economy day by day. The digital economy signifies the use of power electronic equipment like adjustable speed drives, energy-efficient lighting, and programmable logic controllers. These devices are very sensitive to disturbances and cause to complete change of electric load nature [2]. Due to the change of electric load nature, power quality issues occur in the distribution and transmission system. The most common power quality hitches are harmonic distortion, low power factor, unbalancing voltage, and frequency deviation [3].

Power quality issues result in the huge power losses in the system and improper working of electronic equipment. High quality and reliability of power with minimum cost are required for the proper functioning of the system. To achieve this, power electronic controllers are required, which may comprise miscellaneous FACT devices like distributed static compensators (D-STATCOM), dynamic voltage restorer (DVR), and unified power quality controller (UPQC) [4]. D-STATCOM is one of the custom power devices for the improvement of power quality in medium voltage distribution networks. D-STATCOM regulates bus DC-link voltage at estimated voltage and frequency. To increase the limit of power transmission, an electric grid shunted with D-STATCOM is used [5]. D-STATCOM also improves the voltage profile along the compensated line by controlling the reactive power of the system. The load current and voltage are instantaneous quantities and may change frequently. Due to this, the reference component should be real-time quantity. To extract reference components, various controlling techniques have been suggested [6]. The main part of the D-STATCOM operation is a controller. Conventional PI controller and hysteresis current controller (HCC) were used to normalize DC voltage and current harmonics, respectively. However, a conventional PI controller requires a good steady state and dynamic response, which is difficult to obtain under nonlinear load. Recently, fuzzy logic control (FLC) scheme has a great interest in many applications. The fuzzy logic controller provides a good dynamic and steady-state response. It is an accurate mathematical model insensitive with parameter variation and does not require precise input [7]. Also, it offers more flexibility, robustness, and eases to understand. In this paper, a modulating technique originates from the synchronous reference frame theory (SRFT). The VSC control technique is implemented to D-STATCOM system control, which bounds by non-sinusoidal supply conditions [8]. Extraction of non-reference current components is needed to generate gate pulses for VSC of D-STATCOM. For the generation of gate pulses, the current controller is required [9]. This work is proposed in the fuzzy logic controller instead of the PI controller because of many profits [10]. The fuzzy logic control scheme is one of the knowledge base intelligence schemes [11]. Fuzzy controller is maintained in DC-link voltage by determining the switching losses of the IGBT which is used in VSC. For DC-link voltage regulation, active power needs to be controlled.

2 Modeling of D-STATCOM

An arrangement of a three-phase controlled voltage source converter linked with DC capacitor represents the D-STATCOM model [12]. It applies at the utility customer point of common coupling (PCC) of the system to supply reactive power demand by the load. It may operate in two control modes. One is in the current control mode; in this mode, the D-STATCOM compensates the demand for the load current. To balance the unwanted components due to harmonic distortion and load power requirement, a current controller is designed. The other is voltage control mode, in which voltage regulation function is performed through D-STATCOM. For reducing harmonics from voltage waveform, a voltage controller is required to generate references. Three-phase distribution system stability nourishes by D-STATCOM (Fig. 1). D-STATCOM is configured as a universal bridge converter, thus having a constant DC bus voltage.

This voltage is maintained by the DC-link capacitor (C_{dc}), three-phase coupling inductance (L_s), and ripple filter (C_f, R_f). To conserve DC-link voltage ($V_{dc,\,set}$), SRFT is based on an indirect control technique used. To achieve fundamental source current component with positive sequence, active filter feed is compensating current (i_{sc}) to PCC. This will cause a unity power factor at the load end. The KCL at PCC is as follows:

$$i_s + i_{sc} = i_L \tag{1}$$

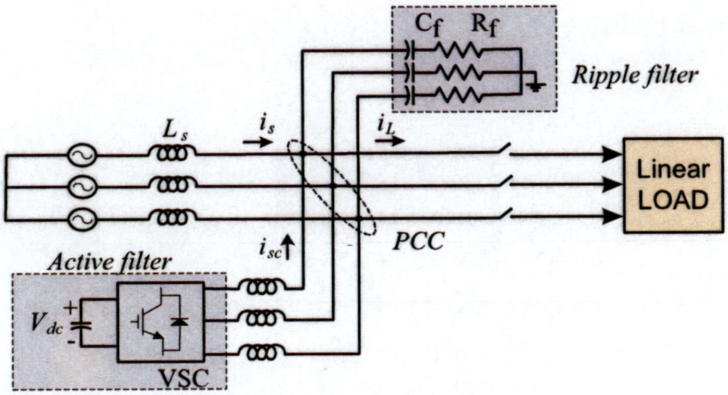

Fig. 1 D-STATCOM connected with three-phase distribution system

3 Synchronous Reference Frame Theory

Figure 2 shows the SRFT-based control technique used to control D-STATCOM by providing gate pulses to VSC. The SRFT technique is based on the indirect current control scheme. The three-phase load current is sensed and provided to the *abc–dq0* frame. By using Park's transformation, '*a–b–c*' phase components are converted into '*α–β–0*' coordinates and '*α–β–0*' coordinates is converted into '*d–q–0*' with the help of Clark's transformation (2). Transformation angle (ωt) is provided through a phased locked loop.

$$\begin{bmatrix} i_\alpha \\ i_\beta \\ i_0 \end{bmatrix} = \frac{2}{3}\begin{bmatrix} 1 & -1/2 & -1/2 \\ 0 & \sqrt{3/2} & -\sqrt{3/2} \\ 1/2 & 1/2 & 1/2 \end{bmatrix}\begin{bmatrix} i_{\mathrm{La}} \\ i_{\mathrm{Lb}} \\ i_{\mathrm{Lc}} \end{bmatrix}; \quad \begin{bmatrix} i_{\mathrm{d}} \\ i_{\mathrm{q}} \\ i_0 \end{bmatrix} = \begin{bmatrix} \sin\omega t & -\cos\omega t & 0 \\ \cos\omega t & \sin\omega t & 0 \\ 1 & 1 & 1 \end{bmatrix}\begin{bmatrix} i_\alpha \\ i_\beta \\ i_0 \end{bmatrix} \tag{2}$$

In this paper, the quadrature component is terminated. The direct component passes through a low-pass filter (LPF) to extract the fundamental component. This signal is added with loss current component for maintaining constant DC voltage of VSC (3). Loss current component is obtained through a PI controller by providing an error signal. The error signal is obtained by substracting the sensed voltage across the DC link capacitor and prespecified reference value.

$$i^*_{\mathrm{active}} = i_{\mathrm{d}} + i_{\mathrm{loss}} \tag{3}$$

Added signal of loss current component and direct component is given to the *dq0–abc* frame. This signal is converted into '*a–b–c*' phasor component (4). The transformation angle (ωt) is estimated through the same PLL. The PLL is used to the synchronization of signals with PCC voltages. The output *abc* current component passes through a hysteresis current controller (HCC). The output of HCC is provided gate pulses for VSC.

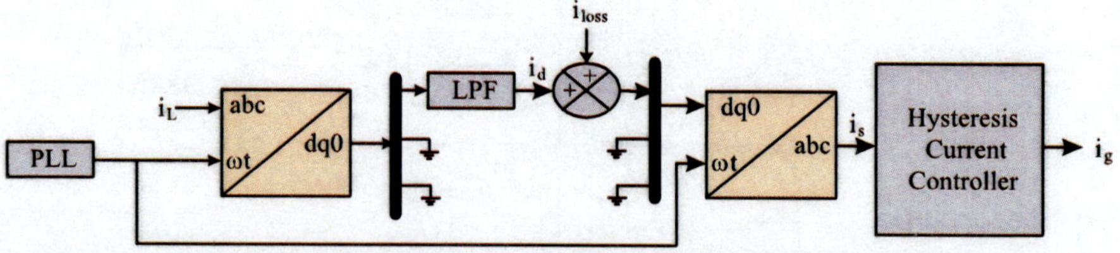

Fig. 2 D-STATCOM control scheme based on SRFT

$$\begin{bmatrix} i_\alpha^* \\ i_\beta^* \\ i_0^* \end{bmatrix} = \begin{bmatrix} \sin\omega t & \cos\omega t & 0 \\ -\cos\omega t & \sin\omega t & 0 \\ 1 & 1 & 1 \end{bmatrix} * \begin{bmatrix} i_\mathrm{d}^* \\ i_\mathrm{q}^* \\ i_0^* \end{bmatrix}; \quad \begin{bmatrix} i_\mathrm{sa}^* \\ i_\mathrm{sb}^* \\ i_\mathrm{sc}^* \end{bmatrix} = \begin{bmatrix} 1 & 0 & 1 \\ -1/2 & \sqrt{3}/2 & 1 \\ -1/2 & -\sqrt{3}/2 & 1 \end{bmatrix} * \begin{bmatrix} i_\alpha^* \\ i_\beta^* \\ i_0^* \end{bmatrix} \tag{4}$$

4 Design of Fuzzy Logic Controller

The fuzzy logic controller provides good dynamic response and steady-state response. This controller stems from simple linguistic variables and eases to understand [13]. Figure 3 shows the process of the fuzzy controller which involves fuzzification, defuzzification, and the formation of rule base and decision-making logic [4]. Fuzzification converts each input variable into linguistic variables using seven membership functions. Membership functions are mapped numerical variables into suitable linguistic values. Linguistic values are viewed as labels of fuzzy sets. All membership functions for FLC input and output are in the range of [−1, 1]. These membership functions overlap each other. Seven membership functions consist of Negative Big, Positive Big, Negative Medium, Positive Medium, Negative Small, Positive Small, and Zero which are denoted by NB, PB, NM, PM, NS, PS, Z, respectively. The two-dimensional knowledge rule base is used to compute the loss DC energy component as shown in Table 1. The formulation of the rule base has controlled the action of the controller. A specific set of rule bases is formed in the support of human thinking and in the form of IF-THEN rule. The system is designed for *n*

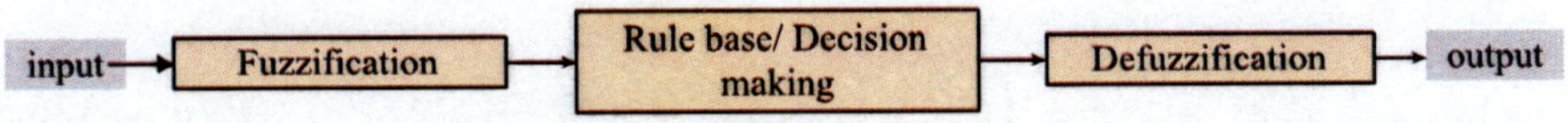

Fig. 3 Simple block diagram of fuzzy logic controller

Table 1 Knowledge rule base

	Δe						
e	NB	PB	NM	PM	NS	PS	Z
NB	NB	Z	NB	NS	NB	NM	NB
NM	NB	PS	NB	Z	NB	NS	NM
NS	NB	PM	NB	PS	NM	Z	NS
Z	NB	PB	NM	PM	NS	PS	Z
PS	NM	PB	NS	PB	Z	PM	PS
PM	NS	PB	Z	PB	PS	PB	PM
PB	Z	PB	PS	PB	PM	PB	PB

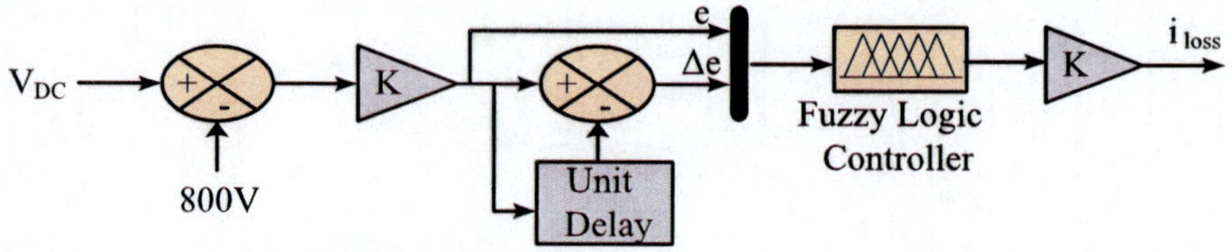

Fig. 4 Schematic diagram of fuzzy logic controller

numbers of functions for each input and m number of inputs, then total $(m * n)$ rules are formed based on combinations of different inputs.

Decision-making logic is required to decide in favor of input error. Defuzzification converts fuzzy set values into control values. In the design of the fuzzy controller, an amplified error signal from the referenced DC voltage and sensed DC voltage across the capacitor is used as an input signal of the fuzzy controller. Figure 4 shows a schematic diagram of FLC. An amplified output of an FLC is considered as the peak of a direct current component. The normalized input of the FLC system is DC voltage error (e), and a change of error is at rth instant (Δe) (5).

$$e = V_{\mathrm{dc,set}} - V^*_{\mathrm{dc}};\quad \Delta e = e(r) - e(r-1) \tag{5}$$

The DC-link voltage error is estimated as the difference between the terminal voltage across the DC-link capacitor and the set reference voltage (800 V). The FLC output is loss current component (i_{loss}) [10]. This system calculates i_{loss} correspond with differential error(Δe_{dc}) and losses in the VSC power circuit (6).

$$i_{\mathrm{loss}} = (2 \times \Delta e_{\mathrm{dc}})/(3 \times V_{\mathrm{sm}} \times T_{\mathrm{x}}) \tag{6}$$

where $\Delta e_{\mathrm{dc}} = e^*_{\mathrm{dc}} - e_{\mathrm{dc}} = \left(C_{\mathrm{dc}} \times \left[(V_{\mathrm{dc,set}})^2 - (V^*_{\mathrm{dc}})^2\right]\right)/2$; $T_x = T_s/6 = 1/6f_s$ V_{sm} is voltage of system and f_s is frequency of supply voltage. There is a possibility of instability and oscillations in the output of FLC [11]. For stabilization, scaling factors tuned while triangular membership functions are constructed. The designed FLC is used Mamdani's min operator for fuzzification and centroid operator for defuzzification.

5 Results and Discussions

The performance evaluation of the distributed network is based on fuzzy control theory presented. In this section, performance evaluation is tested in terms of the following variables: DC-link voltage, source voltage (V_g), source current (i_s), load voltage (V_L), and load current (i_L). Under different conditions of load, simulation results are presented. Parameters of system are listed in an appendix. DC-link voltage (V_{dc}*) is regulated at 800 V.

5.1 Performance Under Linear Load and Nonlinear Loads

Simulation result for the linear balanced load system is shown in Fig. 5. At $t = 0.4$ s, the nonlinear load is interfaced in the system. Nonlinear load introduces harmonics in the power system which is undesirable. Figure 5 shows that grid currents and grid voltages are sinusoidal and balanced throughout the performance. The system is working under a unity power factor even for nonlinear load, and this shows the effective working of the system. The DC-link is regulated at prespecified reference (800 V). It can be seen a sudden increment in DC-link voltage at 0.4 s (at the interfacing instant of nonlinear load). This value is reached 808 V and shows 1% overshoot which is regulated in less than one cycle. There is an increase in settling time due to the rapid change in load.

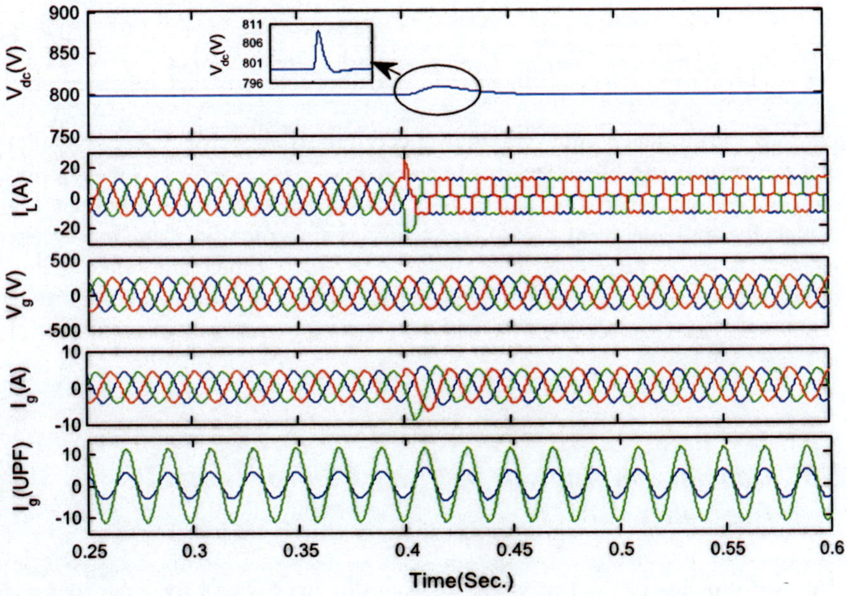

Fig. 5 Performance evaluation for linear and nonlinear loads

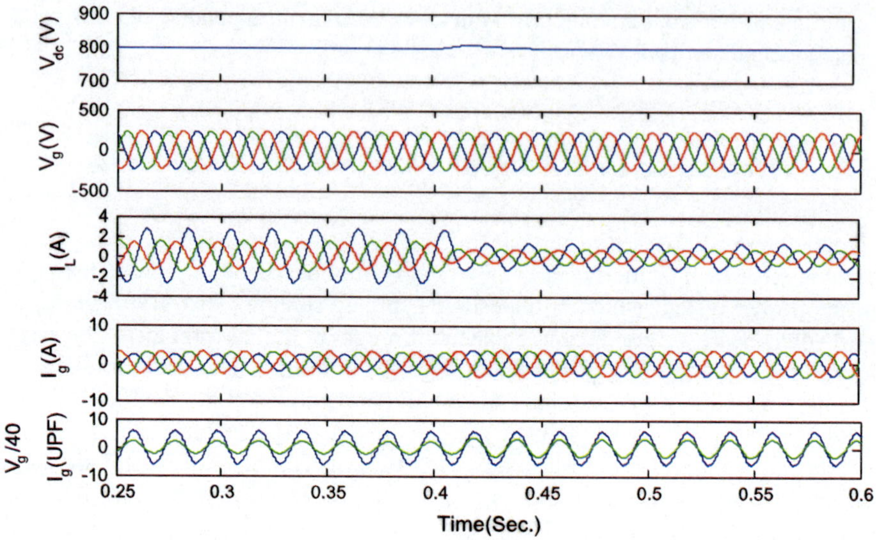

Fig. 6 Performance evaluation under unbalanced load

5.2 *Performance Under Unbalanced Linear Load*

Filtering performance under an unbalanced load is shown in Fig. 6. Results show that the system exhibits a steady-state response. The system is working under an unbalanced linear load, and at $t = 0.4$ s, load is reduced to its half value. Figure 6 depicts the source current and voltage which is sinusoidal in shape and balanced in nature. The fuzzy logic controller regulates DC-link voltage at a steady state. Also, the system is working under a unity power factor. It is experienced that fuzzy control scheme provides less settling time without producing overshoot due to sudden change in load.

5.3 *Comparison Analysis of Fuzzy Controller and PI Controller*

The performance of the fuzzy-based system is investigated for achieving constant DC-link. The behavior of the system is shown in Fig. 7 under transient conditions. The transient response of a fuzzy-based system is compared with the PI controller-based system response. Figure 7 shows that the fuzzy controller takes less initial time as compared to the PI controller during starting. Further, the fuzzy controller provides a fast response to a distributed network. Moreover, the comparative analysis

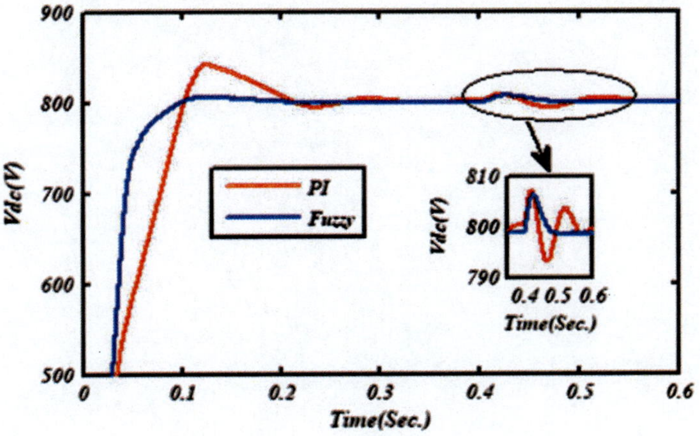

Fig. 7 Performance evaluation for fuzzy logic controller and PI controller

of the FLC and PI controller is given in Table 2. The analysis is based on transient response parameters such as overshoot, undershoot, and settling time. It can be clearly concluded from Table 2 that FLC gives superior performance over the conventional PI controller.

6 Conclusion

The modeling and simulation of D-STATCOM with FLC for linear and nonlinear are presented. The results show the effective performance of the proposed FLC controller over the conventional PI controller under transient conditions. The fuzzy-based system provides a better transient response during load variations. The fuzzy logic controller gives a smaller amount of settling time, peak overshoot, and quick response. Hence, a system based on a fuzzy logic control scheme reduces a 50% error in DC-link voltage by compared with the PI controller during load varying as IEEE standards.

Table 2 Performance costing for fuzzy logic controller and PI controller

	Overshoot			Undershoot			Settling time		
	$t = 0$	$t = 0.4$	$t = 0.6$	$t = 0$	$t = 0.4$	$t = 0.6$	$t = 0$	$t = 0.4$	$t = 0.6$
PI (%)	5	1	1.13	1	0.25	1	0.36	0.1	0.1
Fuzzy (%)	0.63	1	0	0	0	1.13	0.18	0.07	0.09

Appendix

L_s (Coupling Inductance) = 1 mH, V_s (source voltage) = 415 V, f_s (supply frequency) = 50 Hz, filter impedance ($R_f = 10\ \Omega$, $C_f = 1000\ \mu F$), C_{dc} (DC-link capacitor) = 1500 μF, V_{dc} (DC-link) = 800 V.

References

1. Y. Tian, Z. Yu, N. Zhao, Y. Zhu, R. Xia, Optimized operation of multiple energy interconnection network based on energy utilization rate and global energy consumption ratio, in *2nd IEEE Conference on Energy Internet and Energy System Integration (EI2)*, Beijing (2018), pp. 1–6
2. A. Moreno-Munoz, J.J.G. De la Rosa, J.M. Flores-Arias, F.J. Bellido-Outerino, A. Gil-de-Castro, Energy efficiency criteria in uninterruptible power supply selection. Appl. Energy **88**(4), 1312–1321 (2011)
3. N. Alawadhi, A. Elnady, Mitigation of power quality problems using unified power quality conditioner by an improved disturbance extraction technique, in *International Conference on Electrical and Computing Technologies and Applications (ICECTA)*, Ras Al Khaimah (2017), pp. 1–5
4. E. Hossain, M.R. Tür, S. Padmanaban, S. Ay, I. Khan, Analysis and mitigation of power quality issues in distributed generation systems using custom power devices. IEEE Access **6**, 16816–16833 (2018)
5. D. Suresh, K. Venkateswarlu, S.P. Singh, T2FLC based CHBMLI DSTATCOM for power quality improvement, in *International Conference on Computer Communication and Informatics (ICCCI)*, Coimbatore (2018), pp. 1–6
6. M.B. Latran, A. Teke, Y. Yoldaş, Mitigation of power quality problems using distribution static synchronous compensator: a comprehensive review. IET Power Electron. **8**(7), 1312–1328 (2015)
7. F. Hamoud, M.L. Doumbia, A. Chériti, H. Teiar, Power factor improvement using adaptive fuzzy logic control based D-STATCOM, in *2017 Twelfth International Conference on Ecological Vehicles and Renewable Energies (EVER)* (IEEE, 2017), pp. 1–6
8. G. Varshney, D.S. Chauhan, M.P. Dave, Evaluation of power quality issues in grid-connected PV systems. Int. J. Electr. Comput. Eng. **6**(4), 1412 (2016)
9. M. Amirrezai, H. Rezaie, G.B. Gharehpetian, H. Rastegar, A new fuzzy-based current control strategy with fixed switching frequency for improving D-STATCOM performance. J. Intell. Fuzzy Syst. (Preprint), 1–10 (2018)
10. S.R. Reddy, P.V. Prasad, G.N. Srinivas, Design of PI and fuzzy logic controllers for distribution static compensator. Int. J. Power Electron. Drive Syst. **9**(2), 465 (2018)
11. K. Prasad Sharma, N. Baine, Application of a fuzzy logic-based controller for peak load shaving in a typical household, in *2019 IEEE International Conference on Fuzzy Systems (FUZZ-IEEE)*, New Orleans, LA, USA (2019), pp. 1–6
12. V.M. Awasth, V.A. Huchche, Reactive power compensation using D-STATCOM, in *2016 International Conference on Energy Efficient Technologies for Sustainability (ICEETS)*, Nagercoil (2016), pp. 583–585
13. Shikha Gupta, Rachana Garg, Alka Singh, Modeling, simulation and control of fuel cell-based micro grid. J. Green Eng. **7**, 129–158 (2017)

Comparative Study on Machine Learning Classifiers for Epileptic Seizure Detection in Reference to EEG Signals

Samriddhi Raut and Neeru Rathee

Abstract Epilepsy is the key concern of medical practitioners and machine learning researchers since last decade. EEG signals play a very crucial role in early detection of epilepsy as well as cure of epilepsy. The traditional approach to analyze EEG signals includes two main steps: feature extraction and classification. Since multichannel EEG data is chaotic data, selecting optimal features and classifying them are major challenges. There exist a number of feature extraction and classification techniques proposed by researchers which perform well. For feature extraction, wavelets have been proved to perform state-of-the-art performance, but no such state-of-the art performance exists for classification techniques. The classifiers explored in the presented work include random forest classifier, support vector machine, Naïve Bayes, k-nearest neighbor, decision trees, artificial neural network, and logistic regression. Experimental results on the UCI dataset represent that random forest is performing best with 99.78% accuracy.

Keywords Machine learning · Classification · Electroencephalogram · Epileptic seizure detection · Biomedical application

1 Introduction

Seizures are the brain's electrical activities resulting from the excessive electric discharge in a group of brain cells. In neuroscience, recurrent seizures are termed as epilepsy. It produces sudden and ephemeral aberration with several involuntarily body movements, loss of consciousness, and other physical and psychological affects. Epileptic seizures are likely to increase the risk of endemic conditions like neurocysticercosis and other accidental injuries. Affecting over 50 million people, it is one of the serious problems to address [1]. In this context, various developments

S. Raut (✉) · N. Rathee
Department of Electronics and Communication Engineering, Maharaja Surajmal Institute of Technology, New Delhi 110058, India
e-mail: samriddhiraut@ieee.org

P. Bansal et al. (eds.), *Proceedings of International Conference on Artificial Intelligence and Applications*, Advances in Intelligent Systems and Computing 1164,
https://doi.org/10.1007/978-981-15-4992-2_18

are made to analyze the EEG signals of the epileptic patients and to build predictive models for the onset of seizures.

EEG signals represent the electrical activity of brain. Since different parts of brain represent different functionalities, a multi-channel EEG is found to be vital. A multi-channel EEG defines different features in the electrical activities of brain. Analyzing these signals correctly helps us to diagnose different neurological conditions and mental defectiveness. Abnormal seizures are characterized by repetitive high-amplitude brain activities with a combination of spikes and slow waves (SSW) as shown from 1723 in Fig. 1 [2]. Characterizing these signals and extracting beneficiary information visually is a challenging job as it may be misleading. Several researches have been carried out to find computerized, faster, and robust methods for seizure detection hence taking precautionary measures. This paper discusses and compares the performance of various classifiers based on machine learning approaches that make the detection process quite fast, accurate, and reliable.

Several works have been carried out in this field. Gotman et al. [3] have followed the sharp peaks recognition method which detects the high-amplitude brain waves occurring during seizures [4–7]. Shoeb et al. [8] incorporate machine learning models in developing classifiers to classify the signals in epileptic or non-epileptic activities. With support vector machines, it creates nonlinear decision boundaries, and using RBF kernels, a patient-specific model is proposed with a good accuracy and sensitivity. Srinivasan et al. [9] and Guo et al. [10] have applied the concept of neural networks in automated seizure detection, where the present and past values of the signals are compared on the basis of logical parameters like approximate entropy. Acharya et al. [11] explore the application of CNN in EEG signal analysis in reference to epileptic seizures and claims a good accuracy.

Raw EEG signals obtained from the patient are a time series data. It is first preprocessed and then artifacts are removed. Machine learning and deep learning models

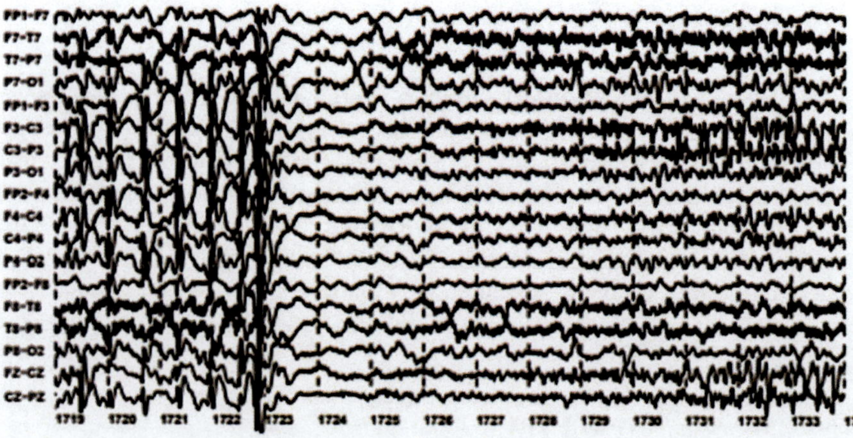

Fig. 1 EEG signal showing seizure activity

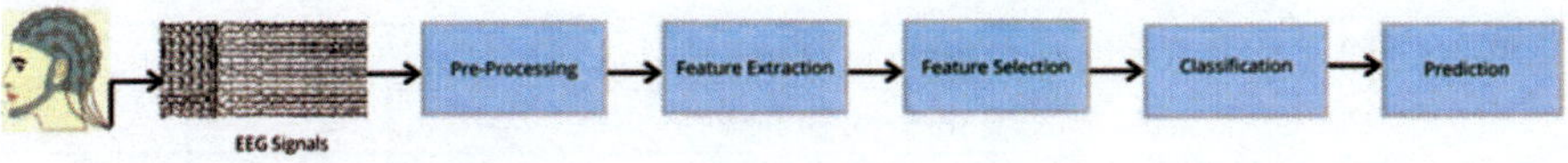

Fig. 2 Steps for EEG signal analysis

in reference to EEG signals are first trained on a set of labeled data of sampled EEG signals and then tested for the predictive analysis.

In this paper, different machine learning and deep neural network techniques are explored, and the accuracies, precision, recall, and F1 score of classifying and predicting the epileptic seizure correctly are compared. From the results, it is found that random forest (RF), support vector machines (SVM), and artificial neural networks (ANN) show better accuracies of 99.78%, 98.18%, and 89.38%, respectively, as compared to other state-of-the-art approaches. Over the past few years, several developments of stand-alone predictive devices and tools have been carried out which employ these classifiers in prediction process.

The outline of the presented work is described in detail in Fig. 1. Section 2 describes the proposed approach. The dataset used to evaluate the proposed work is explained in detail in Sect. 3. Section 4 discusses the classification algorithms used followed with results and future aspects in Sect. 5.

2 Proposed Approach

Standard steps of preprocessing feature extraction and classification are followed in EEG signal analysis, as shown in Fig. 2. The main focus of this study lies in evaluating machine learning models on preprocessed data and predicting the class. Machine learning models and neural networks were modeled in Python3, version 3.7.3. In our approach, the.csv file is first downloaded from the UCI machine learning repository and is loaded into a data frame. The dataset is divided into train and test sets in 80–20 ratio. We train the train sets on different machine learning models, LR, SVM, NB, RF, K-NN, DT, and ANN and then test on the test set. Performance of different classifiers in successful detection and classification in epileptic or non-epileptic classes are compared on the basis of metrics like accuracy, precision, recall, and F1 score.

3 Dataset

The data assessed here is a time-series EEG recordings for 3.3 h. The data is collected via 10–20 electrode system and 128 channel amplifier system following the common frame of reference. It is then sampled at 173.61 Hz to 4097 data points, and subsequently, the band-pass filter was set at 0.53–40 Hz. The 4097 data points

Table 1 Dataset description

Number of attributes	178
Number of subjects	500
Time segment	23.6 s
Number of instances	11,500
Classes	5

are then separated such that 178 data points value represents the EEG signal value for 1 s, at different time instant. The 179th column of the sheet represents the label and can have values 1–5, thus separating the data values in five classes. The class with label 1 records the electric discharge indicating seizures. Second category has recordings from the epileptogenic zone of brain, while third has activities from the hippocampal formations from other hemispheres of the brain. Class with labels 4 and 5 has normal EEG recordings in relaxed state with subject's eyes open and closed, respectively. Among these five classes, current approaches used here treat the classification problem as binary classification problem taking class 1 activity as epileptic and rest classes (2–5) as non-epileptic seizure activity [12]. Table 1 provides a short dataset description.

4 Classification Algorithms

A comparison study was made for the performance and computational accuracies on the dataset with the following machine learning algorithms. All the techniques studied treat the case as binary classification problem.

4.1 Naïve Bayes

It is a probabilistic linear classifier which follows Bayes' theorem. It related the conditional probabilistic theory which defines the occurrence probability of an event *A* based on the certainty that an event *B* happened previously. Here, in this case, Naïve Bayes adjusts the probability of each outcome variable in accordance with the value of other variables in the dataset.

$$P(A|B) = \frac{P(A) * P(B|A)}{P(B)}$$

where

$P(A)$, the probability of the event A to occur.
$P(B)$, the probability of the event B to occur.
$P(B|A)$, the conditional probability of A to occur knowing that B happened.

$P(A|B)$, the conditional probability of B to occur knowing that A occurred.

4.2 Logistic Regression

Logistic regression is a classic machine learning technique to map input into discrete set of classes. Transforming input to output using a sigmoid function and it generates a probability value to map into one of the two or more classes available.

4.3 k-NN

It uses Euclidean distance between the data point and its k neighbor to classify data. Points clustered in good numbers with small distance hence larger neighbors tend to form one class. Having several advantages over other techniques, it is robust and classifies the whole data in one go. Moreover, it has no initial assumptions about the inputs and classes as it classifies purely on the basis of neighbor's distance. It doesn't require training.

4.4 Support Vector Machines

Built on the developments of computational learning, support vector machines are one of the most opted choices in biomedical field. It approaches the classification problem as constructing a hyperplane as decision surface to separate data in high-dimensional space. So, SVM classifier selects a support vector to compute a inner-product kernel between support vector and input vector. A kernel function is principle to the SVM approach.

4.5 Decision Trees

Decision trees are nonparametric supervised learning method wherein an output value is predicted by learning simple decision rules inferred from the data features. Working on if-this-then-that principle, decision trees have a faster implementation and thus are majorly used in a multi-class classification.

4.6 *Random Forest*

Random forest can be treated as a collection of decision trees working on the randomly selected data from the train set, just guarantying that the results are distinct for each trail on a given dataset. This cluster of decision trees works individually by classifying data in each part, and the total number of features in a particular class is counted in the end. This weighted voting technique has several advantages over decision trees like low over-fitting problem and better anomaly detection and isolation.

4.7 *Neural Network*

Neural networks are the biological inspired artificial neural networks wherein it can take input, and through layers of neurons called hidden layers, an output is obtained at the output layer. Following the principle of feature hierarchy, each layer is feeded and trained on the output of previous layer. All the calculations and predictions are made in the individual layers of hidden layer network and finally compiled in last layer. Here, an artificial neural network with five hidden layers is modeled and tested for the 100 epoch on the dataset. 'Binary cross entropy' is chosen as the loss function and 'adam' optimizer is used.

5 Evaluation and Metrics

For comparing the different models tested on the given dataset, accuracy and complexity of models are evaluated. Precision and recall of some classifiers are also discussed. Accuracy can be defined as correct observation to total number of observation. In a more defined way, accuracy is the ratio of number of observations predicted accurately by the model to the total number of observations made by the model, whereas precision is defined as the ratio of true positive to true positive and false positive and recall is the true positive to true positive and false negative. F1 score is the harmonic mean of precision and recall. Mathematically,

$$\begin{aligned}\text{Accuracy} = {} & (\text{True Positives} + \text{True Negatives})/(\text{True positive} \\ & + \text{False Positives} + \text{False Negatives} + \text{True Negative})\end{aligned} \tag{1}$$

$$\text{Precision} = \text{True Positive}/(\text{True positive} + \text{False positive}) \tag{2}$$

$$\text{Recall} = \text{True positive}/(\text{True positive} + \text{False negative}) \tag{3}$$

$$\text{F1 Score} = 2 \times ((\text{Precision} \times \text{Recall})/(\text{Precision} + \text{Recall})) \quad (4)$$

Tables 2 and 3 present the comparison of different approaches in context to seizure detection and classification using machine learning approaches. A confusion matrix is presented in Table 2, in which the rows give the instances of actual class, while columns give that of predicted class. Table 3 derives the precision, recall, and F1 score from Table 2, for all the employed classifiers. Evaluating Table 4, which lists the accuracies of classifiers, it is found that random forest leads with 99.78% accuracy trailing which we have SVM with 98.18% accuracy. RF has a computational complexity of $O(n \log n)$ and is thus easier to train, implement, and interpret. On the other hand, SVM with complexity $O(n^3)$ is undoubtedly a good option for binary classification. Naïve Bayes, K-NN, DT, ANN, and LR give an accuracy of 95.78%, 93.93%, 90%, 89.38%, and 82.80%, respectively. With comparatively lower accuracies, these classifiers also show larger time consumption and space consumption and lower performance in some cases and thus are not a preferred option (Fig. 3).

For the artificial neural network, there are a total of five layers. Backpropagation is employed with 'adam' as the optimizer. The proposed neural network tested gives an accuracy of 89.38% which is still workable and matches the state-of-the-art classifier's performance. In improving the generalization capabilities, complexity and precision periodic training can be carried out. However, with larger number of layers as proposed by other studies, neural network can give promising performance [13–16].

6 Conclusion

The proposed work explored various classifiers on nonlinear deterministic features on the UCI-epileptic seizure dataset. From the comparative analysis, it is found that random forest performs best among other classification algorithms. The computation complexity of the proposed classifiers has not been taken into account. In future, that aspect may be explored. All the classifiers have been explored for the epilepsy, so they may also be explored for other applications so that universally acceptable state-of-the-art performance may be identified.

Table 2 Confusion matrix for different classifiers

Random forest			Support vector machines			Naïve Bayes			k-nearest neighbor		
	TP	FN		TP	FN		TP	FN		TP	FN
FP	1818	21	FP	1492	347	FP	1477	362	FP	1601	238
TN	51	410	TN	374	87	TN	87	367	TN	406	55

Decision trees			Artificial neural network			Logistic regression		
	TP	FN		TP	FN		TP	FN
FP	105	231	FP	1800	39	FP	1837	2
TN	410	51	TN	44	417	TN	410	51

Table 3 Precision, recall, and F1 score for different classifiers

Classifier	Precision	Recall	F1 score
Random forest classifier	0.96	0.94	0.95
Support vector machine	0.97	0.95	0.96
Naïve Bayes	0.94	0.94	0.94
k-nearest neighbor	0.96	0.82	0.87
Decision trees	0.87	0.77	0.80
Artificial neural network	0.97	0.93	0.95
Logistic regression	0.91	0.55	0.53

Table 4 Percentage accuracy of classifiers

Classifier	Accuracy
Random forest classifier	99.78
Support vector machine	98.18
Naïve Bayes	95.75
k-nearest neighbor	93.93
Decision tree classifier	89.38
Artificial neural network	89.38
Logistic regression	82.80

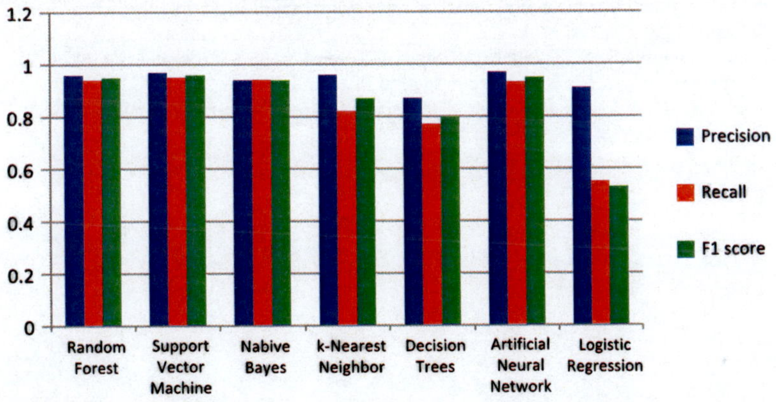

Fig. 3 Precision, recall, and F1 score

References

1. K. Lehnertz, F. Mormann, T. Kreuz, R. Andrzejak, C. Rieke, P. David, C. Elger, Seizure prediction by nonlinear EEG analysis. IEEE Eng. Med. Biol. Mag. **22**, 57–63 (2003)
2. A. Shoeb, J. Guttag, Application of machine learning to epileptic seizure detection, in *Proceedings of the 27th International Conference on International Conference on Machine Learning (ICML'10)*, ed. by J. Fürnkranz, T. Joachims (Omnipress, USA, 2010), pp. 975–982

3. J. Gotman, J. Ives, P. Gloor, Automatic recognition of inter-ictal epileptic activity in prolonged EEG recordings. Electroencephalogr. Clin. Neurophysiol. **46**(5), 510–520 (1979)
4. J. Gotman, Automatic recognition of epileptic seizures in the EEG. Electroencephalogr. Clin. Neurophysiol. **54**(5), 530–540 (1982)
5. J. Gotman, Automatic detection of seizures and spikes. J. Clin. Neurophysiol. **16**(2), 130–140 (1999)
6. D. Koffler, J. Gotman, Automatic detection of spike-and-wave bursts in ambulatory EEG recordings. Electroencephalogr. Clin. Neurophysiol. **61**(2), 165–180 (1985)
7. J. Qu, Gotman, Improvement in seizure detection performance by automatic adaptation to the EEG of each patient. Electroencephalogr. Clin. Neurophysiol. **86**(2), 79–87 (1993)
8. A. Shoeb, Application of machine learning to epileptic seizure onset detection and treatment. Ph.D. thesis, Massachusetts Institute of Technology, 2009
9. V. Srinivasan, C. Eswaran, N. Sriraam, Approximate entropy-based epileptic EEG detection using artificial neural networks. IEEE Trans. Inf. Technol. Biomed. **11**(3), 288–295 (2007)
10. L. Guo, D. Rivero, J. Dorado, J.R. Rabunal, A. Pazos, Automatic epileptic seizure detection in EEGs based on line length feature and artificial neural networks. J. Neurosci. Methods **191**(1), 101–109 (2010)
11. U.R. Acharya, S.L. Oh, Y. Hagiwara, J.H. Tan, H. Adeli, Deep convolutional neural network for the automated detection and diagnosis of seizure using eeg signals. Comput. Biol. Med. **100**, 270–278 (2018)
12. R.G. Andrzejak, K. Lehnertz, C. Rieke, F. Mormann, P. David, C.E. Elger, Indications of nonlinear deterministic and finite dimensional structures in time series of brain electrical activity: dependence on recording region and brain state. Phys. Rev. E **64**, 061907 (2001)
13. Y. Roy, H. Banville, I. Albuquerque, A. Gramfort, T.H. Falk, J. Faubert, Deep learning-based electroencephalography analysis: a systematic review. J. Neural Eng. **16**(5), 051001 (2019)
14. U.R. Acharya, S.L. Oh, Y. Hagiwara, J.H. Tan, H. Adeli, Deep convolutional neural network for the automated detection and diagnosis of seizure using EEG signals. Comput. Biol. Med. 1–9 (2017)
15. M. Golmohammadi, S. Ziyabari, V. Shah, S.L. de Diego, I. Obeid, J. Picone (2017) Deep architectures for automated seizure detection in scalp EEGs. arXiv:1712.09776
16. R. Hussein, H. Palangi, R. Ward, Z.J. Wang, Epileptic seizure detection: a deep learning approach (2018). arXiv:1803.09848

Design Fundamentals: Iris Waveguide Filters Versus Substrate Integrated Waveguide (SIW) Bandpass Filters

Aman Dahiya and Deepti Deshwal

Abstract A waveguide iris bandpass filter is designed in this paper. The proposed filter is created inside the structure by cutting inductive symmetrical windows known as irises. Later, using substrate integrated technology, its equivalent bandpass filter is released. In a dielectric material, SIW is created by adding a top metal over the ground plane and trapping the structure with rows of plated vias on either side. Compared to iris waveguide technique, SIW is an effective and efficient solution. The filters proposed were designed at the center frequency of 5 GHz. The simulated results, when designed using SIW software, show low losses and sharp roll-off characteristics in the passband. SIW has a bandwidth of 3 dB of 2.8 GHz and a fractional bandwidth of 15% in the passband. The filters proposed were suitable for use in applications for microwave communication systems.

Keywords Substrate integrated waveguide · Microstrip · Conventional waveguide · Vias · Iris waveguide · Microwave communication devices

1 Introduction

The innovation of Substrate Integrated Waveguides (SIW) is utilized for electromagnetic wave transmission. This innovation was introduced as a capable solution to accommodate the enormous requirement for good performance and cost-effective resolution for smaller and lightweight devices that operate at extremely high frequencies [1]. Size diminution, low manufacturing cost, and limited losses in SIW devices are the most important aspects to make this technique a strong option over conventional waveguides. These are characterized through a very robust performance;

A. Dahiya (✉) · D. Deshwal
Department of Electronics and Communication Engineering, Maharaja Surajmal Institute of Technology, Janak-Puri, New Delhi, India
e-mail: amandahiya@msit.in

D. Deshwal
e-mail: deshwaldeepti@msit.in

P. Bansal et al. (eds.), *Proceedings of International Conference on Artificial Intelligence and Applications*, Advances in Intelligent Systems and Computing 1164,
https://doi.org/10.1007/978-981-15-4992-2_19

however, they stay massive due to the fact of their stable nature and their bulky size. SIW technique consists of two lines of metal cavities that ensure the electromagnetic waves guided transmission. The rows of metallic cavities act as metallic walls to take part in the confinement of electromagnetic energy [2].

The waveguide embedded in the substrate is widely used as a contact in high-speed switches, power dividers, directional couplers, filters, directional antennas, etc. Filtering is a digital feature that enables us to analyze the band of frequencies by extracting and passing the selected frequencies. For filtering, we must use the appropriate elements, including waveguide design centered on rectangular or circular waveguides, defective ground structures, multimodal structures, metallic cavities, SIWs, etc. [3]. There are several types of waveguide filters [4, 5]: multi modes, E-plane filters, dual-mode resonators, multilayer filters, defective grounds structures, complementary split-ring resonators, etc. In this paper, a bandpass filter functioning in the C-band frequency range, i.e., 3–8 GHz is proposed using both iris waveguide and substrate integrated waveguide (SIW) technology. Simulated and experimental results reveal that SIW bandpass filter is better in response as compared to iris waveguide filter.

2 Design of Waveguide Iris Bandpass Filter

The iris waveguide filter is realized by applying usual discontinuities in the waveguide structure. Iris discontinuities generate a *Wi* ($i = 1, 2, \ldots, n$) symmetrical induction window and the *Li* ($i = 1, 2, \ldots, n + 1$) cavity resonators are inserted between these discontinuities. The cavity resonator coupling is regulated by the iris discontinuity size [6–8] (Fig. 1).

Lumped low-pass filter feature is the standard microwave filter model. The properties of the low-pass prototype transform almost all types of low-pass, bandpass, and band elimination filters. The implementation of Chebyshev low-pass filter is broad, depending on the simple design, wideband, and steep edge. A cross-shaped inductive metal is the identical T-network consisting of inductance, capacitance, and so on lumped components compared to conventional inductive metal posts or diaphragm. Figure 2 represents the equivalent circuit of iris window. For a bandpass filter with Chebyshev frequency response, the impedance inverter normalized characteristic impedance can be determined as the following [9].

$$\Delta = \frac{\lambda_{g1} - \lambda_{g2}}{\lambda_{g0}} \quad (1)$$

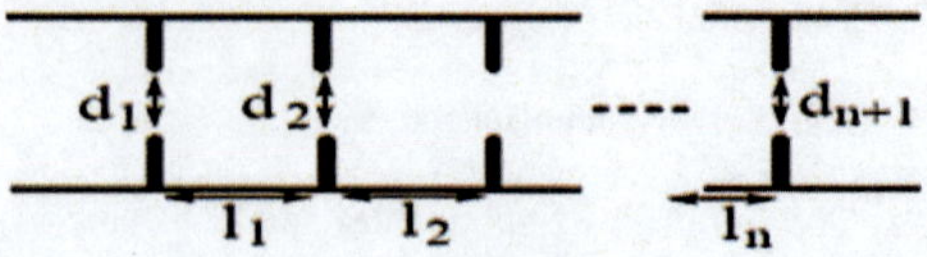

Fig. 1 Prototype of Iris bandpass filter

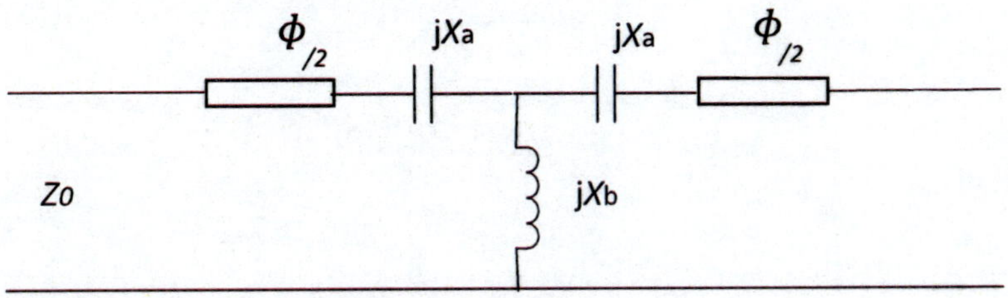

Fig. 2 Equivalent circuit of Iris window

$$\lambda_{g0} = \frac{\lambda_{g1} + \lambda_{g2}}{\lambda_{g0}} \tag{2}$$

$$f_0 = \sqrt{f_1 f_2} \tag{3}$$

$$\lambda_{gi} = \frac{2a}{\sqrt{\frac{(2afi)^2 - 1}{c}}} \quad i = 0, 1, 2 \tag{4}$$

In (4), gi's are the values for the low-pass filter components with Chebyshev response, Ω is the normalized low-pass frequency, Δ is the relative bandwidth, Z_0 is the transmission line characteristic impedance, f_0 the center frequency, λg1, and λg2 are the wavelengths of guided waves in the upper and lower edges of the bandpass filter, *a* is the broadside dimension of the waveguide [10].

The terms used to measure the reactance of the irises and electrical distance are:

$$X_{i,i+1} = \left(\frac{L}{\sqrt{g_i g_{i+1}}}\right) \Big/ \left(1 - \frac{L^2}{g_i g_{i+1}}\right) \quad i = 0, 1, \ldots, n \tag{5}$$

$$\phi_i = \pi - \frac{1}{2}\left[\tan^{-1}(2x_{i-1}) + \tan^{-1}(2x_{i,i+1})\right] \tag{6}$$

where

$X_{i,i+1}$: the reactance of the irises
ϕ_i: the electrical distance between the irises.

Knowing that

$$L = \pi \frac{\lambda_{g1} - \lambda_{g2}}{\lambda_{g1} + \lambda_{g2}} \tag{7}$$

In (7), *L* is the length of iris resonator. The parameters S_i and δ_m are defined as follows:

$$S_{ii} = \mathrm{Sin}\left(\frac{\pi d_i}{2a}\right) \quad i = 1, 2, 3, \ldots, n \tag{8}$$

$$\delta_m = 1 - \sqrt{1 - \left(\frac{2a}{m\lambda_0}\right)^2}, \quad m = 3, 5 \tag{9}$$

Finally,

$$l_i = \frac{\lambda_{go}}{2\pi}\phi_i \quad i = 1, 2, \ldots, n \tag{10}$$

By using the formulations designed above, we determined the initial structure parameters that we used in the design software to execute the structure to optimize the filter to achieve the best possible dimensions.

2.1 Simulated Response of Filter

The return (S_{21}) and transmission (S_{11}) losses are approximately 29 dB and 0.3747, respectively, over the entire frequency spectrum ranging from 3 to 6.5 GHz, indicating a 15% fractional bandwidth as shown in Fig. 3.

3 The Substrate Integrated Waveguide Filter Design

Two lines of metallic vias complete the SIW structure, which is intermittently embedded in the insulator substrate. The Substrate Integrated Waveguide (SIW) is a kind of transmission line that works like a dielectric-filled rectangular waveguide. The

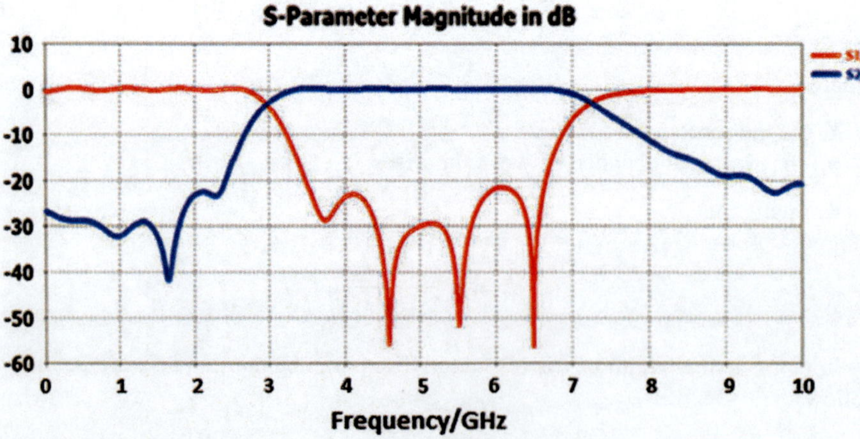

Fig. 3 S_{11} and S_{21} parameters of Iris waveguide bandpass filter

Substrate Integrated Waveguide (SIW) is a category of transmission line that functions as a rectangular waveguide filled with a dielectric substrate. Because of these metallic structures that surround SIW, only TEn0 modes are transmitted within the structure [11]. The distance between adjoining vias and their circumference is the most influential parameter. The integrated waveguide design has the same characteristics as the rectangular waveguide for particular values of p and d, and its radiation losses can be ignored partially. The dominant cut-off frequency mode is just like the TE_{10} rectangular waveguide mode the dominant frequency cut-off mode is the same as the guided mode for rectangular wave TE_{10}. The important features of this mode are the via (d) diameter and the pitch (p) called the gap between two successive vias [12]. The following design equations must be fulfilled in order to decrease the losses:

$$d < \frac{\lambda_g}{5} \ where \ \lambda g = \frac{c}{\sqrt[j]{\epsilon_r}} \tag{11}$$

$$\mathrm{p} < 2\mathrm{d} \tag{12}$$

where λ_g is the guided wavelength in the SIW.

The corresponding SIW parameters are determined after the determination of the physical parameters of the iris waveguide filter by Eqs. (1)–(3) as [13]:

$$L_{\mathrm{SIW}} = L + \frac{d^2}{0.95p} \tag{13}$$

$$W_{\mathrm{SIW}} = W + \frac{d^2}{0.95p} \tag{14}$$

where d is the diameter of cylindrical holes, p is the spacing between adjacent via cavities assuming that that $p < \lambda_O\sqrt{\epsilon_r}/2$ and $p < 4d$ with ε_r relative permittivity, W_{SIW} is the width of SIW and W is the width of the identical rectangular waveguide.

As the propagation attributes are known to be very similar in the metallic rectangular waveguides. Specifically, the distribution of the electromagnetic field is TE_{10} [1], the SIW guide is analogous to that of a traditional rectangular waveguide packed with the same width dielectric. The waveguide's cut-off frequency is described as [14–16]:

$$f_{c10} = \frac{c}{2W\sqrt{\epsilon_r}} \tag{15}$$

Electromagnetic waves are allowed to propagate through designed SIW structures. For effective and lossless transmission, SIW structure needs to be integrated with microstrip tapered transitions. This transition allows the fundamental mode of the waveguide to be adapted, i.e., TE_{10} mode of the Microstrip line [8]. Microstrip transition is designed to transform 50 Ω characteristics impedance with the filter structure (Fig. 4).

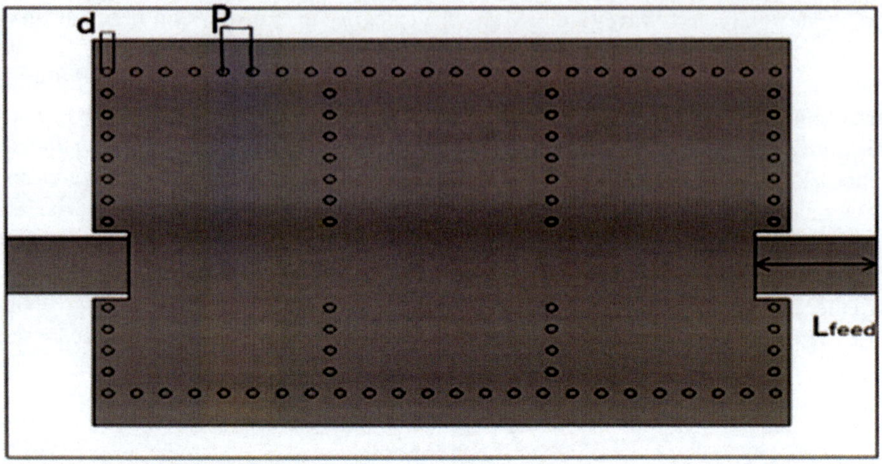

Fig. 4 SIW bandpass filter structure

A SIW-based X-band symmetric window bandpass filter is designed from the above assessment. It is a fifth-order, 0.2 dB ripple Chebyshev bandpass filter with a 0.762 mm thick substrate and permittivity of 3.2. A microstrip transition to SIW structure is designed to achieve 50 Ω characteristics impedance matching. A benchmarking process is performed for the entire filter after an initial CST-MWS model is established.

3.1 *Simulated Response of SIW Bandpass Filter*

The S-parameter simulation results for the proposed SIW filter as shown in Fig. 5 shows good selective filter accuracy with 31 dB return loss and 0.235 dB insertion

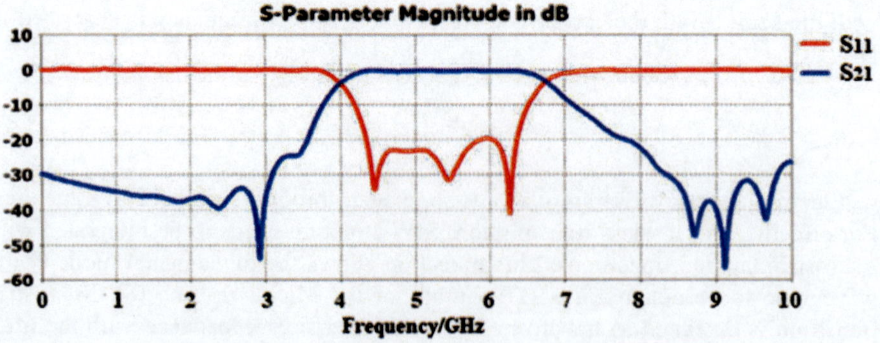

Fig. 5 S-parameters of SIW bandpass filter

loss. The 3-dB bandwidth of the filter obtained is approximately 3 GHz. Simulated responses show that the proposed bandpass filter designed using substrate integrated waveguide technology exhibits sharp roll-off characteristics as compared to filter designed using waveguide iris technique.

4 Conclusion

In this paper, a proposed filter is designed using the two design procedures, one as iris waveguide and the other one as substrate integrated waveguide technology. The simulated results of both filters are compared to the filter developed using SIW technology is more effective, easier to manufacture compared to structures of waveguides and easier to interface with several other printed circuits.

References

1. K. Wu, Substrate Integrated Circuits (SiCs)—a new paradigm for future GHz and THz electronic and photonic systems. IEEE Circuits Syst. Soc. Newsl. **3**(2) (2009)
2. A. Dahiya, A.P.S. Sengar, D.K. Dwivedi, A. Kumar, A critical review of substrate integrated waveguide for microwave applications, in *2016 Second International Conference on Computational Intelligence & Communication Technology (CICT)* (IEEE, 2016), pp. 495–499
3. W. Che, L. Geng, K. Deng, Y.L. Chow, IEEE Trans. Microw. Theory **56** (2008)
4. L. Yechou, A. Tribak, M. Kacim, J. Zbitou, Ku-band waveguide band-pass filter with iris radius, in *Multimedia Computing and Systems (ICMCS)* (2014), pp. 1435–1438
5. G.L. Matthaei, L. Young, E.M.T. Jones, *Microwave Filters, Impedance-Matching Networks, and Coupling Structures*, vol. 5 (McGraw-Hill, New York, 1964)
6. N. Marcuvitz, *Waveguide Handbook* (Peter Peregrinus Ltd, On behalf of The Institution of Electrical Engineers, Polytechnic Institute of New Work, 1986)
7. M. Bozzi, F. Xu, D. Deslandes, K. Wu, Modeling and design considerations for substrate integrated waveguide circuits and components, in *TELSIKS 2007, Serbia, Nis*, 26–28 Sept 2007
8. J.S.G. Hong, M.J. Lancaster, *Microstrip Filters for RF/Microwave Application*, vol. 167 (Wiley, New York, 2004)
9. A. Kumari, D. Kumar, A. Kumar, A Ka-Band bandpass filter with cylindrical cavity resonators using a Substrate Integrated Waveguide (SIW). Int. J. **6**(1) (2017)
10. A. Kumari, A.P. Singh, A. Kumar, D. Kumar, Novel compact wide-band substrate integrated waveguide bandpass filter using conductor backed coplanar waveguide as feed line for W-Band communication applications. Int. J. Comput. Appl. **142**(13) (2016)
11. A. Suntives, R. Abhari, Transition structures for 3-d integration of substrate integrated waveguide interconnects. IEEE Microwave Wirel. Compon. Lett. **17**, 697–699 (2007)
12. D. Deslandes, Design equations for tapered microstrip-to-substrate integrated waveguide transitions, in *IMS* (2010)
13. R. Kumar, S.N. Singh, Design and analysis of ridge substrate integrated waveguide bandpass filter with octagonal complementary split ring resonator for suppression of higher order harmonics. Prog. Electromagn. Res. **89**, 87–99 (2019)
14. A. Kumari, D. Kumar, A. Kumar, Dual-band substrate integrated waveguide (SIW) band pass filter for scientific radar applications. Int. J. Inf. Technol. 1–4 (2018)

15. A. Kumari, A. Pal, D. Kumar, 3.5 GHz microstrip transmission line design for microwave ICs. Int. J. Sci. Res. Rev. **7**(04), 651–654 (2019)
16. A.P.S. Sengar, A. Dahiya, Reconfigurable smart antenna for wireless communication devices, in *Microwave Theory and Technique Society (MTT-S) Project Final Report* (2019)

FPGA Implementation of Recursive Algorithm of DCT

Riya Jain and Priyanka Jain

Abstract Discrete cosine transformation (DCT) is one of an efficient tool which facilitates compression of speech and image signals. This paper focuses on FPGA implementation of discrete cosine transform (DCT-II) which is valid for $N = 2^r$, where N is the length of input sequence and $r > 1$. The architecture used is recursive in nature and implemented with reduced hardware complexity. The structure is designed using Verilog Hardware Description Language (HDL) and Zybo Zynq-7000 Development FPGA board is used for implementation. The design utilized 7.91% LUTs, 6% IOB, 32.5% DSPs, and 1.43% Flip flops and takes 26 clock cycles to compute all the output coefficients of DCT.

Keywords Discrete cosine transform · Signal compression · Recursive structure

1 Introduction

The discrete cosine transform (DCT) is technique for representing a finite sequence in form of weighted sum of cosines. It is a Fourier-related transform similar to that of discrete Fourier transform (DFT) in a sense that it transforms a signal from spatial to frequency domain. Mathematically, DCT is a linear, invertible function which has seven variants with slightly different definitions, that is, DCT I-VII [1].

R. Jain · P. Jain (✉)
Delhi Technological University, New Delhi, India
e-mail: priyankajain@dtu.ac.in

R. Jain
e-mail: riyajain_bt2k16@dtu.ac.in

P. Bansal et al. (eds.), *Proceedings of International Conference on Artificial Intelligence and Applications*, Advances in Intelligent Systems and Computing 1164,
https://doi.org/10.1007/978-981-15-4992-2_20

The DCT finds its applications in transformation technique in signal processing and data compression. Uncompressed digital media and lossless compression have extremely high requirements for bandwidth and memory, which are significantly reduced by using lossy DCT compression technique [2, 3]. The wide adoption of DCT compression standards led to the emergence and proliferation of digital media technologies, such as digital images, digital photos, digital video, streaming media, digital television, streaming television, video-on-demand (VOD), digital cinema, high-definition video (HD video) [4], and high-definition television (HDTV) [5].

In paper [6], coordinate rotation digital computer (CORDIC)-based fast radix-2 algorithm for computation of discrete cosine transformation (DCT) is presented. Efficient implementations of DCT computations are presented based on the shifted discrete Fourier transform (SDFT) [7]. DCT implementation with reduce silicon area is presented in [8]. In the implementation, authors have used fusion of the arithmetic distribution with Loeffler algorithm [8] and the circuit has been implemented in FPGA-Xilinx. Algorithms for computing the DCT on existing VLSI structures of DFT is presented [9]. FPGA implementation of DCT algorithms are also found in [10, 11]; however, algorithms used in these implementations [6–11] require more execution cycles and multipliers. Algorithm presented in [12] reduces the above problem. Practically, proposing a DCT algorithm is not sufficient since it is necessary to validate if they can be embedded physically. Therefore, FPGA implementation is necessary to perform before converting the prototype into a permanent application-specific integrated circuit (ASIC), and thus, it is required to implement these algorithms using Verilog Hardware Description Language (VHDL). This paper proposes new Verilog implementation of a DCT–II algorithm proposed in [12].

Remaining paper is structured as follows: Sect. 2 briefs the proposed algorithm by stating its derivation. Section 3 gives the hardware implementation of DCT algorithm, its architecture and a part of its Verilog code. Further, Sect. 4 mentions the results of hardware implementation, and lastly, Sect. 5 concludes the paper.

2 Proposed DCT-II Algorithm

2.1 Derivation

DCT-II is the most commonly used form of DCT and it is given by following mathematical expression:

If $x(n)$ is the input data sequence such that $n = 0, 1, 2 \ldots (N - 1)$ for $N = 2^r$ [12] then,

$$X(k) = \sqrt{\frac{2}{N}} A_k \sum_{n=0}^{N-1} x(n)\cos\left[\frac{k(2n+1)\pi}{2N}\right] \tag{1a}$$

For $k = 0, 1, 2 \ldots (N-1)$ and,

$$A_k = \begin{cases} \frac{1}{\sqrt{2}}, & k = 0 \\ 1, & \text{otherwise} \end{cases} \tag{1b}$$

According to this algorithm, the DCT coefficients are first divided into two groups: even and odd. For odd group, $k =$ odd and $p = 0$. Even group ($k =$ even) is further divided into $(r-1)$ groups. Thus, for even group, $p = 1, 2 \ldots (r-1)$. $X(0)$ coefficient of DCT is computed separately.

Constant term $\sqrt{\frac{2}{N}} A_k$ is ignored for current calculations and it is multiplied in the end of all mathematical operations. Variable k in (1a) is replaced by $2^p(2i+1)$ so that DCT can be computed through decimation in frequency. Here, $p = 0, 1, 2 \ldots (r-1)$ and $i = 0, 1, 2 \ldots (2^{r-(p+1)} - 1)$, such that

$$X\left(2^p(2i+1)\right) = \sum_{n=0}^{N-1} x(n)\cos\left[2^p(2i+1)\alpha_i\right] \tag{2a}$$

where,

$$\alpha_i = \frac{2^p(2i+1)}{2N} \tag{2b}$$

After following proper procedure and rearranging steps as mentioned in [12], we can come to final solution given as:

$$\begin{aligned} G^p_{ji}(k) = \Big[& (-1)^i \sin\left(\frac{\alpha_{p,i}}{2}\right)\left\{w^p(j) + w^p(j-1)\right\} \\ & + 2G^p_{(j-1)i}(k)\cos\left(\alpha_{p,i}\right) - G^p_{(j-2)i}(k)\Big] \end{aligned} \tag{3a}$$

Such that

$$G^p_{ji}(k) = X(k) = X\left(2^p(2i+1)\right) \tag{3b}$$

where,

$$j = 0, 1, \ldots, \frac{N}{2^{p+1}} - 1 \tag{3c}$$

$$\alpha_{p,i} = \frac{2^p(2i+1)\pi}{N} \tag{3d}$$

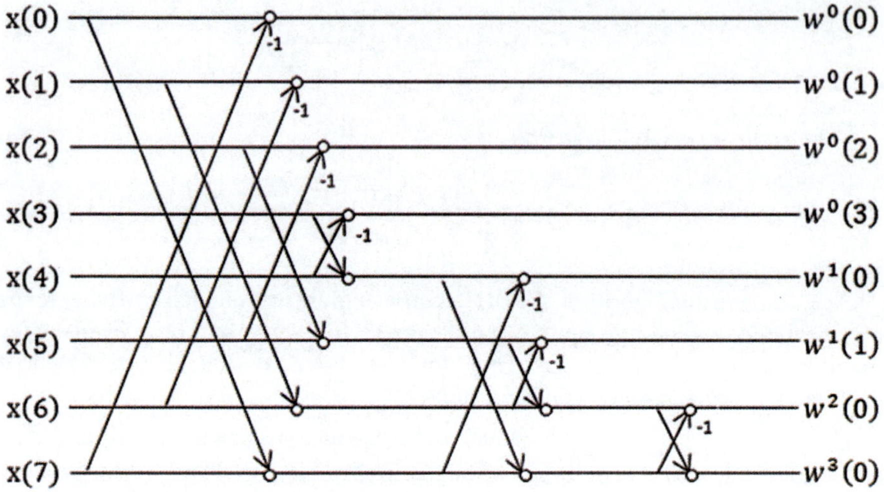

Fig. 1 Preprocessing stage shown for $N = 8$ (2^3)

$$w^0(0) = x(0) - x(7) \tag{4a}$$

$$w^0(1) = x(1) - x(6) \tag{4b}$$

$$w^0(2) = x(2) - x(5) \tag{4c}$$

$$w^0(3) = x(3) - x(4) \tag{4d}$$

And $w^p(j)$ is computed by using a preprocessing stage. Equations for calculating $w^0(j)$ is given in Eq. (4a–d) and rest of the values are computed as shown in Fig. 1.

$w^3(0)$ is used to compute $X(0)$ coefficient of DCT sequence by simply multiplying $w^3(0)$ with $\sqrt{\frac{2}{N}}A_k$, such that $k = 0$. Remaining terms $w^0(j)$, $w^1(j)$, and $w^2(j)$ are used for computation of other DCT coefficients. Equation in (3) can be transformed using Z-Transform and can be implemented using an IIR filter structure. Z-Transform can be represented as follows:

$$\frac{G_{ji}^p(k)}{w^p(j)} = \frac{(-1)^i \sin\left(\frac{\alpha_{p,i}}{2}\right)\left(1 + z^{-1}\right)}{1 - 2\cos\left(\alpha_{p,i}\right)z^{-1} + z^{-2}} \tag{5}$$

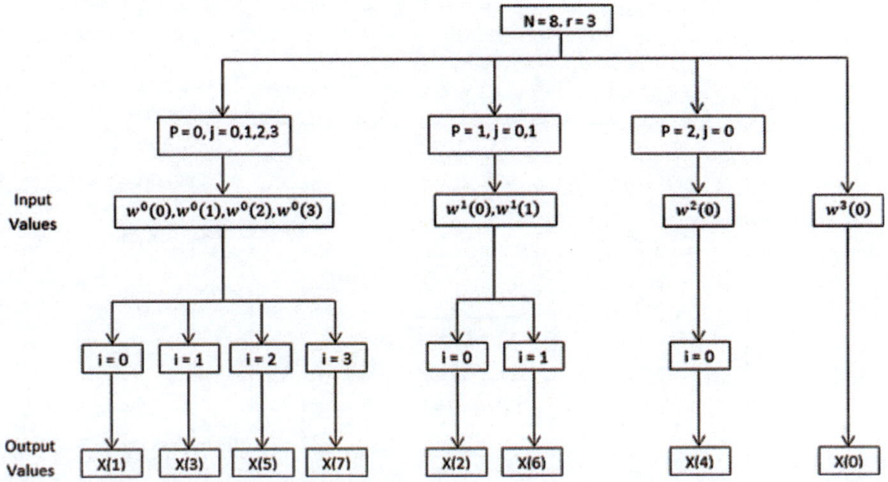

Fig. 2 Computation of DCT coefficients for $N = 8$, $r = 3$

2.2 *DCT-II Algorithm for* N = *8*

Figure 2 shows the stages involved in the algorithm for DCT coefficient computation. For values $N = 8$ and $r = 3$, two groups are divided such that first group has all odd coefficients: $X(1)$, $X(3)$, $X(5)$, and $X(7)$ which are computed for p = 0 corresponding to $i = 0$, 1, 2, and 3, respectively. Second group has even coefficients which are further divided into two subgroups. First subgroup computes $X(2)$ and $X(6)$ for $p = 1$ corresponding to $i = 0$ and 1, respectively. Second subgroup computes $X(4)$ for $p = 2$, $i = 0$. Lastly, $X(0)$ is obtained separately. $w^p(j)$ coefficients are output of preprocessing stage and are fed as input to the second-order IIR filter designed by using (5).

3 Hardware Implementation of DCT Algorithm

3.1 *Architecture*

Equivalent hardware architecture developed for IIR filter given by Eq. (5) is shown by solid lines in Fig. 3. It is a recursive process with feedback and feedforward loop as shown in Fig. 3. It can be seen in the figure that delay element is denoted by a D flip flop. Outputs of preprocessing stage, $w^p(n)$, depending upon values of p, are fed as input into this second-order filter structure in succession, one input in one clock cycle. Therefore, there are three such structures required that run in parallel, one each for $p = 0$, 1, and 2. There are three sections shown in the figure, abbreviated as S1, S2, and S3. Third section (S3) is common for all the filters ($p = 0$, 1, 2). For p

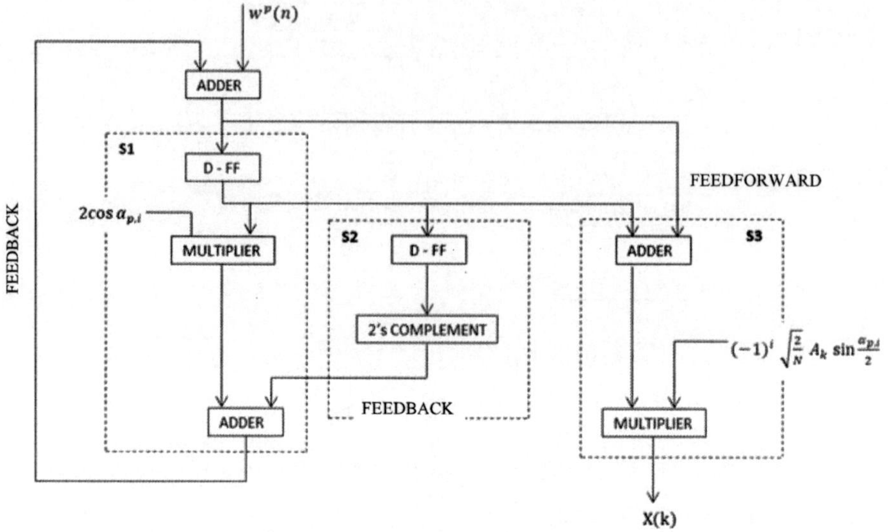

Fig. 3 Hardware circuit for proposed algorithm

= 1, only S1 and S3 are present and for $p = 2$, only S3 is required. Therefore, the hardware for second-order filter for $p = 1$ and 2 is drastically reduced since only 2 and 1 input is required for computation, respectively, as compared to filter required for $p = 0$ where all three sections are required for computation.

Once the inputs are fed in successive clock cycles, the circuit operates in recursive manner and output $X(k)$ is obtained at the edge of clock cycle coming just after last input is fed in the filter circuit.

For above implementation, all numbers are represented in fixed point signed binary such that total word length is 31. One bit (MSB) is sign bit, next 10 bits represent the integer part and least significant 20 bits represent fraction part. Therefore, precision of all the numbers is up to 20 bits. Lastly, the values of sine and cosine functions have 1–4 distinct and constant values (for $N = 8$). Therefore, these values are taken as constant.

3.2 *Verilog Program*

For authenticating the paper, Verilog program for the structure of second-order function designed for $p = 2$ is given below:

```
module SO_filter2 #(parameter S = 1, parameter I = 10 ,
parameter F = 20, parameter N = 8)
(input[S+F+I-1:0] w,
 input clk, rst, en,
 output[S+I+F-1:0] out);

wire[S+I+F-1:0] sin,A;
reg[S+I+F-1:0] T1,out1;
wire[S+F+I-1:0] T2, T3;


assign sin = 31'b10000000000_10110101000001001111;
assign A = 31'b00000000000_10000000000000000000;

always@(posedge clk, negedge rst)
begin
    if(!rst)
        T1 <= 31'b0;
    else if(en)
        T1 <= w;
    else
        T1 <= 31'b0;
end

Multiplier M2(.in1(T1), .in2(sin), .en(en), .out(T2));
Multiplier M3(.in1(T2), .in2(A), .en(en), .out(T3));

always@(posedge clk, negedge rst)
begin
    if(!rst)
        out1 <= 0;
    else if(en)
        out1 <= T3;
    else
        out1 <= 0;
end

assign out = out1;

endmodule
```

4 Results

The given structure of second-order IIR filter was implemented in Verilog HDL using Zybo Zynq-7000 Development FPGA board. An example is taken for ready reference, for $N = 8$, and its simulation result of the implemented design is shown

Fig. 4 Simulation results for the stated example. Input data points $X = [100, 94, 23, 55, 88, 200, 45, 89]$

in Fig. 4. This example is chosen at random without any biasness. It can be seen that all coefficients are successfully obtained in 26 clock cycles. A MATLAB code was also written to verify the output simulation results obtained. For input data points given as $X = [100, 94, 23, 55, 88, 200, 45, 89]$, MATLAB output gave DCT coefficients $Y = [245.3661, -26.6216, 5.1765, 95.7598, -10.6066, -51.9583, 47.6047, -69.9406]$. While output of the simulated Verilog code, after converting fixed point binary representation into decimal representation, came out to be $Y = [245.3656, -26.6216, 5.1765, 95.7596, -10.6066, 95.7597, 47.6046, -69.9403]$.

Moreover, it is clear that Y0 is the coefficient that is computed first which is followed by Y4, Y2, Y1, and so on. For the purpose of compression of a signal, i.e., teleconferencing applications, some specific coefficient has to be calculated and remaining coefficients can be neglected. If we exclude last 4 coefficients, that is, Y6, Y3, Y5, and Y7, then only 8 clock cycles will be required to compute rest of the 4 coefficients.

The Verilog code is simulated exhaustively for 500 different random samples and the output so obtained is verified using MATLAB code. Average relative errors in DCT coefficients for 500 samples were calculated. Average relative errors in Y0, Y1, Y2, Y3, Y4, Y5, Y6, and Y7 were found as $1.6186e^{-6}$, $5.2555e^{-6}$, $1.7930e^{-6}$, $7.5062e^{-7}$, $6.6169e^{-8}$, $1.0970e^{-6}$, $8.6127e^{-7}$, and $2.1361e^{-5}$, respectively. Figure 5 shows the RTL logic of the implemented design. Software used for the same is Vivado 2018.3. In the given RTL logic, at input, we have clock and asynchronous

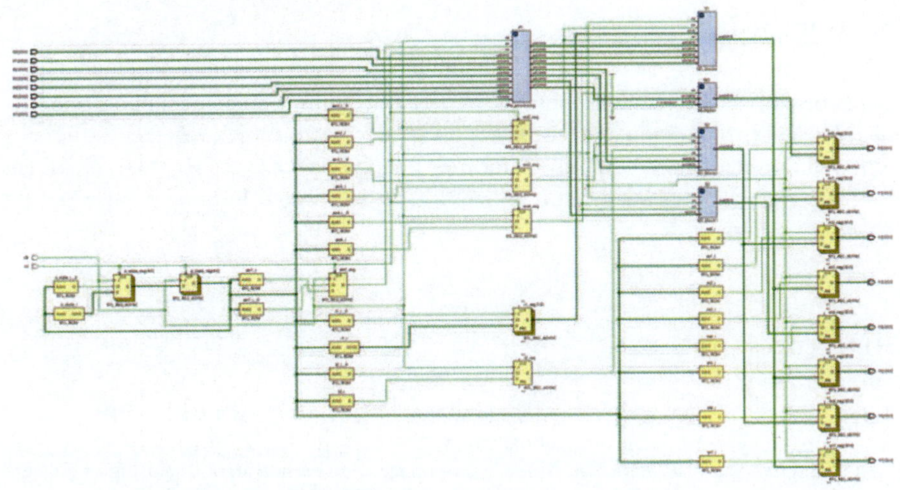

Fig. 5 RTL logic generated for the implemented design in Vivado 2018.3

Table 1 Power analysis of the implemented design

Total on-chip power	0.105 W	Dynamic: 0.012 W
		Static: 0.093 W
Junction temperature	26.2 °C	
Thermal margin	58.8 °C (5.0 W)	
Effective ϑJA	11.5 °C/W	

reset signals along with 8 input data points (X0, X1, X2, X3, X4, X5, X6, and X7). At output, we have 8 output data points (Y0, Y1, Y2, Y3, Y4, Y5, Y6, and Y7).

Lastly, summary of power analysis and slice utilization is given in Tables 1 and 2, respectively.

Table 2 Area utilization of the implemented design on Zybo Zynq FPGA board

	Utilized	Available	Utilization (%)
Slice LUTs	1365	17,600	7.7
DSPs	26	80	32
Bonded IOB	6	100	6
Slice registers	496	35,200	1.4
Slice	505	4400	11.5

5 Conclusion

It can be concluded that the proposed DCT-II algorithm is efficient for compression of video and speech signals. It requires 26 clock cycles to compute all the coefficients with reduced hardware requirement for $p = 1$ and 2. Zybo Zynq-7000 Development FPGA board was used such that it utilizes 7.91% LUTs, 6% IOB, 32.5% DSPs, and 1.43% Flip flops.

References

1. W. Park, B. Lee, M. Kim, Fast computation of integer DCT-V, DCT-VIII and DST-VII for video coding. IEEE Trans. Image Process. **28** (2019)
2. Y. Zhou, C. Wang, X. Zhou, DCT-based color image compression algorithm using an efficient lossless encoder, in *14th Proceeding of ICSP* (2018)
3. R. Kozhemiakin, V. Lukin, B. Vozel, Image quality prediction for DCT-based compression, in *14th CADSM* (2017)
4. U.V. Koc, K.J.R. Liu, DCT-based motion estimation. IEEE Trans. Image Process. **7**, 948–965 (1998)
5. C.T. Chiu, K.J.R. Liu, Real-time recursive two-dimensional DCT for HDTV systems, in *IEEE International Conference on Acoustics, Speech, and Signal Processing, ICASSP-92*, San Francisco, CA, vol. 3, pp. 205–208 23–26 Mar 1992
6. H. Huang, L. Xiao, CORDIC based Fast Radix-2 DCT algorithm, IEEE May 2013
7. S.-F. Hsiao, Y.H. Hu, T.-B. Juang, C.-H. Lee, Efficient VLSI implementations of fast multiplierless approximated DCT using parameterized hardware modules for silicon intellectual property design. IEEE (2001)
8. M.A. BenAyed, L. Dulau, P. Nouel, Y. Berthoumieu, N. Masmoudi, P. Kadionik, L. Kamoun, New design using a VHDL description for DCT based circuits, Dec 1998
9. N.I. Cho, S.U. Lee, DCT algorithms for VLSI parallel implementations. IEEE Trans. Acoustics Speech Sig. Process. **38**(1), 121–127 (1990)
10. M.B. Mutgekar, P.C. Bhaskar, Analysis of DCT and FAST DCT using soft core processor, in *ICOEI* (2019)
11. C.T. Chiu, K.J.R. Liu, Real-time parallel and fully pipelined two-dimensional DCT lattice structures with application to HDTV systems. IEEE Trans. Circuits Syst. Video Technol. **2**, 25–37 (1992)
12. P. Dahiya, P. Jain, Realization of second-order structure of recursive algorithm for discrete cosine transform. Circuits Syst. Sig. Process. **38**, 791–804 (2019)

Classification of EEG Signals for Hand Gripping Motor Imagery and Hardware Representation of Neural States Using Arduino-Based LED Sensors

Deepanshi Dabas, Ayushi, Mehak Lakhani, and Bharti Sharma

Abstract This study aims to classify the multi-frequency EEG signals associated with motor imagery while performing a carefully devised motor imagery task of gripping left/right hand for rehabilitation application. EEG-based brain–computer interface incorporates recording and classification of the transient EEG changes during different imagery tasks. For rehabilitation of gripping and release movement, ten healthy right-handed volunteers (seven males and three females with mean age: 21.3 years) participated in the EEG investigation using 32 Channel Brain product system. The volunteers performed the motor imagery cued tasks in random order. EEG data was processed to calculate relative alpha band power for each motor imagery trial block from channels C3 and C4 to be passed as feature vectors for the classification of the brain states. After a comparative analysis, SVM classification algorithm provided the highest accuracy of 75%, and the binary output was interfaced with Arduino Uno component to reciprocate left and right imaginary hand movement states using LED light bulbs.

Keywords Brain computer interface · EEG · Support vector machines · Rehabilitation · Arduino

D. Dabas (✉) · Ayushi · M. Lakhani · B. Sharma
Maharaja Surajmal Institute of Technology, C-4 Janakpuri, Delhi 110058, India
e-mail: deepanshidabas123@gmail.com

Ayushi
e-mail: ayushi773478@gmail.com

M. Lakhani
e-mail: lakhanimehak23@gmail.com

B. Sharma
e-mail: bhartisharma@msit.in

P. Bansal et al. (eds.), *Proceedings of International Conference on Artificial Intelligence and Applications*, Advances in Intelligent Systems and Computing 1164,
https://doi.org/10.1007/978-981-15-4992-2_21

1 Introduction

A brain–computer interface (BCI) forms a direct link between the human brain and an external computational device. Training the patients suffering from neural disabilities, to initiate brain signals for controlling external devices, provides scope for extensive future development. Establishing a direct communication with external mechanical devices using the brain signals of motor imagery, aids in bypassing the shortcomings imposed on people with neurological disorders [1]. The electrical activity in the brain can be recorded using a non-invasive monitoring method, electroencephalography (EEG). EEG measures voltage fluctuations within the neurons of the brain [2]. Such an EEG-based brain–computer interface (BCI) can be used to develop a simple binary command for the working of a robotic device. The processing pipeline followed by such a system is as follows, acquisition of EEG data, feature extraction, classification of the features and machine control as shown in Fig. 1. In this study, relative alpha band powers are computed for each motor imagery trial from both left channel (C3) and right channel (C4). Alpha band waves fall in the frequency range of 8–12 Hz and are present in a normal wakeful state with minimal activity as associated with motor imagination. These oscillations reflect cognitive

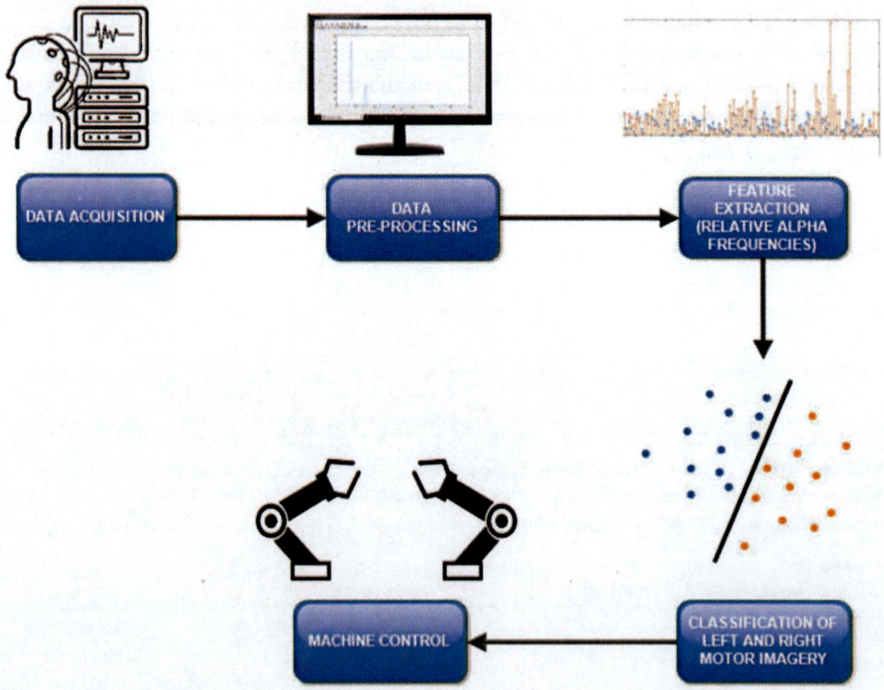

Fig. 1 Timeline of proposed system

states of the brain during imagery and are used as a feature to discriminate between the two classes.

Classification of the two brain states is implemented using Naive Bayes, k-nearest neighbor, support vector machines and linear discriminant analysis. Naive Bayes classification model is used for predictive modeling and is based on Bayes theorem as shown in Eq. (1). It is assumed that all the features are independent of each other and have an equal contribution to the prediction of the result. It calculates the probabilities for the set of input values for each class and classifies as per the outcome with the highest probability.

$$P(A|B) = \frac{P(B|A) \times P(A)}{P(B)}. \tag{1}$$

In the k-nearest neighbor classification model, the combined vote of 'k' neighbors of an object decides the class that the object will fall into. The class of the object is computed locally, and hence, this model is sensitive to the local structure of the data. Support vector machine (SVM) classification model finds a hyperplane in an N-dimensional space that distinguishes between the classes of the data. N is the number of features of the dataset. The objective is to find the hyperplane that maximizes the margin between the classes. Linear discriminant analysis (LDA) computes the mean and variance of the data and makes predictions by estimating the probabilities for a set of input data for each class and chooses the outcome with the highest probability. These four classification models are being used extensively for discrimination of brain states for BCI systems. A comparative analysis for the accuracy of prediction amongst all the models was carried out to choose the model with most accurate predictions based on the relative alpha band power from both left and right channels. For left motor imagery, the output is predicted as 0 and as 1 for right motor imagery. Binary classification is implemented, and the predicted values are passed as input to Arduino microcontroller component. The hardware representation of the left and right brain states is done using LED bulbs. For further implementation, the input digital signal used to illuminate the LED bulbs can be fed to robotic grippers to rehabilitate grip and release motor movement of left and right hands.

2 Literature Review

The reviewed EEG-based BCI articles accommodated a variety of goals as represented in Table 1. These goals can be categorized as (a) development/improvement of existing processing technologies such as developing optimal channel selection methods for motor imagery-based BCIs [3] (b) development of new technologies such as new EEG recording system [4] (c) application-based such as a brain switch for turning on/off a BCI system [5], a real-time drowsiness detection system [6] (d) review articles [7] and (e) evaluation of parameters affecting BCI system such as the design of the electrode layout [8], impact of auditory distraction on BCI performance

Table 1 Literature survey to review EEG-based BCI systems, the algorithms employed and their applications

Author	Year	Algorithm	Key idea	Pros (1) & Cons(2)
J. R. Wolpaw et al.	2000	EEG-based communication	EEG-based communication depends on successful interaction of two adaptive controllers	1.Translates EEG control into device control 2. Depends totally on muscle control
Fabien Lotte et al.	2007	Fuzzy classifier	Briefly presents the commonly employed algorithms and their critical properties	1. Compares and assesses performance 2. Curse of dimensionality and bias-variance trade-off
Jonathan R. Wolpaw et al.	2004	Adaptive algorithm	Identifying and focusing on the EEG features that the person is best able to control and encourages further improvement in that control	1. Might control complex movements 2. Not necessary to implant electrodes for this application
Rajdeep Chatterjee et al.	2016	Support vector machines (SVM) and multi-layered perceptron (MLP)	Focuses on the classification of motor imagery of the left-right hand movements	Slow computation
K. E. Stephan et al.	2016	Bayesian model selection and generative embedding	An overview of the emerging use of computational models for clinically relevant single-subject predictions	1. Avoids erroneous interpretations of neuroimaging data 2. Assumes that features are independent
Jian Kui Feng et al.	2019	CSP rank selection method and linear discriminant analysis(LDA)	Combines the multiband signal decomposition filtering and the CSP rank channel selection methods to select significant channels and then calculate the classification accuracy	1. Improve classification accuracy 2. Large amount of redundant information

(continued)

Table 1 (continued)

Author	Year	Algorithm	Key idea	Pros (1) & Cons(2)
Md.A.M.Joadder et al.	2019	Spatial filtering method and linear discriminant analysis (LDA) classification	A novel SI-based BCI framework is introduced to identify mental states, and a new channel selection concept is also proposed for the same	1. Efficacy of the method is confirmed by statistical and graphical analyses 2. Additional noise
Quadrianto Novi et al.	2007	Sub-band common spatial pattern (SBCSP)	Fine-tuned system is hardly repeatable by other research group so proposed a SBCSP framework, which provides an alternative solution to address this issue	1. Achieves the best accuracy 2. Problem of a time-consuming fine-tuning process in building a BCI for each subject
Benjamin Blankertz et al.	2008	Common spatial patterns (CSP)	To provide comprehensive information about common spatial pattern (CSP), its application and discuss recent variants of CSP	1. Low computational cost 2. Model selection and preprocessing issues or deterioration under outliers
Xiaogang Chen et al.	2015	Joint frequency phase modulation (JFPM) method	Demonstrates that BCIs can provide a truly naturalistic high-speed communication channel using noninvasively recorded brain activities	Low communication speed
M. Z. Ilyas et al.	2015	Regression trees	An algorithm that automatically estimates the class of data as represented by a feature vector	1. Does not require normalization of data 2. Calculation is complex

(continued)

Table 1 (continued)

Author	Year	Algorithm	Key idea	Pros (1) & Cons(2)
Fabien Lotte et al.	2007	k-NN	An algorithm that aims at automatically estimate the class of data as represented by a feature vector	1. Very simple implementation 2. Does not learn anything from the training data and simply uses the training data itself for classification

[9], relationship between motivation and P300 amplitudes [10]. A divergence of orientation toward practical application-based BCI systems is evident. This increase in trend is devoted to assisting people with disabilities, so they can express their intentions with externally controlled devices. BCI researchers have used features derived from PSD, ERP, wavelet coefficients and adopted classification algorithms, such as LDA, SVM, Bayesian, NN. There are broadly two types of BCI systems in terms of functioning: offline and online systems. In the context of this study, the proposed system will operate in an offline environment and work toward the practical application approach.

3 Methodology

3.1 Data Acquisition

Ten healthy right-handed volunteers (seven males and three females with mean age: 21.3 years) participated in the EEG investigation using 32 Channel Brain product system. To improve the discrimination power of the two brain states (left versus right-hand movements), the volunteers were trained for the experiment beforehand using the same paradigm for a duration of 10 min and 30 s. They were asked to perform motor imagery gripping tasks (GO) for left/right hand and resting task (NO-GO) in a randomized manner. Both grip and release cues were provided to the volunteer for a duration of 4 s each for left and right hands. Each trial was of a duration of 14 s as depicted in Fig. 2.

3.2 Data Preprocessing

Following the acquisition of EEG data, the next step was preprocessing of the EEG data using MATLAB. The data was analyzed and processed for motion artifact removal, eye blinking and was down sampled to 250 Hz. It was then band pass

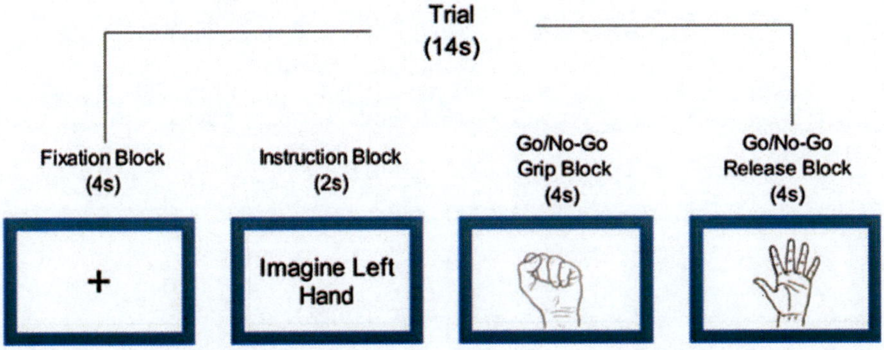

Fig. 2 Illustration of trial block

filtered between 1 and 50 Hz using a basic FIR filter. Re-referencing was carried out without the ECG channel. Wavelet-based analysis gives a suitable base for the evaluation of EEG data. In the context of this study, we use the discrete wavelet transform (DWT) to decompose the preprocessed EEG signal $x[k]$ into its subsequent wavelet coefficients by shifting and scaling the parent signal. The primary task in DWT is to select the number of wavelet decomposition levels for the signal. The output of each level j is represented in the form of two signals: Detail (D_j) and Approximation (A_j), described in Eqs. (2) and (3):

$$D_j[i] = \sum_k x[k] \,.\, h[2\,.\,i - k]. \tag{2}$$

$$A_j[i] = \sum_k x[k] \,.\, l[2\,.\,i - k]. \tag{3}$$

where $h[.]$ and $l[.]$ are the high pass and low pass filters, respectively. The above-mentioned procedure of generating D_j and A_j repeats as long as j does not surpass j_m [11]. The algorithm for decomposition of EEG signal into its subsequent wavelet coefficients and extraction of relative alpha power for each trial block is explained as below:

```
for i = 1 to number of motor imagery trials
    X = EEG.data(channel number, sample points)
    Y = EEG.data(channel number, sample points)
    waveletFunction = 'db8';
    [C,L] = wavedec(X,Y,8,waveletFunction);
    D5 = wrcoef('d',C,L,waveletFunction,5); %GAMMA
    D6 = wrcoef('d',C,L,waveletFunction,6); %BETA
    D7 = wrcoef('d',C,L,waveletFunction,7); %ALPHA
```

```
    D8 = wrcoef('d',C,L,waveletFunction,8); %THETA

    A8 = wrcoef('a',C,L,waveletFunction,8); %DELTA

    D7 = wrcoef('d',C,L,waveletFunction,7); %ALPHA

    POWER_DELTA = (sum(A8.^2))/length(A8);

    POWER_THETA = (sum(D8.^2))/length(D8);

    POWER_ALPHA = (sum(D7.^2))/length(D7);

    POWER_BETA = (sum(D6.^2))/length(D6);

    POWER_GAMMA = (sum(D5.^2))/length(D5);

    Total = POWER_DELTA + POWER_BETA + POWER_ALPHA POWER_THETA
+ POWER_GAMMA;

    RELATIVE_ALPHA = POWER_ALPHA/Total;

end
```

3.3 Feature Extraction

DWT decomposition technique represents EEG data as wavelet coefficients. Their coefficients are converted into features for the classification of brain states. The relative alpha band power was computed for every GO-grip motor imagery trial with 1000 sample points from channels C3 and C4. A clear distinction between the two brain states could be observed upon visualizing the relative alpha powers of each motor imagery gripping trial block from channels C3 and C4 (Fig. 3).

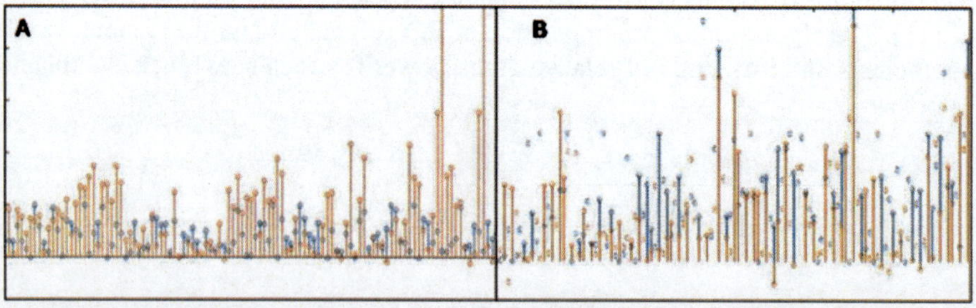

Fig. 3 Plot of relative alpha powers from channel C3 (Blue) and C4 (Red) for trial blocks associated with motor imagery of left hand gripping (**a**) and right hand gripping (**b**)

3.4 Classification

The dataset was modeled using four classification algorithms to perform a comparative analysis on accord of accuracy and recall rate. Accuracy is used as an evaluation metric as it is defined as the ratio of the number of correct predictions made by the classifier to the number of predictions made. It can be computed from the confusion matrix obtained for each model. This metric works well when there are an equal number of samples for the two classes. Also, recall rate has been used as an evaluation metric as well, as it defines the number of true positives obtained in contrast to the total number of predictions. These two metrics have been obtained to cross-validate the hence obtained results and represented by Eqs. (4) and (5).

$$\text{Accuracy} = \frac{\text{Number of correct predictions}}{\text{Total Number of predicitons}}. \tag{4}$$

$$\text{Recall Rate} = \frac{\text{Number of true positives obtained}}{\text{Total Number of predicitons}}. \tag{5}$$

3.5 Hardware Implementation

Two LED bulbs were connected to the Arduino component using a breadboard. The left LED bulb was illuminated on receiving a digital command of 0 and the right LED bulb for a digital command of 1. The hardware implementation represents the two classes of brain states during motor imagery of grip and releases movements, and the digital signals can be used to control robotic grippers to rehabilitate grip movement for people with neurological disorders. The circuit diagram depicting the wired connections of the different hardware components is shown in Fig. 5.

4 Results

The classification accuracies observed were as follows, Naïve Bayes: 61.66%, LDA: 54.16%, k-NN: 54.6% and SVM: 75% using relative alpha power as a feature to separate the two brain states, as shown in Table 2. Accuracy and recall rate have been used as evaluation metrics to perform a comparative analysis between the different models. The plots for accuracy and recall rates for the four classification models are depicted in Fig. 4. As it is evident from the results, SVM classifier gave maximum accuracy and recall rate, its predicted set of output values for left and right motor imagery gripping task was fed as input to Arduino Uno component to represent the two classes using light bulbs as shown in Fig. 5. The representation of motor imagery

Table 2 Classification algorithms accuracy and recall rate results

Classification algorithm	Accuracy (%)	Recall rate
Naïve Bayes	61.66	0.68
LDA	54.16	0.44
SVM	75	1.0
k-NN	54.6	0.44

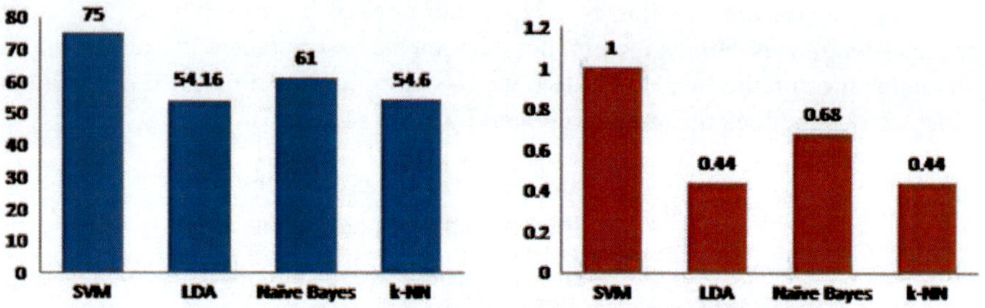

Fig. 4 (Left) Accuracy percentages of the four classification algorithms employed in this study (Right) Recall rates of the four classification algorithms employed in this study

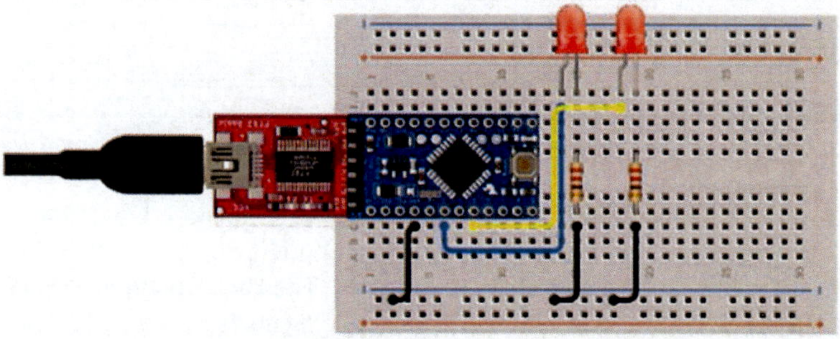

Fig. 5 Arduino Uno microcontroller circuit diagram with breadboard, resistors, USB cable, connecting wires and two LED light bulbs

EEG signals for the gripping task in the form of digital input fed to the Arduino component extends application to further manipulation of hardware devices.

5 Discussion

BCI for neuro-rehabilitation involves the recording and decoding of local brain signals generated by the patient as they try to perform an action-based task or during a mental imagery task. The diversity of applications and range of inclusion of patients with neurological disorders speaks for its adaptable nature and demonstrates the scope of development in this area. Due to the complex nature of the brain data being acquired, constant efforts are made to devise tools and algorithms to improve upon the current systems. At every step, the data is being simplified without compromising its integrity to accurately interpret the imagery action performed by the volunteers and convert it to a binary command for machine control. SVM classifies the two brain states with the highest accuracy of 75% and was used to discriminate between left and right motor imagery state. The performance of the classifiers depends on several factors like the decomposition level for feature extraction using DWT, the relative performance of the subjects during motor imagery task, etc. Upon successful classification of the imaginative states of the human brain, while performing hand gripping task, these two brain states are further represented using the Arduino Uno toolkit with LED bulbs. This representation facilitates visualization of the converted digital signals interpreted from the brain. For future work, the predicted output can be translated as a digital input to control robotic grippers that could rehabilitate grip movement of the hands [12–15].

References

1. Janis J. Daly, Jonathan R. Wolpaw, Brain–computer interfaces in neurological rehabilitation. Lancet Neurol. **7**(11), 1032–1043 (2008)
2. J.C. Henry, Electroencephalography: basic principles, clinical applications, and related fields. Neurology **67**(11), 2092 (2006)
3. W.-K. Tam et al., A minimal set of electrodes for motor imagery BCI to control an assistive device in chronic stroke subjects: a multi-session study. IEEE Trans. Neural Syst. Rehabil. Eng. **19**(6), 617–627 (2011)
4. G. Gargiulo et al., A new EEG recording system for passive dry electrodes. Clin. Neurophysiol. **121**(5), 686–693 (2010)
5. G.R. Müller-Putz et al., Fast set-up asynchronous brain-switch based on detection of foot motor imagery in 1-channel EEG. Med. Biol. Eng. Comput. **48**(3), 229–233 (2010)
6. Gang Li, Wan-Young Chung, A context-aware EEG headset system for early detection of driver drowsiness. Sensors **15**(8), 20873–20893 (2015)
7. H.J. Hwang et al., EEG-based brain-computer interfaces: a thorough literature survey. Int. J. Hum. Comput. Interact. **29**(12), 814–826 (2013)
8. Y. Wang et al., Design of electrode layout for motor imagery based brain–computer interface. Electron. Lett. **43**(10), 557–558 (2007)
9. E.V.C. Friedrich et al., Impact of auditory distraction on user performance in a brain–computer interface driven by different mental tasks. Clin. Neurophysiol. **122**(10), 2003–2009 (2011)
10. S.C. Kleih, F. Nijboer, S. Halder, A. Kübler, Motivation modulates the P300 amplitude during brain–computer interface use. Clin. Neurophysiol. **121**(7), 1023–1031 (2010)
11. Jasmin Kevric, Abdulhamit Subasi, Comparison of signal decomposition methods in classification of EEG signals for motor-imagery BCI system. Biomed. Sig. Process. Control **31**, 398–406 (2017)
12. F. Lotte, A tutorial on EEG signal-processing techniques for mental-state recognition in brain–computer interfaces. in *Guide to Brain-Computer Music Interfacing* (Springer, London, 2014), pp. 133–161
13. A.S. Aghaei, M.S. Mahanta, K.N. Plataniotis, Separable common spatio-spectral patterns for motor imagery BCI systems. IEEE Trans. Biomed. Eng. **63**(1), 15–29 (2015)

14. G. Pfurtscheller et al., Mu rhythm (de) synchronization and EEG single-trial classification of different motor imagery tasks. NeuroImage **31**(1), 153–159 (2006)
15. F. Babiloni et al., Linear classification of low-resolution EEG patterns produced by imagined hand movements. IEEE Trans. Rehabil. Eng. **8**(2), 186–188 (2000)

Bandwidth and Gain Enhancement Techniques of DRA Antenna

Richa Gupta and Garima Bakshi

Abstract This paper presents the investigation done on three DRA geometry to enhance gain and bandwidth. The three geometry consists of Single Rectangular DRA, two-segment Rectangular DRA, and DRA with superstrate structure. The three designs resonants at a frequency of 5.13 GHz, 5.21 GHz and 5.80 GHz, and 6.708 GHz respectively. All the designs are investigated in Ansoft HFSS. The reflection coefficient and radiation pattern of all the designs are presented. The superstrate design provides a maximum bandwidth of 1.08 GHz and gain of 8.80 dB.

Keywords DRA · Gain · Radiation pattern · Bandwidth

1 Introduction

In today's era of wireless communication, a continuous research is required to enhance the gain and bandwidth of the antenna. For long-distance communication and better reception of signal, antenna with high gain and bandwidth is required. DRA antenna is widely used because of its various advantages like radiation efficiency, ease of fabrication [1]. Antenna gain has been enhanced by using MHD technique [2, 3]. The aperture coupled feeding technique is used to increase antenna gain [4].

Numerous techniques have been investigated in the literature to enhance the antenna gain and bandwidth. One of the investigated techniques is stacking DRA [5, 6]. Dielectric resonator antennas of different permittivity are stacked to increase gain and bandwidth of antenna [6]. But it is not easy to fabricate stacked DRA. Many integrated stacking techniques are presented [7, 8] to enhance gain.

R. Gupta (✉)
Maharaja Surajmal Institute of Technology, Delhi, India
e-mail: richaguptamsit@gmail.com

G. Bakshi
School of Engineering & Technology, Ansal University, Gurgaon, India
e-mail: garima.bakshi31@yahoo.com

P. Bansal et al. (eds.), *Proceedings of International Conference on Artificial Intelligence and Applications*, Advances in Intelligent Systems and Computing 1164,
https://doi.org/10.1007/978-981-15-4992-2_22

Investigation has been done on various feeding techniques to generate higher order modes [9, 10]. To generate higher order mode in antenna, a bevel feeding patch structure is demonstrated in [9] to achieve gain up to 7 dB. The technique used in [10] is dual-mode quadrature with high gain results. Both techniques [9, 10] are good but have some fabrication issues. This work presents three DRA designs that have been investigated to enhance gain and bandwidth of antenna. Design 1 is a simple DRA with a microstrip feedline. In design, same DRA is segmented into two dimensions to generate two frequencies and then they are merged for wider bandwidth. In the third design, superstrate layer is used above the DRA to enhance gain. The investigations have revealed that gain has been enhanced using superstrate layer.

2 DRA Mathematical Modeling

The dimensions of DRA can be calculated using the below-mentioned mathematics [1]

$$k_{(x)}^2 + k_y^2 + k_z^2 = \varepsilon_r k_0^2 \tag{1}$$

$$f_0 = \frac{c}{2\pi\sqrt{\varepsilon_r}}\sqrt{k_x^2 + k_y^2 + k_z^2} \tag{2}$$

$$k_x = \frac{\pi}{a};\, k_y = \frac{\pi}{b} \tag{3}$$

$$k_z\left(\tan\frac{k_z d}{2}\right) = \sqrt{(\varepsilon_r - 1)k_0^2 - k_z^2} \tag{4}$$

where ε_r denotes permittivity of DRA, f_0 is resonating frequency velocity of light is represented by c, and k_0, k_x,k_y, and k_z are wave numbers along x, y, and z directions, respectively.

3 Antenna Designs

Three antenna designs are investigated in this paper. The design of Rectangular DRA fed by microstripline is demonstrated in Fig. 1a. The DRA of dimensions, length = 10.5 mm, width = 6 mm, and height = 9.6 mm is placed over an FR4 substrate of dimensions 80 × 80 mm^2 and permittivity, ε_r = 4.4 and thickness 0.84 mm. The dimensions of microstrip line are length = 29.1 mm and Width = 0.33 mm. The permittivity of DRA material is 10.4. DRA has been connected with microstrip line for excitation. HFSS model of two-segment DRA has been shown in Fig. 1b. This technique can be widely used for BW enhancement. A two-segment rectangular

Fig. 1 **a** DRA with direct microstrip feed. **b** HFSS model of two-segment DRA. **c** Superstrate DRA model

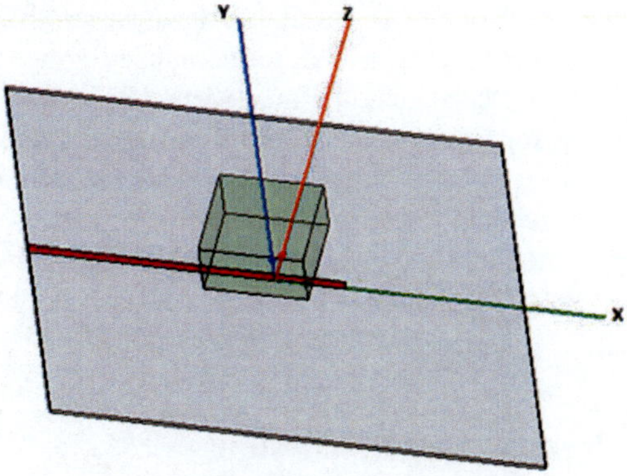

(a). DRA with direct micro strip feed

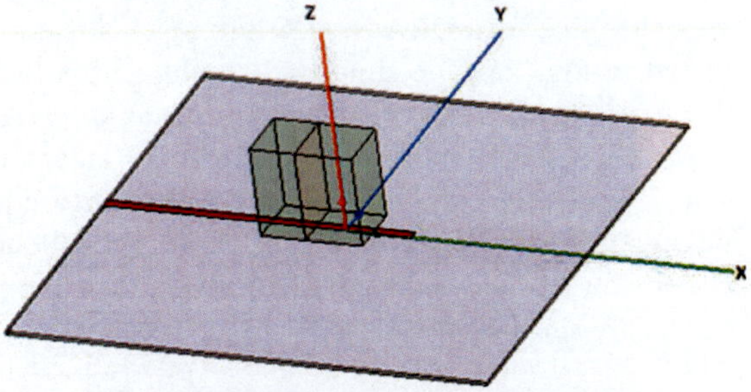

(b). HFSS model of two segment DRA

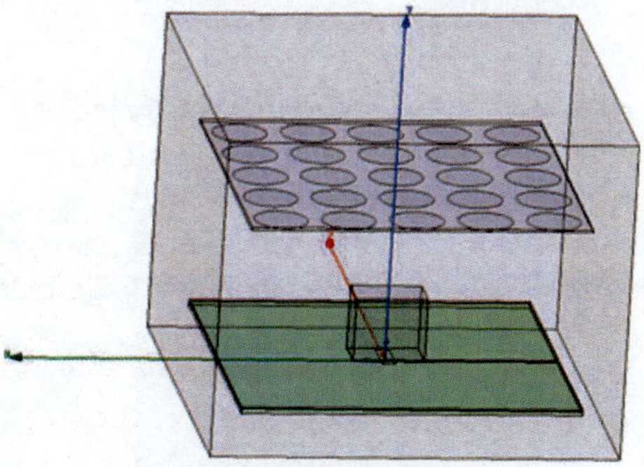

(c). Superstrate DRA model

dielectric resonator in a suitable arrangement has been used to enhance the bandwidth of DRA. This structure is composed of two rectangular sections with different sizes, have been separated by a perfect conducting plate. Thus forming two DRAs, which resonate at two different frequencies. These two frequencies can be merged to obtain wideband characteristics. This technique can be used for wideband applications. Another technique of the superstrate layer above DRA to enhance the gain is presented in Fig. 1c. Superstrate layer consists of circular patches. The superstrate layer above DRA enhances the area of antenna which affects its gain.

4 Results and Discussion

The reflection coefficient results of all three designs are demonstrated in Fig. 2. For microstripline feed DRA S_{11} (Return loss), at frequency 5.13 GHz, is −31.8 dB as shown in Fig. 2a. The simulated results of return loss for two-segment DRA are shown in Fig. 2b. The antenna resonates at two resonant frequencies, i.e., 5.21 and 5.80 GHz. Return loss for the superstrate antenna is shown in Fig. 2c. It shows −30.73 dB return loss at 6.708 GHz resonant frequency. Radiation pattern for all antenna designs are shown in Fig. 3a–c, which shows maximum gain of 8.80 dB at resonant frequency in end fire direction for superstrate configuration. The various results are summarized in Table 1.

5 Conclusion

The investigations have been done on three DRA geometry to enhance gain and bandwidth. The three geometry investigated are Single Rectangular DRA, two-segment Rectangular DRA, and DRA with superstrate structure. The three designs resonants at a frequency of 5.13 GHz, 5.21 GHz & 5.80 GHz, and 6.708 GHz, respectively. The superstrate design provides maximum bandwidth of 1.08 GHz and gain of 8.80 dB.

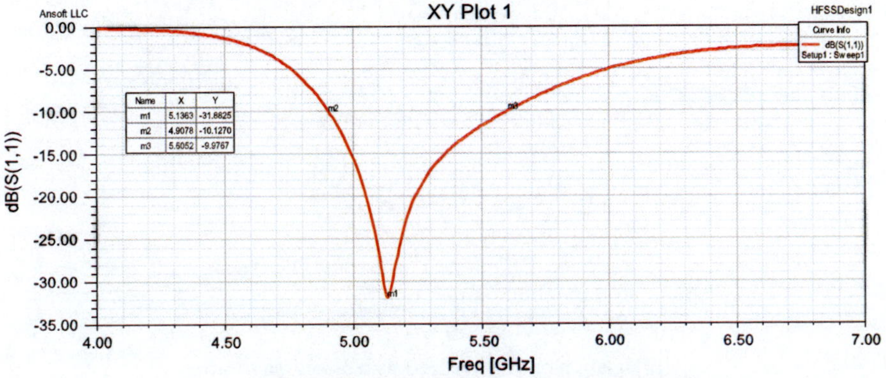

(a) Return loss of DRA feed with direct micro strip

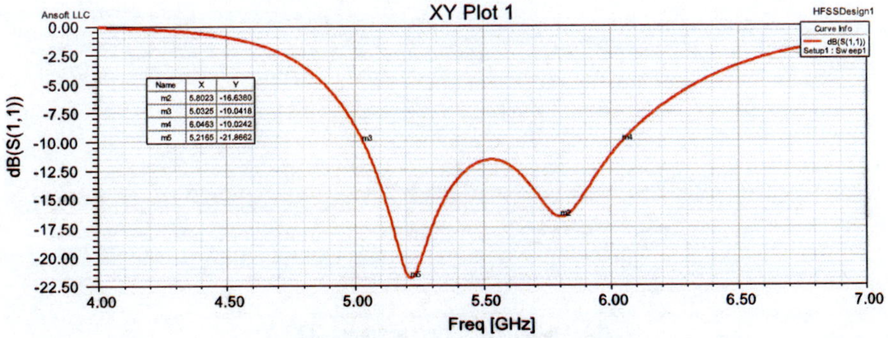

(b) Return loss of two segments DRA

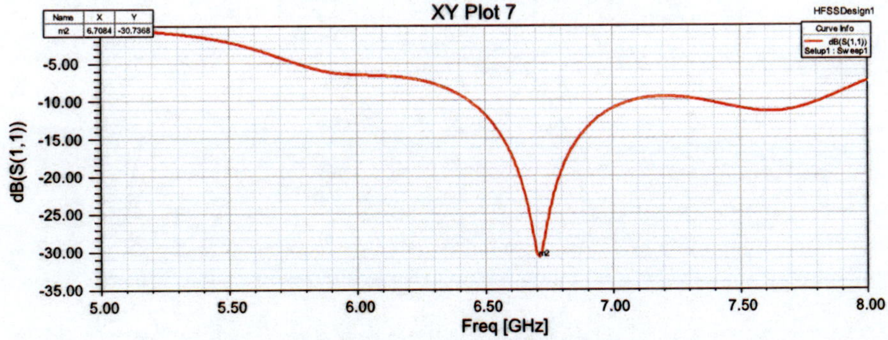

(c) Return loss.of superstrate antenna

Fig. 2 **a** Return loss of DRA feed with direct microstrip. **b** Return loss of two-segment DRA. **c** Return loss of superstrate antenna

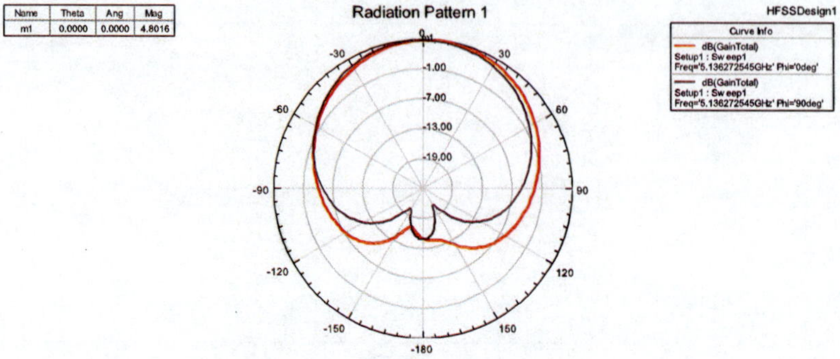

(a). Gain plot of DRA feed with direct micro strip

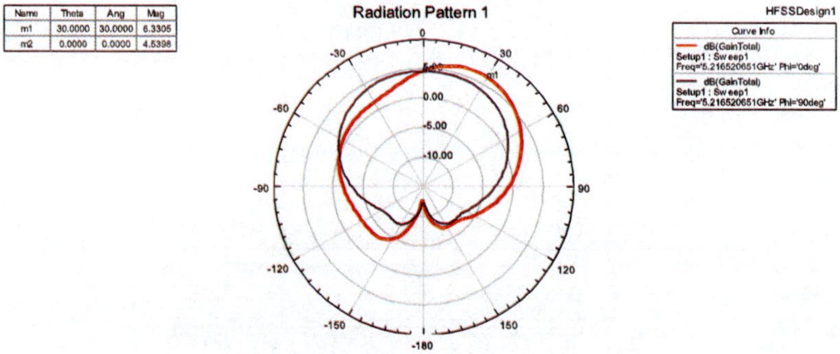

(b). Gain plot of two segments DRA

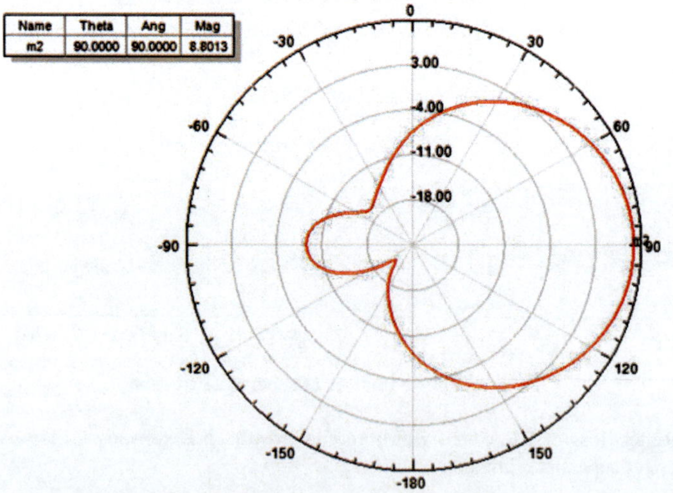

(c). Gain plot of superstrate DRA

Fig. 3 **a** Gain plot of DRA feed with direct microstrip. **b** Gain plot of two-segment DRA. **c** Gain plot of superstrate DRA

Table 1 Result summary of three designs

DRA structure	Single	2 segment	Superstrate
Resonant frequency (GHz)	5.13 GHz	5.2 GHz 1 and 5.80 GHz	6.708 GHz
Bandwidth (GHz)	4.88–5.29 GHz (0.41 GHz)	5.0325–6.046 GHz (1.013 GHz)	6.42–7.50 GHz (1.08 GHz)
Gain (dB)	4.8	4.5	8.80

References

1. R.S. Yaduvanshi, H. Parthasarathy, *Rectangular Dielectric Resonator Antennas* (Springer, New Delhi, India, 2016)
2. R. Gupta, R. Yaduvanshi, Embedded cylindrical magneto-hydrodynamic antenna. Int. J. Ultra-wideband Commun. **3**(2), 68–72 (2015)
3. R. Gupta, R. Yaduvanshi, High gain and wide band Rectangular DRA. Int. J. Ultra-wideband Commun. **3**(2), 107–114 (2015)
4. R. Gupta, R. Yaduvanshi, Reconfigurable aperture coupled rectangular dielectric resonator antenna, in *IEEE International Conference on Computing for Sustainable Global Development* (2015), pp. 859–862
5. A.K. Gautam, M. Singh, Design of gain enhanced stacked rectangular dielectric resonator antenna for C-band applications, in *2016 3rd International Conference on Computing for Sustainable Global Development (INDIACom)* (IEEE, 2016), pp. 119–123
6. Y.M. Pan, S.Y. Zheng, A low-profile stacked dielectric resonator antenna with high-gain and wide bandwidth. IEEE Antennas Wirel. Propag. Lett. **15**, 68–71 (2015)
7. S. Fakhte et al., High gain rectangular dielectric resonator antenna using uniaxial material at fundamental mode. IEEE Trans. Antennas Propag. **65**, 342 (2017). https://doi.org/10.1109/TAP.2016.2627520
8. S. Fakhte, H. Oraizi, L. Matekovits, High gain rectangular dielectric resonator antenna using uniaxial material at fundamental mode. IEEE Trans. Antennas Propag. **65**(1), 342–347 (2016)
9. T.A. Denidni, Z. Weng, Rectangular dielectric resonator antenna for ultrawideband applications. Electron. Lett. **45**(24), 1210–1212 (2009)
10. B. Li, C.-X. Hao, X.-Q. Sheng, A dual-mode quadrature-fed wideband circularly polarized dielectric resonator antenna. IEEE Antennas Wirel. Propag. Lett. **8**, 1036–1038 (2009)

Social Intelligence and Sustainability

TODD: Time-Aware Opinion Dynamics Diffusion Model for Online Social Networks

Aditya Lahiri, Yash Kumar Singhal, and Adwitiya Sinha

Abstract Information diffusion is employed to track the spread of information in the online social network, and hence, is considered extremely relevant to several modern-day applications. It is of pivotal importance for addressing challenges associated with the influence maximization in social media. Our research focuses on the diffusion process to track the temporal extent of information spread in the network. Over time, various models have been suggested in order to best represent the diffusion process. Our paper suggests a novel information diffusion model, time-aware opinion dynamics model (TODD), that derives from opinion dynamics and continuous time models in order to provide a more robust and realistic representation of the real-world problem. A comparative study is conducted between our proposed model and several existing methods.

Keywords Online social networks · Information diffusion · Time aware · Diffusion · Opinion dynamics · Continuous time models

1 Introduction

The diffusion models define the set of rules which allow the infection to spread rapidly and thus serve as a major foundation stone for influencing maximization problem which states that we need to find K nodes which would appropriately cover the whole of the social network. Typically, the approach used by the authors does not take into consideration of both the opinion dynamics and continuous time to model

A. Lahiri · Y. K. Singhal (✉) · A. Sinha
Department of Computer Science Engineering, Jaypee Institute of Information Technology, A-10 Sector 62, Noida 201301, U.P., India
e-mail: yashgyy@gmail.com

A. Lahiri
e-mail: aditya.lahiri2009@gmail.com

A. Sinha
e-mail: adwitiya.sinha@jiit.ac.in

P. Bansal et al. (eds.), *Proceedings of International Conference on Artificial Intelligence and Applications*, Advances in Intelligent Systems and Computing 1164,
https://doi.org/10.1007/978-981-15-4992-2_23

the diffusion process. There has been research in independent studies depicting that content of the tweet as well as nature of the time might be more important in the information diffusion than the influential personality [1].

The opinion dynamics and continuous time spectrum model serve very different purposes and solve different problems. The major factor which required improvement especially in the context of the continuous time model is that the model although is dynamic, there is no data-driven approach to determine the strength of the relationship. On the other hand, while the opinion dynamics model takes into account the suggestions, it is however static and thus not accurately models the changing dynamics [2]. By combining both of them, not only we can make our model dynamic but also take into account the guidelines of the opinion dynamics model. This allows us to model the relationships and explore how power shifts happen in the most accurate way possible. It will also improve the predictions by the influential maximization algorithms which takes into account the underlying diffusion principle as the foundation's, hence finding the person who needs to be targeted in the viral marketing will become highly accurate which may lead to profit to companies. Lastly, the resulting diffusion process can show interesting insights into how the opinions can change the dynamics in the social network.

In the subsequent sections, thus we will do a discussion on the existing diffusion models, the proposed algorithms and its implications followed by the results and conclusions we arrived at.

The process of successful diffusion from a node u to v shall be referred to as v being activated by u in this paper, though it is possible to find other terminology being used in some literature.

Centrality of a node in the social network was considered as the optimum solution to model diffusion process and thus was used to understand how information is propagated in the social network. Most of the early works agreed on measures of centrality such as degree, Eigen vector, betweeness and closeness [3–5]. Each centrality measures have their strength and weakness, and the right measure of centrality is mostly depended on the type of social network you want to model.

Furthermore, to better reflect the opinion dynamics of the social network, a signed approach to reflect trust and mistrust was proposed [6]. Some work on mining the signs from a social network was also done [7]. Negative edges in social networks were examined [8]. There were varying opinions on how to model trust and mistrust. Initially, balance theory was the model proposed and was quite popular [9]. Moreover, it was found that the hierarchical nature was suited to reflect the social dynamics [10].

The influence maximization problem was initially proposed as a problem to find K influential nodes which when selected leads to maximum information spread [11]. The naive solution which was proposed was NP-hard [11]. However, greedy approaches were proposed such as the CELF [12] and CELF++ [13], which both work by exploiting the properties of sub-modularity of the diffusion process. There are inherent inadequacies in how the influence maximization problem was initially modeled and that is due to the way it uses independent cascade models and threshold model as a diffusion model, and they both fail to account for the opinion dynamics of a subject. There has been a substantial amount of work done indicating that the diffusion process should be as more as a function of topic rather than the influence of the person. Furthermore, each node while do have some inherent attractiveness toward the topic and the resistivity of the person to spread the information also matters. Thus, while the in-degree might represent the user popularity, they do not specifically influence the growth of opinions in social networks [14]. The opinion dynamics while helpful in modeling the influence, most of the work considers the opinion dynamics in discrete steps and not considers the progression as a continuous one [2]. This creates a problem as the networks in real life are inherently time continuous but sadly any model which incorporates both opinion dynamics as a diffusion process for the influence maximization has not been still implemented.

2 Existing Diffusion Benchmarks

Some of the early models suggested include the independent cascade model, the linear threshold model and the triggering model. The primary application of these models was the influence maximization problem [11].

The independent cascade model proposes that a vertex u can activate a vertex v with a predetermined probability $p_{u,v}$ provided that u is already activated. It gets to do so only once across all iterations for a given pair (u, v), notwithstanding its success in activating v at that instant [11].

The linear threshold model states that an inactive node u can only be activated if its number of activated neighbors crosses a certain predetermined threshold [15].

The triggering model was suggested as a generalization of the linear threshold and independent cascade model. It states that for a given node u, a subset of influential nodes is chosen randomly according to some predetermined probability distribution. If the set contains a node that is a neighbor of u and was activated in the previous time step, then u is activated [11].

However, none of the aforementioned models make any realistic assumptions about the flow of time or place any time-related constraints whatsoever on the diffusion process. They all flow discretely and continue till no further node can be activated.

Hence, time-aware models were suggested that took into account time-based constraints such as the continuous time model [16] which assumes the activation of v by u to be continuous in time.

The main point of difference between the opinion dynamics model and other existing models is that the state of an individual is liable to oscillate freely, thus correctly simulating how opinion changes in actual world. This is an important correction as an individual opinion is no longer irreversible in nature, i.e., his opinion cannot change once it is activated. Thus, the state of individual is liable to change based on some set of rules by interacting with neighbors [17].

Voter model can be seen as one of the foundation models for opinion dynamics where it is assumed that the opinion of an individual to be plus minus one with the update rule states that the on each round of iteration an infected node would copy his opinion to one of his other neighbors. While this is certainly an improvement over the epidemics model, here it still faces with the inherent flaw where the update rule takes into account the opinion of one individual within its social setting and environment [18].

Majority model is another opinion dynamics model but is an improvement with regard to naive voter model with regard to how the network is processed. Here a group of R individuals is chosen at random, and the group thus takes the opinion of the majority group. The agents here like in the case of the voter model take discrete opinions of $+1$ or -1 [19]. While relatively stable than the voter model, here the model fails in establishing the structure or hierarchy of group selected and formed. There is no rule or constraint in choosing the members of the group. Thus, an individual here might take an opinion which directly contradicts the opinion of his every neighbor seemingly randomly.

Algorithm 1 Proposed TODD Algorithm

```
Input: Graph G, maxIterations, maxT ime Output:
activationStatus[]
  Initialisation : initialize alpha for all pairs, startT ime, activationStatus[]
  LOOP Process
for i = 0 to maxIterations do
  if ((presentTime() - startT ime) > maxT ime) then
     return activationStatus[]
  end if
  for u in G.nodes() do
     if (activationStatus[u] == 0) then
        continue
     end if
     difference = 0
     for v in G.neighbors(u) do
        difference + = 1 − G.alpha(u, v)
     end for
     for v in G.neighbors(u) do
        if (activationStatus[v] == 0) then
           α_{u,v} = G.alpha(u, v)
           β_u = 1 / difference
           t_v = presentT ime()
           t_u = G.activationT ime(u)
           infP rob = 1 − e^(−α_{u,v} *β_u*(t_v −t_u))
           flip = random()
           if (infProb > flip) then
              G.activationT ime(v) = t_v
              activationStatus[v] = 1
           end if
        end if
     end for
  end for
end for
return activationStatus[]
```

Cognitive opinion dynamic model is another relatively complex opinion dynamics model where the individual opinion is modeled in the range between 0 and 1. The opinion here evolves through the combined effect of interacting with the neighbors as well through external information [20]. Furthermore, every individual has its own set of parameters like tendency to inform others, trust in institutions which adversely effect how the diffusion in social network will unfold. This model is marked with improvement over existing opinion dynamics model which considers the opinion of an individual to be purely black and white, while also not providing logical update set based on individual characteristics.

3 Proposed TODD Algorithmic Framework

In our proposed algorithm, time-aware opinion dynamics diffusion (TODD) model was formulated with the object of combining the time awareness and opinion dynamics feature. It uses the exponential distribution to predict the probability of a particular node to infect or diffuse its neighbor(s). It takes into account the flow of time and strength of influence between a given pair of nodes as well as the level of dissent between the given node and its neighbors. The pseudo code is provided in Algorithm 1.

The random() function returns a random number between 0 and 1. The variables maxIterations and maxTime represent the maximum number of iterations that can be run and the maximum time up to which the algorithm will run, respectively.

The presentTime() function returns the time at the moment of the function's call. activationStatus[] is an array like data structure containing the status of activation of all the nodes in the social network graph, G.

G.alpha(u, v) is a function that returns the parameter $\alpha_{u,v}$ which is indicative of the strength of influence from a node u to its neighbor v.

The parameter β_u is a measure of the difference of opinion of u with its neighbors. It increases with decreasing dissent between u and its neighbors.

infProb is the probability of the node u infecting its neighbor v at a time, t_v, given the activation time of u, t_u.

$$\text{infProb} = 1 - \exp\left(-\alpha_{u,v} * \beta_u * (t_v - t_u)\right) \tag{1}$$

It is obtained using a modification of the exponential probability distribution as mentioned in Eq. (1).

4 Discussion

Independent cascade model, SI model, SIS models, Profile model, comes under the broad domain of the epidemics model. The motivation behind the epidemics model is that the spreading process is dependent on the activeness of the neighbor. The reasoning is well sound but one major issue here is the absence of heuristic to signify how much the node is active. SI model and SIS model follow the same trend as depicted in Fig. 1b, c, SIS; however, its slow converges mostly due to its ability to change itself to susceptible at any point of time. Profile model, however, takes into account the profile of an individual as a heuristic to determine the activeness of individual. However, it quickly saturates as depicted in Fig. 1d. The voter model is one of the simplest opinion dynamics models, and truly it shows, the voter model fails in propagating the infection, the trend is stagnant, and there is no change in the structure of the nodes; refer Fig. 2a. The time-aware model has an exponential line trend, and that is, particularly characteristics of time-aware model Fig. 2b [2], in also this case the model quickly saturates even though there is a constraint on time here.

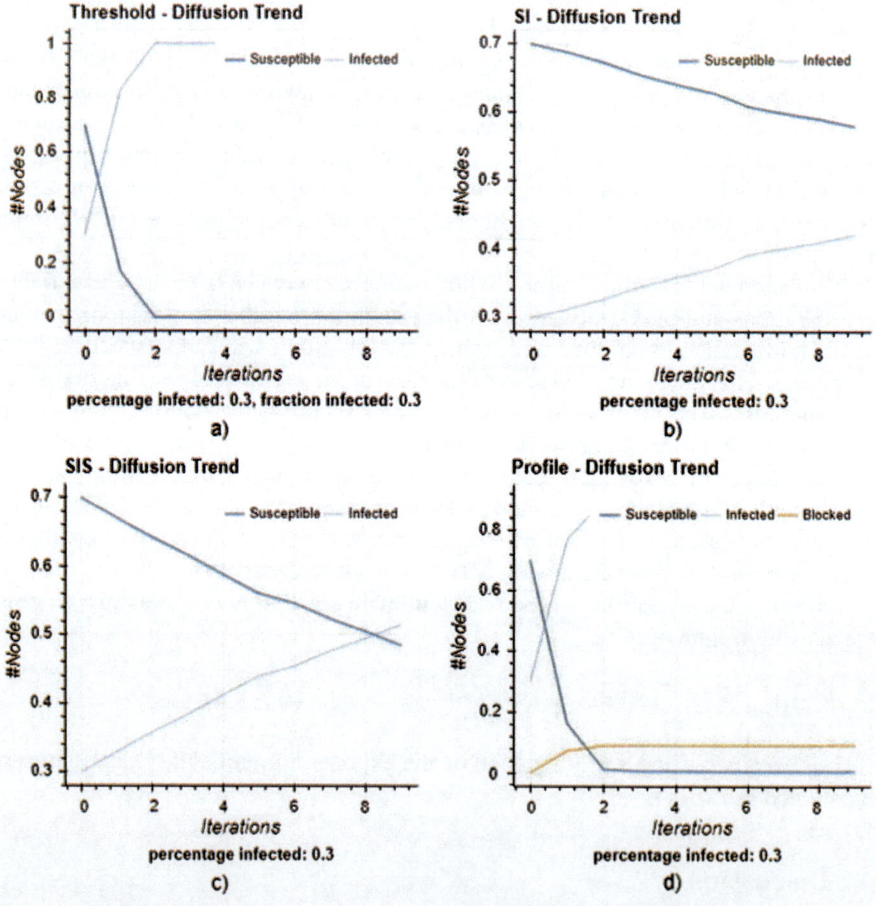

Fig. 1 Comparison of epidemics diffusion models. **a** Independent cascade model, **b** SI model, **c** SIS model, **d** Profile model. All models are here initialized with 20% infection

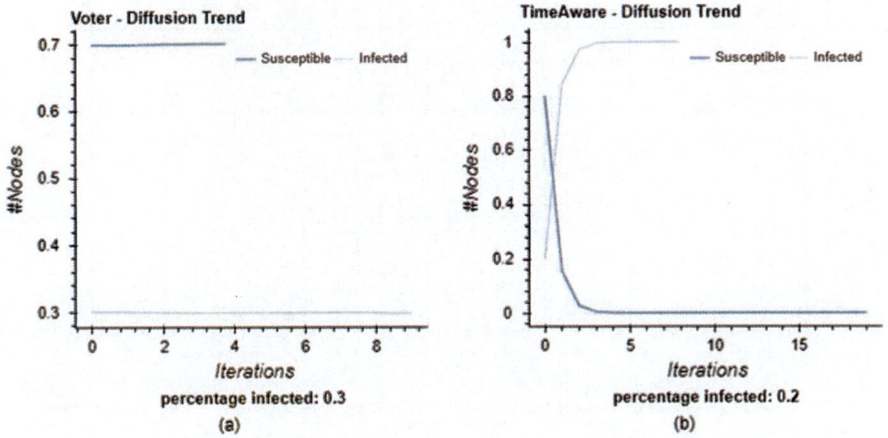

Fig. 2 Opinion dynamics and continuous time model **a** Voter model **b** Time-aware model

The threshold model comes under epidemics model which says that if certain no. of neighbors of uninfected node are infected and proportion of these are above a certain threshold, the node would get infected as well. This is similar to our existing TODD model with respect that here infected node propagates depending upon how strong of an opinion the neighbors of an infected node hold and thus propagate it accordingly. The threshold model saturates quickly within two iterations while the TODD model has a smooth nice ramp structure signifying infection increases here smoothly. Both the modes are based on the opinion dynamics model, so it is interesting how the model differs in propagation Fig. 1a. One reason might be the TODD model has an additional time constraint, and the continuous time model looses their ability to spread the infection quickly with increase in time. The supposed additional constraint mirrors closely to real life [2].

The Sznajd model is the opinion dynamics model based on the theory of social impact, motivated by the fact that a pair of individuals is able to influence a way more than a single person. Hence, in every iteration, a pair of individuals are chosen, and if the opinions coincide, they are able to influence other neighbors with the same opinion as well [21]. Although it sounds reasonable on paper, it does not converge to one opinion as can be observed from Fig. 3d. In contrast, the TODD model not only takes into account the opinion dynamics of an individual, it also takes care of the characteristics of the continuous nature of a time making for a more realistic model.

The majority model is a basic opinion dynamics model that takes into account the opinion of R random individuals at every iteration and thus propagates its opinion accordingly. It slowly converges and follows a parabolic graph. While this model is a better model in general than other opinion dynamics, it continues to suffer from the instability of the opinion dynamics model, here in Fig. 3. The infection rapidly drops after iteration and continues to fall with every iteration. This is partially due to randomized nature of selecting the nodes; no heuristic or algorithm is employed

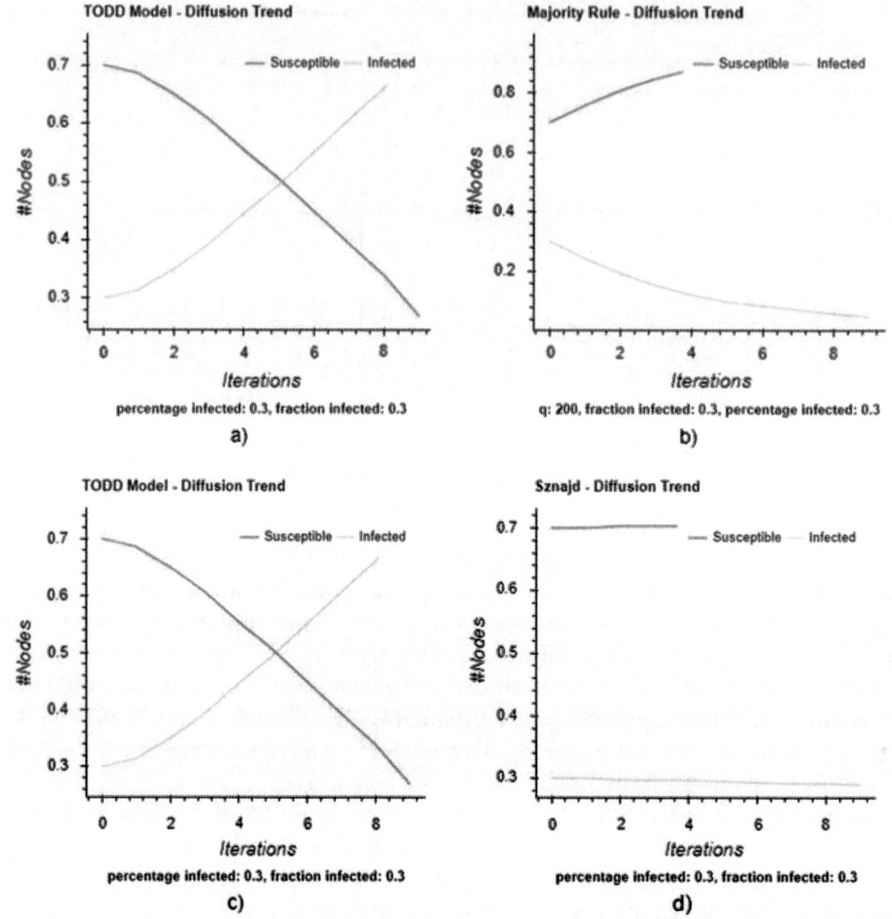

Fig. 3 Comparison of diffusion process in TODD with other models. **a**, **c** TODD model, **b** Majority model, **d** Sznajd model

for selective choosing of some specified nodes. In contrast to this, the TODD algorithm employs heuristic and tends to favor those infected which are surrounded with strongly opinionated individuals as specified in Algorithm 1.

The infection does not spread rapidly here like in epidemics model where almost all the nodes are infected within three iterations, here in all cases of initial infection spread of 30% of above, and 70% of the nodes are infected within eight iterations. Further, Fig. 4a–d shows the infection line is non-decreasing in nature and thus does not show characteristics of opinion dynamics model like majority model where the state of the node once infected is unknown in future iterations. Flat curve on all of the images of Fig. 4 in the beginning of the infection line signifies the rate of

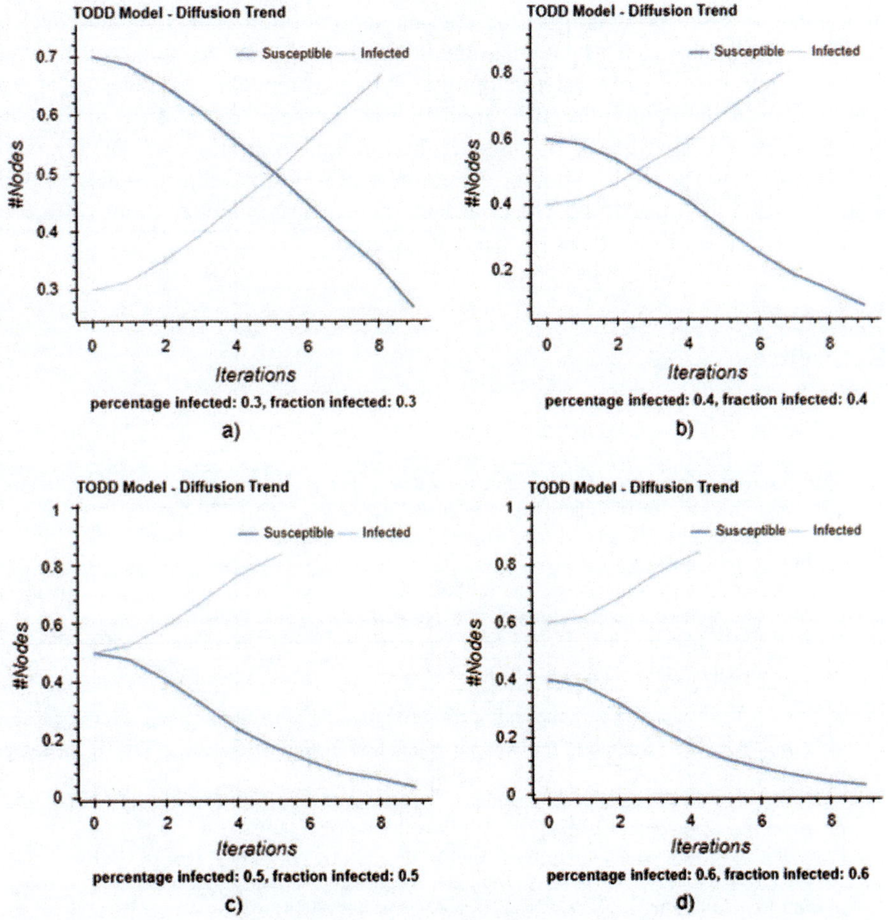

Fig. 4 Comparison of diffusion process in TODD model with different percentage of initial infected nodes. **a** 30% infection, **b** 40% infection, **c** 50% infection, **d** 60% infection

converging is slow, signifying the model takes some nominal time in the beginning before propagating further. The rate of converging seems independent of the no. of infected initial nodes in the beginning; however, with increasing time it ever so slightly resembles to form a plateau.

5 Conclusion

Our proposed model was an attempt to combine two standard diffusion models, the opinion dynamics and continuous time model. Each of the models worked well in isolation; however, there has been little attempt to combine them. One of the things that was sorely lacking in the continuous time models was the absence of the data-driven approach to calculate the strength of the relationship, and thus, it was our intention to incorporate the opinion dynamics model as the measure of the relationship in the continuous time modeling so as the resultant diffusion model mirrors closely to real life as possible. Although the observation of TODD model looks promising, the choice which diffusion model is best is subjective at best, and there is no definite measure at which we can grade diffusion model. Some diffusion model works better than others in different situations.

References

1. O. Tsur, A. Rappoport, What's in a hashtag? in *Proceedings of the Fifth ACM International Conference on Web Search and Data Mining—WSDM '12* (2012)
2. Influence maximization on social graphs: a survey. IEEE Trans. Knowl. Data Eng. (2018)
3. S.P. Borgatti, Centrality and network flow. Soc. Netw. **27**(1), 55–71 (2005)
4. L.C. Freeman, Centrality in social networks conceptual clarification. Soc. Netw. **1**(3), 215–239 (1978)
5. Tore Opsahl, Filip Agneessens, John Skvoretz, Node centrality in weighted networks: generalizing degree and shortest paths. Soc. Netw. **32**(3), 245–251 (2010)
6. J. Leskovec, D. Huttenlocher, J. Kleinberg, Signed networks in social media, in *Proceedings of the 28th International Conference on Human Factors in Computing Systems—CHI '10* (2010)
7. A. Hassan, A. Abu-Jbara, D. Radev, Extracting signed social networks from text, in *Workshop Proceedings of TextGraphs-7 on Graph-based Methods for Natural Language Processing, TextGraphs-7 '12* (Association for Computational Linguistics, Stroudsburg, PA, USA, 2012), pp. 6–14
8. J. Kunegis, A. Lommatzsch, C. Bauckhage, The slashdot zoo, in *Proceedings of the 18th international Conference on World Wide Web—WWW '09* (2009)
9. Fritz Heider, Attitudes and cognitive organization. J. Psychol. **21**(1), 107–112 (1946)
10. R. Guha, R. Kumar, P. Raghavan, A. Tomkins, Propagation of trust and distrust, in *Proceedings of the 13th Conference on World Wide Web—WWW '04* (2004)
11. D. Kempe, J. Kleinberg, É. Tardos, Maximizing the spread of influence through a social network, in *Proceedings of the Ninth ACM SIGKDD International Conference on Knowledge Discovery and Data Mining—KDD '03* (2003)
12. J. Leskovec, A. Krause, C. Guestrin, C. Faloutsos, C. Faloutsos, J. VanBriesen, N. Glance, Cost-effective outbreak detection in networks, in *Proceedings of the 13th ACM SIGKDD International Conference on Knowledge Discovery and Data Mining, KDD '07* (ACM, New York, NY, USA, 2007), pp. 420–429
13. A. Goyal, W. Lu, L.V.S. Lakshmanan, Celf++: optimizing the greedy algorithm for influence maximization in social networks, in *WWW* (2011)
14. M. Cha, H. Haddadi, F. Benevenuto, K. Gummadi, Measuring user influence in twitter: the million follower fallacy (2010)
15. M. Granovetter, Threshold models of collective behavior. Am. J. Sociol. **83**(6), 1420–1443 (1978)
16. M. Gomez-Rodriguez, D. Balduzzi, B. Scholkopf, Uncovering the temporal dynamics of diffusion networks (2011). CoRR, abs/1105.0697
17. Opinion dynamics with disagreement and modulated information. J. Stat. Phys. (2013)
18. A model for spatial conflict. Biometrika (1973)
19. Minority opinion spreading in random geometry. Eur. Phys. J. B Condens. Matter Complex Syst. (1973)

20. Reducing individuals' risk sensitiveness can promote positive and non-alarmist views about catastrophic events in an agent-based simulation
21. Opinion evolution in closed community. Int. J. Mod. Phys. C

Spectral Graph Theory-Based Spatio-spectral Filters for Motor Imagery Brain–Computer Interface

Jyoti Singh Kirar and Ankita Verma

Abstract Motor imagery brain–computer interfaces are one of the widely adopted techniques for imparting basic communication capability to motor disabled patients. The preciseness of a motor imagery BCI task classification is highly dependent on identifying the subject-specific relevant subset of frequency filters. This article proposes a novel approach that utilizes graph theory-based unsupervised feature selection method to determine a reduced set of non-redundant and relevant frequency bands. The empirical analysis of the proposed method is conducted on publicly available datasets, and the obtained results show improvement in classification performance. Further, the performed Friedman statistical test also establishes that the proposed approach surpasses the baseline techniques in classification accuracy.

Keywords Motor imagery · Common spatial pattern · Brain–computer interface · Feature selection · Support vector machine

1 Introduction

A brain–computer interface (BCI) is employed to create an unmediated communication channel between a brain and a device, which interprets neuronal activity for determining user command, with no use of the muscular or peripheral ways [1]. The prime objective of a BCI is to provide communication abilities to the person with severe motor disabilities. In recent years, increased attention has been garnered by EEG-based motor imagery (MI)-based BCI. It uses sensorimotor rhythms (SMRs) which are rhythmic activity generated by visualization of movement of a particular body part. These activities are detected on the scalp by EEG signals and are recorded

J. S. Kirar
Shiv Nadar University, Greater Noida, India
e-mail: kirarjyoti@gmail.com

A. Verma (✉)
Jaypee Institute of Information Technology, Sector 62, Noida, India
e-mail: ankita.verma@jiit.ac.in

P. Bansal et al. (eds.), *Proceedings of International Conference on Artificial Intelligence and Applications*, Advances in Intelligent Systems and Computing 1164,
https://doi.org/10.1007/978-981-15-4992-2_24

over sensorimotor cortex [2]. The momentary reduction in amplitude of SMRs during movement onset is known as "Event-Related Desynchronization" (ERD). Immediately after the movement is performed, there is a subsequent enhancement in the amplitude of SMRs which is called as "Event-Related Synchronization" (ERS) [1, 3–5]. Since, these multichannel EEG signals are extremely dependent of the specific subject. Thus, the selection of relevant subset of subject-specific spatial and spectral filters for classification of motor imagery tasks becomes crucial.

Among several data-driven spatial filtering techniques, common spatial pattern (CSP) [3] is a technique that estimates spatial filters for the analysis of multi-channel EEGs in MI-based BCI by simultaneously maximizing the variance of one class and minimizing variance for the second class [1, 3, 6]. In order to improve the performance of a MI-based BCI, CSP filters are applied to a predefined fixed-sized frequency filter bank. These discriminative frequency bands are dependent upon the specific subject and may differ from subject to subject. However, irrelevant subset of frequency bands may lead to poor performance. To select the relevant subset of frequency bands, supervised filter feature selection method has been suggested in the literature. Novi et al. [3] proposed a method named as subband common spatial pattern (SBCSP) that uses the non-overlapping fixed-sized frequency subbands filtered EEG signals for deriving features. It uses RFE-SVM to rank subbands. Another similar method to SBCSP is filter bank CSP (FBCSP) [5] that employs maximal mutual information for choosing optimal spatio-temporal subband filters. These methods use fixed sized subbands and neglect the significance of exploring other different size subbands in a given frequency bands. Another method called CVSCSP [7] method generates a set of frequency subbands for a given bandwidth and granularity which varies in size. The proposed method utilized these variable-sized filter banks to identify variable-sized relevant frequency bands. In a recent research work, the authors presented CVSTSCSP method for identifying an optimal combination of temporal, spectral and spatial information from different time segments of an EEG trial [8]. However, these features obtained were dependent on static time segments of a given EEG trial. Nonetheless, these methods determine relevance of spectral filters based on class label and fail to capture inherent structure of the data. Recently, unsupervised discriminative feature selection (UDFS) along with CSP has been employed which selects the dominant features for motor imagery task, but this method involves optimization of various parameters for selection of optimal features [9].

Another research work recently utilized kernel PCA-based metric along with CSP to identify the most contributing zones in motor imagery classification [10]. Further, in this direction, the research work utilized the phase lock value (PLV) feature of various channel pairs for classifying motor imagery task [11]. Both these methods reduce the number of electrodes utilized for classification, thereby improving time complexity. However, these methods utilize only spatial features and ignore the spectral content of the data. The research work [12] uses sequential forward feature selection method to identify a subset of frequency bands from a composite filter bank of prior-known and variable-sized frequency bands. However, being a wrapper approach, it is computationally intensive.

In this work, we proposed a model for choosing a subset of frequency bands, which are relevant for the classification of motor tasks in MI-based BCI. The proposed model uses graph theory-based unsupervised filter feature selection method [2]. Experimental study is conducted on the publicly available BCI competition datasets to establish the efficacy of the proposed approach.

2 Proposed Approach: Graph Theory-Based Common Spatial Patterns (GBCSP)

Our proposed method consists of the following four steps as shown in Fig. 1:

(a) **Preprocessing**: Variable-sized frequency band filter bank is generated for a given frequency band (7–32 Hz) using the minimum bandwidth and frequency granularity [7]. Figure 2 shows variable-sized subband filters generation. At last, raw EEG data obtained from each of these subbands is filtered.

(b) **Feature Extraction**: We then obtain the desired features from the filtered data of each frequency band by applying CSP based on the following Raleigh criterion:

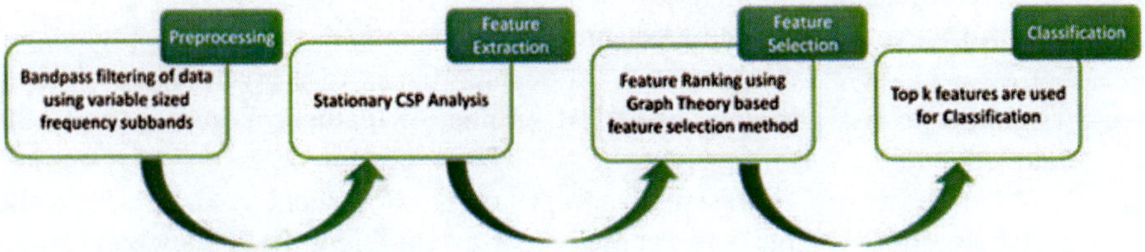

Fig. 1 Proposed Method: GBCSP

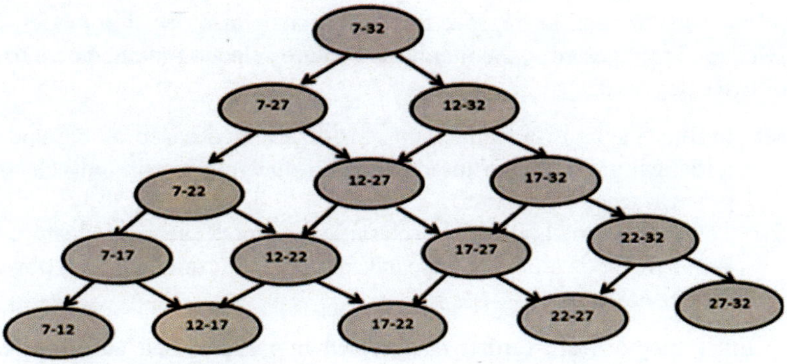

Fig. 2 Variable-sized subband filters

	f_1	f_2	f_3	...	f_M
x_1	x_{11}	x_{12}	x_{13}	...	x_{1M}
x_2	x_{21}	x_{22}	x_{23}	...	x_{2M}
x_3	x_{31}	x_{32}	x_{33}	...	x_{3M}
$\vdots$	$\vdots$	$\vdots$	$\vdots$	$\ddots$	$\vdots$
x_N	x_{N1}	x_{N2}	x_{N3}	...	x_{NM}

Fig. 3 Data matrix

$$R(w) = \frac{W^T \sum_1 W}{W^T \{\sum_1 + \sum_2\} W} \tag{1}$$

where $\sum_1, \sum_2$ denotes normalized average covariance matrices belonging, respectively, to class 1 and class 2 and W is a spatial filter matrix.

(c) **Feature Selection**: The obtained features are ranked using spectral graph theory-based feature selection method.

Graph theoretic Feature Selection:

Let $X = [x_1, x_2, \ldots, x_N]'$ is a $N \times M$ data matrix of a given subject having N number of trials or samples and M number of features. Figure 3 shows the matrix $\boldsymbol{X}$.

Where each row represents a sample $\boldsymbol{x_i} \in R^M$ where $i = 1, \ldots, N$ and each column represents feature column corresponding to frequency band j $f_j \in R^N$ where $j = 1 \ldots M$.

However, the number of features in this data matrix is still high and contains redundant frequency band features. In order to obtain a smaller set of most relevant feature and to improve the performance of a decision model, feature selection is suggested in the literature. Feature selection methods are basically of two categories:

(a) In filter methods, a feature subset is determined based on the underlying statistical properties of the data, and it does not require any classifier to perform feature selection.
(b) The wrapper methods employ classification model selecting feature subset. These methods are more expensive in terms of computational power than former ones.

Filter methods are further categorized into supervised and unsupervised approaches. In supervised method of filter feature selection, class label is associated in determining the relevance of each feature. In case of unsupervised feature selection method, inherent properties like scatterness and separability of data are used to determine the relevance of each feature. Motor imagery EEG data being highly subject specific in nature requires highly subject-specific features for classification.

Among the several unsupervised filter methods, graph theory-based spectral feature selection method [7] is widely used in different applications for reducing the high dimensionality of data. Graph-theoretic feature selection approaches are useful in capturing the inherent geometric structure of data by simultaneously considering both the local dependency and global scale of data. To the best of author's knowledge, graph theory-based feature selection algorithm is not utilized in the domain of motor imagery BCI for feature selection. Motivated by this, in this research work, unsupervised graph theory-based feature selection algorithm [7] is employed for the selection of desired reduced set of features.

In graph-theoretic feature selection method, associations among different samples of the data are defined in terms of the pairwise sample similarity matrix. Only those features that are consistent with the inherent structure of this matrix are considered relevant for classification. To estimate the sample similarity matrix, an undirected weighted graph $G(V, E)$ is defined (V and E denoting the set of vertices and the set of edges, respectively). Every vertex of this graph denotes a sample x of the data matrix. Between every two vertices (samples) $\boldsymbol{x}_i$ and $\boldsymbol{x}_j$, a weighted edge is computed by defining a sample similarity matrix K, where similarity between two vertices $\boldsymbol{x}_i$ and $\boldsymbol{x}_j$ is given by:

$$k_{i,j} = \mathrm{e}^{-\frac{\|x_i - x_j\|^2}{2\sigma^2}} \tag{2}$$

where σ is a tuning parameter for class separabilty. We can then define a degree matrix D as

$$D(i, j) = \begin{cases} d_i & \text{if } i = j \\ 0 & \text{otherwise} \end{cases} \tag{3}$$

where $d_i = \sum_{j=1}^{m} k_{i,j}$ and is considered as an approximation of the density around sample x_i. Similarly, we can also define a Laplacian matrix and normalized Laplacian matrix, respectively by

$$\begin{aligned} L &= D - K \\ \Lambda &= D^{-1/2} L D^{-1/2} \end{aligned} \tag{4}$$

Here, $\phi(f_i)$ that measures the smoothness of every feature f_i for Graph G is given by

$$\phi(f_i) = \frac{f_i^T \Lambda f_i}{1 - \left(f_i^T \xi_0\right)} \tag{5}$$

where spectral decomposition of Λ yields ξ_0(the eigenvector belonging to the least eigenvalue). The lesser value of $\phi(f_i)$ denotes more relevance of the feature f_i for classification. Finally, SVM is used as a classifier over the ranked features.

(d) **Classification**: High-ranked features are then used for learning a decision model.

3 Experimental Results

Two publicly available datasets are utilized for experimental studies:

- Dataset BCI Competition III 4a [13]
- Dataset BCI competition IV 1a [14].

For the Dataset 1, EEG signal matrix of size (118 × 200) per trial is created; here, rows denote channels and columns denote time points per trial. Similar to Dataset 1, data matrix of size 59 × 200 per trial for Dataset 2 was obtained as well. All the experiments have been performed on the variable-sized frequency bands as shown in Fig. 2. For the feature extraction step, CSP yields two features from each trial. SVM is used as a classification model with the parameters specified in Table 1.

Classification error is used as performance metrics. The performance of the proposed method is compared with CSP, SBCSP and FBCSP. Furthermore, we have also seen the performance of the proposed method with different-sized filter banks. For each subject, 10 trials of tenfold cross-validation are performed, and the average value of the classification error is being reported. Table 2 shows the obtained results of competing methods for every subject in Dataset 1. Figure 4 shows the reduction

Table 1 Parameter values for CSP and SVM

Parameters	Value
Number of spatial patterns in CSP r	1
Gaussian Kernel parameters σ	10
SVM regularization parameter C	100

Table 2 Classification error of different methods for Dataset 1

Subject	CSP	SBCSP	FBCSP	GBCSP
aa	0.23	0.19	0.19	0.12
al	0.10	0.01	0.01	0.005
av	0.33	0.28	0.26	0.16
aw	0.16	0.22	0.07	0.08
ay	0.15	0.35	0.37	0.15
Mean error	0.19	0.21	0.18	0.10

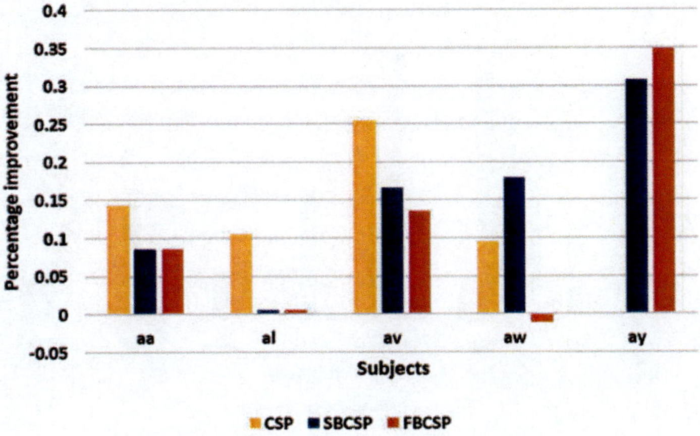

Fig. 4 Improvement in classification error for every subject in Dataset 1

in the classification error by using the proposed GBCSP when compared to the existing methods for every subject in Dataset 1. Similarly, Table 3 and Fig. 5 show the obtained results for all the subjects of Dataset 2. It can be observed from Fig. 5 and Table 3 that the classification error reduces with the uses of the proposed algorithm

We can derive the following conclusions from the obtained results:

- The proposed method GBCSP achieved the best performance.
- For all subjects of Dataset 1, GBCSP outperforms the standard CSP, SBCSP and FBCSP method by achieving minimum classification error (except subject aw).
- For Dataset 1, using the proposed method GBCSP, there is an overall reduction of 11.11, 13.92 and 9.75% in the average classification error as compared to CSP, SBCSP and FBCSP.
- GBCSP outperforms the standard CSP, SBCSP and FBCSP method by achieving minimum classification error for all subjects of Dataset 2.

Table 3 Classification error of different methods for Dataset 2

Subject	CSP	SBCSP	FBCSP	GBCSP
ds1a	0.27	0.30	0.28	0.23
ds1b	0.35	0.47	0.45	0.29
ds1c	0.30	0.41	0.25	0.20
ds1d	0.23	0.20	0.12	0.06
ds1e	0.17	0.16	0.19	0.11
ds1f	0.17	0.24	0.20	0.12
ds1g	0.21	0.11	0.09	0.04
Mean error	0.24	0.27	0.23	0.15

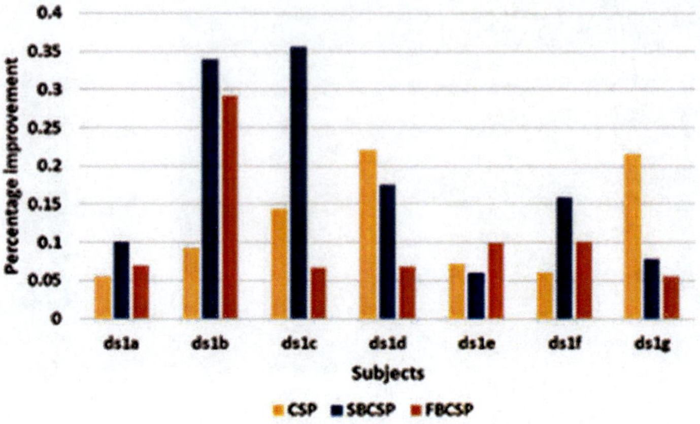

Fig. 5 Improvement in classification error for every subject in Dataset 2

- Using the proposed method GBCSP, there is an overall reduction of **11.84%**, **16.43%** and **10.38%** in the average classification error as compared to CSP, SBCSP and FBCSP for Dataset 2.
- Using the proposed method GBCSP, there is an overall reduction of **11.47%**, **15.18%** and **10.07%** in classification error as compared to CSP, SBCSP and FBCSP over all subjects of both datasets
- By applying feature selection techniques, there is a reduction in classification error.

In Fig. 6, the average classification accuracy of all 12 subjects of both Dataset 1 and Dataset 2 is plotted in the form of a box plot. From Fig. 6, it is evident that the median accuracy, upper quartile and lower quartile of GBCSP are higher as compared to CSP, SBCSP and FBCSP method which is a significant improvement in performance with the proposed method.

For calculating the statistical significant difference between different methods used in this work, Friedman statistical test [15] is conducted. The Friedman ranking is given in Table 4 for all the methods in ascending order. Table 4 shows that the proposed GBCSP method performed the best.

4 Conclusions

This paper aims to suggest an efficient and robust feature selection method which can select a reduced subset of relevant frequency bands from a filter bank of variable-sized frequency subbands. The proposed model GBCSP is a four-phase approach. Band-pass filtering of data from variable-sized filter bank is conducted in the first phase.CSP is used for feature extraction from each of the frequency bands in the second phase. Features from each frequency bands are then ranked using spectral graph

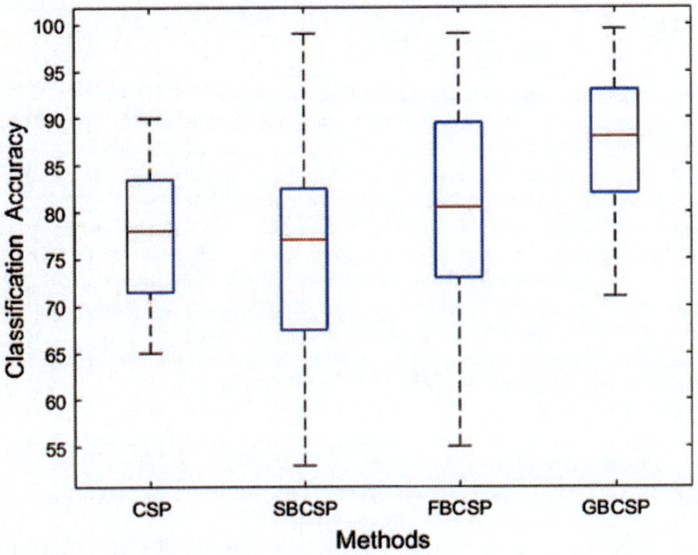

Fig. 6 Boxplot of various methods overall subjects

Table 4 Friedman statistical ranking of different algorithms

Algorithm	Rank
GBCSP	1.125
FBCSP	2.583
CSP	3.0417
SBCSP	3.25
p-value	1.8452E−4

ranking method. Top ranked features are then used to build the decision model. Two publicly available datasets are utilized for conducting empirical study and comparative analysis. The proposed model outperforms the existing methods (CSP, SBCSP and FBCSP) as depicted by the results obtained from both the datasets. Our model also achieved the best ranking in the Friedman statistical test as well.

The proposed work performs efficiently for binary classification. In the future, we will extend the proposed work for multiclass classification. The current work is limited to only spectral content and does not utilize temporal and spatial information. We will develop algorithms that can utilize spectral, temporal and spatial content simultaneously to determine optimal set of features for classification.

References

1. J.R. Wolpaw, N. Birbaumer, D.J. McFarland, G. Pfurtscheller, T.M. Vaughan, Brain–computer interfaces for communication and control. Clin. Neurophysiol. **113**(6), 767–791 (2002)
2. B. Rana, A. Juneja, M. Saxena, S. Gudwani, S.S. Kumaran, M. Behari, R.K. Agrawal, Graph-theory-based spectral feature selection for computer aided diagnosis of Parkinson's disease using T1-weighted MRI. Int. J. Imaging Syst. Technol. **25**(3), 245–255 (2015)
3. B. Blankertz, R. Tomioka, S. Lemm, M. Kawanabe, K.R. Muller, Optimizing spatial filters for robust EEG single-trial analysis. IEEE Sig. Process. Mag. **25**(1), 41–56 (2008)
4. Q. Novi, C. Guan, T.H. Dat, P. Xue, Sub-band common spatial pattern (SBCSP) for brain-computer interface, in *3rd International IEEE/EMBS Conference on Neural Engineering, CNE'07* (2007), pp. 204–207
5. K.K. Ang, Z.Y. Chin, H. Zhang, C. Guan, Filter bank common spatial pattern (FBCSP) in brain-computer interface, in *Proceedings of The 2008 International Joint Conference on Neural Networks (IJCNN)* (2008), pp. 2390–2397
6. J.R. Wolpaw, E.W. Wolpaw, *Brain-Computer Interfaces: Principles and Practice* (Oxford University Press, Oxford, 2012)
7. J.S. Kirar, R.K. Agrawal, Optimal spatio-spectral variable size subbands filter for motor imagery brain computer interface. Procedia Comput. Sci. **84**, 14–21 (2016)
8. J.S. Kirar, R.K. Agrawal, Relevant feature selection from a combination of spectral-temporal and spatial features for classification of motor imagery EEG. J. Med. Syst. **42**(5), 78 (2018)
9. L.F. Velasquez-Martinez, D. Luna-Naranjo, D. Cárdenas-Peña, C.D. Acosta-Medina, G.A. Castaño, G. Castellanos-Dominguez, Relevance of common spatial patterns ranked by kernel PCA in motor imagery classification, in *International Conference on Brain Informatics*(Springer, Cham, 2019), pp. 13–20
10. V.K. Benzy, A.P. Vinod, Classification of motor imagery hand movement directions from EEG extracted phase locking value features for brain computer interfaces, in *TENCON 2019-2019 IEEE Region 10 Conference (TENCON)* (IEEE, 2019), pp. 2315–2319
11. A. Al Shiam, M.R. Islam, T. Tanaka, M.K.I. Molla, Electroencephalography based motor imagery classification using unsupervised feature selection, in *2019 International Conference on Cyberworlds (CW)* (IEEE, 2019), pp. 239–246
12. J.S. Kirar, R.K. Agrawal, Relevant frequency band selection using sequential forward feature selection for motor imagery brain computer interfaces, in *2018 IEEE Symposium Series on Computational Intelligence (SSCI)* (IEEE, 2018), pp. 52–59
13. G. Dornhege, B. Blankertz, G. Curio, K.-R. Muller, Boosting bit rates in noninvasive EEG single-trial classifications by feature combination and multiclass paradigms. IEEE Trans. Biomed. Eng. **51**(6), 993–1002 (2004)
14. B. Blankertz, G. Dornhege, M. Krauledat, K.-R. Müller, G. Curio, The non-invasive Berlin brain–computer interface: fast acquisition of effective performance in untrained subjects. NeuroImage **37**(2), 539–550 (2007)
15. R.L. Iman, J.M. Davenport, Approximations of the critical region of the fbietkan statistic. Commun. Stat. Theory Methods **9**(6), 571–595 (1980)

Discovering Mutated Motifs in DNA Sequences: A Comparative Analysis

Rajat Parashar, Mansi Goel, Nikitasha Sharma, Abhinav Jain, Adwitiya Sinha, and Prantik Biswas

Abstract Locating motifs in DNA sequences is a traditional conjunctional and challenging problem in the discipline of bioinformatics. Motif refers to the biologically functional short, recurring common sequence pattern in DNA strands involved in important processes taking place at the genetic level in an organism. Motifs are the small set of immunity genes present in the DNA sequences as a binding site and turn on whenever the organism gets infected. Motif discovery is integral to problems such as antibody biomarker identification and transcription factor binding sites (TBFS) in the field of genetics and holds greater importance to enable advances in understanding human genetics, social intelligence, biology and health. This problem has been broadly studied over the years, but the complexity and optimality of most of the existing algorithms are still very high. Recently, this field of bioinformatics has grown significantly, and many algorithms have been proposed to solve this problem. However, high complexity is the most challenging aspect of this problem which still grabs the attention of many researchers. This paper analyzes the several algorithms mainly on the basis of their computing models and optimality of the solution and presents the critical reviews of the methods adopted for finding the planted motif.

R. Parashar (✉) · M. Goel · N. Sharma · A. Jain · A. Sinha · P. Biswas
Department of Computer Science Engineering, Jaypee Institute of Information Technology, Noida, India
e-mail: rajatparashar2910@gmail.com

M. Goel
e-mail: goel.mansi23@gmail.com

N. Sharma
e-mail: nikkisharma827@gmail.com

A. Jain
e-mail: iamabjain@gmail.com

A. Sinha
e-mail: adwitiya.sinha@jiit.ac.in

P. Biswas
e-mail: prantik.biswas@jiit.ac.in

P. Bansal et al. (eds.), *Proceedings of International Conference on Artificial Intelligence and Applications*, Advances in Intelligent Systems and Computing 1164,
https://doi.org/10.1007/978-981-15-4992-2_25

These algorithms can also be used for string matching, data mining and pattern detection, etc.

Keywords MapReduce · Transcription factors · Genetic algorithm · DNA sequences · Clustering · Sustainability · Motif finding problem · Mutation · Heuristic · Clique · Social intelligence

1 Introduction

A sequence motif can be described as an amino acid sequence or nucleotide pattern that is ubiquitous and has a biological essence. The major goal here is to recognize the binding sites prevailing in deoxyribonucleic acid (DNA) for transcription factors [1]. DNA sequences contain subsequences that are nearly comparable to each other containing only a handful of mismatches. The repeated occurring patterns in those sequences are called motifs [2]. We have to analyze the frequencies of patterns in nucleotide sequences. We do not have the complete dictionary of the motif, i.e., the starting position of the motif is unknown, while motif length is given provided motif is mutated. Motif discovery will help to detect the complex disease easily; it will be a great help in social intelligence and sustainability. The main task is to find out length, starting index of the mutated motif. Binding sites can help us in finding mutations in the DNA dataset. Once we know the sequence pattern of the motif, then we can apply many search algorithms like anchor-based algorithms [2], clustering, genetic algorithm-based [3, 4] and probabilistic-based algorithms [5]. Recent advancement in genome sequence availability has led to the expansion of computational techniques for motif finding. As an outcome, many unique algorithms for finding motif have been proposed in a large amount over the past decade. Researchers proposed numerous algorithms to discover motifs in promoter progression from homologous or correlated genes, either mainly on position weight matrices in ChIP-seq dataset or consensus sequences [6].

2 Problem Definition

A list of DNA sequences says '*n*' is given of length '*m*'. The purpose of the mentioned motif finding problem for the mutated version is to find the implanted sequence of length '*l*' having exactly '*d*' mutations in each line of every DNA sequence. Figure 1 shows the implanted mutated motif pattern in each line of DNA sequence exactly once and represents a set of five DNA sequences having length 11. This problem can be broadened to many other adaptations such as gapped motif finding, mutable motif finding, and many others. Here, in Fig. 1, we have $d = 1$ and $l = 4$ and the implanted mutated motif is CTCA. It can be seen that the motif does not have to be available in its precise form or pattern in any of the considered DNA sequences. In

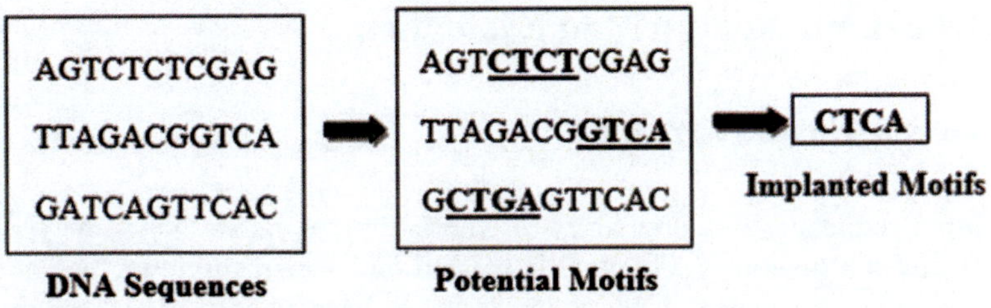

Fig. 1 Example of motif problem where $n = 3$, $l = 4$ and $d = 1$

the generic form of the planted motif finding (PMP), motif '*M*' exists in at least q out of the '*n*' sequences, and it can be referred to as qPMP 'Quorum Planted Motif Search Problem'. It is an NP-complete problem.

3 Literature Survey

The motif finding problem entails an in-depth hunt of the motif in DNA sequences and hence is a computationally hard problem. The main challenge here is that there is no preliminary knowledge of the specific pattern of the motif and also due to mutations in the motif has become a very challenging task. Since it is not known how the biological elements function in nature, researchers have been working to propose algorithms to work in this area and simulate motif finding efficiently. The complexities of all the algorithms proposed are exponential [2]. Most approaches suggest using advanced data structures and different programming paradigm, while others suggest different heuristics and probabilistic approaches to give approximate results. Motif finding problem can also be solved as a word-based problem using position weight matrices [7]. For providing more accurate and faster results, some existing algorithms have also been modified to find the implanted motif [5]. Efficient machine learning techniques can also be used [8]. An excellent review of well-known algorithms has been provided by Dai and Das et al. [9] that can be immensely used to resolve the motif finding problem. Researchers have tried to solve an easier version of the motif finding, i.e., non-mutated variant of the problem, while the harder variant, i.e., the mutated motif finding problem is a bigger challenge for all.

4 Algorithmic Analysis and Evaluation

4.1 Graph-Based Algorithm

Let n input sequences be denoted by Si where $1 \leq i \leq n$. Say Sij is the substring of length 1 from position j. Graph G is created on l-mers as follows. The nodes in the graph will be sequences. Two nodes Si and Sj have an edge only if the distance ('hamming distance') between the two nodes is less than $2d$. The graph 'G' defined would be a multipartite graph having n partitions, each of which would contain nodes matching to inimitable l-mers from Si. It is not hard to observe that if there is a motif M implanted in all the n sequences, then the nodes in G would form a clique. It is the heart of this approach. Note that there may occur cliques having size 'n' in the graph 'G', lacking an implanted motif. So, that would give a false-positive result [7]. This situation may arise when a probable motif having a distance greater than d from respective 2 l-mers, but these 2 l-mers have a distance $\leq$2d and so share an edge. The problem is that finding cliques is quite a hard problem too, though knowing the maximal clique significantly prunes the search space. Now, all the cliques discovered before are checked for valid motif discovery. It works on the observation that if there exists a motif, it would have be 'd' distance away from any node in that clique. Hence, a search tree of all sequences at distance d from T is created. Every traversal from root to leaf maps to a possible motif. Each of these possible motifs is then checked for validity. For the traversal of the search tree, the tree is categorized into layers. The tree is created ensuring that the count of unique subtrees is few in each layer. A DFS traversal is used to traverse the tree in a bound and branch paradigm. To find cliques of size N, firstly, the cliques of size $N - 1$ are discovered, which are recursively discovered by smaller cliques' sizes. The base case for recursion is that every node in 'G' forms 1-cliques. The $(i + 1)$-cliques are formed using i-cliques in ith iteration. Note that to just store the cliques would require processing capability. Thus, at every stage, all the redundant and useless i cliques are eliminated. The concept of bit masking is applied to store all the i cliques which are formed using the i nodes. It is stored as a binary number whose ith bit is set if the clique has ith node, else it is 0. The binary number formed is then converted to decimal number and that clique is stored in a bucket with that index. Every individual bucket is granted an exclusive bucket number, calculated as defined before. To prune the search space, it is workable to identify nodes that cease to exist in 'n-clique', and therefore, it would be removed from upcoming considerations.

Algorithm 4.1: To find 'm'-cliques in graph 'G'.

```
Graph_Based_Algorithm (g, l, d);
form_clique()
        Form cliques of size I+1 by adding S to all cliques that S is a
        neighbour of.
        Compute bucket indx for new cliques by setting the jth least
        significant bit to
        return

delete_clique()
        Eliminate S and all cliques which contain S from the search space.
        return

 for I in range (1 to m-1) do
          for each node v belongs to Pj ∀ k: 1<=k<=m do
                 if S is adjacent to all the buckets in
                 this iteration then
                        form_clique()
                 end
                 else
                        delete_clique()
                 end
                 if bucket == NULL then
                             Stop and declare no cliques found
                 end
 end
 return all available cliques from the nth bucket
```

4.2 Anchor-Based Clustering Algorithm

The algorithm has the following features: (1) They initiate a modern clustering technique for finding the motif in DNA sequences quickly by pre-treating sequences. They break whole DNA sequences into several trivial clusters, and then on each existing cluster, various algorithms are applied to get the preferred motif. They said that motifs can be lost, so randomly sampling cannot be done. Another method is to cluster sequences on the basis of their comprehensive resemblance, but this was not beneficial as motifs can be seen only in subregion of sequences, for example, edit distance as a resemblance in K-means. (2) The first anchor (gapped g-grams)-based clustering algorithm is initiated and termed as ASC [2]. (3) The present theoretical visualization for an anchor-based resemblance measures and demonstrates that no two sequences should hold the identical motif with the same precision. They initiate a link between anchor-based resemblance and DNA sequences containing preferred motif to perform anchor-based clustering. Two random DNA sequences may have some resemblance anchors. The chance for any arbitrary sequence of length k to contain N g-anchors is very less when q-anchors are selected. As a cluster center, any randomly chosen sequence is picked, and then, all other sequences are compared for similarity. If the similarity is above a threshold, then it is captured and included in this cluster. This process is repeated until every sequence is a part of a cluster. Though this process has some limitations. The worst-case time complexity of this approach is $O(N * N)$ which occurs when only those sequences are chosen that do not have any motif as cluster centers [2]. Also, many small redundant clusters may also be formed that do not contain any motif, but still, these clusters have to be traversed as well to complete the algorithm's process. ASC can detect motifs in protein sequences faster than any existing algorithm. The algorithm which works well usually does not go in the worst case.

```
Algorithm 4.2: Sequence Clustering based on Anchor
Input: A set of sequences S, the number of anchors d in a cluster center,
       the number of clusters K.
Output: K clusters of sequences
     Initialize K anchor sets to empty.
     for all Si € S do
        Categorize Si in a set of anchors
     end
     Compute the fitness score of all the anchors.
     Consider d*k anchors having maximum scores and divide
     them into K anchor sets in any arbitrary order; each
     has d anchors.
     repeat
             Map each sequence to a cluster
             Recompute each cluster's centre
     until termination condition
     return the clusters containing sequences.
```

4.3 Speed-up Technique

In this method, the two exact algorithms PMS0 and PMS1 have been referred. It creates all the possible l-mers from the set of sequences. The set of l-mers having a distance of less than or equal to d from u is called as 'd-neighborhood of u'. Now all the l-mers are generated which are in the d-neighborhood. While to check the validation of candidate motifs (*l*, *d*), it takes $O(m * n * l)$ complexity. Thus the running time of PMS0 is $O(m^2nl * (l\ C\ d) * 3^{\wedge}d)$ [10]. There is speed-up part which is mentioned in the paper. The Algorithm PMS0 is run not for all the *n* input sequences but for randomly chosen *K* sequences. The value K chose empirically. This way, the candidate motifs are generated for those *K* sequences. Now, each of these candidate motifs is iterated and checked for presence in all the remaining (*N-K*) Sequences as well. The speed-up part is that since the candidate motifs are generated only on K sequences. The count of all candidate motifs is not that much large; hence the process speeds up. If the value of d is larger, then the count of all candidate motifs that arise by random chance may be huge. A probabilistic analysis of this effect can be conducted which shows that the speed-up part of this algorithm works as it prunes down the search space by a large factor.

```
Algorithm 4.3: Planted Motif Search

Run the algorithm PMS0 on n input sequence. Pick randomly k sequences, and
so on. Let S be the set of (l, d) motifs found in these k sequences set.
       for all l-mer in V: S do
       verify
       if V is true
               motif in [O(m*n*l)] time complexity
        Output is v.
```

4.4 Heuristic Algorithm Using Topic Models

It is a type of heuristic algorithm and follows stochastic behavior, which means it gives an approximate solution not the optimal one, so when it runs each time with the same set of sequences, it will probably give different outputs [5]. It works only for ungapped motifs or says continue substring, so the patterns that have slits can be divided into distinct motifs. Additionally, in disparity with further different algorithms, which generally take the assumption that in the dataset minimum an occurrence of the motifs is present. The algorithm is executed as an exemplary genetic algorithm. In the first step, it starts generating a population of probable solutions; then, it iterates one by one above them and stores the best possible solution of each iteration, deleting the worst ones and again forming the new possible solutions on the basis of previously stored solutions in the iterations, till the optimal solution not obtained or maximum iterations will not be reached. Thus, the one and only criteria that want to be determined that how the generated population is signified, how the evaluation will be done, how in each iteration, we can select the individuals with the best fit, how the new entities are being formed from the existing ones and lastly when will the algorithm terminates to provide the concluding solution. Every individual that belongs to the population is a set of '*m*' l-mers. The l-mer has a fixed length. The initialization step is as follows: Given a set of 'n' DNA sequences, a l-mer, $C_{\min}$ stands for the minimum count of reappearances for each l-mer, and a count of words for each individual in set of n. The iteration runs till the initial set of '*m*' l-mers would not reach.

Algorithm 4.4: Heuristic Algorithm using Topic Models

- i. Select any arbitrary length l in the range of (lmin, lmax).
- ii. Select an arbitrary Sequence S from the Input Set.
- iii. Select any position p in that sequence S between 0 and l-k.
- iv. Select the substring w of length l starting from the position p.
- v. Compute the frequency of the substring w in the Sequence S, ensuring that max 25% of ill matches are permitted.
- vi. Rearrange w into w_s and compute the frequency of c' of the word w_s in the given sequence, permitting a max of 25% of mismatches.
- vii. If (c-c') >= $c_{\min}$, insert the l-mer to the set for the given individual.

4.5 Using Genetic Algorithm

The authors [3, 4] here suggested a very elegant approach for solving the mutated motif finding problem. They have used a genetic algorithm-based approach. Initially, they take all the possible 3 length l-mers which can be at most 64 since the sequences are composed of only four alphabets—A, C, G, T. Now, for all these 64 initialized l-mers, the fitness score is computed. A 3-length sequence is given the score of the number of matches with the N sequences. Then, the 3-length sequence with the highest fitness will be considered, and with it, 4-mers are generated and so on. Then, the length of individuals is increased by 1 after each iteration until it reaches a standard maximum length. The operators used in this process are mutation, addition and deletion of nucleotide. Mutation: In order to keep the variety of the individual's candidate motifs, the mutation is done ensuring that the initial 3 length sequence is not altered during the whole computation. So, the algorithm always gives the global optimum given that the mutation probability value is chosen optimally. Addition: To generate a motif of length $l + 1$ from a l-mer, a new nucleotide is added to the front as well as to the backside of the existing l-mer. So, at each stage, 2 new $(l + 1)$ mers are generated. Out of both, the one with the maximum fitness score is considered. Deletion: To generate a l-mer, the reverse of addition is performed, and the last nucleotide added is deleted.

```
Algorithm 4.5: Genetic Algorithm Based Technique
Input: {A_1,A_2,···,A_m}, maxLength, maxiloop, pm
Output: Best Motif, Best Score Initialization:{B_1^3,B_2^3,...B_64^3}, compute the
fitness score {TFS_1,TFS_2 ,..,TFS_64 }, Len=3
   while Len <= maxLength
       for I in range (1 to 64) do
             get B_i1^{Len+1} by performing Addition on B_Len^3
             calculate the TFS_i1^{Len+1} from B_i1^{Len+1}
         for jloop in range (1 to maxiloop) do
                 get B_i1^{Len} by performing Deletion onB_i1^{Len+1}
            if Len > 3 then
                 get B_i2^{Len} by performing Mutation on B_i1^{Len}
                 get B_i2^{Len+1} by performing Addition on B_i2^{Len}
            end
            else
                 get B_i2^{Len+1} by performing Addition on B_i1^{Len}
            end
         end
             calculate the TFS_i2^{Len+1} from motif B_i2^{Len+1}
             if TFS_i1^{Len+1} < TFS_i2^{Len+1} then
                 TFS_i1^{Len+1} ← TFS_i2^{Len+1}
                  B_i1^{Len+1} ← B_i2^{Len+1}
             end
       end
          TFS_i^{Len+1} ← TFS_i1^{Len+1}
          B_i^{Len+1} ← B_i1^{Len+1}
     end
     Let Len = Len + 1
 end
```

4.6 MCES Algorithm for Large Dataset

The algorithm progresses through the following three stages:

i. *Mining Parameter*: In the mining step, there are mainly two parameters; the first one is the minimum evolution rate, and the second one is threshold frequency. To decrease the anomalies, the minimum evolution rate is set to 2.
ii. *MapReduce Strategy*: When the input dataset is very large, then the storage area will overreach the single machine's storage capacity. To deal with this issue, the authors proposed a method based on the MapReduce technique which uses the 'emerging substrings mining' method.
iii. *Combining Method*: This step comprises two parts: 'clustering emerging substrings' (CESs) and 'clustering position weight matrix'(CPWMs). To ensure efficient performance, the evolving substrings having similar length are clustered iteratively. If any obtained clusters contain ≤ 5 substrings, then that cluster is termed as an invalid cluster and is discarded. All the substrings residing in the cluster are aligned by forming a spanning tree with maximum weight [10]. The position weight matrix is created, and then according to that matrix, useless redundant substrings are eliminated. The final motifs are obtained by passing the high information sections of collective PWMs.

```
Algorithm 4.6: Mining and Combining Emerging Substrings (MCES)

Input: a data P for testing and a control data Q of DNA patterns
Output: the set of the required motifs K
Initialize: set the parameters for mining step
        if |P|+ |Q| ≤ z' then
                Apply mining technique in a single machine
        else
                Apply mining technique separately using MapReduce
        Perform the step to combine and get K
        if k = Φ then
                p(th) ← p(th) / 2
        Conduct mining and combining step again
        end
 return K
```

5 Comparative Analysis

The graph-based algorithm proposed by Indranil Roy and Srinivas Aluru [7] is an exact algorithm. Out of all the instances in the published literature, the two most challenging instances (24, 10) and (25, 11) are solved by it in the same 46 h. The idea is here to prune or minimize the search space. The maximum possible number of nodes that can be present in graph g is (20 * 600), assuming that no repetition of l-mers occurs in the sequence. The complexity of each node is $(2d + 1) * l - 2d\hat{}2$. The author makes an unrealistic assumption that all nodes sustain in all the repetitions. For each iteration, the parameters like the bucket size and the cliques' size are fixed. Here is Table 1 containing some execution results of the approach mentioned in the paper.

The anchor-based similarity clustering [2] reduces the execution time as compared to others. It is a faster motif finder than other approximate algorithms. It can detect motifs in protein sequences faster than any other existing algorithms. The only limitation is that this algorithm works great but for short sequences. The results given by this algorithm are purely random as it is based on clustering. So, it may work faster for some sequences, while it may be slow for other sequences of the same length. But, it works fast for short sequences for sure. We also analyzed this algorithm on the basis of effectiveness and efficiency. (1) Effectiveness: If the two patterns comprise the same motif but they are stored in distinct clusters by the algorithm, then that duo of sequences is called a misplaced duo. (2) Efficiency: Tests were conducted to compute the effectiveness of ASC. The results were that the ASC worked better for short sequences as compared to the larger sequences because the clustering is taking a larger time to terminate. Speed-up technique [10] is based on two exact algorithms PMS0 and PMS1. The author reviewed these two algorithms and then added a speed-up technique of his own which further reduces the running time by pruning the search space. So, the number of candidate motifs generated by this algorithm is reduced which speeds up the overall process. In the speed-up process (PMS4), the author suggests that the algorithms PMS0 and PMS1 are applied only for some sequences and not for all the sequences. Then, the candidate motifs found in that are then checked for the presence in all the remaining sequences. Although this checking part is a costly operation, it is fast because checking has to be done only for the candidate motifs which are already small in number due to the speed-up process. Joseph Basha and Kenta Nakai modified genetic algorithm to propose the solution [5]. The improvement is done by using the perplexity measurement. The algorithm has some steps. The first step is to create the candidate instances with GA. Then following this step, the instances are clustered using the hamming distances. The main limitation of this approach is that it results in a huge number of false positives. The reason behind it is that the results of the confidence could not be measured. So, a new measure of perplexity had to be added, and the CTM was built on the basis of perplexity measure. The last step was to include only those motifs having perplexity less than 100. The conclusion is that no statistics analyzed should be taken as a fixed estimate of the methods. Since we still do not know how the subjacent biology works. A criterion to sort out the high confidence outcomes is required. The approach suggested by Fan and Wu [4] was tested on three types of datasets, the simulated data, biological data and data CRP. The algorithm works like a parallel

Table 1 Running time of Algorithm 4.1

(l, d)	Nodes count per iteration	Time to find clique (min)
(24, 10)	5736	12.267
(24, 10)	5241	3.958
(24, 10)	4538	17.337

random search. Since this is implemented using the parallel computing paradigm, its computational efficiency is increased dramatically. On comparing the data CRP results with other famous approaches like GARPS, GA and Gibbs sampler motif, this algorithm had the highest score. The algorithm is effective as it avoids getting stuck in the local optima. The mutation part of the GA is made ensuring that the initial three nucleotides remain unaltered during the whole process. This keeps the variety of the individuals and so makes it more effective than others. A probabilistic algorithm that uses concepts of data mining which includes MapReduce and clustering techniques has also been proposed [3]. It does not give an optimal answer. It uses the concept of the threshold value which is only an assumption and does not provide stability to the algorithm. Its complexity is very high, and running time is also very high for the test cases presented in the paper as compared to different other algorithms. In the worst case, it may take infinite time because of the self-loop.

After analysis and discussion on the above algorithm, we find that if we apply the combination of the graph-based algorithm with an anchor clustering algorithm, the results are better and are accurate.

6 Conclusion

The algorithms mentioned in this paper are efficient as provided their results, and they work perfectly fine for both long and short length of DNA sequences as specified. Although every algorithm has its strong and weak points. In the field of pattern and string recognition and data mining, these algorithms will provide to generate good results. Being a problem of biological importance, motif finding in DNA sequences remains scope for further future work as the existing algorithms work fine, but the execution time needs to be improved and the accuracy of the solution needs to be optimized. The work till date mostly uses a probabilistic approach, so in the future with the help of better computing models, more accurate results can be achieved.

References

1. S. Prado, Motif discovery using optimized suffix tries. Master thesis, Faculty of Engineering and Architecture, Department of Information Technology—IBCN, Ghent University, Ghent, Belgium (2012)
2. H. Liu, F. Han, H. Zhou, X. Yan, K.S. Kosik, Fast motif discovery in short sequences, in *2016 IEEE 32nd International Conference on Data Engineering (ICDE)*, Helsinki (2016), pp. 1158–1169. https://doi.org/10.1109/icde.2016.7498321
3. Q. Yu, H. Huo, X. Chen, H. Guo, An efficient motif finding algorithm for large DNA data sets, in *2014 IEEE International Conference on Bioinformatics and Biomedicine* 2014
4. Y. Fana, W. Wua, R. Liua, W. Yangb, An iterative algorithm for motif discovery, in *17th Asia Pacific Symposium on Intelligent and Evolutionary Systems, IES2013*
5. J.B. Gutierrez, K. Nakai, A study on the application of topic models to motif-finding algorithms, in *From 15th International Conference on Bioinformatics*, Queenstown, Singapore (2016)

6. Y. Zhang, P. Wang, A fast cluster motif finding algorithm for ChIP-Seq data sets. Biomed. Res. Int. **2015**, 1–10 (2015). https://doi.org/10.1155/2015/218068
7. I. Roy, S. Aluru, Discovering motifs in biological sequence using micron automata processor, in *Conference Paper in IEEE/ACM Transactions on Computational Biology and Bioinformatics*, Jan 2016
8. M.K. Das, H.K. Dai, A survey of DNA motif finding algorithms, in *Proceedings of the Fourth Annual MCBIOS Conference. Computational Frontiers in Biomedicine*, Nov 2007
9. S. Prakash, H. Agarwal, U. Agarwal, P. Biswas, S. Dawn, Discovering motifs in DNA sequences: a suffix tree based approach, in *8th International Advance Computing Conference (IACC 2018)*
10. S. Rajasekaran, H. Dinh, A speed-up technique for (l, d)- motif-finding algorithms. BMC Res. Notes **4**, 54 (2011)
11. H.C.M. Leung, F.Y.L. Chin, An efficient motif discovery algorithm with unknown motif length and a number of binding sites. Int. J. Data Min. Bioinform. **1**(2), 201 (2006). https://doi.org/10.1504/ijdmb.2006.0108562006
12. A. Majumdar, Finding DNA motifs: a probabilistic suffix tree approach. Ph.D. dissertation, Computer Science and Engineering, Graduate College, University of Nebraska, Lincoln, 2016. Accessed 18 Apr 2018
13. H.R. Hassanzadeh, P. Kolhe, C.L. Isbell, M.D. Wang, MotifMark: finding regulatory motifs in DNA sequences, in *2017 39th Annual International Conference of the IEEE Engineering in Medicine and Biology Society (EMBC)*, Seogwipo (2017), pp. 3890–3893. https://doi.org/10.1109/embc.2017.8037706
14. J.S. Fink, M. Verhave, S. Kasper, T. Tsukada, G. Mandel, R.H. Goodman, The CGTCA sequence motif is essential for biological activity of the vasoactive intestinal peptide gene AMP-regulated enhancer. Proc. Natl. Acad. Sci. U.S.A. **85**(18), 6662–6666 (1988)
15. A. Jain, R. Parashar, A. Goyal, P. Biswas, S. Dawn, A. Nanda, Discovering motifs in DNa sequences: a candidate motifs based approach, in IEEE *5th International Conference on Parallel, Distributed and Grid Computing*, Solan, India (2018)
16. U. Draisbach, F. Naumann, A generalization of blocking and windowing algorithms for duplicate detection, in *2011 International Conference on Data and Knowledge Engineering (ICDKE)* (IEEE, 2011), pp. 18–24
17. P. Pevzner, N.C. Jones, *An Introduction to Bioinformatics Algorithms* (The MIT Press, Cambridge, MA, 2004)
18. F. Zambelli, G. Pesole, G. Pavesi, Motif discovery and transcription factor binding sites before and after the next-generation sequencing era. Brief. Bioinform. **14**(2), 225–237 (2013)
19. C. Angkawidjaja, A. Paul, Y. Koga, K. Takano, S. Kanaya, Importance of a repetitive nine-residue sequence motif for intracellular stability and functional structure of a family I.3 lipase. FEBS Lett. **579**(21), 4707–4712 (2005)
20. J.T. Ballew, J.A. Murray, P. Collin, M. Mäki, M.F. Kagnoff, K. Kaukinen, P.S. Daugherty, Antibody biomarker discovery through in vitro directed evolution of consensus recognition epitopes. Proc. Natl. Acad. Sci. **110**(48), 19330–19335 (2013)
21. J. Davila, S. Balla, S. Rajasekaran, Fast and practical algorithms for planted (l, d) motif search. IEEE/ACM Trans. Comput. Biol. Bioinform. **4**, 544–552 (2007)
22. T.L. Bailey, C. Elkan, Fitting a mixture model by expectation maximization to discover motifs in bipolymers, in *Proceedings of 2nd International Conference on Intelligent System for Molecular Biology*, Menlo Park, California (1994), pp. 28–36
23. C.E. Lawrence, S.F. Altschul, M.S. Boguski, J.S. Liu, A.F. Neuwald, J.C. Wootton, Detecting subtle sequence signals: A Gibbs sampling strategy for multiple alignment. Sc. **262**, 208–214 (1993)
24. U. Keich, P.A. Pevzner, Finding motifs in the twilight zone, in *Proceedings of 6th Annual International Conference on Computational Biology* (2002), pp. 195–204
25. A. Price, S. Ramabhadran, P.A. Pevzner, Finding subtle motifs by branching from sample strings. Bioinformatics **19**(suppl 2), ii149–ii155 (2003)

26. A.M. Carvalho, A.T. Freitas, A.L. Oliveira, M.-F. Sagot, Highly scalable algorithm for the extraction of cis-regulatory regions, in *Proceedings of Asia-Pacific Bioinformatics Conference* (2005), pp. 273–283
27. P.A. Evans, A.D. Smith, Toward optimal motif enumeration, in *Algorithms anRd Data Structures* (Springer, New York, NY, USA, 2003), pp. 47–58
28. L. Marsan, M.F. Sagot, Extracting structured motifs using a suffix tree-Algorithms and application to promoter consensus identification, *in Proceedings of 4th Annual International Conference on Computational Molecular Biology* (2000), pp. 210–219
29. M. Nicolae, S. Rajasekaran, Efficient sequential and parallel algorithms for planted motif search (2013). arXiv preprint arXiv:1307.0571
30. N. Sharma, M. Goel, A. Verma, A. Jain, R. Parashar, P. Biswas, Discovering non-mutated motifs in DNA sequences: a HashMap based binary search motif finder, *4th International Conference on Computing Communication and Automation 2018*, Galgotias University, India, 14–15 Dec 2018

Classification of S&P 500 Stocks Based on Correlating Market Trends

Minakshi Tomer, Vaibhav Anand, Raghav Shandilya, and Shubham Tiwari

Abstract Stock market is a correlated entanglement of companies whose prices depend upon various factors including how well each individual company is doing, their stock's performance and the general trend of the stock market. A simple linear prediction based upon the past performances of the stock is not enough to declare whether a particular stock will perform in a linear way. To predict the behavior of stocks, it is required to look at all of its correlated counterparts, compile their performance numbers and take a logical guess on whether the stock's performance will be on the positive side or on the negative side. For that measure, data of all the stocks from Standard and Poor's 500 Ratings (S&P 500) was collected. The novel idea in this paper is to classify a stock's performance based upon its correlated features and predict that whether it should be bought, sold, or held.

Keywords Supervised classification · Support vector machines · Random forest classifiers · Stock market · Correlation

1 Introduction

The stock market is volatile. On one side, it will remain stable and perform linearly while on the other it would abruptly rise or crash all contingent on some vital bit of information or news being released [1]. An example of this trend was displayed

---

M. Tomer · V. Anand (✉) · R. Shandilya · S. Tiwari
Department of Information Technology, Maharaja Surajmal Institute of Technology, C-4 JanakPuri, New Delhi 110058, India
e-mail: supercd93@gmail.com

M. Tomer
e-mail: tomer.minakshi@gmail.com

R. Shandilya
e-mail: raghavshandilya18@gmail.com

S. Tiwari
e-mail: iamshubhamfc@gmail.com

P. Bansal et al. (eds.), *Proceedings of International Conference on Artificial Intelligence and Applications*, Advances in Intelligent Systems and Computing 1164,
https://doi.org/10.1007/978-981-15-4992-2_26

during the trade war situation between USA and China, another trend that is prevalent in-stock trading is the first-fall trend which happens when a specific report is released due to which the prices of a leading company's stock are affected. Now witnessing the sudden surge in the market, all the other stocks of similar areas of development garner interest from the investors, money starts getting invested in or being pulled out from these stocks and eventually the market trend shifts. In the analysis of such trends, it is found that even though the other companies were not the target of the initial change in the sentiment of the market, but still there is a significant amount of difference in their price. This analysis goes on demonstrating a significant amount of correlation present between the S&P 500 stocks [2]. This relationship is observed since each of the stocks present in the S&P 500 ratings belong to companies that are considered leaders in their respective fields of operations. Since these companies have such immense reputation and prestige in their domain that if their stock prices undergo any change, then these changes are felt across the industry. Even if a negative trend starts developing for the specific stock, the investors sense that its adverse effects would be reflected upon the whole industry. By taking this trend as a reference, it was deduced that the price change in all the major stocks present in the S&P 500 ratings is highly linked together [3].

To predict this trend, in our paper the data of all the S&P 500 stocks were taken into account and the model extracted the daily change in the price of the stocks to compute the per cent change in the price of the stock which paper predicts. If the per cent change in the price of the stock hits a certain threshold then the stock is labeled according to the price change and a prediction is made regarding the action an investor should take in the given situation. The main strategy in the classification of the model is that it doesn't predict the absolute price of the selected stock rather it computes the per cent change in the price of the stocks over the last seven days and then classifies the stock into one of the three categories—buy, sell, and hold [4]. These categories are defined based on whether the per cent change has hit two per cent, if the change is a negative of 2% then the stock is classified as a sell, if the change is positive then the stock is classified as a buy and if the 2% threshold is not met then the stock is classified as a hold. The whole model design is distributed into steps as shown in Fig. 1.

2 Related Work

For a long time, the stock market has been the playground for investors and scientists to develop algorithms to make accurate predictions over its behavior. Market trends are constantly changing so using a linear model will not provide accurate results on real-time data. Another observation was how each stock is correlated with each other, this results in a corresponding price shift upon a sudden market correction. Due to this process, the whole market gets heavily affected by a general trend which explains why researchers have found the stock market to be highly volatile in nature.

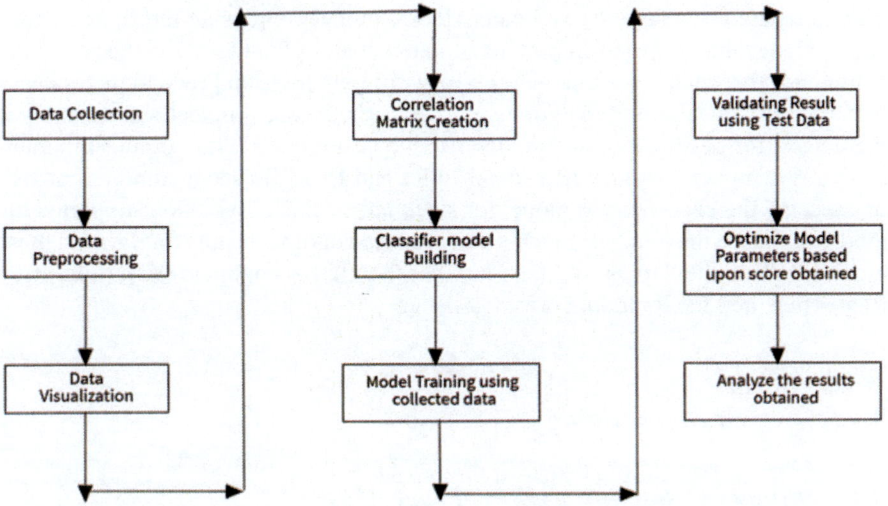

Fig. 1 System design

Data visualization done by Liu and Dashti [3] on S&P 500 stocks gave insight on how data is distributed over some time and how data points are spread over a linearly fit line [4]. It was deducted that a prediction model should understand the correlating stock features and make predictions on a stock by observing how the other stock's behavior and trends. From further analysis, it was observed how major stocks get affected by a sudden change in market trends and apart from special cases, tend to perform similarly.

In their paper, Chaudhuri and De [4] proposed to use a classification model that classifies stocks into two labels of buy and sell. These labels are based upon the past performances of each stock. A similar approach was applied by Fu and Du [5] when they performed stock selection based on their return price and label them into "good" or "bad" classes. Both approaches work best upon some peculiar cases as the former approach worked best on a single stock index and latterly worked best in-stock selection. This analysis gave the inference to derive a model that takes advantage of both kinds of situations.

Another observation inferred by Kalyani [1] in her research paper was the difficulty to predict future trends in a stock market with just the past performance data of the stock, this feature is observed due to how each stock's performance can observe a sudden dip in performance due to some unexpected conditions or news that directly affect the company. Since the whole system of the stock market is highly correlated, the effects can be seen throughout the market and the whole market trend changes abruptly.

Lu and Zhao [6] also highlighted the correlation factor between the major stocks and how market depends upon certain key factors for its trends in their research. This further enhanced the deductions on how to optimize the model and dataset to accommodate these features in the classifier. Similarly, choosing the right classification algorithms is also as important as stated in all of the literature, the reasoning behind it is that each algorithm works with a different logic and each logic produces better results in different conditions, on the basis of that a model should analyze the dataset for desired output and then use the best models with optimized parameters to get higher accuracy and score. Elliot and Hsu [7] used a similar approach in choosing the classification model for stock labels, this observation supported the model to choose the four best models that produced optimal results for the task, these models include the Support Vector Classifier (SVC), K-Neighbors (KNN), Logistic Regression, and the Random Forest Classifier.

3 Research Methodology

3.1 Dataset Collection and Preprocessing

For the paper, initially, the dataset was collected from the list of all the stocks present in the S&P 500 ratings. The model scraped the stock names in ticker format from Wikipedia. Sequentially, the actual stock price was collected for each of the S&P 500 stocks using the Yahoo Finance API, which collected each stock's daily data over a span of the last 20 years. The data format consists of the stock's Opening Price, closing price and volume of the shares traded daily. The collected data was cleaned for the removal of all the missing values, and the data was finally stored into a master CSV file for further processing.

3.2 Data Visualization

After data collection, the dataset was analyzed for the key aspects and trends observed in the collected S&P 500 stock dataset. To further analyze the correlation between both the stocks, a correlation matrix was generated that identifies the relationship between the stocks and how significant is the impact of the price change of one is on the other stock [8]. The matrix calculates the correlation by using Eq. 1.

$$R = \frac{n(\sum xy) - (\sum x)(\sum y)}{\sqrt{\left[n\sum x^2 - (\sum x)^2\right]\left[n\sum y^2 - (\sum y)^2\right]}} \tag{1}$$

In Fig. 3, a plot between the stock data of Google and Amazon over the course of the last 20 years is generated. Since both Amazon and Google work in the same domain, it is expected that a relationship must be present between them. From Fig. 2, it is observed that both Google and Amazon stocks are highly correlated with each other and this correlated behavior was confirmed in Fig. 3 as a surge in the price of the former tends to impact the prices of the latter.

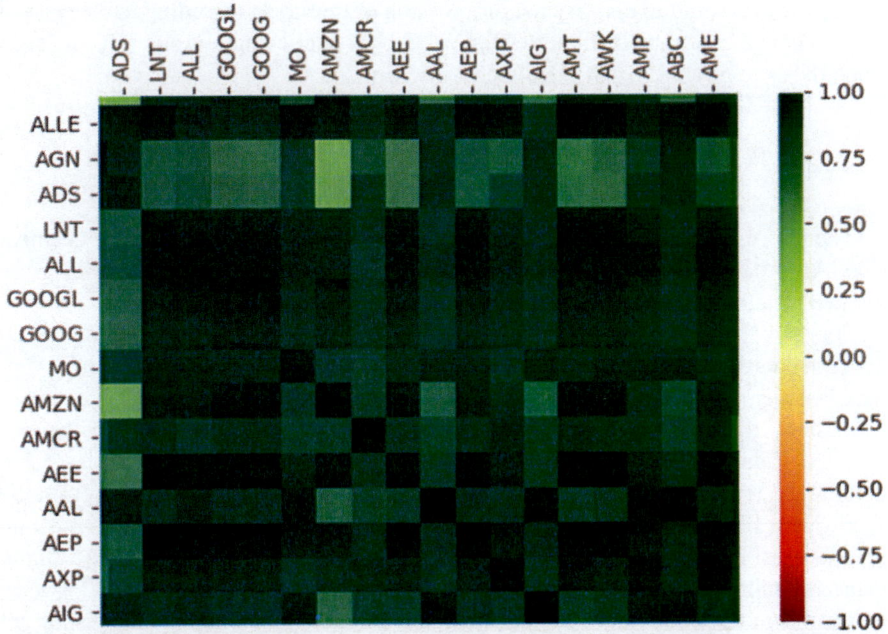

Fig. 2 Part of the correlation matrix between S&P 500 stocks

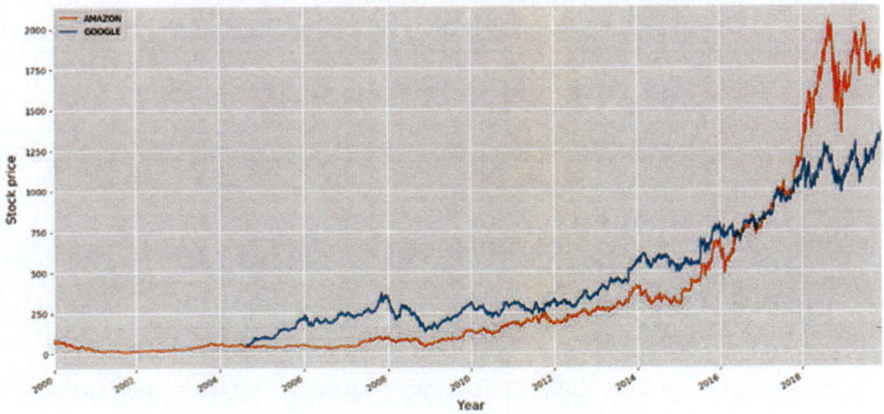

Fig. 3 Google versus Amazon price trend over the last 20 years

3.3 Classifier Modeling and Implementation

Identifying the trends and the correlation between our dataset, the Machine Learning model has been built to classify the progression of the stock according to the criteria of 2% change. For model implementation, the per cent change in the price of each stock over the past 7 days is calculated using Eq. 2.

$$\text{Percent Change} = \frac{(\text{Stock}[\text{price}].\text{shift}(-7) - \text{Stock}[\text{price}])}{\text{Stock}[\text{price}]} \quad (2)$$

The shift function signifies that the price of the stock has shifted to the negative side. After calculation, the stocks were classified based on the results. They were marked as either −1, 0, and 1 for buy, sell, and hold, respectively.

For prediction, different sets of Supervised Classification models [9] were trained over the dataset. The algorithms implemented for classification were—Support Vector Classifier [10], Logistic Regression, Random Forest Classifier, and K-Neighbors Classifier. To select the most optimized parameters for each model Gridsearchcv function from the scikit-learn library was used. Then, the training data was partitioned into Training Data and Test Data using the train_test_split function from the scikit-learn library. Using this function, the dataset was then sliced into a training set and a test set in the ratio of 80:20. The model is initially trained over the training dataset. Finally, the dataset is evaluated over the test dataset. This model is assessed on various metrics such as Precision, Recall, F1, and Support [11] and the results are mentioned in Table 1.

Table 1 Classification reports of the used classification models

Model	Label	Precision	Recall	F1-score	Support	Average score
KNN	−1	0.62	0.58	0.60	249	0.6339
	0	0.42	0.36	0.39	73	
	1	0.68	0.73	0.71	350	
SVC	−1	0.72	0.52	0.60	243	0.6711
	0	0.69	0.24	0.36	83	
	1	0.65	0.88	0.75	346	
Logistic regression	−1	0.63	0.73	0.67	234	0.6979
	0	0.49	0.48	0.49	66	
	1	0.79	0.78	0.75	372	
Random Forest Classifier	−1	0.75	0.78	0.77	234	0.7470
	0	0.64	0.49	0.56	66	
	1	0.76	0.78	0.77	372	

4 Results

After testing the classification models on the test set, insight has been gained on how the model performs over the dataset. The model was tested on each of the S&P 500 stocks and found that due to their correlating features, the classification models yielded similar results. As for scoring, each model was given an average weighted score based upon various parameters such as Precision, Recall, and F1-score. From a broader point of view, it was observed that Random Forest classifiers gave the most accurate predictions but to understand the reason it performed the best, each scoring parameter is reviewed on how they affect the final scores.

The Precision determines how much of your positively predicted classifications were classified correctly while the recall score determines how much data of a label was correctly classified for the total data of a label. The F1 score determines the weighted mean of both the precision and recall score of a classifier. The analysis of similar results revealed that an optimized model must maintain a balance between the precision and recall [12]. From the results in Table 1, models with a high Recall score for -1 label (Hold) gave higher accuracy as compared to other models, this was demonstrated by the Random Forest Classifier [13] as it gave a high Recall Score which resulted in total accuracy of 74.70%. From the results, we observed the low Recall and F1 score of the model the 0-label (hold) classification. This behavior of the model can be analyzed as an aberration of the model used in the paper to label stocks as a 0-label(hold) over a longer period (greater than 7 days). There is also factor of less training data available as some of the S&P 500 companies have listed their stocks quite recently which means these companies don't have the daily stock data for the period of last 20 years due to which its data isn't present in our stock market dataset [14].

5 Conclusion and Future Scope

In this paper, the critical relationships between the stock prices and the general trends of a stock market were highlighted. It has been found that the daily prices of the stocks depend upon a variety of features and data, a linear prediction model is not enough to predict real-time stock data, even though it performs well on the training set. The model demonstrated how each of the S&P 500 stocks is correlated with each other and how a simple impact on a major stock can shift the current market trends.

The classification of stocks was done according to the percentage price change of all the correlated stocks. The supervised classifier models used were trained for each of the S&P 500 stock's daily dataset and they produced results with accuracy ranging from 63 to 75%. From the results, an anomaly was detected that the models were unable to correctly classify the stocks whenever there was a sudden surge in the price of a stock due to some major news or information being released. The S&P 500 dataset used in the paper produced variable results upon classification. The model produced the highest accuracy in classification with the Random Forest Classifier to generate a score of 0.7470.

The model produced results with high accuracy over real-time stock market data, the labels generated by the classification model are identifying the stock market trends accurately. On analysis of the results, an insight was formed on how to improve its performance in future work. The insights produced were as follows:

- Increase the training dataset.
- Add more features to identify more market trends.
- Perform analysis of daily news to identify any major market shift.

References

1. J. Kalyani, P. Bharathi, P. Jyothi, *Stock Trend Prediction Using News Sentiment Analysis* (2016). arXiv:1607.01958
2. M. Poženel, D. Lavbič, *Discovering Language of the Stocks* (2019). arXiv:1902.08684
3. Z. Liu, M.D. Moghaddam, R.A. Serota, *Distributions of Historic Market Data—Stock Returns* (2017). arXiv:1711.11003
4. A. Chaudhuri, K. De, D Chatterjee, *Discovering Stock Price Prediction Rules of Bombay Stock Exchange Using Rough Fuzzy Multi Layer Perception Networks* (2013). arXiv:1307.1895
5. X.Y. Fu, J.H. Du, Y.F. Guo, M.W. Liu, T. Dong, X.W. Duan., *A Machine Learning Framework for Stock Selection* (2018). arXiv:1806.01743
6. S Lu, J. Zhao, H. Wang, *The Emergence of Critical Stocks in Market Crash* (2019). arXiv:1908.07244
7. A. Elliot, C.H. Hsu, *Time Series Prediction: Predicting Stock Price* (2017). arXiv:1710.05751
8. Z. Kakushadze, W. Yu, *Stock Market Visualization* (2018). arXiv:1802.05264
9. J. Liu, F. Chao, Y.-C. Lin, C.-M. Lin, *Stock Prices Prediction using Deep Learning Models* (2019). arXiv:1909.12227
10. S. Arik, S.B. Eryilmaz, A. Goldberg, *Supervised Classification-Based Stock Prediction and Portfolio Optimization* (2014). arXiv:1406.0824
11. A. Filos, *Reinforcement Learning for Portfolio Management* (2019). arXiv:1909.09571
12. M.S., Lauretto, B.B.C. Silva, P.M. Andrade, *Evaluation of a Supervised Learning Approach for Stock Market Operations* (2013). arXiv:1301.4944
13. L. Khaidem, S. Saha, S.R. Dey, *Predicting the Direction of Stock Market Prices using Random Forest* (2016). arXiv:1605.00003
14. M. Velay, F. Daniel, *Using NLP on News Headlines to Predict Index Trends* (2018). arXiv:1806.09533

Blockchain and Industrial Internet of Things: Applications for Industry 4.0

Mahesh Swami, Divya Verma, and Virendra P. Vishwakarma

Abstract Internet of Things (IoT) is commanding a great role in connecting electronic appliances through the cloud. It is an expanding area of research and is finding association with other emerging technologies like 3D printing and machine vision. The growing popularity of the technology is an extension of the Internet boom. An expanding Internet user base, cheaper smart devices and greater on-cloud memory are encouraging the spread of IoT. From fridges to water geysers, any electronic appliance may be connected to the IoT network. The advantages of remote access, master control and cloud storage give it unprecedented advantage over conventional means of communication. In this paper, we attempt to review some latest applications of IoT across various sectors: location tracking, sports, smart devices and blockchain-based security. IoT remains a rapidly developing topic with significant interdisciplinary research and development potential. Further integration of sensors for inputs will propagate the spread of IoT.

Keywords IoT · IIoT · Cloud computing · Sensors · Computer vision · Smart devices · MQTT · Industries · Blockchain · Cryptocurrency · Ethereum · P2P · Ethernet · Hashing

1 Introduction and a Brief History of IoT

Internet of Things or IoT is the interconnection of domains, devices and data through the physical media of electronic devices. It involves connections through the use of cloud computing and existing Internet protocols and servers. With the further integration of computer vision, the technology has scopes of reaching great heights

M. Swami (✉) · D. Verma · V. P. Vishwakarma
University School of Information Communication and Technology, Guru Gobind Singh Indraprastha University, New Delhi, Delhi 110078, India
e-mail: mahesh23312@gmail.com

D. Verma
e-mail: dverma0732@gmail.com

P. Bansal et al. (eds.), *Proceedings of International Conference on Artificial Intelligence and Applications*, Advances in Intelligent Systems and Computing 1164,
https://doi.org/10.1007/978-981-15-4992-2_27

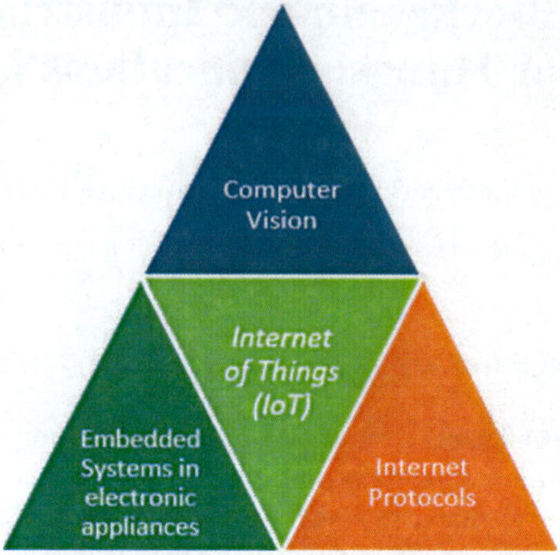

Fig. 1 IoT framework

of an interconnected world previously not possible through human intervention. IoT was first conceived by writers in the 1930s as they imagined the day when manual tasks would not be required to take care of repetitive work. A breakthrough introduction to the idea happened with the invention of barcode by John Woodland in 1932. Barcodes were assigned to each product and soon became common in the US supermarkets. The ARPANET, predecessor of the modern Internet, came around in 1969 [1]. The first IoT prototype was a cola vending machine at Carnegie Mellon University (USA) which kept a track of the inventory and whether cola bottles were cold or not [1].

The technology has grown at a rapid pace since then, due to the following reasons:- miniaturization of electronics, improved Internet and Communication Technology (ICT) innovations, globalization and advancement of package-based software and coding and automation. Thus as we write this paper, several hundred devices get added to the vast Internet of Things around the globe. This paper presents a review of industrial IoT in the intelligent monitoring segment, utilities and water utilization management segment in Sect. 2. Section 3 presents a blockchain-based IoT system. In Sects. 4 and 5, we attempt to present future proposals of systems based on summarization of the systems (Fig. 1).

2 IoT for Industry: Applications for Industrial Internet of Things

There are huge changes in industries during the trend of automation across all over the world. IoT helps to transform in various aspects like smart cities, healthcare, etc. Due to a sudden increase in the use of machines and vehicles in the past few years, it causes a huge effect on the atmosphere and leads to diseases like heart problems, asthma and bronchitis [2, 3]. IoT in industries will lead to an increase in the production and output's efficiency of any plant, but the atmosphere concern is the primary interest. In chemical industries, there are huge chances of gas leakage like methane, butane, CO_2, etc. which can prove lethal for life. Smart sensors help to detect these gases and can be used to ease the problem [4]. A microcontroller-based gas monitoring system was designed to detect and alert the system when the threshold level is crossed [5].

2.1 Intelligent Industry Monitoring Systems: Toward Industrial Internet of Things (IIoT)

This system was designed by using various smart real-time monitoring sensors that are deployed on the shop floors. These sensors help into detect gas leakage, boiler temperature, fuel leakage when crossing the threshold level and alert the floors manager via various means like cloud data and raise alarm. This system is combined with Raspberry Pi to sense the various leakages and used for real-time monitoring of the shop floor. In this setup, MQ-6, MQ-7 and MQ-135 are used to monitor the smoke, carbon monoxide, LPG, methane, etc. [6]. The DHT-11 is used to monitoring the boiler temperature and humidity level. This data is monitoring continuously and uploaded to the Google Cloud from which it is possible to introduce to IoT shield and Raspberry Pi. The connection between Raspberry Pi and server is established by the application program interface (API). The edge devices, application-level security and security of gateways are ensured by using IoT shield [6] (Fig. 2).

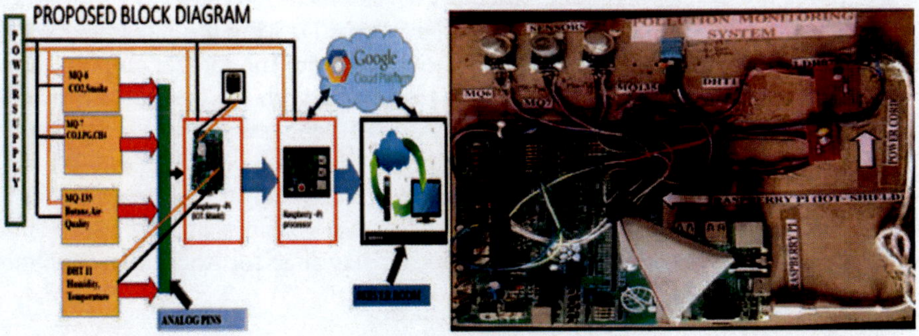

Fig. 2 Proposed system versus actual prototype [6]

2.2 IoT Paradigm for Supervision in the Utilities Industry

In the current technology, there are lots of scope of IoTs applications such as smart wireless sensors, universal computing and communication via mobile. However, there are lots of options in this field because of a huge variety of diversity of hardware and software, and some are under study [7]. The middleware (broker) platform plays a vital role, and so, there are lots of options for these platforms and protocols, selecting based on requirement meets. In this device, we use the Message Queue Telemetry Transport (MQTT) protocol. The reason for choosing this protocol since it is available as open source for both publish and subscribe a message and it is also designed for limited no. of devices having a high latency, less bandwidth and untrusted networks [8]. There are lots of applications of MQTT, and some include water utilization metering and to monitor the energy generator systems.

Energy generator systems monitoring based on IoT

Due to the desired features of MQTT, this system is designed by using MQTT protocol and IoT because of its less hardware cost, less communication bandwidth, high latency network and more instability. This type of system has the capability to monitor the data as well as to describe the type of signal which is monitored.

The generator set which usually is diesel operated is controlled or driven by programmable logic controller (PLC), and then, Modbus RTU protocol helps to communicate with the management module which helps to communicate with broker with the help of MQTT operating in cloud computing [9]. The broker helps into make the data available to Web application which is monitored by personal computers (PC) or from mobiles which communicates by MQTT. The Wemos D1 Mini Development Kit is used by this type of protocol which features the "ESP8266EX" chip [10, 11]. This chip helps to integrate with Arduino's development interface [12], and it is a low cost and has built-in Wi-Fi modules. The Wemos D1 Mini has integrated Wi-Fi antenna, micro USB port, 16 pins out of which some are for general purpose, reset button and a micro USB port which helps to connect with the computer via cable [9]. Data collection from the PLC is done by using RS485 transceiver via a protocol. The general purpose input/output (GPIO) helps to direction control of RS-485 half duplex. The setup of the prototype is done by the computer or mobile using Wi-Fi network. There are also three LEDs to indicate the system state [9]. The mobile application named as MQTT dash is available on play store for Android to configure the prototype using mobile browser as shown [13] (Fig. 3).

Water utilization metering for the large area using IoT [9]

In daily life, most of the water utilization measuring devices uses a personal protocols or technologies which are used as private and do not interact with others in a friendly way, and it causes a problem at big enterprise level. Thus we need to design a device based on IoT which is capable of collecting the data and have a tendency to interact with others as open-source. The purposed device utilizes the low-power microcontroller system having a radio frequency information transmission and uses cloud services by MQTT protocol for water consumption measuring [9]. It aims to

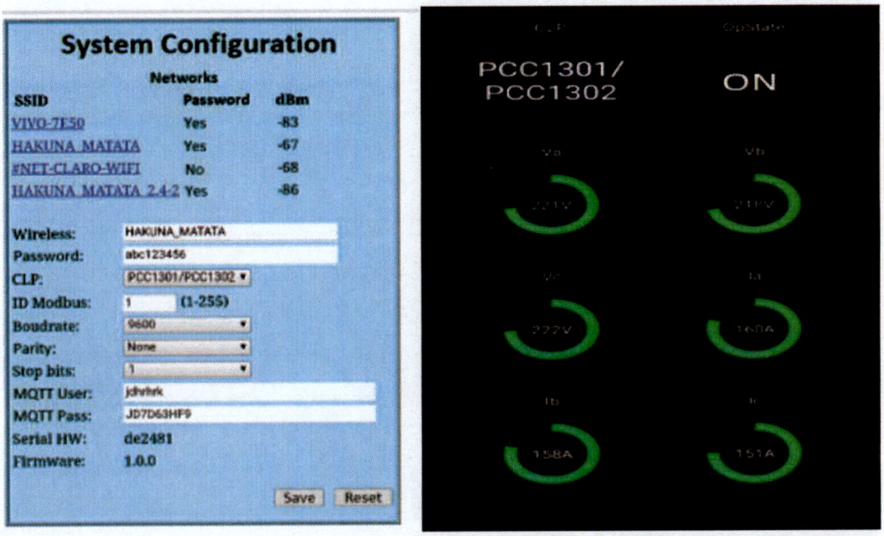

Fig. 3 LHS shows configuration page and RHS shows monitoring dashboard [9]

minimize the cost of setup and increases the coverage area by expanding the Wi-Fi coverage. This setup uses a CC1200DK development kit which integrates with ESP8266 Wi-Fi device and together of this sends data to a server. For monitoring the data of transmitting and collecting volume consumed by the concentrator, we use another CC1200DK development kit with an output water meter's pulse [9]. This kit has the capability to stand with different transmission ranges, and it is possible to work by fluctuating output powers that vary from 14, 10, 5, 0, −5, −10 and −16 dB [9]. The test is conducted on the large distance which is obstacle free (Fig. 4).

The distance and the location of test conducted as shown below [14].

To check the capability of MQTT, two hardwares are designed. The first hardware uses a Mosquitto broker and Ethernet network-based BeagleBones Black boards, and in the second hardware, BeagleBone is replaced by the ESP8266-12F to establish the network by using Wi-Fi module as shown in Fig. 5 (LHS and RHS). This setup measures the time interval between the Wi-Fi network and the Ethernet network which is used for data transmission. Grafana dashboard application is used for measuring the water consumption reading, and in this, it is possible to represent the data in graphical form and also consolidate the data, and we can also use MQTT dashboard in smartphone application [9] (Fig. 6).

3 Blockchain for IoT: Applications

The Internet of Things is gaining wider acceptance for day-to-day transactions to establish a machine–machine interaction with minimal human interaction. The era

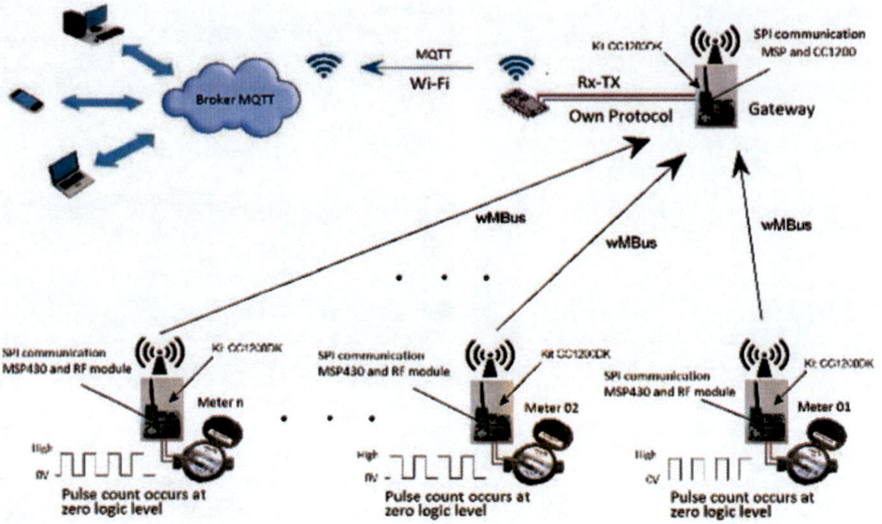

Fig. 4 Communication system [9]

Fig. 5 Local and test distance [9]

of bots is upon us, and soon the need for human labor shall be minimized and changed to fit into the needs of the Fourth Industrial Revolution. The aspect of security in an increasingly connected world becomes ever important, as cyberattacks become common. The need for securing IoT networks can be listed as a conjugation of one or more of the following reasons, namely ransomware attacks by blocking crucial

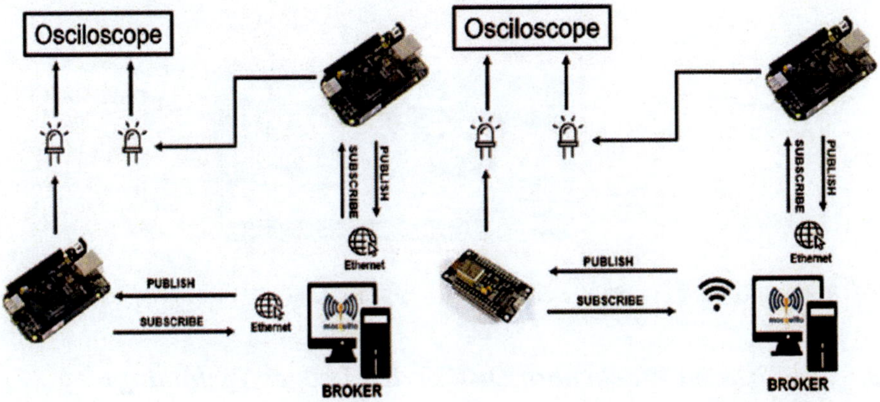

Fig. 6 Ethernet to Ethernet and Ethernet to wireless (LHS and RHS, respectively) [9]

financial infrastructure (for instance, "WannaCry" malware), informational leaks and hijacking of entire systems for criminal purposes. In this respect, blockchain can be seen as a major research field application for IoT systems [15]. It is a relatively new concept, and countries are wary to implement it for lack of central authorities, but the idea can certainly be explored (Fig. 7).

A blockchain-based IoT system works on distributed ledger technology (DLT) that performs the following functions named as Guarantees the integrity of stored data, Maintains each user as an integral part, avoids centralization of server data and supports data distribution. Blockchain became a common word through the introduction of bitcoin introduced by Satoshi Nakamoto back in 2008 [16]. The ordered blocks in the ledger system are interconnected through hashes from the previous block. This ensures that transaction history cannot be deleted without deleting the characteristic part of the system. This is the reason for the rapid popularity gained by bitcoin amid investors. Security issues remain a challenge to be sidestepped for researchers and designers. Some common issues may be spoofing and insecure physical interface attacks on low level which involve physical link layers of communication; buffer attacks, routing attacks and authentication-based privacy violation on intermediate levels and insecure firmware and middleware security for interfaces [15] (Fig. 8).

Fig. 7 Distributed and decentralized network for blockchain-based IoT system

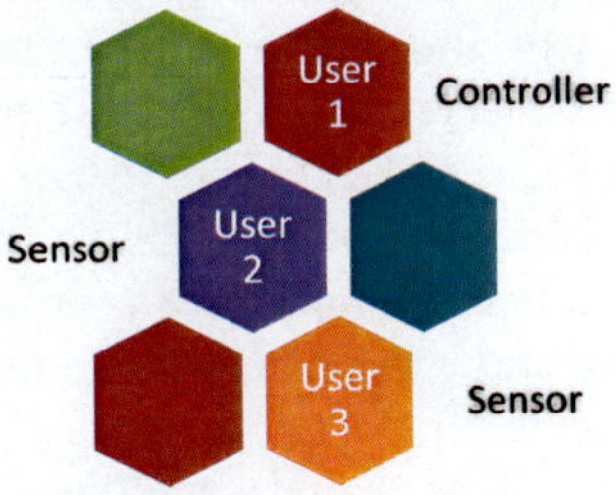

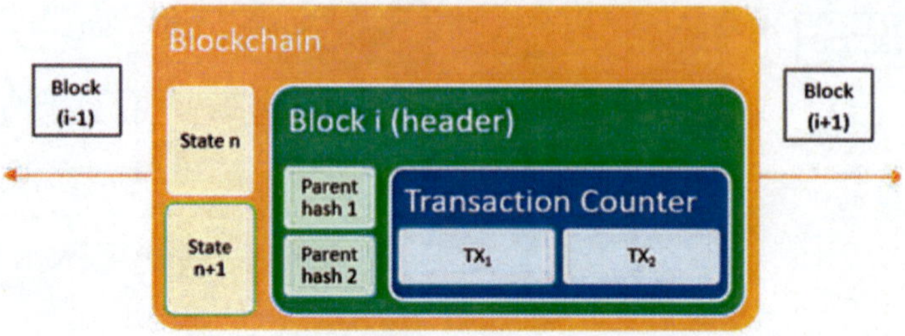

Fig. 8 Representation of a single block of a financial blockchain [17, 18]

3.1 *IoT-Based Blockchain Distributed Ledger Technology (DLT) System: Decentralized Networking Using Blockchain*

In a blockchain-based IoT network, the communication protocol is replaced with a blockchain network. As viewed in figures above, data processed in IoT systems will only grow over time; hence, robust handling systems are needed. A way to handle these is through peer-to-peer messaging using distributed ledgers, similar to peer-to-peer networking (PPN) systems [15–22]. The blockchain platform is Ethereum, which has an advantage over bitcoin as being Turing-complete, i.e., being compatible with all types of systems through their unique identities. The Ethereum Virtual Machine (EVM) with Golang programming language is used. Additionally, Keccak-256 hash function and encoding are used on Ethereum [16]. Tests reveal that systems with and without blockchain have almost similar security levels of encryption algorithms. To minimize data leaks, a proper linkage between protocols is necessary. This can be made through a feedback loop between analytical machines and sensor networks. Communication is possible through ZigBee and/or Bluetooth. Edge servers may be used which have the advantage of high power computing in case of arduous hash value verification calculation results. The blockchain nodes may be interconnected together to help store data. Broadcast of data occurs as node i transmits data to the nearest node, i.e., "$i - 1$" or "$i + 1$" (refer Fig. 2). This chosen node then broadcasts the data further to all P2P nodes. Following this, a task is delivered by the network which needs all nodes to calculate a hash value. Time consumption between different nodes is compared for local computation offloading levels and other considerations. The node carries the choice to either compute locally or offload to edge servers following digital signatures. The result is validated by the home node and if found to be correct, is transmitted to all nodes. If all nodes validate the result and if legal as per encoding, the block is added to the blockchain [20]. Dataset sharing among communities is an upcoming topic in IoT applications, which further uses a Reference Integrity Metric (RIM) for each dataset on a blockchain ledger system (Fig. 9).

Remote updates to the system can be made possible since the system is always online [21]. Self-healing of the code can also be done through data redundancy.

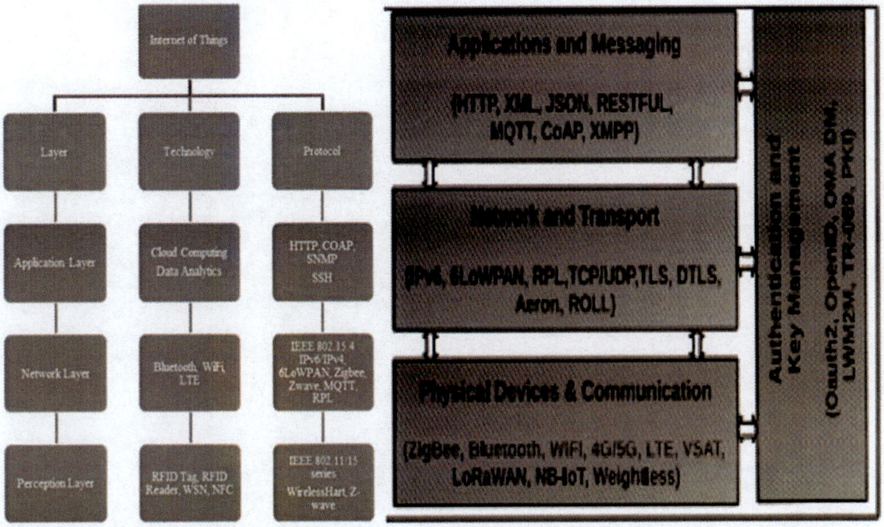

Fig. 9 Architecture for IoT [16]

3.2 *Cryptocurrency Based on Blockchain: Currency of the Future*

Blockchain-based cryptocurrency like bitcoin is making news around the world. The technology works as a decentralized ledger system that keeps adding to the list of transactions much like a stack. The transactions remain anonymous.

The proposed IoT network could work on the basics of blockchain such as host configuration and BIOS details on ledger, requiring every device on that IoT cloud to download latest configurations at specific time intervals and encryption to preserve network privacy. The system could be an effective "means and end" solution to the problem of achieving cloud/fog security, prevent unauthorized access and maintain a decentralized system [22] (Figs. 10 and 11).

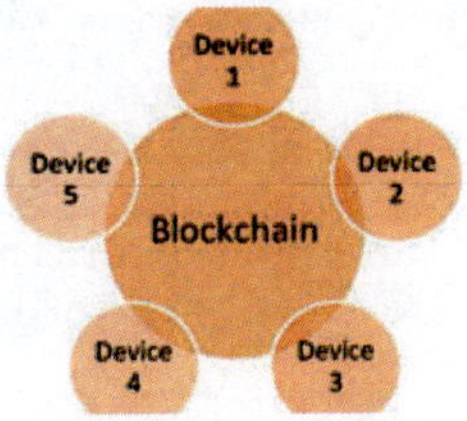

Fig. 10 Protocols and standards for common IoT systems [18]. Communication is possible through ZigBee and/or Bluetooth

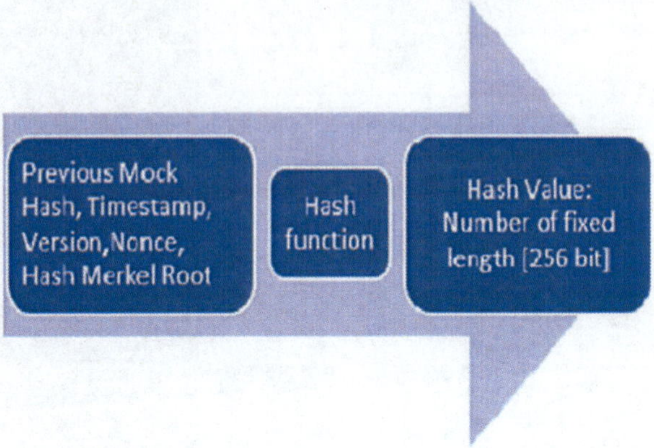

Fig. 11 Hashing in blockchain technology [22]

4 Summary: IoT Applications and Key Pointers

MQTT is the preferred choice of open-access messaging protocol for the present IoT systems, except in ledger-based systems as seen. In summary, the following features of IoT are utilized for smart applications: (a) connectivity; (b) single master control; (c) cost effective; (d) single server control. Single server control may be replaced with blockchain technology (Table 1).

Table 1 Summary of IoT applications

Application	Specifications
IoT industry sensor	• Uses Raspberry Pi to monitor leakage • Uses DHT-11 sensor to monitor temperature and humidity levels. MQTT protocol as the open-source messaging protocol for limited devices
IoT-based energy generator systems	• PLC monitored • Modbus RTU protocol helps communicate with the management module • ESP8266 microprocessor integrates with Arduino's interface
IoT-based water metering for wide area	• Builds on MQTT protocol for consumption measurement • ESP8266 Wi-Fi device to transmit data to server • "MQTT Dash" application can be used
IoT-based blockchain technology	• Proposes to use ledger-based database technology • Reference Integrity Metric for each dataset on the ledger system to identify each node • Updates and self-healing are easy due to online nature of system

Table 2 Future system proposals

S. No.	System integration	Application
1	Localization and tracking sensor-based IoT network and ledger	• Cellphone-based linkage • To facilitate Direct Benefit Transfer to farmer beneficiaries and pensioners; enables citizens to access government schemes at single place • Ledger nature of system would ensure that no records can be tampered with or without tampering the entire system
2	Industrial IoT system	• Linkage of entire industries through on-floor mapping and connections, interconnections on IoT, etc. • To link water tanks in residential societies, automated refilling and pilferage alarms
3	Blockchain ledger linkage for financial IoT	• To link cryptocurrency transactions in a small regional area (can be further extended to states) • To hinder dark Internet-based terror financing by engaging more and more users • Linkage to the existing banking infrastructure at a later stage • Establishment of common organizations on global and regional levels to monitor ledger-based transactions

5 Conclusion: Future System Proposals Based on Systems Reviewed

IoT presents keychain technologies coming up for humankind. In view of the changing climatic conditions and globalization, online database management is becoming the need of the hour. Governments and enterprises world over are setting up committees to integrate frameworks of IoT linkages relevant to our need. It is the time for researchers to develop systems which can be integrated into a global network, which may someday cross the domains of space and time as we know today. Some applications relevant to the sensors reviewed and pertinent to global problems are presented below (Table 2).

References

1. *A Very Short History of the Internet of Things*. https://www.forbes.com/sites/gilpress/2014/06/18/a-very-short-history-of-the-internet-of-things/#5c66685b10de

2. A.M. Leman, Nor Hidayah, *Occupational Safety and Health: Workers and Industrial Safety Monitoring for Sustainable Work Environment Development* (Health and Safety, AET, 2013)
3. Wu-Zhao, L. Lei-Hong, H. Yue-shan, W. Xiao-ming, A community health service architecture based on the internet of things on health-care, in World Congress on Medical Physics and Biomedical Engineering, vol. 39, pp. 1317–1320 (Springer, Heidelberg, 2013)
4. R. Rushikesh, S. Chandra Mohan Reddy, S. Reddy, Development of IoT based vehicular pollution monitoring system, in *International Conference on Green Computing and Internet of Things (ICGCIoT) proceedings* (IEEE Press, 2015)
5. A. Banik, B. Aich and S. Ghosh: Microcontroller based low-cost gas leakage detector with SMS alert, in *Emerging Trends in Electronic Devices and Computational Techniques (EDCT)* (IEEE, 2018). https://doi.org/10.1109/edct.2018.8405094
6. B.C. Kavitha, R. Vallikannu, IoT based intelligent industry monitoring system, in *2019 6th International Conference on Signal Processing and Integrated Networks (SPIN)*, pp. 63–65 (IEEE, 2019). https://doi.org/10.1109/spin.2019.8711597
7. L. Atzori, A. Iera, G. Morabito, The internet of things: a survey, in *Computer Networks* (Elsevier, 2010)
8. G.C. Hillar (eds.), MQTT Essentials—A lightweight IoT Protocol. First. (Packt, Birmingham, 2017)
9. F.V. Schreiber, R.A. Dias, B.D. Borba, V. Noll, V.Z.M. Gallo, Application of the IoT paradigm for supervision in the utilities industry, in *13th IEEE International Conference on Industry Applications (INDUSCON)* (IEEE, 2018)
10. D1 mini. https://wiki.wemos.cc/products:d1:d1_mini
11. ESP8266EX Datasheet. https://www.espressif.com/sites/default/files/documentation/0a-esp8266ex_datasheet_en.pdf
12. ESP8266 core for Arduino. https://github.com/esp8266/Arduino
13. MQTT Dash (IoT, Smart Home)—Apps on Google Play. https://play.google.com/store/apps/details?id=net.routix.mqttdash&hl=en
14. Google Maps. https://www.google.com/maps/@-27.4344032,-48.450146,900m/data=!3m1!1e3
15. N.M. Kumar, P.K. Mallick, Blockchain technology for security issues and challenges in IoT, in *International Conference on Computational Intelligence and Data Science (ICCIDS 2018) proceedings*, Procedia Comput. Sci. **132**(2018) 1815–1823 (Elsevier, 2018)
16. D. Fakhri, K. Mutijarsa, in *Secure IoT Communication using Blockchain Technology* (IEEE Press, IEEE, 2018). 978-1-5386-6670-8/18
17. Z. Zheng et al., An overview of blockchain technology: architecture, consensus, and future trends, in *6th International Congress on Big Data (BigData Congress) Proceedings* (IEEE, 2017)
18. M.A. Khan, K. Salah, IoT security: review, blockchain solutions, and open challenges. Future Gener. Comput. Syst. **82**, 395–411 (2018)
19. M. Son, H. Kim, Blockchain-based secure firmware management system in IoT environment, in *International Conference on Advanced Communications Technology (ICACT), Proceedings* (IEEE, 2019). ISBN 979-11-88428-02-1
20. C. Fang et al., Edge computing and caching based blockchain IoT network, in *Proceedings of 2018 1st IEEE International Conference on Hot Information-Centric Networking (HotICN 2018)* (IEEE, 2018). 978-1-5386-4870-4/18
21. M. Banerjee, J. Lee, K.R. Choo, A blockchain future for internet of things security: a position paper, in *Digital Communications and Networks* (Elsevier, 2017). 20172352-8648
22. M. Singh, A. Singh, S. Kim, Blockchain: a game changer for securing IoT data. in *Proceedings of IEEE 4th World Forum on Internet of Things (WF-IoT)*, pp. 51–55 (IEEE, Singapore, 2018)

Opinion Mining to Aid User Acceptance Testing for Open Beta Versions

Rohit Beniwal, Minni Jain, and Yatin Gupta

Abstract Social media enables the sharing of opinions, ideas, interests, thoughts, hobbies, etc., by creating social platforms such as Twitter, Facebook, TripAdvisor, Yelp, Rooter, Goodreads, etc. The above sharing generates a lot of information over various social media platforms, which is used in different research areas. Among the many research areas, one is opinion mining, which identifies and extracts subjective information from source materials by using text analysis, natural language processing, and computational linguistics. The application of opinion mining is abundantly seen in the areas of marketing, governance, tourism industry, etc. However, the application of opinion mining is not much explored in the area of user acceptance testing. Therefore, in this research article, we propose a model for opinion mining and use it as a supporting tool for user acceptance testing. Specifically, we use the opinion mining on "Twitter" tweets to analyze the response of an Open Beta, i.e., Public Beta versions of the software to quantify acceptance criteria qualities. Hence, we try to inculcate opinion mining in earlier stages of software development to aid user acceptance testing in order to get the user feedback as accurately as possible to make change in the concerned software before its final version is released. The model is implemented and evaluated for Open Beta versions of IOS 13 and AndroidQ using machine learning techniques such as decision tree, K-nearest neighbor, multilayer perceptron, and logistic regression. Our evaluation shows that we achieved the best result with K-nearest neighbor having an accuracy of 97.12%, which is 14.37% higher as compared to the highest accuracy ever obtained in this area.

R. Beniwal · M. Jain · Y. Gupta (✉)
Department of Computer Science and Engineering, Delhi Technological University, New Delhi, Delhi 110042, India
e-mail: yatingupta7777@gmail.com

R. Beniwal
e-mail: rohitbeniwal@yahoo.co.in

M. Jain
e-mail: minnijain91@gmail.com

P. Bansal et al. (eds.), *Proceedings of International Conference on Artificial Intelligence and Applications*, Advances in Intelligent Systems and Computing 1164,
https://doi.org/10.1007/978-981-15-4992-2_28

Keywords Opinion mining · Social media · User acceptance testing · Open beta version

1 Introduction

Social media has caught the attention of the entire world by providing social platforms such as Twitter, Facebook, TripAdvisor, Yelp, Rooter, Goodreads, etc., and thus connecting the countless number of individuals. All the content such as text, images, audio, video that is available on social media is generated by its users who access its services through various platforms. Twitter is one such renowned social media platform, which receives around 500 million tweets per day. Twitter users post or tweet every day about numerous subjects like places, pictures, services, products, interests, etc. Twitter also provides the facilities to let its users remain informed about the various products or services of their interest and get direct comments about them. Tweets containing opinions and reviews from potential users, critics, etc., can heavily influence the demand of the concerned product or service.

Pertinent literature has demonstrated the use of social media for strategic, focused, and collaborative development. The social, mobile, analytics, and cloud (SMAC) is trending among both practitioners and researchers. There are many researchers, who used the opinion mining on various social media platforms such as GitHub [1–3], stack overflow [4, 5], app reviews [6], and Twitter [7]. Moreover, opinion mining has been used in the field of software testing also. Opinion mining is a topic in text analysis, natural language processing (NLP), and in data mining domain. The opinion is defined as a thought, attitude, or a view that is based on emotion and not on logical reason. The objective of opinion mining is to give computers the ability to identify and show emotions [8].

As far as software testing is concerned, it is defined as a process with the intent to find errors so that quality software can be developed [9]. There are several sorts of software testing that exist in Software Engineering, and among them, one is user acceptance testing (UAT) also known as (a.k.a.) "Beta testing," which involves testing the business functionality of the software system being developed. UAT is that stage of software development where the audience or the business agents test the software in the "real world" [10]. There are many advantages of UAT such as increasing the usability of the software, rising the happiness among end-users, providing the opportunities to identify and repair issues, and offering a vision to end-user. Figure 1 shown explains the various types of testing in the life cycle of software development.

Beta testing is generally divided among two classes, open and closed. Closed Beta Testing also referred to as private testing in which a specific group of people is identified who can contribute to this type of testing. In Open Beta Testing also known as public testing, software products are publicly released and are openly available to the worldwide community. Thus, this type of testing involves feedback from anyone available in the public domain. Though the main purpose of Open Beta Testing is generally not to get actionable feedback from the "testers" but to generate

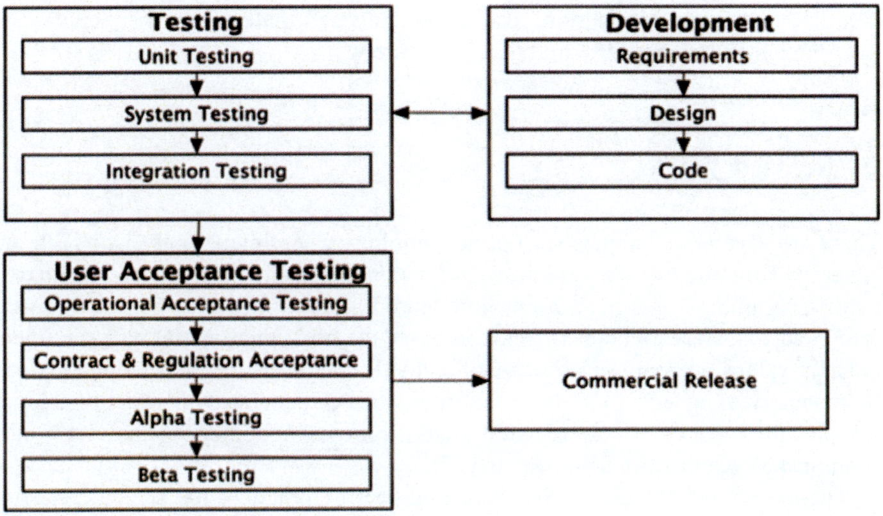

Fig. 1 Testing in software development life cycle

buzz and awareness about the product [11]. Open Beta Testing is often preceded by the intense Closed Beta Testing. For example, before releasing the Windows 8 software, Microsoft conducted one of the largest all-time Open Beta Testing for the concerned operating system. In another example of Open Beta Testing, Apple's iOS 13 beta 1 and iPadOS beta 1 are available in public domain since July 2019, and their availability allowed the users to test the latest operating system features for free of cost.

Similarly, there are abundant examples where software development organizations release the Beta version of the software, which provides ample opportunities for the users to test Open Beta versions. Moreover, the above opportunities allow researchers to perform opinion mining, sentiment analysis, rumor analysis, current trend analysis, etc., on the feedback of the users that aids in sensing the mood of the users for Open Beta versions. Among them, opinion mining has already been heavily used for business intelligence purposes [11]. We have especially seen that in user acceptance testing (UAT), opinion mining can be proven to be a great resourceful tool to check the acceptance of the Open Beta release of the software. Therefore, in this research paper, we perform the opinion mining on Open Beta version of the software that aids the user acceptance testing. Moreover, we provide an opinion mining model that extracts and, respectively, scores opinion markers (adjectives, verbs, and adverbs) and finds the emotion value attached with a tweet. Our objective is to analyze and mine the public mood quantified by sentiment measures such as happiness, anger, sadness, fear, and disgust that can act as metric to characterize the quality of an Open Beta version of a software and thus examine the goals of a business.

The remaining paper is organized as follows: Sect. 2 deliberates the related work; Sect. 3 examines the opinion mining model for Open Beta version followed by Sect. 4

that deliberates implementation, result, and analysis; finally, Sect. 5 concludes the research paper and provides direction for future work.

2 Related Work

There are several techniques for opinion mining or sentiment analysis, which are generally split into two classes known as lexicon-based and machine learning-based methods. Opinion classifiers in machine learning-based methods are trained using the extracted features, whereas in lexicon-based methods, dictionaries with sentiment polarity values are required instead of trained labeled data. The dictionaries can be manually assigned or can be semi-automatically generated in which machine-readable dictionaries or corpus-based methods are used. Sentiment analysis is also examined by Kumar and Sebastian in [12].

The use of social media in Software Engineering practices has seen a great rise in recent times. Social media is been integrated with a variety of tools such as Web-based portals, IDEs, issue trackers, and code editors. Storey et al. [13] presented an evaluation of the present-day use of social media in Software Engineering and thus advocated its use.

Social media has altered the mechanism of interaction between people for working together and sharing information with each other. Begel et al. in [14] discussed the impact of empowering fresh methods for software development teams to work together. People would be enabled to organize within and outside organizational boundaries, and lone individuals would be able to use social media in developing successful product lines.

User-generated microblogs are becoming more crucial to many fields and present an attractive opportunity for researchers to study. Kim et al. [15] analyzed smartphone-related reviews on social media site "Twitter" to evaluate which factors affect a user's interests and attitudes using opinion mining.

Opinion mining to aid user acceptance testing for Open Beta version is also examined by Kumar and Abraham in [16]. Kumar and Abraham used machine learning techniques such as Naive Bayesian, Bernoulli Naive Bayesian, multinomial Naive Bayesian, Gaussian Naive Bayesian, and support vector machines to determine the accuracy of the opinion polarity classifier where multinomial Naive Bayesian gave the highest accuracy of 82.74%. However, in this paper, we extend their work by increasing the accuracy from 82.74 to 97.12% using machine learning techniques such as multilayer perceptron, K-nearest neighbor, decision tree, and logistic regression for opinion mining to aid user acceptance testing for Open Beta versions IOS 13 and AndroidQ where K-nearest neighbor provided the highest accuracy.

3 Opinion Mining Model for Open Beta Versions

We propose a model as shown by Fig. 2 that determines the emotion vector of an Open Beta version of the software by mining opinions on Twitter tweets. The main purpose is to check the acceptance of the developed software by acquiring feedback of the users from Twitter. The following subsections explain the details.

3.1 Pre-processing Phase

This phase will create the desired transaction file that will contain the opinion markers, viz. verb, adverb, and adjective. The desired data is obtained by downloading the publicly available tweets of Twitter using the Twitter API. Further, the data is cleaned. The cleaning of data involves removing usernames, hashtags, punctuations, emoticon replacement, non-English characters, re-tweet symbols etc. Then POS tagger is applied on the extracted cleaned terms where only adjectives, adverbs, and verbs are retained, and rest data is discarded.

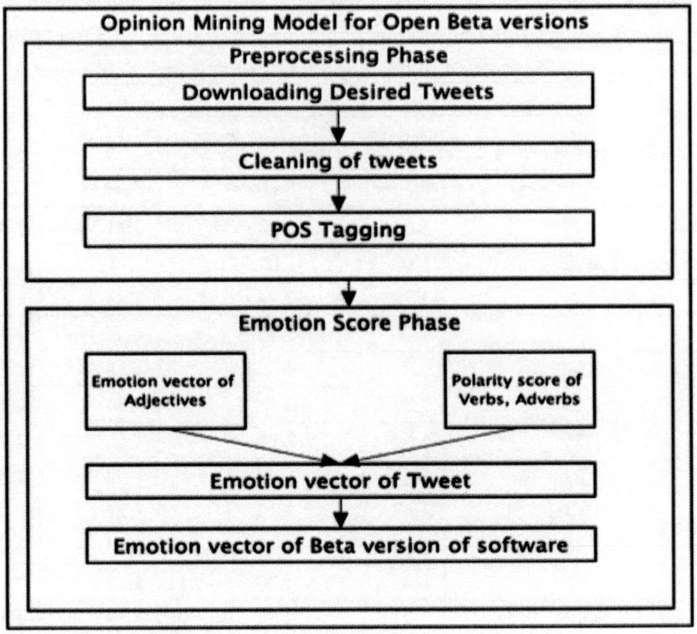

Fig. 2 Opinion mining model for Open Beta versions

3.2 Emotion Scoring of Tweets

Once the file with the required data is prepared, the next step is to calculate the emotional value of each tweet. Words present in a tweet are either scored individually (adjective) or as a group (verbs or adverbs with subsequent adjective) [16]. Each adjective is associated with an emotion vector whose values are calculated using a hybrid technique, i.e., corpus and machine learning-based. Adverb–verb polarity scores are estimated using a dictionary-based method. To calculate the emotional value, five basic emotion scores viz. happiness, anger, sadness, fear, and disgust are determined using the hybrid technique combining lexicon and machine learning method. The emotion values are initially assigned manually and then used as features to train the classifier using the WEKA tool [17] to implement and compare multilayer perceptron, K-nearest neighbor, decision tree, and logistic regression [18, 19].

The file containing a list of adjectives is prepared by conducting a survey online with 350 undergraduate students to score the emotions between 0 and 5. The file had 1000 adjectives with their respective emotion score vectors. A part of this survey outcome is shown below in Table 1.

For adverbs and verbs, a list of positive and negative adverb and verbs was created and then grown using WordNet. Values considered by Kumar and Sebastian [16] for most frequently used verbs and adverbs are used here as well as seen in Table 2.

Finally, the tweet's emotion score is calculated based on the conception that the presence of a verb or adverb prior to any adjective intensifies the emotion of adjective.

Table 1 Sample adjectives with their respective emotion score vectors

Word	Happiness	Anger	Sadness	Fear	Disgust
Abduction	1.09	3.54	3.93	4.11	3.47
Adventure	3.72	1.08	1.09	2.23	1.05
Bake	3.23	1.20	1.14	1.23	1.09
Bland	1.3	1.46	1.72	1.27	1.80
Contempt	1.8	2.42	1.65	1.86	1.70
Crash	1.19	2.82	3.38	3.64	2.17
Deceit	1.15	3.52	2.93	2.57	2.91
Desire	3.84	1.62	1.47	1.53	1.09
Errand	1.74	1.69	1.29	1.21	1.41
Exercise	3.52	1.58	1.47	1.53	1.40
Fame	3.61	1.30	1.26	1.42	1.29
Fear	1.12	2.27	2.52	4.37	1.64
Grin	4.14	1.32	1.21	1.30	1.16
Insolent	1.41	2.02	2.24	1.93	1.82
Invader	1.07	3.38	2.40	3.65	2.68
Hawk	1.92	1.20	1.14	1.82	1.17

Table 2 Polarity scores for adverbs and verbs

Adverb	Strength	Verb	Strength
Most	0.9	Enjoy	0.7
Quite	−0.3	Reject	−0.2
Much	0.1	Suck	−0.9
Never	−0.9	Hate	−1
Hardly	−1	Attract	0.4
Pretty	0.3	Like	0.8
Too	0.6	Adore	0.9
Complete	1	Dislike	−0.7
Less	−0.6	Impress	0.5
Any	−0.2	Smile	0.6

If an adverb or verb is encountered after another adverb or verb then we add it. We repeat this process till an adjective is encountered

Now of the added value of adverbs or verbs is less than 0 i.e., negative, then for the upcoming adjective, we subtract its value from 5.

Or if added value of verbs or adverbs is positive and >=0.5 then multiply it with the upcoming adjective else multiply 0.5 with upcoming adjective.

Fig. 3 Algorithm to calculate emotion vector

The algorithm used for this purpose is shown in Fig. 3, which is the same as given in [20].

4 Implementation and Analysis

To pertinently demonstrate the efficaciousness of the proposed model, we picked the tweets available on IOS13 and AndroidQ.

4.1 Preprocessing of Tweets

Using the Twitter Application Program Interface (API), we downloaded the publicly available tweets on IOS13 and AndroidQ using Python programming language. After downloading all the available tweets, we preprocessed the data for cleaning, removed all the unnecessary characters like hashtags, username indicator (@), hyperlinks, re-tweet symbol (RT), punctuations as well as Non-English characters. Next, by using natural language toolkit, we tagged the adverbs, verbs, and adjectives. The resultant file contained a list of tweets that only have the desired words, namely adverbs, verbs, and adjectives. For example, a tweet "Totally excited to try out #androidQ #androidBeta!!" is converted to "totally excited."

4.2 Emotion Scoring

Once the preprocessing steps are done, the tweets are scored using the algorithm described in Sect. 3. For example, scoring of the above POS-tagged tweet is done as:

1. "Totally" is an adverb, and "excited" is an adjective
2. The adjective emotions of excited are represented by a vector [4.56, 1.25, 1.13, 1.36, 1.13] (<Happiness>, <Anger>, <Sadness>, <Fear>, <Disgust>)
3. In the list of adverbs, the value of totally is 0.8
4. Using the algorithm described in Sect. 3, we calculated the emotion score of the tweet.
5. The adverb value, i.e., "Totally" is 0.8, the emotion scores of the tweet are multiplied by 0.8, and then, the respective scores are divided by 5
6. Resultant score of tweet is vector [0.729, 0.2, 0.18, 0.217, 0.18].

Table 3 shows the sample tweet emotion values. We repeat the process for approx. 1000 tweets for different tags of Open Beta versions of software like #ios13Beta, #androidQ. Figures 4 and 5 show the emotion scores found for two Beta versions, namely IOS13 and AndroidQ.

Table 3 Sample tweets with their emotion values

S. No.	TWEET	Happiness	Anger	Sadness	Fear	Disgust
1	#AndroidQ latest beta is incredible on my #Pixel3! Battery is so much better now and everything is fluid and smooth	0.504	0.038	0.037	0.036	0.038
2	One Drive crashing immediately after opening. Has not worked under the last 3 developer beta updates	0.154	0.534	0.208	0.090	0.188
3	#iOS13Beta is such a success! feels good	0.8653	0.229	0.238	0.298	0.222
4	Fun times ahead!! #AndroidQ	0.759	0.257	0.272	0.293	0.225

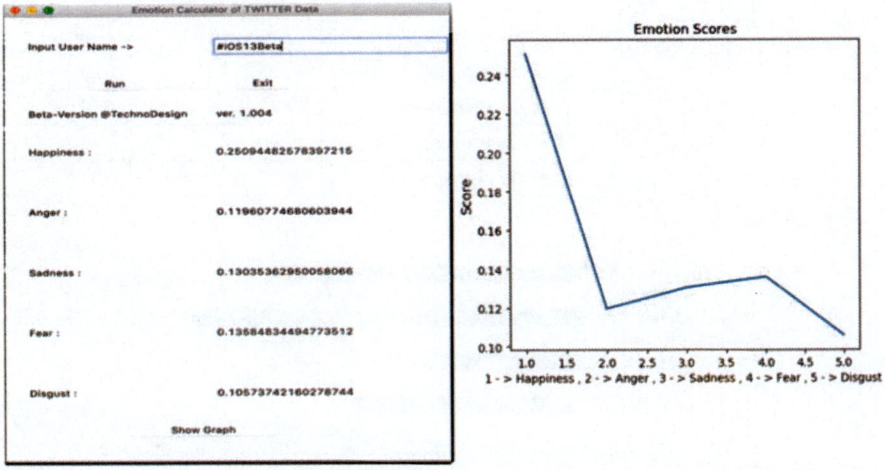

Fig. 4 For 352 tweets with hashtag #iOS13Beta

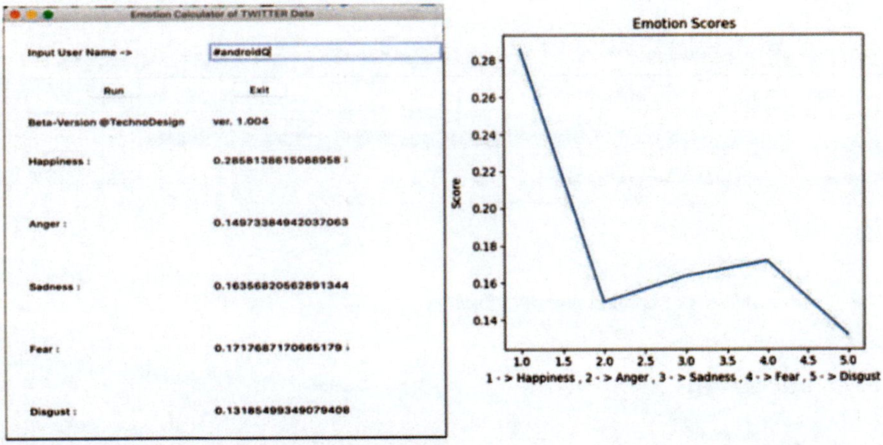

Fig. 5 For 546 tweets with hashtag #androidQ

4.3 Performance of Opinion Polarity Classifier

We used the machine learning techniques such as multilayer perceptron, K-nearest neighbor, decision tree, and logistic regression to determine the performance of our system on the basis of accuracy and time taken. Table 4; Figs. 6 and 7 show the results. We measured the performance of the proposed model using ten-fold cross-validation.

Table 4 Accuracy and time taken by the machine learning techniques

Algorithm	Accuracy (%)	Time (s)
Decision tree	96.55	0.01
K-nearest neighbor	97.12	0.01
Multilayer perceptron	95.40	1.81
Logistic regression	95.97	0.1

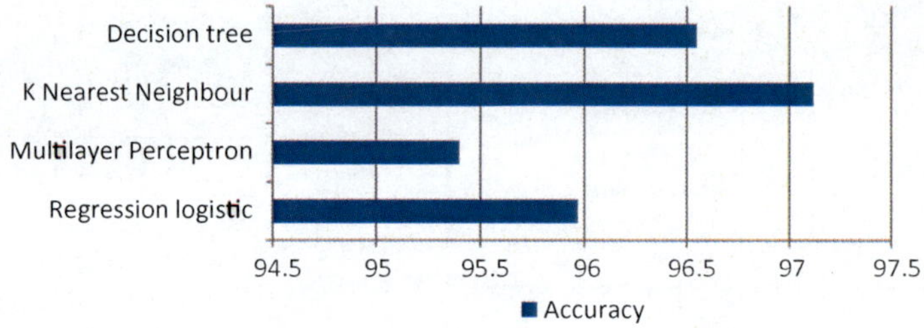

Fig. 6 Accuracy analysis of opinion polarity classifier

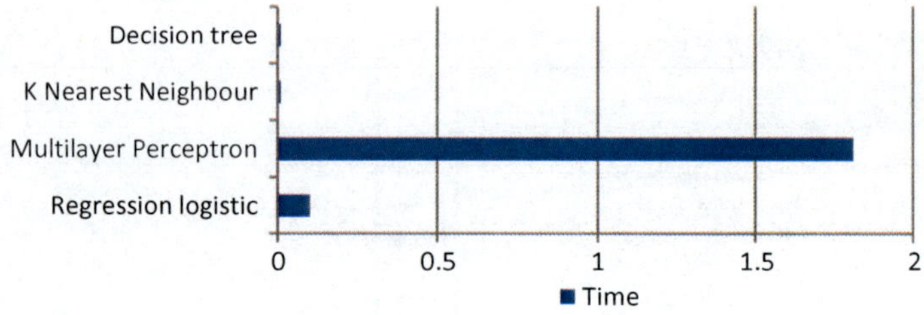

Fig. 7 Time analysis of opinion polarity classifier

5 Conclusion

Opinion mining is extensively researched in areas of marketing, governance, but the field of beta testing of software has not been explored. This research tries to inculcate opinion mining in earlier stages of development to ease the process and get user feedback as accurately as possible to make changes in the software in the Beta stage itself. Our evaluation shows that we achieved the best result with an accuracy of 97.12%, which is 14.37% higher as compared to the best accuracy result of [20].

References

1. V. Sinha, A. Lazar, B. Sharif, Analyzing developer sentiment in commit logs, in *Proceedings of the 13th International Conference on Mining Software Repositories*, pp. 520–523 (ACM, 2016)
2. E. Guzman, D. Azócar, Y. Li, Sentiment Analysis of commit comments in GitHub: an empirical study, in *Proceedings of the 11th Working Conference on Mining Software Repositories*, pp. 352–355 (ACM, 2014)
3. E. Guzman, B. Bruegge, Towards emotional awareness in software development teams, in *Proceedings of the 2013 9th joint meeting on foundations of software engineering*, pp. 671–674 (ACM, 2013)
4. M.M. Rahman, C.K. Roy, I. Keivanloo, Recommending insightful comments for source code using crowdsourced knowledge, in *2015 IEEE 15th International Working Conference on Source Code Analysis and Manipulation (SCAM)*, pp. 81–90 (IEEE, 2015)
5. F. Calefato, F. Lanubile, M.C. Marasciulo, N. Novielli, Mining successful answers in stack overflow, in *Proceedings of the 12th Working Conference on Mining Software Repositories*, pp. 430–433 (IEEE Press, 2015)
6. W. Maalej, Z. Kurtanović, H. Nabil, C. Stanik, On the automatic classification of app reviews. Requirements Eng. **21**(3), 311–331 (2016)
7. E. Guzman, R. Alkadhi, N. Seyff, A needle in a haystack: What do twitter users say about software?, in *2016 IEEE 24th International Requirements Engineering Conference (RE)*, pp. 96–105 (IEEE, 2016)
8. S. Kumar, B. Reddy, An analysis on opinion mining: techniques and tools. Indian J. Res. **5**(8) (2016)
9. C. Kaner, Exploratory testing, in *Quality Assurance Institute Worldwide Annual Software Testing Conference* (2006)
10. V. Bordo, *Overview of User Acceptance Testing (UAT) for Business Analysts (BAs)*. (n.d.) Retrieved 10 Oct 2019, from https://www.develop.com/useracceptancetests
11. J. Perino, *Private & Public Betas: What's the Difference (2014)*. Retrieved 10 Oct 2019, from http://www.betabound.com/private-beta-tests-vs-pUblic-beta-tests-whats-difference
12. A. Kumar, T.M. Sebastian, Sentiment analysis: a perspective on its past, present and future. Int. J. Intell. Syst. Appl. **4**(10), 1–14 (2012)
13. M.A. Storey, C. Treude, A. van Deursen, L.T. Cheng, The impact of social media on software engineering practices and tools, in *Proceedings of the FSE/SDP Workshop on Future of Software Engineering Research*, pp. 359–364 (ACM, 2010)
14. A. Begel, R. DeLine, T. Zimmermann, Social media for software engineering, in *Proceedings of the FSE/SDP Workshop on Future of Software Engineering Research*, pp. 33–38 (ACM, 2010)
15. J. Kim, D. Choi, M. Hwang, P. Kim, Analysis on smartphone related twitter reviews by using Opinion Mining techniques, in *Advanced Approaches to Intelligent Information and Database Systems*, pp. 205–212 (Springer, Cham, 2014)
16. A. Kumar, T.M. Sebastian, Sentiment analysis on twitter. Int. J. Comput. Sci. Issues (IJCSI) **9**(4), 372 (2012)
17. G. Holmes, A. Donkin, I.H. Witten, *Weka: A Machine Learning Workbench* (1994)
18. M. Jain, R. Beniwal, A. Ghosh, T. Grover, U. Tyagi, Classifying question papers with bloom's taxonomy using machine learning techniques. Commun. Comput. Inf. Sci., pp. 399–408 (2019)
19. M. Bhatia, A. Kumar, R. Beniwal, An optimized classification of apps reviews for improving requirement engineering, in *Recent Advances in Computer Science and Communications*, vol. 13(1) (2020)
20. A. Kumar, A. Abraham, Opinion mining to assist user acceptance testing for open-beta versions. J. Inf. Assur. Secur. **12**(4) (2017)

Feature Extraction and Learning on Image and Speech Data

A Genesis of an Effective Clustering-Based Fusion Descriptor for an Image Retrieval System

Shikha Bhardwaj, Gitanjali Pandove, and Pawan Kumar Dahiya

Abstract The tremendous advancements in digital technology pertaining to diverse application areas like medical diagnostics, crime detection, defense, etc., have led to an exceptional increase in the multimedia image content. This bears an acute requirement of an efficacious recovery system to cope up with human demands. The content-based is among the renowned retrieval systems which uses color, texture, shape, edge and other spatial information to extract basic image features. Here, in this paper to enhance the performance of the image retrieval system, a unique and unexcelled color fusion descriptor which combines color moment and color histogram is being proposed. A hybrid feature vector (HFV) is formed after combining features of these two color techniques, and this HFV is given as input to the clustering algorithm. This clustering algorithm performs an efficient class prediction of the given query image by the user. This clustering-based system is also very effective in solving issues related to retrieval time of desired images. Various benchmark datasets like Corel-1K, Corel-5K, Corel-10K and COIL-100 have been tested on the proposed system, and average precision of 91.35%, 77.6%, 71.5% and 97.7% has been obtained, respectively, which is comparatively higher than many state-of-the-art-related techniques.

Keywords Color moment · Color histogram · K-means clustering

1 Introduction

Digital imaging has nowadays evolved into a paramount section in assorted areas of image evaluation and processing [1]. Due to this ever increasing digital image content, the retrieval or recovery of desired images becomes too strenuous for users. For the

S. Bhardwaj (✉) · G. Pandove · P. K. Dahiya
ECE Department, DCRUST, Murthal, Sonepat, Haryana, India
e-mail: shikpank@yahoo.com

S. Bhardwaj
UIET, Kurukshetra University, Haryana, India

P. Bansal et al. (eds.), *Proceedings of International Conference on Artificial Intelligence and Applications*, Advances in Intelligent Systems and Computing 1164,
https://doi.org/10.1007/978-981-15-4992-2_29

ease of users, there is a cardinal necessity of a technology for the effective retrieval of images. And the solution to this dilemma is known by a technique, designated as content-based image retrieval (CBIR) system. It extracts basic visual attributes like texture, shape, color, etc., both from the query image and total database images.

Images can be recovered in two basic ways: (1) Text-based image retrieval (TBIR). (2) Content-based image retrieval (CBIR). TBIR is the traditional method of retrieving images and is based on text or keyword manual annotation of images. But, it suffers from vast errors like use of synonyms, homonyms, misspellings, etc. [2]. In contrast to TBIR, content-based is an efficient technique of retrieving images from huge repository of collected images. This system eliminates majority of the pitfalls present in TBIR system. In this system, image recovery is based on the presence of various image features. A basic CBIR system consists of a query image given by the user, and different image features are extracted from this user input image, and feature vectors are created. The similar procedure is adopted for the whole dataset. These calculated feature vectors are matched in the similarity matching stage of the system by using a specific distance metric, resulting in visually similar images [3]. Both local and global type of features can be withdrawn from an image.

Among all the basic image features, color is the utmost indispensable attribute that can be easily perceived by human eye. Each color and every object associated with color has a specific wavelength related to it, which gets reflected and forms an image in human retina. For color feature extraction, there are numerous techniques [4] which can be applied to CBIR system. But, the challenge here is to find the optimal technique for feature recovery and also the best amalgamation of these techniques. Thus, the main contributions of this paper are as follows:

- To form a hybrid color descriptor by using the best two color feature extraction techniques namely color moment (CM) and color histogram (CH).
- To enhance the accuracy of the proposed color descriptor by using K-means clustering with the developed fusion-based color descriptor.
- Precision, recall and retrieval time of the proposed system have been compared with many related techniques.
- Four benchmark datasets have been tested on the developed image retrieval system.

The remaining arrangement of the paper is as follows: Related work is presented in Sect. 2, Sect. 3 gives a detailed information about employed methodology, in Sect. 4, experimental analysis and results are given, and finally, conclusion and future trends are presented in Sect. 5.

2 Related Work

In the above section, it is already stated that on the basis of various image attributes like texture, color, shape, edge, etc., images can be searched, indexed and retrieved in a CBIR system [3]. This section confers to different techniques which are used for color feature extraction in CBIR system.

Liu et al. [5] describe about a system in which color difference histogram (CDH) is calculated. The technique applied here is used to evaluate the color difference amidst duo points which exhibits perceptual uniformity above the HSV color space. Color auto correlogram is also an eminent descriptor for the withdrawal of color features where calculation is based on the spatial relationship between two similar pixels. Color auto correlogram [6] in conjunction with color moments is applied as a fusion technique to extract color features. Support vector machines (SVM) is also employed as a classifier, in order to increase the performance of the system [7, 8]. But, all these systems are incapable to compensate the retrieval time of the system as all these systems lack the inclusion of any cluster-based technique.

Color moment is defined as a technique which describes the distribution of color in an image. Ranklet transform is used here with color moment for final recovery of images. This transform is used to minimize the effects of scaling and rotation [9].

Chen et al. [10] propose a color feature extraction system based on thresholding, where binary quaternion-moment-preserving (BQMP) is used in combination with different retrieval techniques. The methods used for color extraction are variable cardinality (VC) and fixed cardinality (FC). Color space is a model for the representation of intensity values of a given image. Varied types of color spaces can be employed in a CBIR system like RGB, HSV, L*u*v*, CIE, L*a*b*. Database of different pathology images, dermatological images, hematological images, dental images has been used for experimentation in medical-based CBIR systems [11].

Dominant color descriptor (DCD) is another color descriptor used in CBIR systems. Here, partitions are obtained out of color space, and every partition has its center and percentage. A combination DCD, gray level co-occurance matrix (GLCM) [12] and Fourier descriptor is the deployed techniques of the constituted hybrid system [13]. Color coherence vector (CCV) and improved color coherence vector (ICCV) [14] have also been used as color extraction techniques. Many drawbacks of color histogram are replaced by this method. But, CCV suffers from exhaustive computation time and more feature vector space.

Many types of classifiers and clustering techniques can also be embedded with CBIR systems. Pandey et al. [15] proposed a CBIR system based on the usage of agglomerative clustering which forms the groups of images within the total dataset of images. Another clustering approach known as density clustering [16] is employed with t-distributed stochastic neighbor embedding (t-SNE) to reduce the drawbacks of storage space and time.

But, this agglomerative clustering has a long computational time, and due to change in densities, density clustering does not provide accurate results. Thus, most of the above-described color feature descriptors suffered from one or another limitation.

So, a methodology is proposed in this paper which is an amalgamation of color histogram and color moment feature extraction techniques. Both the methods jointly contribute in producing a highly effective system with reduced time consumption by the inclusion of K-means clustering with the fusion descriptor.

3 Proposed Methodology

Feature can be defined as a localized, pertinent and perceptible attribute of an image. Feature extraction is a salient process in CBIR systems. In this paper, a combination of color histogram and color moment has been done for the withdrawal of color features. Also, K-means clustering technique has been utilized to solve the issue of time complexity and improve the overall performance of the system.

3.1 Color Histogram

Color histogram is the most commonly used color feature extraction technique in CBIR systems. In this technique, overall pixels in the image are considered and are represented in the form of bar graphs. Pixels are computed in fixed intervals known as bins. It has two axes where X-axis denotes tonal colors and Y-axis signifies the number of pixels in a particular bin. A basic form of color [17] histogram can be denoted by Eq. (1).

$$h_{P,Q,R}(p, q, r) = N.\text{prob}\,(P = p, Q = q, R = r) \tag{1}$$

where the three color channels of an image are given by P, Q and R and N signifies the number of pixels of an image. This technique can also be used for the calculation of mean and standard deviation of images [18].

Color histogram is invariant to scale, translation and rotation, but they suffer from lack of spatial information. To overcome this limitation, many other descriptors like color moment, color signatures, etc., can be used.

3.2 Color Moment

Color moments are metrics which describe the color distributions in an image. Its distribution can be effectively described by moments, according to probability theory. It eliminates the drawback of color histogram by including spatial information from images [19].

If I_{ij} specifies, the ith color channel, jth image pixel, total number of pixels in an image are N, then index entries of the color channel and region r are given by first color moment, denoted by Eq. (2).

$$\text{Mean}(E_{r,i})\frac{1}{N}\sum_{j=1}^{N} I_{ij} \tag{2}$$

Now, standard deviation is the next moment and is calculated by taking square root of variance of color distribution in an image. It is given by Eq. (3).

$$\text{Standard deviation}(\sigma_{r,i}) = \left(\frac{1}{N}\sum_{j=1}^{N}(I_{ij} - E_{r,i})^2\right)^{\frac{1}{2}} \tag{3}$$

Next moment is called as skewness and signifies that how asymmetric is the color distribution. It is given by Eq. (4).

$$\text{Skewness }(S_{r,i}) = \left(\frac{1}{N}\sum_{j=1}^{N}(I_{ij} - E_{r,i})^3\right)^{\frac{1}{3}} \tag{4}$$

3.3 K-Means Clustering Algorithm

K-means is a partitioning algorithm which uses a two-step iterative process to find the user specified K clusters. If the set of observations $(\boldsymbol{y}_1, \boldsymbol{y}_2, \boldsymbol{y}_3, \ldots, \boldsymbol{y}_n)$ are given, then the K-means clustering algorithm disintegrates the n observations into K parts $(K \leq n)$ denoted by $S = (\boldsymbol{S}_1, \boldsymbol{S}_2, \boldsymbol{S}_3 \ldots \boldsymbol{S}_k)$ in order to minimize the variance. Thus, the main objective of this algorithm [20] can be given as

$$\min_{S_k} \sum_{k} \sum_{y \in S_k} \pi_y \text{dist}(y, m_k) \tag{5}$$

In Eq. (5), S_k denotes k clusters, m_k denotes the centroid (centroid is the mean of the various cluster members) of S_k, $\pi_y > 0$ denotes the weight of y and dist is the distance function used.

3.4 Fusion of Feature Extraction Techniques

In this paper, an exclusive technique which integrates color histogram and color moment has been proposed. Both these techniques are jointly inexplicable. Formally, the proposed technique can be divided into two parts: (1) Cluster creation stage (2) Testing/cluster prediction stage.

(1) **Cluster creation stage**

This stage consists of the following sub-steps:

A. **Feature extraction stage:** It is the first stage of this phase and is based on the extraction of features from the utilized dataset using CM and CH to extract color features. After this stage, two independent feature vectors are formed for each of the attribute.
B. **Fusion stage:** The two feature vectors are fused together in this stage for the formation of a hybrid feature vector (HFV). These two feature vectors are normalized independently by using minimum–maximum normalization for the creation of HFV.
C. **Cluster creation stage:** The selected HFV is given as an input to the clustering algorithm. Suitable centroid has been obtained by using K-means clustering algorithm, which is an image, based on which matching with the query is performed. Main steps are as follows:

1. Initialize the initial value of K randomly.
2. K clusters are formed by connecting every given input with the nearest mean.
3. Then, the arithmetic mean (Centroid) of each of the K cluster becomes the new mean, and step 2 and 3 are repeated unless a convergence is reached, and finally, desired clusters are obtained.
4. End.

Then, finally desired clusters are obtained which successfully divides the utilized dataset into several parts. The architecture of this part of the proposed system is shown in Fig. 1a.

(2) **Testing/Cluster prediction stage**

In this stage, the different clusters which are created by using clustering approach in the previous section are utilized to retrieve an image by using a specific distance metric. The main subsections of this stage are as follows:

A. **Query formulation stage**: The user enters the query image to test the working of the proposed system. Here, each and every image is utilized as a query image.
B. **Feature extraction and Fusion stage**: Again, the features of the query image are extracted using CM and CH, and finally, a hybrid feature vector (HFV) is created.
C. **Cluster Prediction and Similarity matching stage**: In this stage, the matching of the query image is done with the already created cluster, and the best cluster is chosen. Then, by using a specific distance metric, matching of the query image

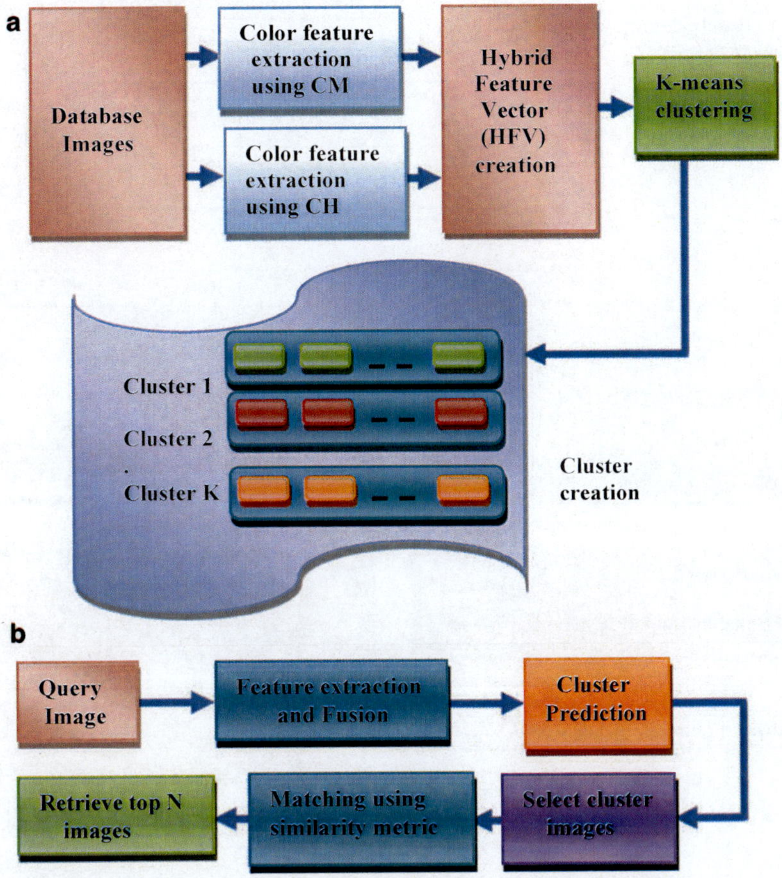

Fig. 1 **a** Framework of cluster creation stage. **b** Framework of cluster prediction stage

is performed with images in that selected cluster and finally top N images are retrieved based on the minimal distance. The basic architecture depicting this part is given in Fig. 1b.

4 Experimental Setup and Results

In order to obtain accurate and standardized results, the setup for experimentation is as follows.

Table 1 Briefly description of utilized datasets

Image database	Total images	Total classes	Size of images
Corel-1K	1000	10	256 × 384 or 384 × 256
Corel-5K	5000	50	256 × 384 or 384 × 256
Corel-10K	10,000	100	256 × 384 or 384 × 256
COIL-100	7200	100	128 × 128

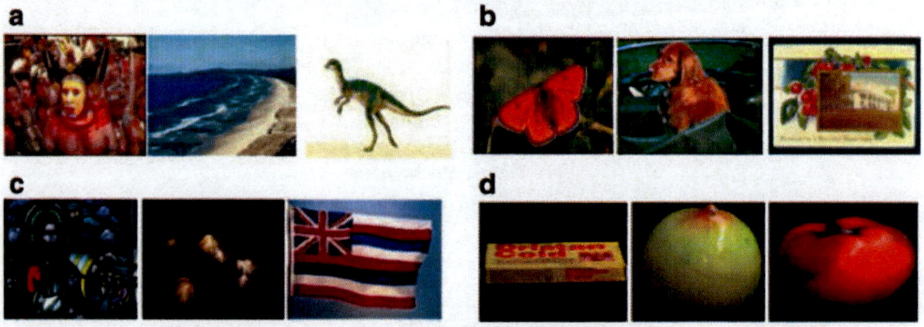

Fig. 2 **a** Sample images from Corel-1K. **b** Sample images from Corel-5K. **c** Sample images from Corel-10K. **d** Sample images from COIL-100

4.1 Experimental Setup

Here, the experimental analysis is being done over four datasets namely Corel-1K, Corel-5K, Corel-10K [21] and COIL-100 [12]. These experiments are performed in MATLAB R2017a, core i3 processor, 4 GB memory and 64 bit windows. A brief description of these datasets is given in Table 1, and sample images from each dataset are shown in Fig. 2a–d.

4.2 Results

There are different evaluation parameters which can be utilized for calculating the capability of a particular CBIR system. But, the prominent metrics are precision and recall [22, 23].

$$\text{Precision} = \frac{\text{Number of relevant images retrieved}}{\text{Total number of images retrieved}} \tag{6}$$

Table 2 Average precision using three distance metrics

Dataset	Euclidean	Manhattan	Minkowski
Corel-1K	91.35	88.56	84.3
Corel-5K	77.6	72.4	67.7
Corel-10K	71.5	66.5	65.2
COIL-100	97.7	90.65	89.64

$$\text{Recall} = \frac{\text{Number of relevant images retrieved}}{\text{Number of relevant images in database}} \quad (7)$$

Many types of distance metrics [24] are utilized for the purpose of calculating similarity. The results obtained by using varied distance metrics on the proposed system are given in Table 2. The average precision obtained by using Euclidean distance metric outperforms the other distance metrics. Manhattan produces innumerable false negatives and does not yield accurate results. Euclidean distance metric is based on weighted and normalized attributes and has speedy computational performance.

The implemented system has been initially compared to many state-of-the-art-related techniques regarding average precision obtained on Corel-1K, Corel- K and Corel-10K datasets. This comparison has been shown in Table 3.

Again, the comparative analysis for COIL-100 dataset with the related techniques is given in Table 4.

Hence, from Tables 3 and 4, it can be seen that the proposed technique has superior results as compared to many related techniques. In order to verify the retrieval time efficiency of the proposed system, the retrieval time of the proposed system is also

Table 3 Comparison of the proposed system with the related techniques

Dataset	Semantic name	Ref. [24]	Ref. [19]	Ref. [21]	Ref. [23]	Proposed system
Corel-1K	Africa	95	81	72.4	81	80.5
	Beach	60	66	51.5	92	97
	Building	55	78.75	59.55	79	78.5
	Bus	100	96.25	92.35	93	85
	Dinosaur	100	100	99	99	100
	Elephant	90	70.75	72.7	79	100
	Flower	100	95.75	92.25	99	96
	Food	100	98.75	72.35	88	95
	Horse	100	67.75	96.6	80	96.5
	Mountain	75	77.25	55.75	85	85
	Average	87.5	83.225	76.5	87.5	91.35
Corel-5K	Average	72.56	68.6	55.25	75.4	77.6
Corel-10K	Average	65.45	59.98	49.58	68.9	71.5

Table 4 Comparison of the proposed system with the related techniques

Database	Average precision (%)			
COIL-100	Ref. [12]	Ref.[14]	Ref. [25]	Proposed
	95	82.5	96.5	97.7

compared to many techniques, and its performance is given in Fig. 3. It can be seen that the proposed system has less retrieval time as compared to related techniques. Separate graphical user interface (GUI) [25] has been designed for each of the utilized dataset in order to show the retrieval results the GUI for two utilized datasets have been shown in Fig. 4a, b.

Thus, it can be seen from Fig. 4a, b that top ten images have been retrieved and the retrieved images belong to the same category of the given query image.

Precision versus Recall (P vs R) curves by varying the number of images from 10 to 50 for the utilized datasets are shown in Fig. 5a, b.

Thus, from the curves given in Fig. 5a, b, it can be seen that as there is a decrease in the value of precision, and there is an increase in the value of recall.

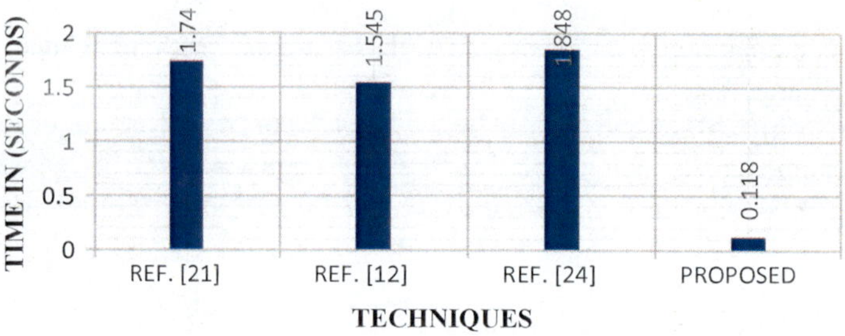

Fig. 3 Comparison of retrieval time with the related techniques on Corel-1K dataset

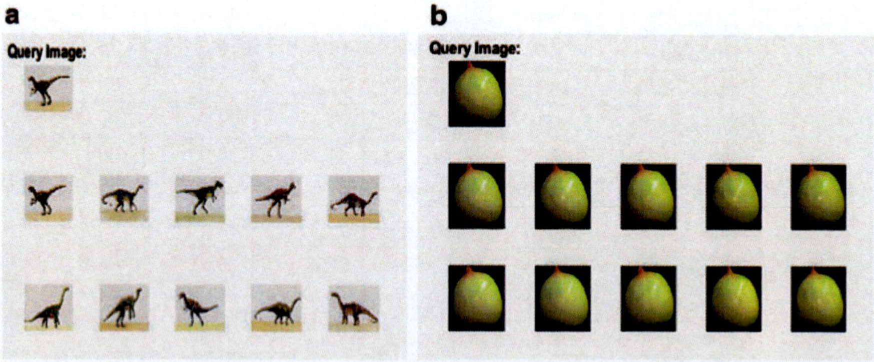

Fig. 4 **a** Retrieval results from Corel-1K. **b** Retrieval results from COIL-100

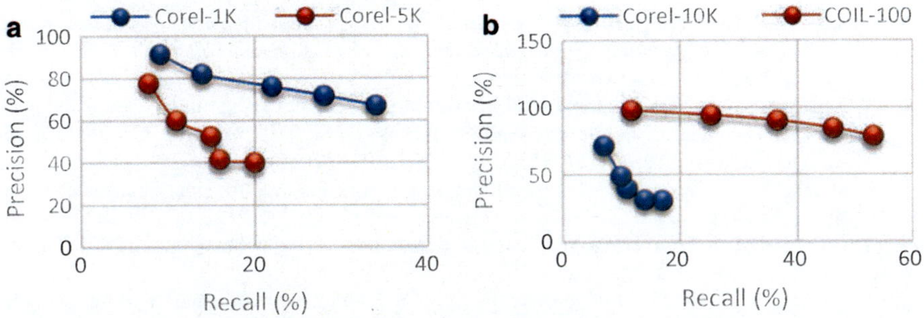

Fig. 5 **a** P versus R curve on Corel-1K and Corel-5K. **b**. P versus R curve on Corel-10K and COIL-100

5 Conclusion and Future Trends

Customary traditional techniques for color feature extraction suffer from one or the other disadvantage. Therefore, this paper presents a unique fusion color descriptor which combines color histogram and color moment. Color histogram has very fast computation time and is easy to calculate. Similarly, in color moment, different moments can be calculated, and it also includes the spatial information of the input image. These advantages of the two techniques contribute in producing an efficient retrieval system. To validate these facts, simulation results are given which have been compared with many related techniques. Further, to improve the performance of the proposed system, K-means clustering is also been applied. The proposed method has been successfully tested on four benchmark datasets namely Corel-1K, Corel-5K, Corel- K and COIL-100. Further, in future, the proposed work can include texture and shape features and can be applied to some applications areas like crime detection, medical diagnosis, face recognition and many other pattern recognition applications. Moreover, techniques like deep learning and Internet of things (IoT) can be added to the proposed system.

References

1. N. Jhanwar, C. Subhasis, S. Guna, Z. Bertrand, Content based image retrieval using motif cooccurance matrix. Image Vis. Comput. **22**(14), 1211–1220 (2004)
2. A. Alzu'bi, A. Amira, N. Ramzan, Semantic content-based image retrieval: A comprehensive study. Journal of Vis. Commun. Image Represent. **32**, 20–54 (2015)
3. A.K. Naveena, N.K. Narayanan, Image retrieval using combination of color, texture and shape descriptor, in *International Proceedings on Next Generation Intelligent Systems (ICNGIS)*, pp. 1–5 (IEEE, 2016)
4. L. Chuen-Horng, C. Rong-Tai, C. Yung-Kuan, A smart content based image retrieval system based on color and texture feature. Image Vis. Comput. **27**(6), 658–665 (2009)
5. G.H. Liu, J.Y. Yang, Content-based image retrieval using color difference histogram. J. Pattern Recogn. **46**(1), 188–198 (2013)

6. J. Huang, S.R. Kumar, M. Mitra, W.J. Zhu, R. Zabih, Image indexing using color correlograms, in *Proceedings of International Proceedings on Computer Vision and Pattern Recognition*, vol. 191, no. 3–4, pp. 762–768 (1994)
7. N. Tripathi, A new technique for cbir with contrast enhancement using multi- feature and multi class SVM classification, in *International Proceedings on Signal Processing, Communication, Power and Embedded System (SCOPES)*, pp. 2031–2036 (2016)
8. A. Shahbahrami, Comparison between color and texture features for image retrieval. J. Iran **27648**, 1–9 (2008)
9. A.J. Afifi, W.M. Ashour, Image retrieval based on content using color feature. ISRN Comput. Graph. **2012**, 1–11 (2012)
10. W.T. Chen, W.C. Liu, M.S. Chen, Adaptive color feature extraction based on image color distributions. IEEE Trans. Image Process. **19**(8), 2005–2016 (2010)
11. M.S. Banu, K. Nallaperumal, Analysis of color feature extraction techniques for pathology image retrieval system, in *International Proceedings on Computational Intelligence and Computing Research*, pp 1–7 (IEEE, 2010)
12. E. Mehdi, E. Aroussi, N. El. Houssif, *Content-Based Image Retrieval Approach Using Color and Texture Applied to Two Databases (Coil-100 and Wang)*, pp. 49–59 (Springer International Publishing, 2018). https://doi.org/10.1007/978-3-319-76357-6_5
13. M.V. Lande, P. Bhanodiya, P. Jain, An effective content-based image retrieval using color, texture and shape feature, in *Intelligent Computing, Networking, and Informatics, Advances in Intelligent Systems and Computing*, pp. 1163–1170 (Springer, 2014)
14. H.A. Elnemr, Combining SURF and MSER along with color features for image retrieval system based on bag of visual words. J. Comp. Sci., pp. 213–222 (2016)
15. S. Pandey, P. Khanna, Content-based image retrieval embedded with agglomerative clustering built on information loss. Comput. Electr. Eng. **54**, 506–521 (2016)
16. H. Lacheheb, S. Aouat, A density clustering approach for CBIR system, in *13th International Proceedings on Computer Systems and Applications, AICCSA*, pp. 1–8 (IEEE/ACS, Agadir, 2016)
17. J.M. Patel, A review on feature extraction techniques in content based image retrieval, in *International Proceedings on Wireless Comm, Signal Processing and Networking*, pp. 2259–2263 (IEEE, 2016)
18. S.M. Singh, K. Hemachandran, Content -based image retrieval using color moment and gabor texture feature. Int. J. Comput. Sci. Issues **9**(5), 299–309 (2012)
19. L.K. Pavithra, T.S. Sharmila, An efficient framework for image retrieval using color texture and edge features. J. Comput. Electr. Eng. **0**, 1–14 (2017)
20. J. Cao, Z. Wu, J. Wu, W. Liu, Towards information-theoretic K-means clustering for image indexing. Sig. Process. **93**(7), 2026–2037 (2013)
21. S. Fadaei, R. Amirfattahi, M.R. Ahmadzadeh, New content-based image retrieval system based on optimised integration of DCD, wavelet and curvelet features. IET Image Process. **11**(2), 89–98 (2017)
22. C. Singh, K.P. Kaur, A fast and efficient image retrieval system based on color and texture features. J. Vis. Commun. Image Represent. **41**, 225–238 (2016)
23. Y. Mistry, D.T. Ingole, M.D. Ingole, Content based image retrieval using hybrid features and various distance metric. J. Electr. Syst. Inf. Technol. **2016**, 1–15 (2017)
24. J. Pradhan, S. Kumar, A.K. Pal, H. Banka, A hierarchical CBIR framework using adaptive tetrolet transform and novel histograms from color and shape features. Digit. Signal Process. A Rev. J. **82**, 258–281 (2018)
25. S. Liu et al., Perceptual uniform descriptor and ranking on manifold : a bridge between image representation and ranking for image retrieval. J. Latex. arXiv:1609.07615v1 [cs.CV] 24 Sept 1–14 (2016)

MR Image Synthesis Using Generative Adversarial Networks for Parkinson's Disease Classification

Sukhpal Kaur, Himanshu Aggarwal, and Rinkle Rani

Abstract Parkinson's disease (PD) is one of the most common types of neurological diseases triggered by progressive degeneration of dopaminergic neurons in the brain. Although this neurodegenerative disease does not have a known cure, early detection accompanied by earlier care may help patients to have a better quality of life. Throughout recent years, Magnetic Resonance Imaging (MRI) has been one of the most common diagnostic tools because it removes harmful radiation. Deep learning approaches have shown high applicability for diverse databases of clinical objects. But, there is huge sum of data that is required to train deep networks limiting the application of some activities like medical image classification thereby lacking sufficient training samples for each class. In this paper, a solution focused on the Generative Adversarial Network (GAN) is proposed for the augmentation of training data to improve the quality of the diagnosis of Parkinson's disease. A Generative Adversarial Network is designed to generate MR images of high quality. A mixing training methodology is then implemented through a mixture of GAN-based data augmentation strategies to further improve the performance of the Deep CNN classifier. The MR images of healthy and PD patients obtained from Parkinson Progression Markers Initiative (PPMI) are used to test the proposed approach. A total of 185 samples are collected and 95 sample images were actually applied for data augmentation. This model's improved data set is as many as 600 samples, and it is found that the images produced are of good quality using this data set for the detection of maximum signal-to-noise ratio (PSNR), which is of revolutionary value in the real

S. Kaur (✉) · H. Aggarwal
Department of Computer Engineering, Punjabi University, Patiala 147002, India
e-mail: sukhpal91@gmail.com

H. Aggarwal
e-mail: himanshu.pup@gmail.com

R. Rani
Computer Science and Engineering Department, Thapar Institute of Engineering and Technology, Patiala 147004, India
e-mail: raggarwal@thapar.edu

P. Bansal et al. (eds.), *Proceedings of International Conference on Artificial Intelligence and Applications*, Advances in Intelligent Systems and Computing 1164,
https://doi.org/10.1007/978-981-15-4992-2_30

test norm. On synthetic and real MR images of healthy and PD patients, the Deep CNN Model reaches an AUC of 88.4%. The proposed approach shows promising in PD diagnosis.

Keywords Generative adversarial networks · Parkinson's disease · Convolutional neural networks

1 Introduction

Following Alzheimer's disease (AD), Parkinson's disease (PD) is the second most important neurodegenerative disease impacting middle-aged and elderly people. PD induces a gradual loss of neuron-generating dopamine in the brain which occurs in two forms of effects, namely motor and non-motor. Although some experimental techniques are available to identify and assess the development of PD, the outcomes of these studies are empirical and rely on the experience of the clinicians, on the other side, are costly and the treatment phase is time-consuming [1]. There are different types of neuroimaging methods, one of the most common of which is Magnetic Resonance Imaging (MRI) because it is an inexpensive and non-invasive process. Individuals with PD develop signs if they lose nearly 80% of their brain dopamine [2]. In the area of clinical image analysis, machine learning has shown remarkable results throughout recent years. The proposed neurodisease diagnostic CAD system uses multiple imaging data types, including single-photon computed tomography (SPECT) [3], diffusion tension imaging (DTI), positron emission tomography (PET) and MRI. Artificial intelligence (AI) use has drawn attention in medical imaging activities in recent years [4]. Small data datasets are however major obstacles, in general for controlled machine learning and for rare conditions for which only a small number of cases can occur even in large databases [5, 6]. Many strategies for information expansion have been proposed in this respect, but minor changes such as overfitting in learning processes or structural transformations have not fulfilled the immediate need for larger data sets [7]. Generative adversarial networks (GANs) may help meet the need for a rise in synthetic information. The idea of GAN has been extended to medical imaging in several reports. The MR images of healthy patients and PD patients are used as data sets in this paper and the GAN is used to maximize results for the classification of Parkinson's disease.

2 The Proposed Approach

The proposed approach demonstrates our three-component design architecture: Generator network, discriminator network, and classification network. The network of generators and the network of discriminators is a GAN. The random noise z is used as an input to the generator. The generator network is used to produce synthetic images while the discriminator network filters these images. The DNN classifier provides a mechanism for the identification of healthy and Parkinson's disease patients based on synthetic and original MR images.

2.1 Dataset

T1-weighted brain MR images collected from the PPMI registry are used in researching this paper. PPMI is a global, large-scale community experiment designed to identify biomarkers for PD progression [8]. The evidence used in analysis includes the original 598 reference T1 MR images of 75 Parkinson's disease (PD) and 110 healthy controls (HC).

2.2 Generator

Six-layer architecture is chosen as the generator network and is displayed in Fig. 1a.

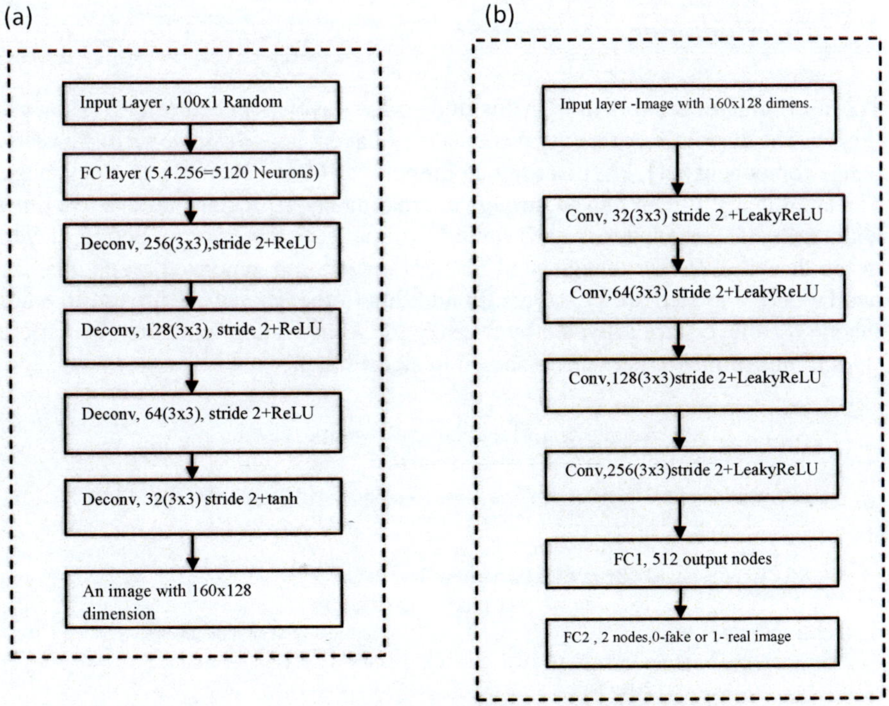

Fig. 1 Structure diagram of **a** Generator **b** Discriminator

The generator input is a vector sampled from a uniform distribution of 100 random values in the [0; 1] array [9]. Every time a random noise is added to the input variable on being exposed to the model. This process of adding noise to the input samples can be called data augmentation [10]. The input feeds directly into a fully connected layer of 5120 units and is then converted into 16 × 16 × 256 sizes. Finally, MR images with a width of 160 × 128 are produced after five deconvolution layers. Algorithm 2 gives the detail of various steps involved in the generator network.

```
Algorithm 2.Generator network
1.Input:-Random noise Z.
2.Output:-Real and fake images.
3.define_generator(l_dim)
4.for i in rande do
5.model.add(dense(nodes,l_dim=100)     //addition of random noise
6.model.add(Conv2DTranspose(channel
    no,kernel_size,stride,padding)
7.model.add(ReLU)                       //addition of activation function
8.return model.
```

2.3 Discriminator

The discriminator network used in this study is five-layer architecture and is shown in Fig. 1b. The discriminator input layer consists of a real image sample set and a vector image component [11]. The discriminator input is a 160 × 128 single-channel image. The input image is then moved through alternating layers of convolutions five times with stride of 2; the latter is used for sub-sampling as there are no pooling layers in the design [12]. The weight and bias parameters are generated randomly. The two layers completed are FC layers. In addition to the last layer with no activation function, all layers are activated by a Leaky ReLU activation function. The various steps of discriminator network is shown in algorithm 3.

```
Algorithm 3.Discriminator network
1.Input:-Real and fake images
2.Output:-class label 1 for real and 0 for fake.
3.define_discriminator(lmage size)
4.for i in range do
5.model.add(Conv2D(channel
    no,kernel_size,stride,padding)
6.model.add(LeakyReLU)          //addition of acvitation function
7.model.add(dense(no of neurons,activation_fn)
    // fully connected layer
8.return model
```

The back propagation method supports both the generative and the Discriminator models to change the parameters.

In this paper, the learning rate of the neural network is set to 0.0002, the number of iterations is 1000, and the batch size is 8. In the test, the convolution kernel is set to 3 * 3 [5, 13]. We use the antagonistic effect of GAN to produce objects that are close to the actual information in the iteration phase following iteration of the generative network [14].

2.4 *Classification Network*

The famous Le-Net-5 architecture is used as a DNN classifier [15]. The pre-trained Le-Net based framework contains five layers such as two convolution layers, two Max-pooling layers; one fully connected layer. Algorithm 5 explains the various steps of DNN classifier. The ReLU activation function is used on all layers except the last fully connected layer on which sigmoid function is used. After data augmentation, the DNN classifier is trained on synthetic and real MR images of healthy and PD patients [16–18]. The model is trained for a total of 200 epochs with the Adam optimization algorithm.

Thus, we get a new output of around $160 \times 128 \times 96$. The local accessible field layer is occupied by a max-pool layer with kernel size 2 and a stride of 1 having zero paddlings. The second part is equal to the first convolution part with a parallel max-pool layer. The last layer is a softmax layer used for data categorization. This layer contains a fully linked layer of two neurons. Neuron one corresponding to healthy controls, neuron two corresponds to PD patients.

3 Results and Discussion

All the experiments are conducted using Theano deep learning library in python 2.7.0 software running on a laptop with Intel® Core™ i5-6700K microprocessor and 6 GB RAM. The MR image dataset used in this work involves subjects from the PPMI database regarded by the healthy individuals and Parkinson's disease patients. The complete collections of MR images are shown in Fig. 2a, b for healthy Individuals and PD patients. The MR images can demonstrate the degeneration of brain systems. A substantial reduction of gray matter intensity is noted in PD patients relative to healthy individuals with modifications in the striatum region.

The analysis aimed to demonstrate that the model based on data augmentation by using GANs beats the one without. Through measuring its accuracy, specificity, and sensitivity, the efficiency of the GAN classifier is achieved. The specificity is calculated by the percentage of the true negative where the nature of the negative situations is determined correctly. To give the sensitivity of the classification procedure, the proportion of the true positive is determined. The accuracy is determined

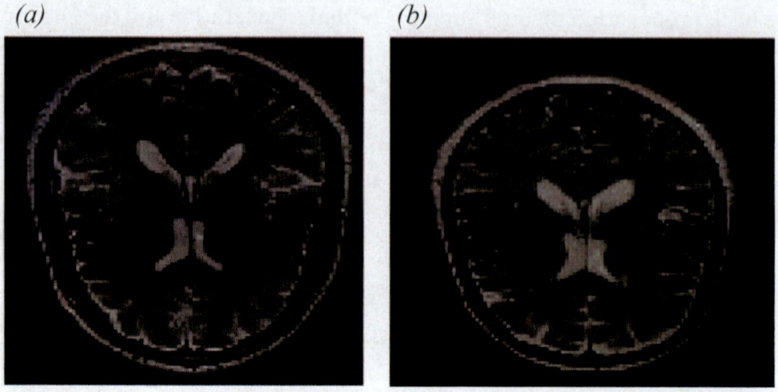

Fig. 2 MR images of a Healthy individuals, b Parkinson's disease patients

by the ratio between true positive or false negative for the total number of correct predictions [19, 20]. The situation under which training is performed included a lack of bias against the discriminator to achieve 0. This is accomplished after approximately 200 epochs, as seen in Fig. 3, which indicates the training losses of the GAN on the trained subset in a PPMI control group (i.e. healthy controls). Similar training error curves are found in the PD subset.

In the results, the diversion in the loss function can be seen through the below-given trend chart as seen from Fig. 3a, b. The change in the acquired model is seen as same but the loss function for the discriminator model is high. This means that both have separate spaces and degree for improvement for the degree of simulation and discrimination for the existing information that will prove to be advantageous for the entire model. It is visible from the variation figure for loss function generative model seen from Fig. 3a. It is seen from the generated image of the generative framework that it's very unstable at the start of the iteration process. After constant 150 iterations,

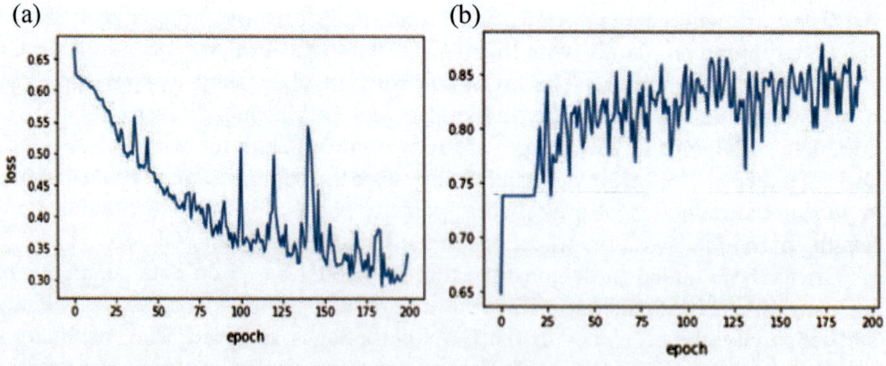

Fig. 3 Loss function value of a Generator network. b Discriminator network

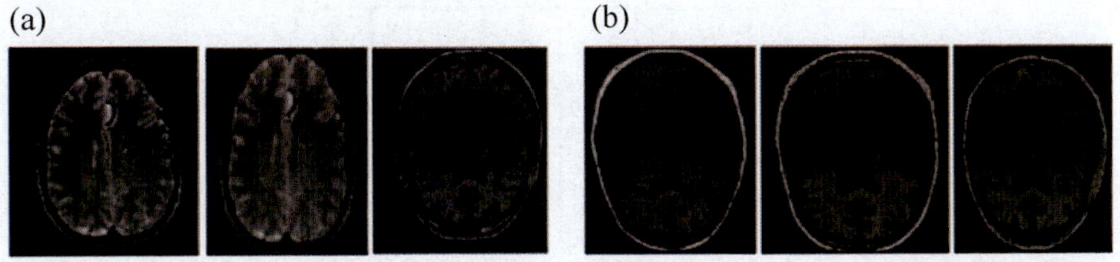

Fig. 4 Synthetic images after **a** 100 iteration, **b** 190 iteration

it is starting to go down and then with time it become stable. The total stability is attained after a total of 150 iterations. From the findings and analysis of the time figure of iteration Fig. 3b, it is observed that the value of change in the loss function in the model of adversarial begins from 0.05 when it rises. From the change value of the loss function, it starts to rise quickly. But, from the above two figures, it has been observed that the training of GAN is not stable and the generator and discriminator may drop or rise constantly during the entire process.

In the 100th version, the images are distorted and the simple tone is faulty as shown in Fig. 4a. Figure 4b shows the Contours and textures appeared during the 150th version. Following 190 runs, the proposed approach can be balanced. We can see in Fig. 4b that the customized characteristics are enhanced after 190 iterations which are somewhat similar to the real data set. The images produced are thus chosen for 190 iterations in this experiment. Nonetheless, if we expand the images, the overall created image is similar in style and in tone to the original image but has defects as well. The image's granulation is more extreme, and the image is noisy.

In order to measure the effectiveness of the proposed approach, the following study was devised: after being thoroughly trained, the GANs model generates 500 fake images for each class. These were merged with images from the initial database in four independent hybrid datasets, each with a specific false/real image proportion (i.e. 25, 50, 75, and 100%). Subsequently, synthetic images are applied to the training data, creating a hybrid dataset, which is again used to train the same network with and without the use of the conventional augmentation. In this paper, to illustrate the ability of our proposed approach, we present a number of images generated by the trained generator. Most of the images produced proved to be of high quality (Fig. 4b) and displayed all the characteristics of a real MRI scan.

Figure 5 shows the performance of DNN classifier on mixed MR images of healthy and PD patients over different iterations. The pre-trained Le-Net framework has been chosen to produce a reasonable performance in the identification of the two classes (i.e. Parkinson's disease patients, healthy controls). The real MR images obtained from PPMI dataset are used to test the DNN classifier, which is a multi-layer CNN. This network has been equipped with and without the use of data augmentation techniques. The accuracy gradually increased with an improvement in the synthetic training sample size. For 100 synthetic sample sizes, the accuracy is 81.61%. When the synthetic training sample size is 200, the accuracy is greater than 0.8 (84.52%) and when the synthetic training sample size is 500, the accuracy is 87.74%. After that,

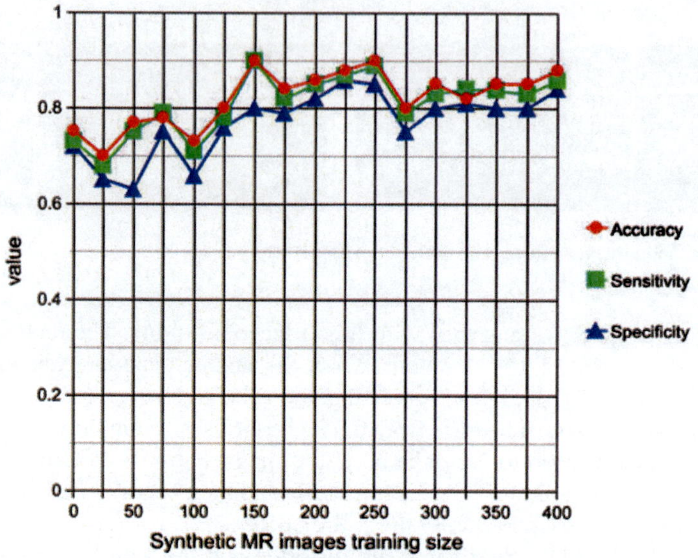

Fig. 5 Performance of DNN classifier over different training synthetic MR images

increasing the synthetic training sample size, however, did not result in an increase in the accuracy, and it tends to fluctuate about 88%. When the sample size of the synthetic training is 600, the performance remained stable at around 88%.

This work aims to improve the capacity of a classification model to classify images of healthy and PD patients via the application of synthetic images. The classification results obtained using various Deep CNN performance with and without data augmentation techniques on image synthesis outlined in Table 1. All the image synthesis models and techniques can significantly improve the classification outcomes as opposed to classification results with the original images. Furthermore, with the introduction of synthetic images, the SD findings for identification in accuracy, sensitivity, specificity and AUC are considerably decreased. In other terms, increasing the training samples improves the robustness of the classification method. With both 300 healthy controls and 300 images of Parkinson's disease patients, the Deep CNN model reaches the best accuracy of 88%.

The Deep CNN Model using synthetic samples lets the CNN classification system to obtain up to 3% higher accuracy compared to classification results utilizing original

Table 1 Deep CNN performance with and without data augmentation

Methodology	Accuracy (%)	Specificity (%)	Sensitivity (%)
Deep-CNN (transfer-learned) with data augmentation	88	87.14	87.92
Deep-CNN (transfer-learned) without data augmentation	84.67	83.76	84.13

Table 2 Comparison of proposed approach with existing techniques on same dataset

Authors	Methodology	Accuracy (%)
Amoroso et al. [21]	Random forest + SVM	82.67
Cigdem et al. [22].	PCA + PDF + SVM	83.23
Sivaranjini et al. [1]	AlexNet + Transfer Learning	86.9
The proposed approach	Pre-trained Deep-CNN with data augmentation	88

images. However, higher accuracies are obtained with more synthetic images. There were 185 samples obtained and 95 sample images are used for data augmentation. The updated data set of this model contains as many as 600 MR images, and it is found that the generated image is of good quality by using this data set for the peak signal-to-noise ratio (PSNR) detection, which is of innovative value in the standard of real samples.

Several trials have explored computer training approaches that use MRI information to diagnose Parkinson's disease individuals. The efficacy of the suggested technique is contrasted to other state-of-the-art research using PPMI database MR pictures as shown in Table 2. Amoroso et al. [21] used Random Forest and SVM to develop a therapeutic aid scheme. The complex network measurements obtained from brain lesions were used as PD measurements in this research. The effectiveness of the main element assessment (PCA) and likelihood allocation model (PDF) with SVM for PD evaluation was evaluated by Cigdem et al. [22]. These methods could, therefore, benefit from information loss. These techniques are also inherently influenced by segmentation mistakes and would have an impact on the strength of discrimination. The suggested CNN model centered on complete brain assessment takes advantage of the temporal framework without depending on hand-crafted characteristics and is therefore deemed solid. Previous trials revealed similar efficiency rates on multiple.

Datasets of Parkinson's disease, using multiple methods of dimension reduction [22] and computer teaching algorithms [23]. Similarly, PD evaluation was held out using distinct CNN settings on MR and SPECT images [23, 24]. The preliminary findings indicated by this technique, however, indicate an accuracy of 88%.

4 Conclusion

This article, utilizing Generative Adversarial Networks, provides a novel technique for data augmentation. For each category of the initial dataset, this requires training a GAN and then using it to produce a number of synthetic images. The use of a powerful image generation method has many benefits over conventional augmentation methods. The most significant is the quality of the images generated and the potential to generalize new patterns beyond the boundaries of the original data collection. We would increase further the number of samples generated in the future and explore how the classification quality is further enhanced to ultimately outperform a hand-made feature-based process. The use of GANs as an enhancement on the quality of imbalanced datasets is another study path.

References

1. S. Sivaranjini, C.M. Sujatha, Deep learning based diagnosis of Parkinson's disease using convolutional neural network. Springer J. Multimedia Tools Appl., pp. 1–13 (2019)
2. S. Shinde, S. Prasad, Y. Saboo, Predictive markers for Parkinson's disease using deep neural nets on neuromelanin sensitive MRI. Elsevier J. NeuroImage Clin.22(3), 1–8 (2019)
3. D. Saranyaraj, M. Manikandan, S. Maheswari, A deep convolutional neural network for the early detection of breast carcinoma with respect to hyper-parameter tuning. Springer J. Multimedia Tools Appl., pp. 1–26 (2019)
4. D. Long, J. Wang, M. Xuan, M. Zhang, Automatic classification of early Parkinson's disease with multi-modal MR imaging. Plos One **11**(7), 1–9 (2012)
5. X. Wu, K. Xu, P. Hall, A survey of image synthesis and editing with generative adversarial networks. Tsinghua Sci. Technol. **3**, 1–16 (2019)
6. S. Soltaninejad, I. Cheng, A. Basu, *Towards the Identification of Parkinson's Disease Using only T1 MR Images*, pp. 145–156 (Springer Nature, 2018)
7. J. Islam, Y. Zhang, Early diagnosis of alzheimer's disease: a neuroimaging study with deep learning architectures. IEEE Explore, pp. 1994–1996 (2017)
8. www.ppmi-info.org
9. V. Wegmayr, M. Hörold, J.M. Buhmann, Generative aging of brain MRI for early prediction of MCI-AD conversion, in *IEEE International Symposium on Biomedical Imaging*, pp. 1042–1046 (2019)
10. T. Iqbal, H. Ali, Generative adversarial network for medical images (MI-GAN). Image Sign. Process. **42**(231), 1–11 (2018)
11. Y. Liu, Y. Zhou, X. Liu, F. Dong, C. Wang, Wasserstein GAN-based small-sample augmentation for new-generation artificial intelligence: a case study of cancer-staging data in biology. Engineering **5**, 156–163 (2019)
12. M. Febrian, S. Makin, J. Wardlaw, T. Komura1, *Predicting the Evolution of White Matter Hyper-intensities in Brain MRI using Generative Adversarial Networks and Irregularity Map*, pp. 1–9 (Taylor and Francis, 2019)
13. Q. Kong, B. Tong, M. Klinkigt, Y. Watanabe, N. Akira, T. Murakami, *Active Generative Adversarial Network for Image Classification*, pp. 1–8 (2019)
14. G. Wang, W. Kang, Q. Wu, *Generative Adversarial Network (GAN) based Data Augmentation for Palmprint Recognition*, pp. 1–7 (IEEE, 2018)
15. T. Han, C. Liu, W. Yang, D. Jiang, A novel adversarial learning framework in deep convolutional neural network for intelligent diagnosis of mechanical faults. J. Knowled. Based Syst., pp. 1–19 (2018)
16. A.M. Dawud, K. Yurtkan, H. Oztoprak, Application of deep learning in neuroradiology: brain haemorrhage classification using transfer learning. Hindawi Comput. Intell. Neurosci., pp. 1–12 (2019)
17. D. Kollias, A. Tagaris, A. Stafylopatis, Deep neural architectures for prediction in healthcare. Springer J. Complex Intell. Syst. **4**, 119–131 (2018)
18. Mikołajczyk, A., Grochowski, M.., Data augmentation for improving deep learning in image classification problem,IEEE,2018,pp. 117–122
19. S. Shams,R. Platania, J. Zhang, J. Kim, K. Lee, *Deep Generative Breast Cancer Screening and Diagnosis*, pp. 859–867 (Springer Nature, 2018)
20. X. Wang, N. Yubo, Unsupervised domain adaptation for facial expression recognition using generative adversarial networks. Hindawi Comput. Intell. Neurosci., pp. 1–10 (2018)

21. N. Amoroso, M. Rocca, A. Monaco, R. Bellotti, Complex networks reveal early MRI markers of Parkinson's disease. Medical Image Anal. **48**, 12–24 (2018)
22. O. Cigdem, A. Yilmaz, I. Beheshti, H. Demirel, Comparing the performances of PDF and PCA on Parkinson's disease classification using structural MRI images, in *26th Signal Processing and Communications Applications Conference SIU2018*, pp. 1–4
23. D. Long, J. Wang, M. Xuan, Automatic classification of early Parkinson's disease with multi-modal MR imaging. PLoS One **7**(11) 2012
24. F. Nemmi, U. Sabatini, O. Rascol, P. Péran, Parkinson's disease and local atrophy in subcortical nuclei: insight from shape analysis. Neurobiol. Aging **36**, 424–433 (2015)

Chest X-Ray Images Based Automated Detection of Pneumonia Using Transfer Learning and CNN

Saurabh Thakur, Yajash Goplani, Sahil Arora, Rohit Upadhyay, and Geetanjali Sharma

Abstract Pneumonia is also known as Bronchopneumonia is a condition of the lungs which primarily affects the air sacks called alveoli. It results in dry and productive cough, fever, weakness, chest pain, and difficulties in breathing. Pneumonia is caused by infections brought upon by microorganisms such as bacteria and viruses. The diagnosis of pneumonia is compassed by scrutinizing X-ray images of the chest. This diagnosis is subjective to factors such as the unclear appearance of the disease in the X-ray image. Hence, artificial intelligence-based systems are required to beacon the clinics. In this study, we have implemented a very popular convolutional neural network model known as VGG16 for the diagnosis of pneumonia. In the training stage, we have used transfer learning and fine-tuning. We were able to achieve an accuracy of 90.54% with a 98.71% recall and 87.69% precision using the aforementioned convolutional neural network model.

Keywords Pneumonia · VGG16 · Transfer learning · Deep learning · Convolutional neural network

1 Introduction

Pneumonia is a type of infection which occurs in one or both lungs of a person. It results in inflammation (swelling) in the affected region of the lungs. It is primarily caused by microorganisms such as viruses and bacteria and to a lesser extent by drugs and other microorganisms. As per a study conducted by U.N. in 2010, about 3.97 lakh children, under the age of 5, died of pneumonia in India [1]. It is estimated that 20% of the deaths caused by pneumonia worldwide are from India. Thanks to the recent developments in the field of medical science, this disease can now be treated without complications which lead to death. However, this requires early detection of the disease in the patient [2]. There are several methods by which doctors detect the

S. Thakur (✉) · Y. Goplani · S. Arora · R. Upadhyay · G. Sharma
Maharaja Surajmal Institute of Technology, New Delhi, India
e-mail: thakurksaurabh@gmail.com

P. Bansal et al. (eds.), *Proceedings of International Conference on Artificial Intelligence and Applications*, Advances in Intelligent Systems and Computing 1164,
https://doi.org/10.1007/978-981-15-4992-2_31

presence of infection in the lungs. Analysis of X-ray images of the chest is a reliable method for the detection of pneumonia. This method of detecting pneumonia is difficult as images are hard to distinguish. The semblance of pneumonia is unclear and often is mistaken by radiologists for another disease [3].

Thus, to curtail these failings and inconsistencies, a computerized software program is needed which will help radiologists in the diagnosis of pneumonia. Recent developments in the field of machine learning, more specifically deep learning has shown great promise and success in recognition of images [4–9]. Convolutional Neural Networks (CNNs), a class of deep neural networks, can classify images with high accuracy [10]. CNN's are made just like the connections present in the human brain's visual cortex. They are made to resemble biological connectivity patterns of neurons [11]. They can extract more significant features over handmade features from the entire image [12].

Several deep networks have been developed by researchers which are based on CNNs. These networks have been successfully implemented in the detection of brain tumor segmentation [9], Alzheimer's [13], Breast Cancer [14], etc. Several studies have been done about the detection of pneumonia with the deep learning approach. In the year 2017, P. Rajpurkar et al. [15] devised an algorithm called CheXNet which trained a 121-layer convolutional neural network on the dataset ChestX-ray14. ChestX-ray14 is one of the largest publicly available dataset of chest x-ray images. It contains over 100,000 frontal view x-ray images of the chest. They achieved an f1 score of 0.435 (95% CI 0.387, 0.481), which is quite higher about 0.048 than the 0.387 (95% CI 0.330, 0.442) that is the radiologist average score.

Considering the above information as our basis, we updated and trained a well-known network for the detection of pneumonia from chest X-ray images. This neural network is based upon the model VGG16. The fine-tuned VGG16 model is shown in Fig. 1. The authors have also employed data augmentation, transfer learning, and fine-tuning methods.

2 Dataset and Proposed Model

In this section of the paper, the chest X-Ray Image dataset and the proposed model has been described in detail.

2.1 Dataset

In this study, we have trained our model on a dataset of chest X-ray images which contained a total of 5863 images. This dataset was shared by Paul Mooney on kaagle.com. The images were divided into two categories—normal and pneumonia. All

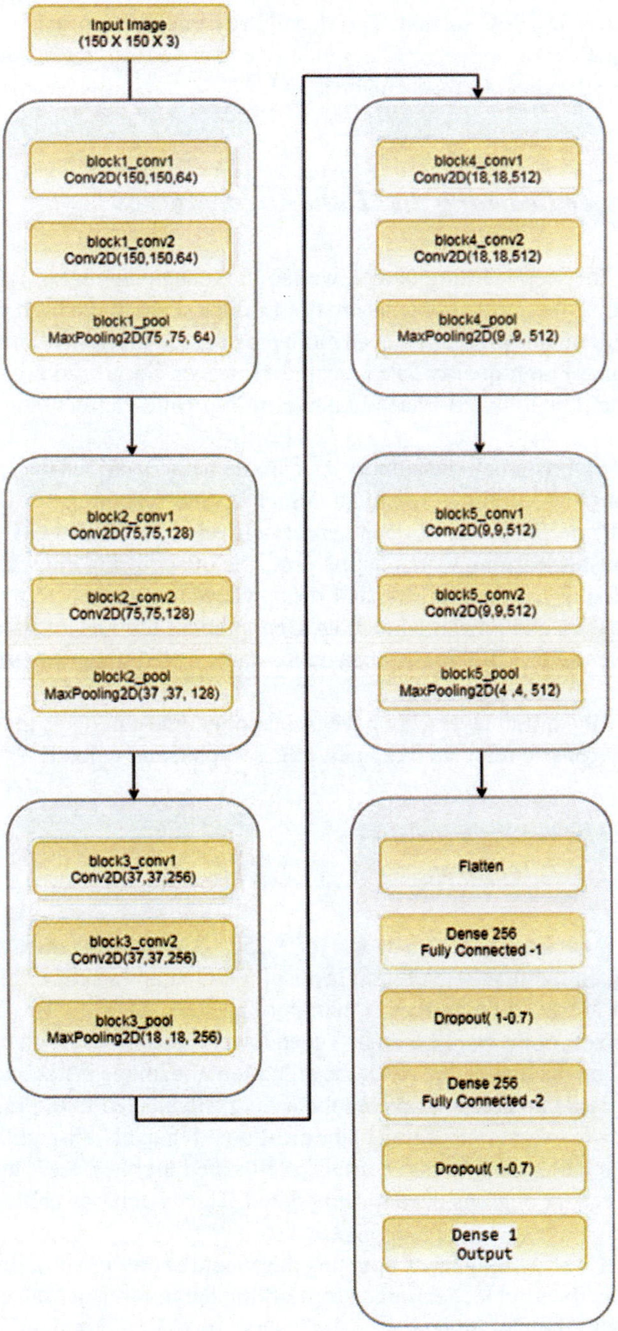

Fig. 1 Fine tuned VGG16 network

the images were in JPEG format. The chest X-ray images were taken from a group of pediatric patients who were all less than five years old from Guangzhou Women and Children's Medical Centre, Guangzhou [16].

2.2 Transfer Learning and Data Augmentation

For a successful deep learning model, we require a huge amount of reliable data. In many situations, these requirements are not fulfilled. Due to the high costs involved and time required, this problem is even more prevalent in the field of medical science.

The amount of high-quality data is scarce. However, some workarounds are available which help us to avoid issues like overfitting. Data augmentation is one such method.

In this proposed model, the authors have implemented various data augmentation techniques such as flipping, zooming, rescaling and resizing. Another technique to improve the performance of deep neural networks, especially CNNs, is transfer learning. Transfer learning works on the principle of acquiring knowledge from one task and utilizing it in solving related tasks. Currently, only a minute amount of people work on CNN's from scratch as a gargantuan amount of data is required. Thus, we have used a CNN pre-trained on a dataset called ImageNet which contains more than a million images [17].

In some of the initial layers, we have also applied fine-tuning as these layers tend to have many generic features like lines, edges, blobs, and colors.

2.3 CNN Architectures

In this study, input images of dimension 150 * 150 * 3 have been taken. After processing the image in the first convolution layer of VGG16 architecture, the dimensions get reduce to 75 * 75 * 64 and reach max-pooling state. Max-pooling helps in compression of pixels of the image data into even lower dimensions. It gradually reduces to 4 * 4 * 512 by the final layer. After the reduction, the image is flattened. Flattening converts the data into a 1-D array and the array is connected to the fully connected layers. Its weight gets set according to the pre-trained ImageNet weights. Here, transfer learning is being used to make use of pre-trained architecture. Hence, the fully connected layers or neurons need to be updated. This is done by calling the VGG16 architecture.

This model is not simply fancy updating the weight of the convolution layer. Therefore, the model does not let get trained by marking the training as "false". In VGG16, before setting the convolution layer "false", it has 16,828,607 trainable parameters to train and afterward, we are left with 14,714,688 parameters. The detailed information of trainable parameters is given in Table 1. The VGG16 model has 16 convolutional

Table 1 Complete details of trainable parameters

Layer (type)	Output shape	Parameter #
Vgg16 (Model)	(None, 4, 4, 512)	14,714,688
Flatten_1	(None, 8192)	0
Dense_1	(None, 256)	2,097,408
Dropout_1	(None, 256)	0
Dense_2	(None, 64)	16,448
Dropout_2	(None, 64)	0
Dense_3	(None, 1)	65
Total params: 16,828,609		
Trainable params: 2,113,921		
Non-trainable params: 14,714,688		

Table 2 Distribution of chest X-ray image dataset

Classes	Training	Validation	Testing
Normal	1341	234	234
Pneumonia	3876	390	390
Total	5217	624	624

layers. It also has three fully connected layers, five max-pooling layers and a final layer that has a soft-max activation function.

3 Experimental Results

The proposed model has been trained in 1341 X-ray images of healthy people and 3876 X-ray images of people suffering from pneumonia. The model was then evaluated on 234 X-ray images of healthy people and 390 X-ray images of people suffering from pneumonia. The detailed information about the data is presented in Table 2.

The model has achieved an accuracy of 90.54%, a recall of 98.7%, a precision of 87.7%, and an f1 score of 92.9% by training our model on the VGG16 architecture. The loss graph and accuracy graph are shown in Figs. 2 and 3, respectively. The parameters set for training are—Adam as optimizer with default learning rate, Adam handles learning rate optimization, loss function as categorical cross-entropy, epochs 30, steps per epochs 50, drop rate 0.3, and batch size as 32.

4 Conclusion

In this proposed work, the authors have tried to obtain the state of art results for the detection of pneumonia disease from X-ray images of the chest by using a transfer

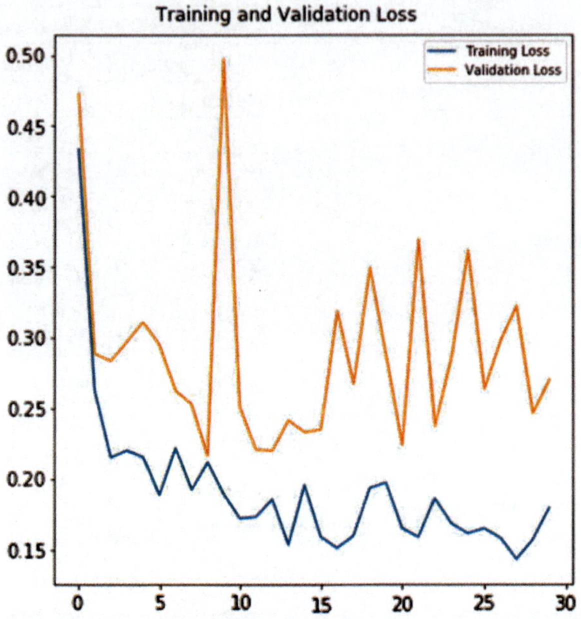

Fig. 2 Loss graph of training and validation data

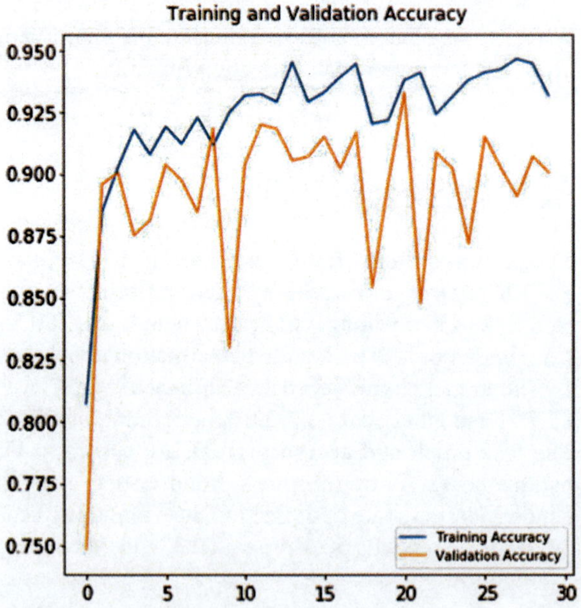

Fig. 3 Accuracy graph of training and validation data

learning algorithm. We have collected and labeled a total of 5863 images. The model fetched highly favorable results by achieving an overall accuracy of 90.54% and a recall value of 98.7%. The f1 score reached a mark of 92.9. Therefore, it can be inadvertently accepted in the clinical industry to correctly identify the chest X-rays with disorders from the normal ones for a weighted decision.

References

1. National Health Portal, Pneumonia, obtained from https://www.nhp.gov.in/disease/respiratory/lungs/pneumonia
2. M. Aydogdu, E. Ozyilmaz, H. Aksoy, G. Gursel, N. Ekim, Mortality prediction in community-acquired pneumonia requiring mechanical ventilation; values of pneumonia and intensive care unit severity scores. Tuberk Toraks **58**(1), 25–34 (2010)
3. M.I. Neuman, E.Y. Lee, S. Bixby, S. Diperna, J. Hellinger, R. Markowitz, S. Servaes, M.C. Monuteaux, S.S. Shah, Variability in the interpretation of chest radiographs for the diagnosis of pneumonia in children. J. Hosp. Med. **7**(4), 294–298 (2012)
4. J. Ker, L. Wang, J. Rao, T. Lim, Deep learning applications in medical image analysis. IEEE Access **6**, 9375–9389 (2018)
5. G. Litjens et al., A survey on deep learning in medical image analysis. Med. Image Anal. **42**, 60–88 (2017)
6. O. Ronneberger, P. Fischer, T. Brox, U-net: convolutional networks for biomedical image segmentation, in *International Conference on Medical Image Computing and Computer-Assisted Intervention*, pp. 234–241 (Springer, 2015)
7. J. Redmon, A. Farhadi, *Yolov3: An Incremental Improvement*. arXiv preprint arXiv:1804.02767 (2018)
8. D.A. Ragab, M. Sharkas, S. Marshall, J. Ren, Breast cancer detection using deep convolutional neural networks and support vector machines. PeerJ **7**, e6201 (2019)
9. S. Pereira, A. Pinto, V. Alves, C.A. Silva, Brain tumor segmentation using convolutional neural networks in MRI images. IEEE Trans. Med. Imaging **35**(5), 1240–1251 (2016)
10. A. Krizhevsky, I. Sutskever, G.E. Hinton, Imagenet classification with deep convolutional neural networks, in *Advances in Neural Information Processing Systems*, pp. 1097–1105 (2012)
11. Brain Corporation, Neocognitron, obtained from http://www.scholarpedia.org/article/Neocognitron
12. Y. LeCun, Y. Bengio, G. Hinton, Deep learning. Nature **521**(7553), 436 (2015)
13. Radiology society of North America, Deep learning algorithm detects Alzheimer's up to six years before doctors obtained from https://newatlas.com/ai-algorithm-pet-scan-alzheimers-diagnosis/57138/
14. S. Albarqouni, C. Baur, F. Achilles, V. Belagiannis, S. Demirci, N. Navab, Aggnet: deep learning from crowds for mitosis detection in breast cancer histology images. IEEE Trans. Med. Imaging **35**(5), 1313–1321 (2016)
15. P. Rajpurkar, J. Irvin, K. Zhu, B. Yang, H. Mehta, T. Duan, D. Ding et al., *Chexnet: Radiologist-Level Pneumonia Detection on Chest X-Rays with Deep Learning*. arXiv preprint arXiv:1711.05225 (2017)
16. Chest X-ray dataset was obtained from https://www.kaggle.com/paultimothymooney/chest-xray-pneumonia/download
17. O. Russakovsky et al., Imagenet large scale visual recognition challenge. Int. J. Comput. Vis. **115**(3), 211–252 (2015)

Relative Examination of Texture Feature Extraction Techniques in Image Retrieval Systems by Employing Neural Network: An Experimental Review

Shefali Dhingra and Poonam Bansal

Abstract In this paper, the performance of the four most significant texture feature extraction techniques of content-based image retrieval (CBIR) systems are technically analyzed and compared in two phases. In the first phase, simply texture features are extracted by various techniques on three important CBIR datasets which are Wang, Corel-5K and Corel-10K and retrieval parameters are evaluated. In the next phase, artificial neural network, i.e., cascade forward back propagation neural network (CFBPNN) is employed as a classifier after the step of feature extraction which increases the accuracy and average precision of the system. The main issue with CBIR systems is the proper selection of techniques for the extraction of low-level features of the images which comprises of color, texture, and edge. Among these features, texture is one of the most influential features. This selection of features completely depends upon the type of images to be retrieved from the database. The four important texture techniques i.e. Grey level co-occurrence matrix (GLCM), discrete wavelet transforms (DWT), Gabor transform and local binary pattern (LBP) are experimented here and performance parameters are evaluated. So this analysis will help the researchers in selection of proper texture feature techniques depending upon the type of images for CBIR systems. During phase 1, which is without classifier the LBP technique has the highest precision which is around 82, 70 and 66% on Wang, Corel-5K, and Corel-10K respectively. The results spectacles a significant improvement with the use of CFBPNN and these results now comes up to around 94, 83, and 79%, respectively, on the same datasets.

Keywords Deep learning · F-measure · Precision · Relevant feedback · Cascade feed forward back propagation network (CFBPNN)

S. Dhingra (✉)
Guru Gobind Singh Indraprastha University, New Delhi, India
e-mail: munjalshefali@gmail.com

P. Bansal
Maharaja Surajmal Institute of Technology, New Delhi, India
e-mail: poonam.bansal@msit.in

P. Bansal et al. (eds.), *Proceedings of International Conference on Artificial Intelligence and Applications*, Advances in Intelligent Systems and Computing 1164,
https://doi.org/10.1007/978-981-15-4992-2_32

1 Introduction

Large image datasets have been created with the advancement in the technologies in the domain of image acquisition and data storage. But dealing with these databases is a very challenging and tedious task. Therefore, for searching and indexing images from these databases, efficient and fast retrieval systems are needed. Traditionally, retrieval of images is accomplished by text-based image retrieval (TBIR), where textual description helps to retrieve the images. But a very famous proverb is there "A picture is worth one thousand words". So to overcome the limitations of the TBIR systems content-based image retrieval system (CBIR) was developed in which an image itself is used for the retrieval of relevant images [1].

So in order to make this type of system the first step is the extracting of productive and efficient low-level features of the image and after that similarity algorithm is applied for retrieval of images. The three common well known visual features are color, shape, and texture [1]. The contents of the image are directly related to these low-level features. Almost every image annotation and retrieval systems have been developed successfully by using these features. When all the three features are integrated the retrieval results obtained are much better. But individually the color and texture are the two most important and widely used features due to their easier extraction process and higher accuracy as compared with shape [2].

The procedure of recovering the images from the group of database on the basis of the feature of images is called CBIR system. The block diagram of CBIR system is shown in Fig. 1 Inconsistent automated annotation, high-priced manual annotation and inaccuracy are the main problems suffered by textual-based image retrieval systems [3]. Due to these problems, the efficient and dominant image retrieval algorithms have been proposed which retrieves the images on the basis of its visual attributes only. Presently, CBIR is the backbone of the image retrieval systems and is able to retrieve the images from the large repositories [4, 5].

Feature extraction and similarity matching play a crucial role in the performance of the CBIR systems. Features can be extracted by various techniques, mostly used color and texture extraction techniques are color moment, color histogram, dominant color descriptor, LBP, DWT, GLCM, etc. Feature matching is the process of matching

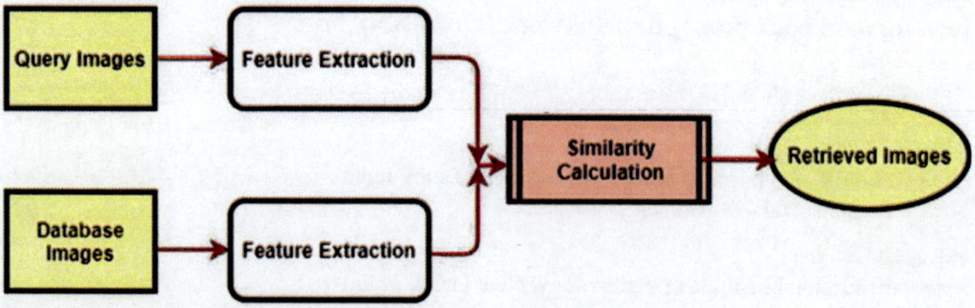

Fig. 1 Basic block diagram of CBIR system

the extracted query image features with the features of the database images. This similarity measurement can be performed by using various distance metrics such as Euclidean distance, Manhattan distance, city block and many more. Lesser is the distance more is the similarity between the image [6, 7].

With large datasets, CBIR systems have to face many challenges. The main issue with these systems is accuracy which goes on decreasing with an increase in datasets. According to the latest studies various deep learning algorithms have been applied in the areas of image and video processing [8]. The most efficient deep learning architecture used for the classification of large image datasets are feed forward neural networks and cascade forward backpropagation neural networks. These functions are further used as deep representations for the purpose of classification, retrieval or clustering [9]. Another major concern with CBIR systems is the problem of semantic gap. The semantic gap is the difference in the low-level features of the image captured by the machine and human perception. To blown away this gap relevant feedback techniques are exploited [10, 11].

The proposed framework here is just the experimental survey of different texture feature extraction techniques on various datasets in three phases. Firstly, the average precision is calculated simply from texture techniques. Secondly, the CFBPNN is employed after the step of feature extraction which is modeled as a classifier that helps in the classification of images.

This paper is structured as following way: Sect. 1 gives the introduction about Image Processing, Image Retrieval, CBIR technique, CFBPNN and the simple overview of the proposed approach. Section 2 reveals about the literature which is interrelated to the proposed work based on different texture based CBIR system.

In the next section, the main contributions of this proposed work are discussed in detail. Section 4 is the proposed framework and fifth section provides the results and discussions. The last section, i.e., Sect. 6 is the final concluding section.

2 Related Works

In this section, the state of the art schemes based mostly on the texture techniques in CBIR system are discussed. The curvelet approach was proposed in [12] which provides good accuracy in texture retrieval. In this scheme, from the transformed images, low order statistics were computed. Results here proved that it provides better results than Gabor texture feature. By keeping this in mind Jin et al. designed the shape based CBIR system which can then be combined with texture and color to provide better results [13].

The two features were combined i.e. discrete wavelet transforms and edge histogram descriptors for texture and shapes respectively [14]. The edge histogram was enforced on selected wavelet coefficients which come after the decomposition of the input image. This combination had increased the performance of the system. Sometimes only the single texture feature provides good results such local tri-directional pattern was proposed in [15] which was the texture feature and uses the pixel intensity in three directions.

To overcome the limitation of semantic gap relevance feedback was used. In [16] the work was done to decrease the semantic gap by the process of feature selection and adaptation. In [17] different types of relevant feedback techniques were analyzed such as Support Vector Machine, query refining, deep learning with their performance and the various issues related to these RF techniques. Different types of neural-based algorithms are there such as feed forward networks, cascade forward backpropagation networks, Multi-Layer Perceptrons, Convolutional Neural Networks, Autoencoders, Sparse coding, etc. [18]. The basic operations performed by these algorithms are classification, regression, clustering, etc. [19, 20].

3 Main Contributions

The main write-off in this experimental survey is the various texture feature extraction techniques that are mainly used in CBIR systems and the classifier CFBPNN which is used here with these techniques to increase the efficiency of the system. All these are described in detail in the below sections.

3.1 Texture Feature Extraction Techniques

Several approaches are used for extracting the texture, such as structural, model-based, statistical and transform-based. But the most used methods reported in the literature as seen in the above-described table are statistically based and transform-based because of their good performance [21]. The types of texture techniques are shown in Fig. 2, some of which will be further described and experimented in this paper.

3.1.1 Gray-Level Co-occurrence Matrix (GLCM)

This technique was proposed by Haralick. It is one of the most used and accurate feature extraction methods and evaluates the different second-order statistical properties of the images. [22]. The four most useful and important statistical parameters of GLCM are given in Eqs. (1)–(4)

Energy

$$E = \sum_u \sum_v (k(u, v)) \tag{1}$$

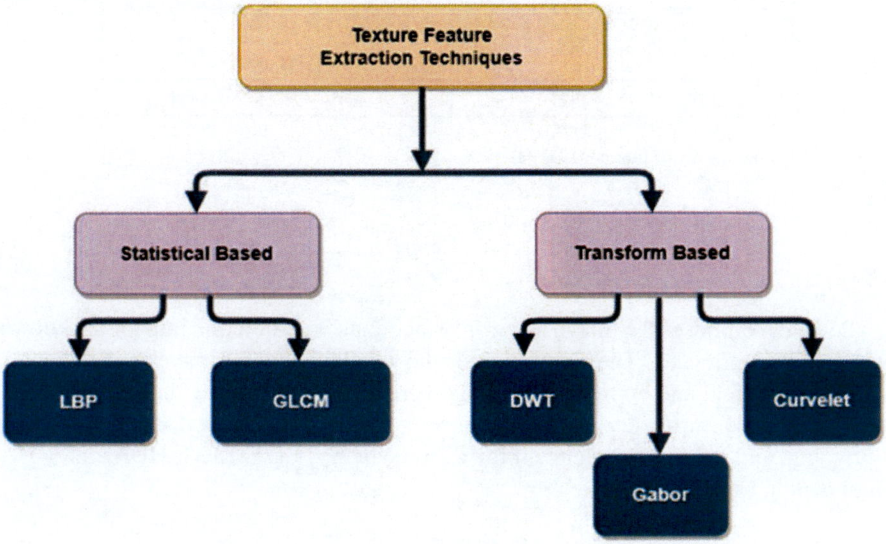

Fig. 2 Type of texture techniques

Contrast

$$c = \sum_{u}\sum_{v}(u-v)^2 k(u,v) \tag{2}$$

Entropy

$$T = \sum_{u}\sum_{v}(k(u,v))\log(k(u,v)) \tag{3}$$

Inverse Difference

$$I = \sum_{u}\sum_{v}\frac{1}{1+(u-v)^2}k(u,v) \tag{4}$$

where u and v are the co-occurrence matrix coefficients.

3.1.2 Discrete Wavelet Transform (DWT)

The wavelet transforms having frequency and spatial characteristics obtain the multi-scale resolution of the images [23]. For extracting the texture features, the coefficient distribution of the mother wavelet is computed. This wavelet is translated by b and scaled by a is given in Eq. 5. The wavelet transform mathematically stated as,

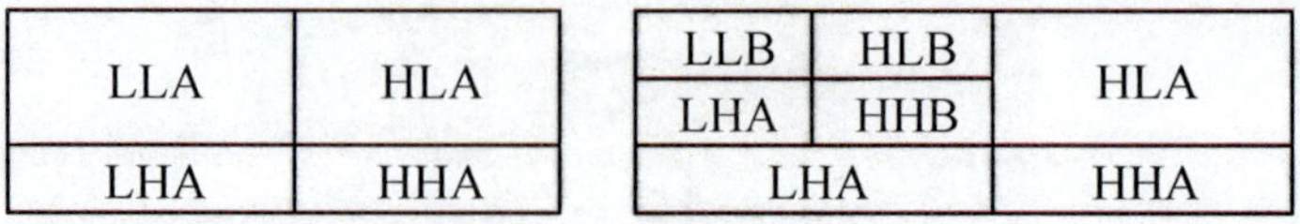

Fig. 3 Decomposition of bands of DWT

$$\Psi_{a,b}(t) = \frac{1}{\sqrt{a}}\Psi\left(\frac{t-b}{a}\right) \tag{5}$$

DWT decomposes the image using an orthogonal basis function into the set known as wavelet coefficients. These sets provide the details of four parts such as horizontal, diagonal, vertical and approximation and therefore it can provide the fine details of the image [24]. After decomposition four sub-bands are formed containing both horizontal and vertical frequency information which is LLA, LHA, HLA, and HHA as shown in Fig. 3.

3.1.3 Gabor Transform

Gabor wavelet transforms efficiently extracts the texture feature from the images and had been widely used in the literature. These filters are a group of wavelets and every wavelet is capturing energy on a specific orientation and frequency. These functions produce the optimized resolution in both frequency and spatial domains so this wavelet transform extracts the local features due to multi-orientation properties [25]. The 2D Gabor function is described in Eq. (6).

$$G(x, y) = \exp\left[\left(\frac{x^2}{\sigma_x^2} + \frac{y^2}{\sigma_y^2}\right) + 2\pi j W_x\right] \tag{6}$$

W = centre frequency, σ_x and σ_y are scaling functions.

3.1.4 Local Binary Pattern

LBP feature extraction technique for the texture is widely used in various applications of image processing due to its simplicity, performance, and implementation. Before applying this technique the RGB images are converted into grayscale. This operator is based on the difference between the centre value of the pixel and neighbor pixels. By thresholding of neighbor pixel with the centre pixel, the binary code is produced for every pixel of the image. In the next step, histogram is constructed for the determination of frequency of binary patterns. The bins number in the histogram tells about the number of pixels used for the LBP calculation. The calculations of LBP are shown in Eqs. (7) and (8).

$$\mathrm{LBP}_N = \sum_{i=0}^{N-1} f(P_i - CP)2^i; \tag{7}$$

$$f(p) = \begin{cases} 1; \; P \geq 0 \\ 0; \; P < 0 \end{cases}, \; N \text{ is total neighboring pixels.}$$

$$\mathrm{Hist}_k = \sum_{x=1}^{M}\sum_{y=1}^{M} f_2(\mathrm{LBP}(x, y), k) \tag{8}$$

The survey has been done on these texture techniques and various advantages and limitations of these techniques have been noticed. Like the LBP technique provides highlighted results but it is more time-consuming. The GLCM technique captures approximately all the properties of an image but its cost is quiet high. So each and every feature extraction technique has its own pros and cons, they can be used according to the type of images.

3.2 Cascade Forward Back Propagation Neural Network

This artificial intelligent model comprises of input, output, and hidden layers. During the training time of these networks, the computations are carried out from the input to the output layers of the network and the errors are propagated to the previous layers. So this is known as back propagation network The CFBPNN is basically employed for the prediction of the data at the output [26]. For the training of this neural network, initially, the images are stored in database. The input of the network is the feature vectors of the images extracted from the texture techniques and the purpose of this network here is the classification of the texture features of the images into different classes. The architecture of CFBPNN is shown in Fig. 4.

4 The Proposed Framework

This system is proposed basically to check the working of various decisive texture techniques of CBIR systems with and without the neural network based classifier. The classifier used here for testing is CFBPNN as it provides much better results as discussed in the literature section. The proposed systems will help the researchers to select the best texture feature extraction technique depending on the type of the images. It works in two separate phases.

Phase 1: In the first step, preprocessing of images is performed i.e. they are converted to the required color spaces. Then, features are extracted individually by all the texture techniques displayed in the above section on all the datasets. After that, the features of the query image are matched directly with the features of the

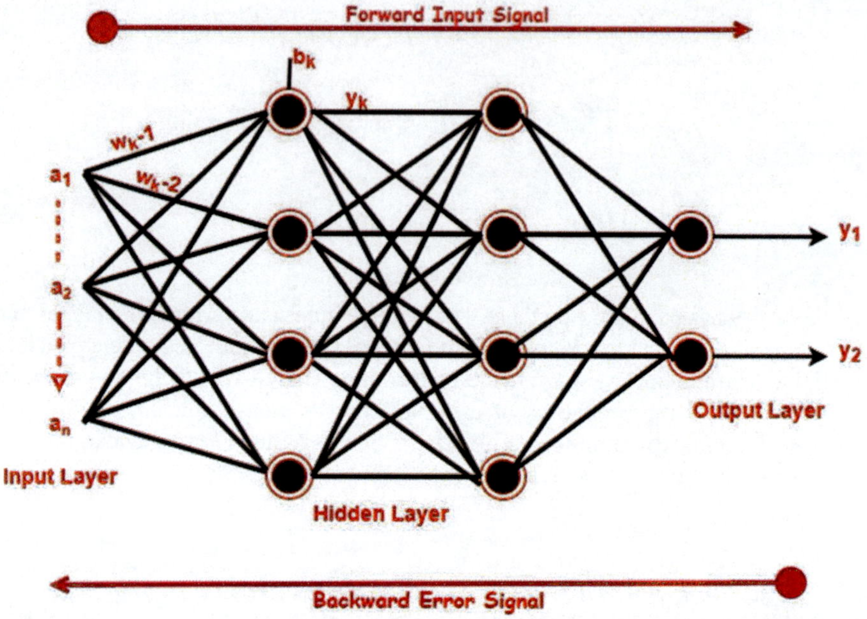

Fig. 4 Architecture of CFBPNN

database images. Features of both query image and database images are extracted by the same technique.

Phase 2: Similar to the above, preprocessing of the images is done and the features are extracted by all the leading texture techniques. After this step, CFBPNN is trained as a classifier which makes the various classes of the feature vectors of the database images. Now the query image features enter into the classifier and it predicts its best matching class. The similarity of the query image is only calculated from its best matching class only and not from the whole image dataset which in turn decreases time and increases the accuracy of the system. The complete block diagram showing phase 1 and phase 2 is shown in Fig. 5.

5 Experimental Set up and Results

5.1 Experimental Set up

Experiments for the proposed system have been conducted on three benchmark datasets which are Wang, Corel-5K and 10K. These are the standard datasets used for the testing of CBIR systems. The implementations are done on MATLAB-15A with 64-bit windows and 2 GB memory.

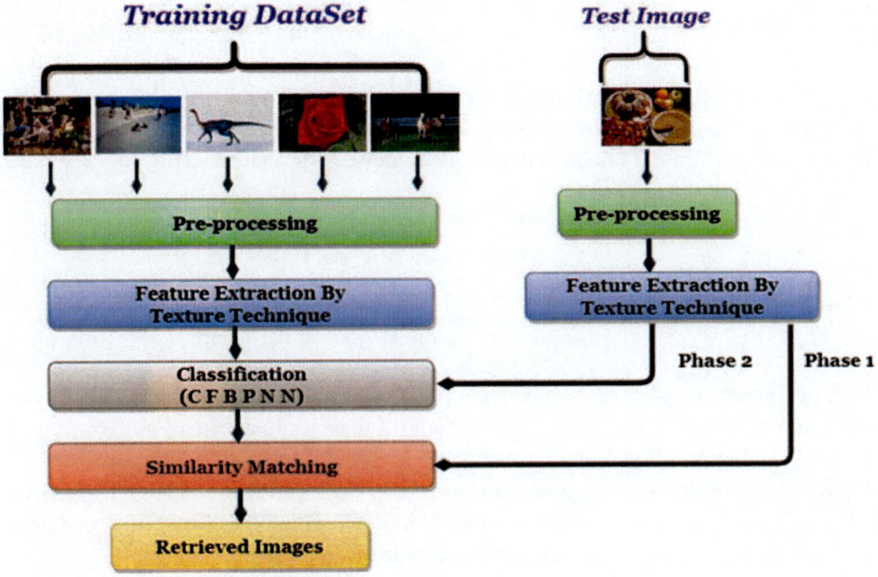

Fig. 5 Framework during phase 1 and phase 2

Description of Wang, Corel-5K and Corel-10K dataset
The Wang dataset comprises of 1000 images. The images are categorized into 10 different groups such as buses, dinosaurs, flowers, beaches, etc. The image size is 256 × 384 otherwise 384 × 256. The Corel-5K datasets contain around 5000 images. These images are categorized into fifty groups such as animals, cars, ships, butterflies, etc. The image size is 128 × 187 or 187 × 128. The Corel-10K datasets contain 10,000 images of same size as Corel-5K. Some sample images of these datasets are shown in Fig. 6.

Fig. 6 Sample images of three databases used in experiments

5.2 Experimental Results

In order to evaluate the performance of the proposed framework the most important parameters of the CBIR systems which are precision, recall and accuracy are computed for different benchmark datasets which are described in the above section.

$$\text{Precision} = \frac{\text{No. of relevant images Retrieved}}{\text{Total No. of images Retrieved}}$$

$$\text{Recall} = \frac{\text{No. of relevant images retrieved}}{\text{No. of relevant images in database}}$$

Experiments in phase 1: Experiments are conducted on all the texture feature extraction techniques individually which are all described in the above sections. In this framework, all the database images are considered as query image and the average value of precision and recall are calculated. Euclidean distance is used here for the similarity measurement between the images. The top 10 images are retrieved and the value of precision, recall, and *f*-measure is calculated.

$$D_{\mathrm{E}} = \sqrt{\sum_{i=1}^{n} (|q_i - D_i|)^2} \tag{10}$$

where q_i and D_i are the feature vectors of query and database image respectively.

Experiments in phase 2: In these experiments, firstly the features are extracted by different texture techniques just like in experiments 1 but after the step of feature extraction CFBPNN is modeled as a classifier that makes the different classes of the images based on their similar features. Like the above case, all the database images are taken as the query images and the average value of precision, recall is calculated. The view of the network that is obtained during the training process is shown in Fig. 7. This is the network formed when the features are extracted from the LBP

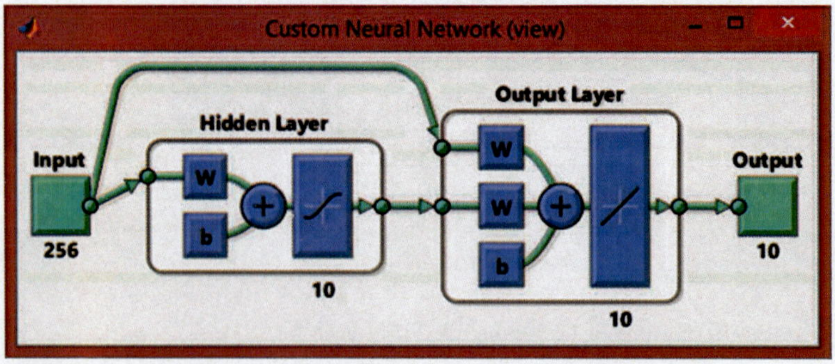

Fig. 7 CFBPNN framed during training with the extraction of LBP features in Wang database

Table 1 Comparison of average precision value of texture techniques in Wang database

Texture technique	Without classifier	With classifier
LBP	82	94
DWT	70	90
GLCM	75	88
Gabor	66	78

Table 2 Comparison of average precision value of texture techniques in Corel-5K database

Texture technique	Without classifier	With classifier
LBP	70	83
DWT	60	78
GLCM	63	80
Gabor	58	70

Table 3 Comparison of average precision value of texture techniques in Corel-10K database

Texture technique	Without classifier	With classifier
LBP	66	79
DWT	52	70
GLCM	58	72
Gabor	51	66

technique in the case of Wang database. Similar types of networks are framed by other techniques also during the training process of CFBPNN. The difference with others is only in the number of nodes.

The comparison of average precisions in both experimental phases of all the techniques is shown in Tables 1, 2, and 3. The results in tables reveal that the LBP shows the highest rate of precision value in all the databases with and without the application of CFBPNN. The Gabor technique shows the least results in these types of images.

It is also very clear from the results that the DWT technique makes a very high jump of almost 20% with the application of classifier which is much better than others and its results become almost similar to the results of LBP results. So, it will help the researchers in the field of CBIR to select the suitable texture technique according to the conditions such as resolution changes, illumination changes, etc. Proper selection of technique also depends upon the type of images. LBP inherits the steadiness regarding changes in illumination and rotation while DWT technique is multi-resolutional and multi-oriented with higher directionality.

The image retrieval results of LBP technique in Wang database by choosing any random image with classifier is shown in Fig. 8 respectively. In Fig. 8 it is clearly observed that all the images that are retrieved with the classifier are similar to the query image and hence the precision is 1 or 100%.

Fig. 8 Images retrieved by LBP technique with classifier

6 Conclusion

In this experimental survey paper, the texture feature extraction techniques which are primarily used in CBIR systems are analyzed and compared. Initially, all the texture techniques are tested individually and the average value of precision is evaluated. In this case, LBP technique provides the superlative results for these types of images. After that, to see how the precision rate varies with the classifier, neural network CFBPNN is applied. The simulation results show that there is a significant improvement in the rate of precision values with the use of classifier and it is the maximum in the case of DWT technique. So this experimental survey will help the researchers to choose the best texture extraction technique for image retrieval depending on the image conditions as it is the most influential feature among all.

References

1. P. Sharma, Improved shape matching and retrieval using robust histograms of spatially distributed points and angular radial transform. Optik **145**, 346–364 (2017)
2. K. Hemachandran, A. Paul, M. Singha, Content-based image retrieval using the combination of the fast wavelet transformation and the colour histogram. IET Image Process. **6**, 1221–1226 (2012)
3. S. Murala, R.P. Maheshwari, R. Balasubramanian, Local tetra patterns: a new feature descriptor for content-based image retrieval. Int. J. Multim. Inf. Retriev. 191–203 (2012)
4. S. Dhingra, P. Bansal, An overview on the progress of content based image retrieval systems: from 1990's to 2000's, in *Proceedings of 5th International IEEE Conference on Computing for Global Sustainable Development, INDIACOM* (2018)
5. S. Fadaei, R. Amirfattahi, M.R. Ahmadzadeh, A new content-based image retrieval system based on optimized integration of DCD, wavelet and curvelet features. IET Image Process. 1–19 (2017)
6. J. Prabhu, J.S. Kumar, Wavelet based content based image retrieval using color and texture feature extraction by gray level coocurence. J. Comput. Sci. **1**, 15–22 (2014)

7. A. Alzu'bi, A. Amira, N. Ramzan, Semantic content-based image retrieval: a comprehensive study. J. Vis. Commun. Image Represent. **32**, 20–54 (2015)
8. B. Warsito, R. Santoso, H. Yasin, Cascade forward neural network for time series prediction. J. Phys. 1–9 (2018)
9. N.N. Van, A. Boucher, J.M. Ogier, S. Tabbone, Cluster-based relevance feedback for CBIR: A combination of query point movement and query expansion. J. Ambient Intell. Humaniz. Comput. **3**, 281–292 (2012)
10. M. Mosbah, B. Boucheham, Distance selection based on relevance feedback in the context of CBIR using the SFS meta-heuristic with one round. Egypt Inf. **18**, 1–9 (2017)
11. I. Sumana, M. Islam, Content based image retrieval using curvelet transform, in *IEEE Conference on Signal Process* (2008), pp. 11–16
12. C.J.S. Ke, *Content-Based Image Retrieval Based on Shape Similarity Calculation*. 3D Research Express Springer (2017)
13. M. Verma, B. Raman, Local tri-directional patterns: A new texture feature descriptor for image retrieval. Dig. Signal Process. A Rev. J. **51**, 62–72 (2016)
14. R. JVCI, S. Murala, Q.M.J. Wu, Expert content-based image retrieval system using robust local patterns. J. Vis. Commun. Image Represent. **25**, 1324–1334 (2014)
15. M.K. Alsmadi, An efficient similarity measure for content based image retrieval using memetic algorithm. Egypt J. Basic Appl. Sci. **4**, 112–122 (2017)
16. A. Sheshasaayee, C. Jasmine, CBIR: current trends and issues. Int. J. Eng. Trends Tech. **10**, 166–175 (2014)
17. Y. Guo, Y. Liu, A. Oerlemans, S. Lao, S. Wu, M.S. Lew, Deep learning for visual understanding: a review. Neurocomputing **187**, 27–48 (2016)
18. D. Jaswal, V. Sowmya, K.P. Soman, Image classification using convolutional neural networks. Int. J. Adv. Res. Tech. **3**, 1661–1668 (2014)
19. L. Wang, H. Wang, Improving feature matching strategies for efficient image retrieval. Signal Process. Image Commun. **53**, 86–94 (2017)
20. S. Dhingra, P. Bansal, A competent and novel approach of designing an intelligent image retrieval system. EAI End. Trans. Scalable Inf. Syst. (2019). https://doi.org/10.4108/eai.13-7-2018.160538
21. Y. Liu, D. Zhang, G. Lu, W.Y. Ma, A survey of content-based image retrieval with high-level semantics. Pattern Recognit. **40**, 262–282 (2007)
22. A. Shahbahrami, Comparison between color and texture features for image retrieval, in *2008 6th International Symposium of Applied Machine and Intelligence Informatics* (2008), pp. 221–224
23. S. Agarwal, *Content based image retrieval using discrete wavelet transform and edge histogram descriptor* (Int. Conf. Inf. Syst. Comput. Netw., IEEE, 2013), pp. 19–23
24. P. Srivastava, A. Khare, Integration of wavelet transform, local binary patterns and moments for content-based image retrieval. J. Vis. Commun. Image Represent. (2016)
25. S.M. Singh, K. Hemachandran, Content-content—based image retrieval using color moment and gabor texture feature. **9**, 299–309 (2012)
26. D.S. Badde, A. Gupta, V.K. Patki, Cascade and feed forward back propagation artificial neural network models for prediction of compressive strength of ready

Machine Learning Based Automatic Prediction of Parkinson's Disease Using Speech Features

Deepali Jain, Arnab Kumar Mishra, and Sujit Kumar Das

Abstract Parkinson's disease is a severe neurodegenerative disease where primarily the motor system of the human body gets affected. It is currently one of the leading causes of disability around the world. Although currently there does not exist any cure for the disease, early detection of the disease can lead to better treatment and better disease management, which can improve the quality of life of the patients drastically. Researchers have shown that speech data can be very helpful in the early diagnosis of the disease. Clinical decision support systems built using speech features from previous patients could potentially be of very high utility for early diagnosis of the disease. In this work, a machine learning based automatic prediction framework is presented, which can be used to build such effective decision support systems. Various evaluation metrics like prediction accuracy, sensitivity, specificity and Area Under the Curve of the Receiver Operating Characteristics have been considered for evaluating the applicability of various machine learning algorithms. It has been observed that ensemble methods like Random Forests and Gradient Boosting classifiers can be used to effectively perform predictive modelling, achieving average prediction accuracy of 86.5%. Oversampling strategy is also used in order to increase the size of the training data, so that deep neural network based approaches can be studied. Significant improvements have been observed post oversampling, achieving average prediction accuracy up to 91.5, which suggests the potential for applicability of the approach as a decision support system in real diagnostic scenarios.

Keywords Parkinson's disease · Machine learning · Predictive modelling · Clinical decision support system

---

D. Jain (✉) · A. K. Mishra · S. K. Das
CSE Department, NIT Silchar, Silchar, India
e-mail: jaindeepali010@gmail.com

A. K. Mishra
e-mail: arnab.mishra.1992@gmail.com

S. K. Das
e-mail: dassujit88@gmail.com

P. Bansal et al. (eds.), *Proceedings of International Conference on Artificial Intelligence and Applications*, Advances in Intelligent Systems and Computing 1164,
https://doi.org/10.1007/978-981-15-4992-2_33

1 Introduction

Parkinson's disease is a neurodegenerative disease that affects the central nervous system and causes degradation in motor system control. Currently, such neurological diseases are the main causes of disability around the world. According to a study conducted by [1], the number of cases of Parkinson's in 2016 has nearly doubled from 2.5 million in 1990 to 6.1 million. This number is expected to rise to 12 million by 2050. Although medical researchers are currently working towards finding a cure for the disease very actively, no such cure has been found yet. The current approach for dealing with the disease is to try to manage the symptoms by treatment, as well as physical therapy. Usually, the detection of the disease is done by examining motor symptoms, however by this time significant neurological damage may have already been done. If the disease can be detected at an early stage by examining non-motor symptoms, then early treatment can be started and the disease can be potentially managed better, thereby improving the quality of life of the patient. Researchers have proposed the use of speech data for the purpose of early detection of the disease [2–4]. The idea of using speech data from Parkinson's disease patients and healthy controls for performing predictive modelling is explored in this work.

In today's digital era, an abundance of data is usually available for most fields, which upon proper utilization can provide an exciting new understanding of the particular field. Speech data is one such source of information, which can reveal a lot of hidden knowledge for early diagnosis of Parkinson's disease. With the advancements in the fields of Machine Learning (ML) and Artificial Intelligence (AI), it is prime time now, to make use of the data available to build effective clinical decision support systems. Moreover, recent advancements in the field of Deep Learning can open new avenues towards building more effective data-driven decision support systems.

In this work, a publicly available dataset [5] is used which contains speech features extracted with the help of various speech signal processing techniques from 188 patients and 64 healthy controls. Various baseline ML classification algorithms, as well as ensemble methods, have been studied and evaluated on the basis of multiple valuation metrics. Furthermore, oversampling technique has also been applied to increase the size of the dataset, so that the applicability of Deep Neural Networks (DNN) algorithms can be studied. From the experimental results, it has been observed that with proper preprocessing of the data and oversampling, DNN algorithms can provide a very efficient prediction.

The rest of the paper is organized as follows. In Sect. 2, various related works are mentioned and how the current work relates to them is discussed. The proposed methodology is discussed in the following section. Section 4 provides a description of the speech dataset used in this work. Following which the experimental results are discussed in Sect. 5. Finally, the paper is concluded in Sect. 6.

2 Related Work

Automatic detection of Parkinson's disease using machine learning based approaches is a problem that has been studied by many researchers in the past [5–13]. In this section, the related works are discussed, by first briefing about the use of non-speech data, and then speech data based approaches.

In [6], the use of non-motor symptoms like Rapid Eye Movement sleep Behaviour Disorder (RBD), Olfactory Loss data, cerebrospinal fluid measurements and dopeminergic imaging markers, for predicting the presence of the disease is proposed, along with machine learning classifiers like Naive Bayes, Support Vector Machines (SVM) and Boosted trees. Whereas, in [7], the authors have proposed the use of patient questionnaires from the Movement Disorder Society-Unified Parkinson's Disease Rating Scale (MDS-UPDRS), for the purpose of developing predictive models. Multiple classifiers like Logistic Regression (LR), Boosted Trees, and SVMs are applied on the questionnaire data.

Among all the different types of non-motor data based predictive modellings, the use of speech data is very popular, since speech is one source of information which can be very helpful in the early detection of the disease. In [8], the application of Fuzzy K Nearest Neighbours approach is proposed, after applying PCA based dimensionality reduction on features extracted from patient and healthy control speech data. It is observed in [8] that the application of PCA can improve the prediction system's accuracy significantly. The use of multiple unique neural networks in parallel is proposed by [9], where the researchers found that the use of 9 NNs in parallel gives the best performance improvement of about 8 over single NN based method. In [10], the use of four feature selection techniques combined with binary classification algorithms like Random Forest and SVMs is studied for selecting more predictive dysphonia measures from speech data containing sustained vowel pronunciations. The researchers in [11] presented a detailed study where the predictive modelling tasks have been carried out on speech data from both English speaking patients from the USA and German-speaking patients from Germany. According to the observations drawn from this multilingual setting, separate features dominate the predictions when the predictive modellings for both the languages are considered separately. The accuracy of the system reduces when the training and testing of the ML models are done on different language speech data.

In [12, 13], the authors conducted studies for determining the severity of the progress of Parkinson's disease, using voice data. The predictive features from the voice data are extracted using various speech processing approaches. Comprehensive statistical analysis is done on these features to find out value ranges that correspond to various levels of severity of the disease.

The most relevant of the related works to the current work is presented in [5], where the authors have demonstrated the usefulness of Tunable Q-factor Wavelet Transformation (TQWT) technique for extracting speech features, which can improve the performance of ML-based predictive models significantly. This is, to the best of our knowledge, the only work done on the dataset which is considered in the current work also. The researchers in [5] presented a comprehensive analysis of various other speech feature extraction techniques as well. The best models as demonstrated in [5] were able to achieve a maximum accuracy of around 86%. In our current work, the experimental results on the same dataset, using the proposed approach, show significant improvements and achieves above 90% accuracy, while achieving impressive results with respect to the other performance measures as well.

3 Proposed Methodology

In this work, two approaches are proposed for performing machine learning based automatic prediction of Parkinson's disease. This section describes both of these approaches in detail.

3.1 Predictive Modelling Without Oversampling

The first approach is depicted in Fig. 1. In this approach, the raw dataset is firstly passed through a Data Preprocessing phase, where all the features are transformed via feature scaling which transforms the data for each feature to have the data mean as 0 and standard deviation as 1. This sort of transformation is important because it brings the variances of each feature in a similar range. If this is not done, then some high variance features may dominate the classifier's performance, thereby degrading the quality of the prediction model learned.

After feature scaling, the dataset is split randomly into 80:20 ratio, where 80% of the data is considered for training the ML algorithms and the rest is kept aside as the testing data to be used later during the model evaluation phase.

Now the training dataset is used for training various ML algorithms. In this work, multiple kinds of classifiers are used, like Logistic Regression (LR), K Nearest Neighbours (KNN), Decision Trees (DT), Support Vector Machines (SVM), AdaBoost (ADB), Gradient Boosting (GB), Random Forest (RF) and Deep Neural Networks (DNN). The reason for choosing these particular classifiers is to perform experimentation using baseline models, bagging and boosting based models as well as complex models like DNNs.

Once the ML models are trained, then the next task is to evaluate the trained models on the testing data that was kept aside earlier, so that the models' ability to generalize can be studied. This entire process from splitting the dataset to evaluating the models is repeated multiple times so that the average behaviour of the models can be observed because, in each of the iterations, the models are learned on different random splits of the data.

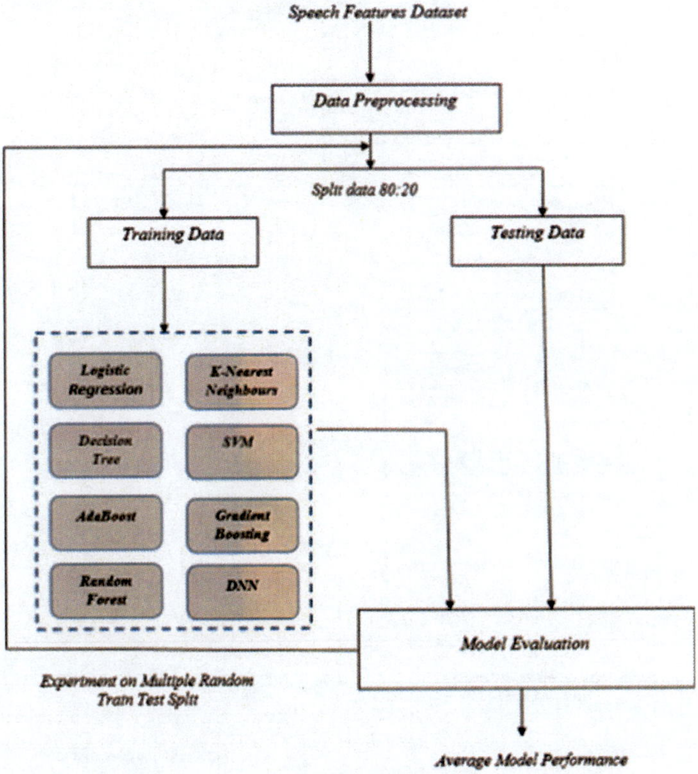

Fig. 1 Predictive modelling approach without oversampling

3.2 Predictive Modelling with SMOTE Oversampling

The second approach is depicted in Fig. 2. This approach is almost similar to the approach depicted in Fig. 1, the only difference is that here Synthetic Minority Oversampling TEchnique (SMOTE) oversampling phase is also included. If the dataset under consideration for any machine learning task has many instances of a class, and much fewer instances of the other, then such a dataset is called as a class imbalanced dataset.

SMOTE oversampling solves this problem of class imbalance, by creating synthetic samples from the minority class. SMOTE considers each minority instance and chooses one of its k closest minority neighbours, following which SMOTE synthesizes a new minority instance between the neighbour and itself. This process gets rid of the class imbalance problem. Once the oversampling phase is over, then the approach proceeds the same as the first approach, to find the average performance of the trained models on the testing data which has been kept aside earlier.

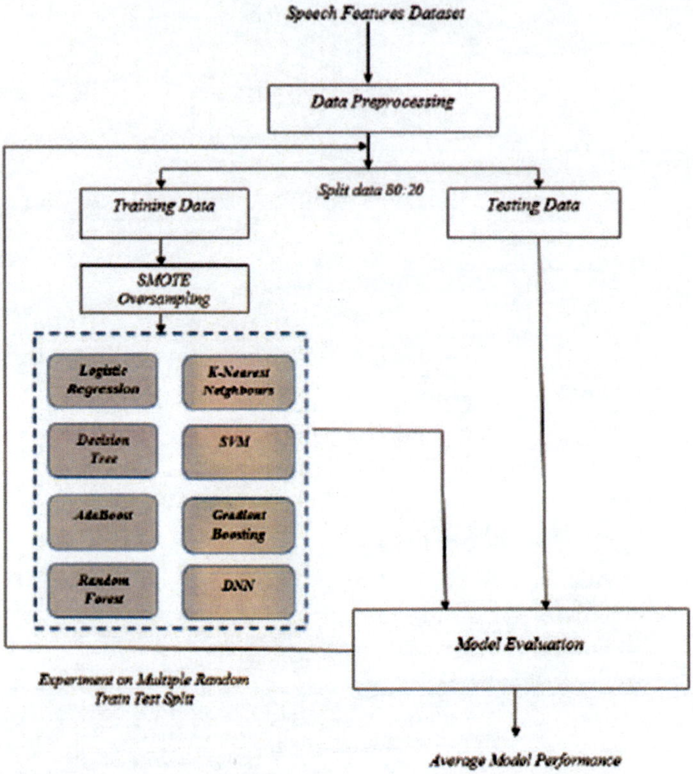

Fig. 2 Predictive modelling approach with SMOTE oversampling

4 Dataset Description

In this work, a dataset containing features derived from speech signals, which is first introduced by [5] is used in order to perform predictive modelling. The dataset is derived from speech signals by 188 Parkinson's disease patients (107 men and 81 women) and 64 healthy controls (23 men and 41 women). The patient ages range from 33 to 87, whereas the control group's ages range from 41 to 82. The dataset is publicly available at https://archive.ics.uci.edu/ml/datasets/Parkinson%27s+Disease+Classification. Various speech signal processing techniques were applied to derive speech features like Mel Frequency Cepstral Coefficients (MFCCs), Time Frequency Features, Wavelet transformation based features, etc. from the speech signal consisting of three repetitions of the sustained phonation of the vowel /a/. The sustained vocalization of phoneme /a/was recorded with a microphone set to 44.1 kHz. Via these preprocessing steps on the speech signals, this dataset consisting of 756 instances, with 754 features each was created. The number of positive samples for Parkinson's disease in this dataset is 564, and number of negative samples is 192.

5 Experimental Results and Discussion

In this section, the experimental setting is first discussed, following which the experimental results are presented. Multiple performance metrics have been considered in this work, for the purpose of evaluating the ML models, like prediction accuracy, sensitivity, specificity and area under the curve (AUC) of the receiver operating characteristic (ROC) curve.

Some classifier specific settings are also important to be noted here. The bagging and boosting classifiers used in this work all are parameterized by 50 individual internal learners. For the Deep Neural Network classifier used in this work, which is characterized by a learning rate of 0.0001, batch size of 10 and number of epochs as 50, consider the structure of the network as depicted in Fig. 3 below. During training of the network, an early stopping criterion was also applied, based on 10% validation data kept aside from the training dataset. This was done to avoid over fitting by the model.

Using the above-defined setting, experiments have been conducted in this work using the approaches mentioned in Sect. 3.

The results and observations drawn from these experiments are discussed in the following subsections.

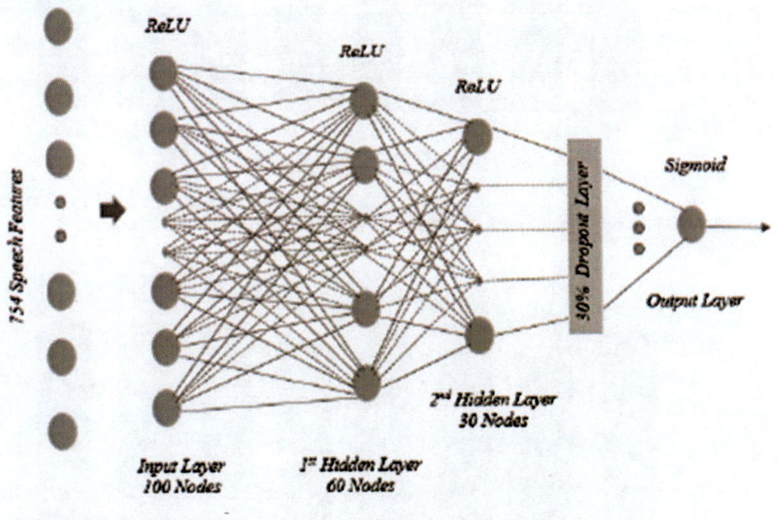

Fig. 3 Deep neural networks model considered for experimentation

5.1 Experiments Without Oversampling

In these sets of experiments, in order to find the average accuracy, sensitivity, specificity and AUC values, experiments are conducted on 50 different random 80:20 splits of the dataset, for each individual ML model. Consider the results using this approach as depicted in Figs. 4, 5, 6 and 7.

One common observation from the above results is that among the baseline models, Decision Tree classifier performs better with respect to all the performance metrics, achieving an average prediction accuracy of 84.7%, average sensitivity of 93.19%, average specificity of 66.04% and an average AUC value of 78.97%. Also,

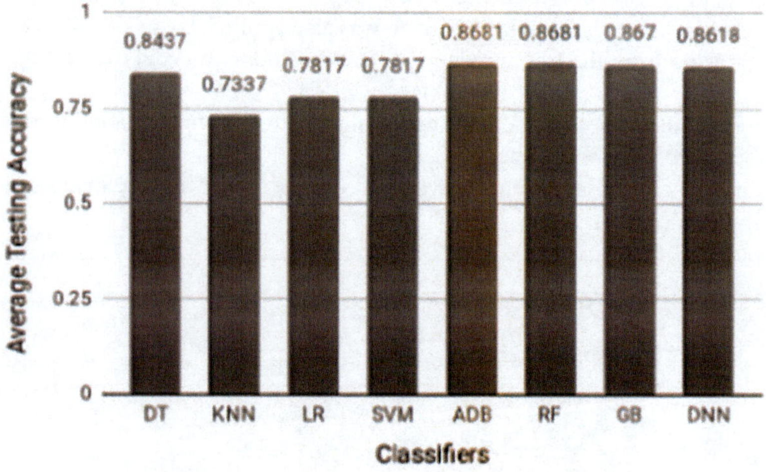

Fig. 4 Average prediction accuracies of classifiers without using oversampling

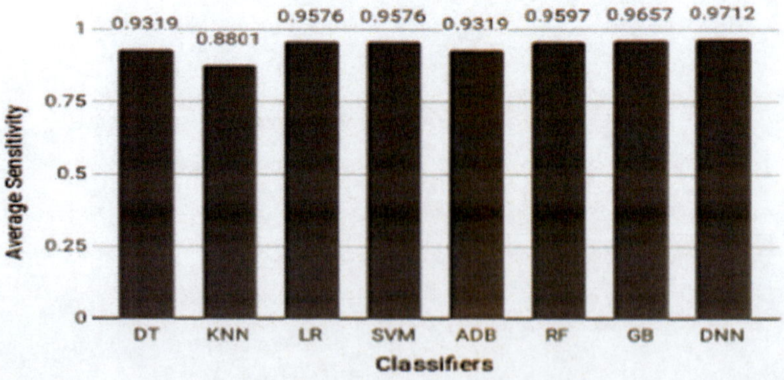

Fig. 5 Average sensitivities of classifiers without using oversampling

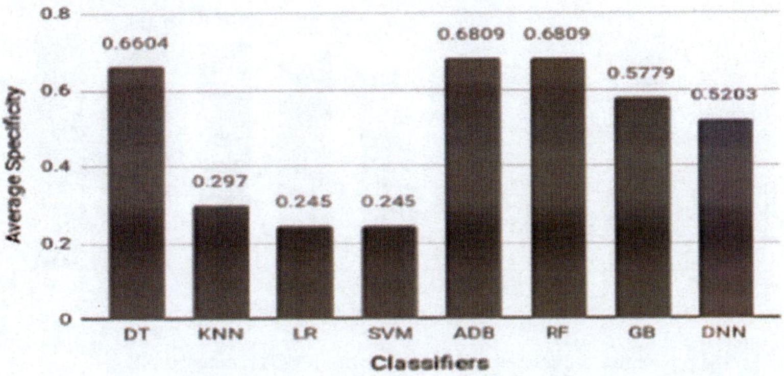

Fig. 6 Average specificities of classifiers without using oversampling

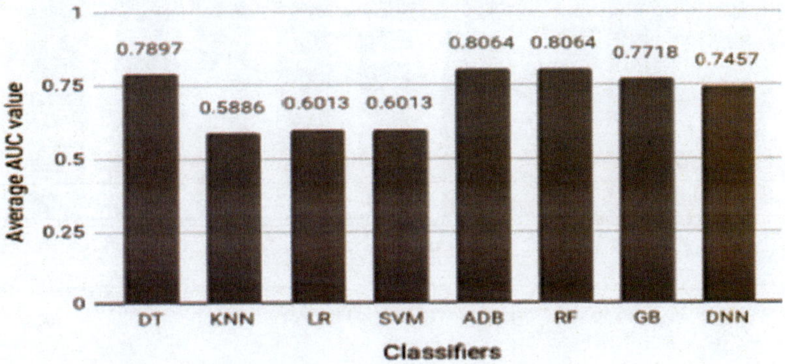

Fig. 7 Average AUC values of classifiers without using oversampling

another important observation is that all the ensemble methods perform comparatively better than the baseline models. This is expected since it has been proven previously by many researchers that an ensemble of simple models can perform better when combined together. Note that without any oversampling, DNN model doesn't perform as well as the ensemble models, this is due to the small size of training data and also the class imbalance problem in the dataset. The best model for predicting the presence of Parkinson's disease, according to these set of experiments is the Random Forest model, with respect to all the performance metrics.

5.2 *Experiments with SMOTE Oversampling*

In these set of experiments also, the same experimental setting is considered. The results using this approach are depicted in Figs. 8, 9, 10 and 11.

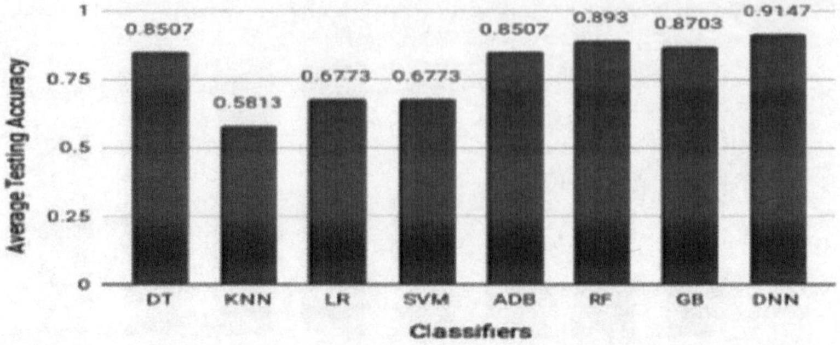

Fig. 8 Average prediction accuracies of classifiers with SMOTE oversampling

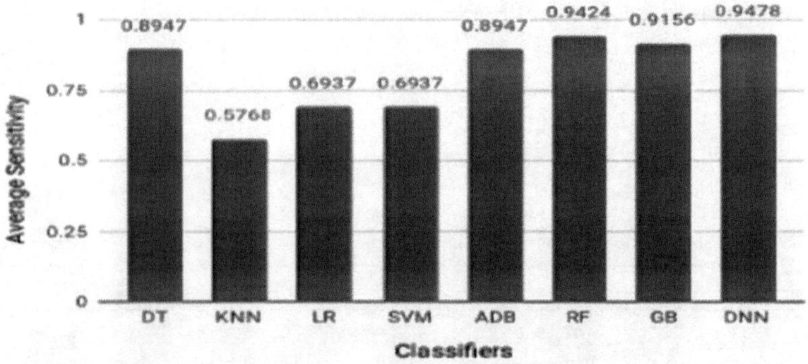

Fig. 9 Average sensitivities of classifiers with SMOTE oversampling

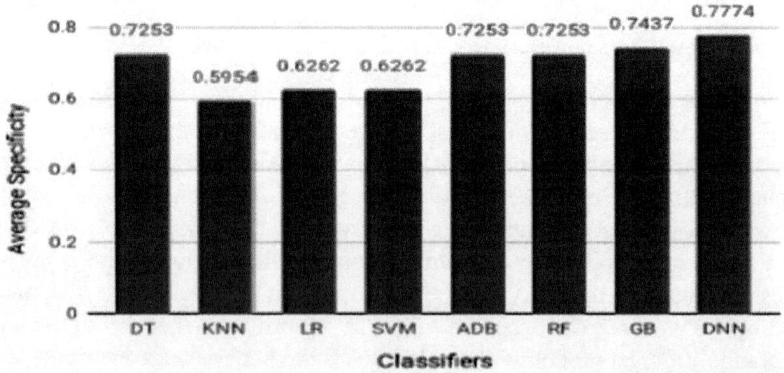

Fig. 10 Average specificities of classifiers with SMOTE oversampling

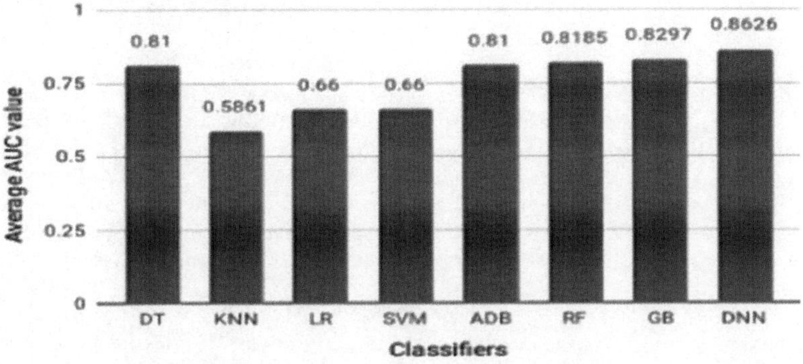

Fig. 11 Average AUC values of classifiers with SMOTE oversampling

Similar trend can be observed here as well, with all the ensemble learners doing better than the baseline learners, except for Decision Trees, which show similar performance as the ensemble learners. However, with oversampling one important change has occurred in the experimental results, which is DNNs has outperformed all the other classifiers, across all the performance metrics. This demonstrates the benefit of oversampling strategy. With SMOTE oversampling, best results are achieved using DNN, with an average prediction accuracy of 91.47%, an average sensitivity of 94.78%, average specificity of 77.74% and an average AUC value of 86.26%.

From the experimental results it can be concluded that with enough class balanced data, DNNs can outperform all the other classifier, for the purpose of detection of Parkinson's disease.

6 Conclusion and Future Work

In this work, approaches for automatic prediction of Parkinson's disease using various ML techniques have been discussed. Experimental results suggest that ensemble learners, in general, are good at predicting the presence or absence of the disease. Also, it has been observed that DNN based approach can be improved for this prediction task with the application of proper oversampling. The best approach for automatic prediction of Parkinson's disease has been found to be DNN with SMOTE oversampling, achieving above 90% accuracy, and also consistently better performance with respect to the other performance metrics as well. These observations demonstrate the potential applicability of the proposed approaches in real diagnostic settings. It is important to note here however, that due to the lack of publicly available datasets, this work only demonstrates results on one dataset. A more comprehensive analysis will be done in the future, considering multiple datasets. Such a comprehensive analysis is expected to provide a more detailed understanding of the applicability of the proposed approaches.

References

1. E.R. Dorsey, E. Alexis, E. Nichols, F. Abd-Allah, A. Abdelalim, J.C. Adsuar, M.G. Ansha et al., Global, regional, and national burden of Parkinson's disease, 1990–2016: a systematic analysis for the global burden of disease study 2016. Lancet Neurol. **17**(11), 939–953 (2018)
2. R.K. Sharma, Early detection of Parkinson's disease through Voice, in *2014 International Conference on Advances in Engineering and Technology (ICAET)* (IEEE, 2014), pp. 1–5
3. J. Rusz, R. Cmejla, H. Ruzickova, E. Ruzicka, Quantitative acoustic measurements for characterization of speech and voice disorders in early untreated Parkinson's disease. J. Acoust. Soc. Am. **129**(1), 350–367 (2011)
4. B.T. Harel, M.S. Cannizzaro, H. Cohen, N. Reilly, P.J. Snyder, Acoustic characteristics of Parkinsonian speech: a potential biomarker of early disease progression and treatment. J. Neurolinguist. **17**(6), 439–453 (2004)
5. C.O. Sakar, S. Gorkem, A. Gunduz, H.C. Tunc, H. Nizam, B.E. Sakar, M. Tutuncu, T. Aydin, M. Erdem Isenkul, H. Apaydin, A comparative analysis of speech signal processing algorithms for Parkinson's disease classification and the use of the tunable Q-factor wavelet transform. Appl. Soft Comput. **74**, 255–263 (2019)
6. R. Prashanth, S.D. Roy, P.K. Mandal, S. Ghosh, High-accuracy detection of early Parkinson's disease through multimodal features and machine learning. Int. J. Med. Inf. **90**, 13–21 (2016)
7. R. Prashanth, S.D. Roy, Early detection of Parkinson's disease through patient questionnaire and predictive modelling. Int. J. Med. Inf. **119**, 75–87 (2018)
8. H.-L. Chen, C.-C. Huang, X.-G. Yu, X. Xin, X. Sun, G. Wang, S.-J. Wang, An efficient diagnosis system for detection of Parkinson's disease using fuzzy k-nearest neighbor approach. Expert Syst. Appl. **40**(1), 263–271 (2013)
9. F. Åström, R. Koker, A parallel neural network approach to prediction of Parkinson's Disease. Expert Syst. Appl. **38**(10), 12470–12474 (2011)
10. A. Tsanas, M.A. Little, P.E. McSharry, J. Spielman, L.O. Ramig, Novel speech signal processing algorithms for high-accuracy classification of Parkinson's disease. IEEE Trans. Biomed. Eng. **59**(5), 1264–1271 (2012)
11. H. Hazan, D. Hilu, L. Manevitz, L.O. Ramig, S. Sapir, Early diagnosis of Parkinson's disease via machine learning on speech data, in *2012 IEEE 27th Convention of Electrical and Electronics Engineers in Israel* (IEEE, 2012), pp. 1–4
12. M. Asgari, I. Shafran, Extracting cues from speech for predicting severity of parkinson's disease, in *2010 IEEE International Workshop on Machine Learning for Signal Processing* (IEEE, 2010), pp. 462–467
13. M. Asgari, I. Shafran, Predicting severity of Parkinson's disease from speech, in *2010 Annual International Conference of the IEEE Engineering in Medicine and Biology* (IEEE, 2010), pp. 5201–5204

Local Binary Pattern Based ELM for Face Identification

Bhawna Ahuja and Virendra P. Vishwakarma

Abstract In this chapter, local binary pattern (LBP) based extreme learning machine (ELM) is presented for identification of high-dimensional face images of different resolution. In this scheme, LBP is used for the representation of micro-regions of face image in the form of feature vectors which are further concatenated to be utilized as face descriptors for representation of face image. The obtained face descriptors are classified using ELM which is a simple and fast learning algorithm. Extensive simulations on AT&T and Yale databases of face images illustrate the effectiveness of the proposed scheme.

Keywords Face identification · Face descriptor · ELM · LBP

1 Introduction

Face recognition is a very active and interesting subject in computer vision due to its utility in various applications like human–machine interaction, visual surveillance and biometric identification [1, 2]. Face recognition is a very challenging problem as the appearance of face images vary due to changes in lightning conditions, expressions, pose and age. The main issue in face recognition is finding efficient features (descriptors) to represent face images. In the literature of face recognition, different linear techniques such as linear discriminant analysis [3], independent component analysis [4], principal component analysis [5] had been explored which maps the high-dimensional face image to low dimensional image. As the face recognition problem is highly non-convex, so these techniques are not efficient to represent the non-convex variations necessary to differentiate intra and inter-class individuals.

B. Ahuja (✉) · V. P. Vishwakarma (✉)
University School of Information Communication Technology, Guru Gobind Singh Indraprastha University, New Delhi 110078, India
e-mail: Bhawana.ahu@gmail.com

V. P. Vishwakarma
e-mail: vpv@ipu.ac.in; virendravishwa@rediffmail.com

P. Bansal et al. (eds.), *Proceedings of International Conference on Artificial Intelligence and Applications*, Advances in Intelligent Systems and Computing 1164,
https://doi.org/10.1007/978-981-15-4992-2_34

Pentland et al. [6] introduced a eigen features approach based on the local region descriptors in which principal component analysis is applied to sub-region of the image. Obtaining efficient descriptors to represent local regions in face image is an open issue. These descriptors should have less discrepancy in intra-class (i.e. different images of the same subject) and high variations in extra class with less computational cost. A lot of research has been conducted in texture analysis to develop a range of distinct descriptors for the representation of image patches.

Ojala et al. [7] introduced the concept of local binary pattern (LBP) features, for depiction of local spatial structure of face images, producing outstanding results. LBP is a unifying method to variant structural and statistical framework of texture analysis. It summarizes the local structure of images by comparing each pixel with its neighboring pixels. It is used in numerous real-world problems due to its robustness to grayscale variations like changes in illumination conditions. It has highly discriminative and computational efficient property, which promotes it to be used in image analysis.

After feature extraction, the next step in face recognition is face matching or classification of faces with different faces in database. The traditional classification algorithms are iterative and inefficient when a large database is used. To overcome the problem associated with these algorithms like local minima, slow learning, overfitting, underfitting ELM, and its variants were introduced in literature [8–12]. ELM belongs to the class of a single-hidden layer feed forward neural network. It is non-iterative in nature as the input layer parameters are randomly assigned. ELM randomly projects the input samples without using the input data charters tics, which leads to non-deterministic output solution [13, 14]. The hidden layer neurons in ELM are experimentally calculated to achieve the desired output. ELM is proved to be a fast algorithm than traditional gradient-based algorithms when used for the classification problem of high dimension, due to arbitrary assignment of input weights instead of iteratively updation of weights as in case of traditional classification algorithms.

In the proposed approach for face identification, the local features of the face (eyes, nose, lips, etc.) are calculated by applying LBP on micro-patterns of face image which are further concatenated to obtain a face descriptor. The descriptors are classified using ELM. The efficiency of proposed algorithm is compared with ELM on face databases.

The remaining paper is structured in the following way. Section 2 briefs the concept of LBP. The proposed work is explained in Sect. 3. In Sect. 4 experimental results are evaluated on the proposed scheme. Section 5 draws our conclusions.

2 Local Binary Pattern (LBP)

LBP is a simple and very powerful texture operator that tags the pixels of a face image by thresholding the proximity of each pixel with the value of the center pixel element and acknowledges the result as a binary number. Due to its distinguished power and computational easiness, LBP texture-operator has to turn out as a popular

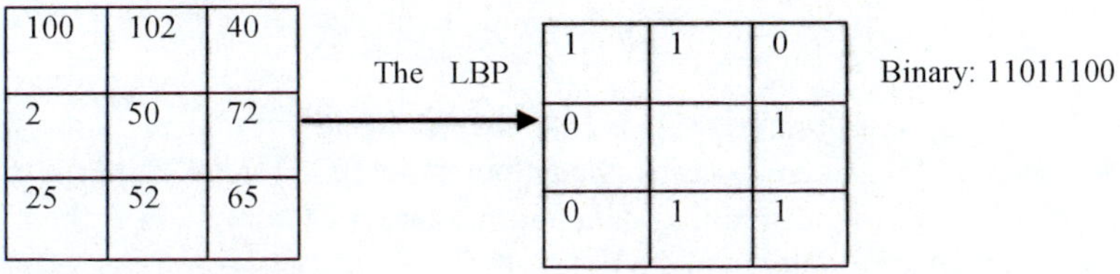

Fig. 1 Basic LBP operator

methodology in numerous applications. It can be viewed as a combined approach to the traditionally variant structural and statistical models of texture analysis [15, 16]. Perhaps the utmost essential proprietary of the LBP operator in real-world problems is its robustness to monotonic changes caused by variations in grayscale intensity level of pixel element, for example, variations in illumination conditions [17, 18]. Another significant property of LBP is computational simplicity, which helps it in analyzing images in challenging real-time conditions. The basic LBP operator is explained in Fig. 1, in which neighborhood is of size 3×3. Considering every pixel as threshold, it transferred its 3×3 neighborhood into an 8-bit binary code. The LBP code for considered neighborhood is obtained by concatenation of all 0 and 1 elements from the top left corner in clockwise direction. Later on, LBP is extended to arbitrary number of bilinear interpolated pixels on a circle with arbitrary size are used as neighbor pixels instead of its 3×3 neighborhood.

The potential of features obtained with the help of LBP operator depends on the size of micro-region. The feature vectors associated with these regions are local features which are concatenated to be used for classification. Unlike the different statistical feature extraction methods that tuned a large number of parameters, the LBP scheme is very efficient due to its simple matching approach and easy to compute features extraction operation.

3 Proposed LBP Based Extreme Learning Machine (LBP-ELM)

Face recognition is non-linear classification problem in which the accuracy rate depends on the feature vectors used to represent the face images. In our proposed approach LBP method is adopted for feature extraction. The face image is partitioned into micro-regions of same size and then LBP is applied on these regions to find the feature vectors associated with these regions. The obtained feature vectors are classified using ELM. ELM is fast and less complex learning algorithm, which is based on feedforward neural network concept [19, 20]. In ELM the weights associated with input data are assigned randomly. The output weight matrix (between hidden layer and output layer) in ELM is evaluated by the following Equation

$$\beta = \Omega^{\dagger} Z \tag{1}$$

where β is output weight matrix, $\Omega^{\dagger}$ is M-P inverse of output matrix corresponding to hidden layer and Z is target matrix. The output target function for ELM for training set $\{(y_k, z_k)\}_{k=1}^{N}$ (y is input training data and z is target vector and N is the number of instances of training data) is given by

$$F = \Omega\beta = \Omega(y)\Omega^T\left(\frac{C}{Q} + \Omega\Omega^T\right)^{-1} Z \tag{2}$$

where C is identity-matrix and Q is regularization coefficient. The class of the input data is determined by the index of the output node with the largest value.

In LBP-ELM the input data vector is replaced by the final feature vector obtained by concatenation of features vectors of micro-regions of face image. The output function for LBP-ELM is represented as

$$F = \Omega\beta = \Omega(y_f)\Omega^T\left(\frac{C}{Q} + \Omega\Omega^T\right)^{-1} Z \tag{3}$$

Here y_f is the final feature vector.
The proposed LBP-ELM is explained as follows:

Step 1: Calculate the LBP of face image to extract the edges, lines, flat area and spots which are further combined to obtain the face descriptor.
Step 2: Determine the output weight matrix using Eq. (1).
Step 3: Evaluate the output function for LBP-ELM using Eq. (3) for classification of input face images.

4 Experimental Evaluations

The experimental results are calculated on popular face databases for calculating the performance of proposed algorithm LBP-ELM. The percentage recognition rates have been evaluated on these databases for estimating the improvement in performance of presented scheme. The experiments are performed using MATLAB 8.1.

4.1 Face Databases

The databases considered for experimental calculations are AT&T [21] and Yale [22]. The AT&T database includes images with distinct facial details, pose, illumination conditions and expressions. There are 400 images of 40 different individuals, with each image of size 112 × 92. The Yale database contains grayscale images for 15 persons, with 11 different face images per person. The dimension of each face image is 320 × 243. The image of this database has a lot of variations in terms of expressions (happy/normal) and lightning conditions. The above mentioned databases are not pre-processed for performance calculation.

4.2 *Results and Discussions*

The number of face images per individual used for training are considered sequentially and are increased from 1 to 7. For example, in the case of AT&T database, the training dataset consists of minimum of 40 images (i.e. one image per person) and maximum of 280 images (seven images per person). Corresponding to training database testing database consists of 360 and 120 images, respectively. The obtained training and testing datasets are disjoint. For results calculation using ELM different values of hidden neurons (500,1000, 1500, 2000) and regularization coefficients (1, 10, 20, …, 100) have been tried, but the best results are achieved by considering hidden neurons equal to 500 and regularization coefficient equal to 10. The activation function applied to hidden layer neurons is sigmoidal function. For feature vector calculation using LBP, the neighborhood size is taken as 3 × 3. For comparative analysis, the results of LBP-ELM are evaluated on the same parameters as used for the ELM. Tables 1 and 2 displays the experimental results for ELM and LBP-ELM for AT&T and Yale databases respectively. It can be seen from the experimental results that the percentage testing accuracy in case of AT&T database is more than Yale database in most of the cases, this is because of the fact that face images in case of Yale database contain a lot of variations due to lightning conditions. The performance results clearly reveal that LBP-ELM outperforms ELM.

Table 1 Comparison of percentage accuracy of ELM and LBP-ELM on AT&T database

Algorithm	No. of images per individual used for training						
	1	2	3	4	5	6	7
ELM	64.27	75.18	77.63	79.74	82.10	86.62	91.15
LBP-ELM	66.10	77.20	79.10	80.72	83.50	88.70	92.20

Table 2 Comparison of percentage accuracy of ELM and LBP-ELM on yale database

Algorithm	No. of images per individual used for training						
	1	2	3	4	5	6	7
ELM	45.56	61.21	78.16	84.72	87.79	85.14	93.32
LBP-ELM	46.71	62.35	79.28	86.10	88.02	86.70	94.00

5 Conclusions

This paper introduced a framework for face identification which is integration of LBP and ELM. In the proposed scheme the LBP is used to describe the texture description of micro-region of the face image and the concatenation of all-region description represents the global geometry of face image. The classification of calculated face descriptors is done using ELM. The proposed scheme is certified on face databases. The simulation results reveal the significant improvement in percentage recognition rate by the proposed approach. In the future, more sophisticated feature extraction methods will be explored that works efficiently in uncontrolled conditions.

References

1. W. Zhao, R. Chellappa, P.J. Phillips, A. Rosenfeld, Face recognition: a literature survey. ACM Comput. Surv. **35**, 399–458 (2003)
2. W. Chen, M.J. Er, S. Wu, Illumination compensation and normalization for robust face recognition using discrete cosine transform in logarithm domain. IEEE Trans. Syst. Man Cybern Part B **36**, 458–466 (2006)
3. K. Etemad, R. Chellappa, Discriminant analysis for recognition of human face images. Josa a **14**, 1724–1733 (1997)
4. M.S. Bartlett, J.R. Movellan, T.J. Sejnowski, Face recognition by independent component analysis. IEEE Trans. Neural Netw. **13**, 1450–1464 (2002)
5. B. Heisele, P. Ho, J. Wu, T. Poggio, Face recognition: component-based versus global approaches. Comput. Vis. Image Underst. **91**, 6–21 (2003)
6. A. Pentland, B. Moghaddam, T. Starner, View-based and modular eigenspaces for face recognition (1994)
7. T. Ojala, M. Pietikäinen, T. Mäenpää, Gray scale and rotation invariant texture classification with local binary patterns, in *European Conference on Computer Vision*, pp. 404–420 (2000)
8. G.-B. Huang, Q.-Y. Zhu, C.-K. Siew, Extreme learning machine: a new learning scheme of feedforward neural networks, in: 2004 IEEE International Joint Conference on Neural Network, pp 985–990 (2004)
9. G.-B. Huang, Z. Bai, L.L.C. Kasun, C.M. Vong, Local receptive fields based extreme learning machine. IEEE Comput. Intell. Mag. **10**, 18–29 (2015)
10. Y. Miche, A. Sorjamaa, P. Bas et al., OP-ELM: optimally pruned extreme learning machine. IEEE Trans. Neural Netw. **21**, 158–162 (2009)
11. L. Zhang, D. Zhang, Evolutionary cost-sensitive extreme learning machine. IEEE Trans. Neural Netw. Learn. Syst. **28**, 3045–3060 (2017)
12. S. Ding, X. Xu, R. Nie, Extreme learning machine and its applications. Neural Comput. Appl. **25**, 549–556 (2014)
13. W. Zong, G.-B. Huang, Y. Chen, Weighted extreme learning machine for imbalance learning. Neurocomputing **101**, 229–242 (2013)
14. X. Wang, M. Han, Improved extreme learning machine for multivariate time series online sequential prediction. Eng. Appl. Artif. Intell. **40**, 28–36 (2015)
15. T. Ahonen, A. Hadid, M. Pietikäinen, Face recognition with local binary patterns. In: *European Conference on Computer Vision*, pp. 469–481 (2004)
16. S. Liao, X. Zhu, Z. Lei et al., Learning multi-scale block local binary patterns for face recognition, in *International Conference on Biometrics*, pp. 828–837 (2007)
17. W. Zhang, S. Shan, W. Gao et al., Local gabor binary pattern histogram sequence (lgbphs): a novel non-statistical model for face representation and recognition, in: Tenth IEEE International Conference on Computer Vision, vol. 1, pp. 786–791 (2005)

18. C.-H. Chan, J. Kittler, K. Messer Multi-scale local binary pattern histograms for face recognition, in *International Conference on Biometrics*, pp. 809–818 (2007)
19. G.-B. Huang, Q.-Y. Zhu, C.-K. Siew, Extreme learning machine: theory and applications. Neurocomputing **70**, 489–501 (2006)
20. G.-B. Huang, L. Chen, Convex incremental extreme learning machine. Neurocomputing **70**, 3056–3062 (2007)
21. F.S. Samaria, A.C. Harter, Parameterisation of a stochastic model for human face identification, in *Proceedings of 1994 Second IEEE Workshop on Applications of Computer Vision*, pp. 138–142 (1994)
22. P.N. Belhumeur, J.P. Hespanha, D.J. Kriegman, Eigenfaces vs. fisherfaces: recognition using class specific linear projection. IEEE Trans. Pattern Anal. Mach. Intell. **19**, 711–720 (1997)

Optimization Techniques and its Applications in Machine Learning

Binary Particle Swarm Optimization Based Feature Selection (BPSO-FS) for Improving Breast Cancer Prediction

Arnab Kumar Mishra, Pinki Roy, and Sivaji Bandyopadhyay

Abstract Breast cancer is currently one of the leading causes of cancer-related deaths among women around the world. Although the severity of the disease is undeniable, an efficient early diagnosis of the disease can lead to a much higher chance of survival for the patients. Effective clinical decision support systems (CDSS) could potentially be of very high utility for medical practitioners, in this regard. In this paper, a binary Particle Swarm Optimization (BPSO) based feature selection approach is presented, which can be used to improve the performance of automatic breast cancer prediction CDSS. The key idea is to formulate the problem of feature selection in terms of a discrete optimization problem, with appropriate data-driven objective function. The average cost of bad feature selection is considered as the objective function to minimize, in this work. Multiple evaluation metrics like average prediction accuracy, sensitivity, specificity and area under the curve (AUC) of the receiver operating characteristics (ROC) curve have been considered in this work, to evaluate the performance of the prediction system. Average prediction accuracies of 80.83% and 98.24% have been observed respectively for the Breast Cancer Coimbra Dataset (BCCD) and Breast Cancer Wisconsin Diagnostic Dataset (BCWDD) after feature selection is performed. When pre-feature selection and post feature selection performances are compared, overall improvements of 4–8% and 1–4% have been observed across all the evaluation metrics for BCCD and BCWDD respectively, suggesting the potential applicability of the proposed approach in real diagnostic settings.

Keywords Breast cancer · Feature selection · Machine learning · Clinical decision support system · Binary particle swarm optimization

A. K. Mishra (✉) · P. Roy · S. Bandyopadhyay
CSE Department, National Institute of Technology, Silchar, Assam 788010, India
e-mail: arnab.mishra.1992@gmail.com

P. Roy
e-mail: pinkiroy2405@gmail.com

S. Bandyopadhyay
e-mail: directornits@gmail.com

P. Bansal et al. (eds.), *Proceedings of International Conference on Artificial Intelligence and Applications*, Advances in Intelligent Systems and Computing 1164,
https://doi.org/10.1007/978-981-15-4992-2_35

1 Introduction

Breast cancer is the most common type of cancer among women, according to the World Health Organization (WHO) [1]. Nearly 2.1 million women get affected every year by this disease. According to WHO estimates, in 2018, over 600,000 women died due to the disease, constituting nearly 15% of all deaths related to cancer among women. However, research works have shown better treatment can be achieved if an early diagnosis of breast cancer is done [2]. Recent advancements in the field of Artificial Intelligence (AI) and Machine Learning (ML), it is nowadays possible to achieve such a highly effective diagnosis of the disease. Past breast cancer patient data can be utilized for building such kind of effective decision support systems, which has been shown in many research works in this area [3–6]. In this work, the application of a feature selection technique is studied for improving the performance of such a system. More specifically, the binary variant of nature-inspired approach of optimization, Particle Swarm Optimization (BPSO) is adapted in this work for the purpose of performing feature selection, thereby improving the performance of automatic ML-based breast cancer prediction systems.

In today's world, there is an abundance of data in almost every field. This data, if utilized properly, can unravel important insights into the currently unsolved problems of the field. The problem of automatic breast cancer prediction is also such kind of a problem. A lot of different types of data are usually collected in the form of doctor consultation reports, blood test reports, histopathological slide images, results of mammography, etc., during the course of diagnosis and treatment of breast cancer. All of these data can help build very efficient predictive models, which can act as decision support systems for physicians. The fields of AI and ML have recently shown great promise in this area. Inspired by this, various kinds of ML approaches have been considered in this work, for experimentations on multiple publicly available data sets. These breast cancer prediction ML approaches can be improved by the application of feature selection techniques. From the comprehensive experimental study, it has been observed that BPSO-based feature selection can be applied to improve the performance of ML-based decision support systems significantly.

The rest of the paper is organized as follows. In Sect. 2, a detailed review of the previous related work presented. In Sect. 3, the proposed methodology for developing effective predictive models for the task of breast cancer prediction is explained. Section 4 discusses the data sets used in this work for the purpose of experimentation, which is then followed by a section on the experimental results and discussions on the observations obtained from the results. Following this, the paper is concluded in the final Section where the findings of the paper are summarized and future work directions are suggested.

2 Related Work

In this section, the main focus is on the related works that specifically deal with the task of building and improving automatic breast cancer prediction decision support systems; a detailed review of ML techniques applied for solving the more general problem of automatic prediction of various types of cancer is discussed in [3].

Researchers have studied this problem of applying ML techniques for performing automatic breast cancer prediction extensively, with most of the recent works focusing on working with image-based data. There are two main reasons for this (i) in the course of breast cancer diagnosis and further treatments, a lot of different kinds of image data is usually collected, like mammography images, histopathological slide images from tissue samples, etc., and (ii) recent advancements in the field of deep learning, especially in the field of computer vision has paved new ways towards getting more insights from image data. The application of deep learning approaches on such image data sets has been explored by many researchers in recent years [7–10]. However before this change of trend, researchers used to focus on the application of more traditional approaches like Support Vector Machines [7, 11–16], Ensemble methods like Random Forest and Adaboost classifiers [15, 17], and Artificial Neural Networks [11, 18].

Breast Cancer Wisconsin (Diagnostic) Dataset (BCWDD) is one of the data sets used in this work for the purpose of an experimental study, which is such a data set that has been explored by many researchers in the past. The power of combining ANN and fuzzy inference systems for the prediction task has been demonstrated by Addeh et al. [19]. Pipelining of association rules based feature selection with adaptive neuro-fuzzy inference systems (ANFIS) is proposed in this hybrid approach, in order to improve the predictive model's performance. In Baneriee et al. [20], the researchers have suggested the use of bagging based approaches, instead of other types of ensemble learners, where a detailed comparative analysis of various ensemble learners on the data set is presented. Aličković et al. [21] proposed a rotation forest-based approach in combination with the Genetic Algorithm based feature selection approach, demonstrating the utility of feature selection approaches for improving the performance of breast cancer prediction systems. Whereas the researchers in [22] proposed a hybrid knowledge-based system that can achieve the highest tenfold cross-validation accuracy of 0.932. It is important to note here that the performances observed from the experimentations based on the proposed approach for feature selection are also very competitive, compared to the results of these recent works.

Another data set considered in this work is the Breast Cancer Coimbra Dataset (BCCD), which is a much newer data set, first introduced by Patrício et al. [23]. This data set consists of breast cancer patient data, that can be easily acquired using routine blood analysis. In [23], the researchers have suggested the usage of the feature subset [Age, BMI, Glucose, Resistin], as attributes to be used by SVM, for performing predictive modelling. This approach achieved a range of [0.87–0.91] for the 95% confidence interval of the AUC value. The sensitivity range of 82–88%

and specificity range of 85–90% was achieved by the approach. It is important to note here that the proposed BPSO-based feature selection technique also finds a similar feature subset, with Leptin found as another important attribute. Li et al. [24] compared various ML-based methods on the same data set, suggesting the application of random forest classifier. In [24], prediction accuracy, AUC value, and *F*-measure metrics were used for the evaluation purpose, attaining 0.743, 0.785 and 0.78 values for the evaluation metrics respectively. It is noteworthy here that the proposed BPSO approach was able to achieve much better performance as compared to Li and Chen [24].

3 Proposed Methodology

In this section, the proposed methodology is described in a detailed manner. There are essentially three major phases involved, BPSO-based feature selection, ML-based predictive modelling, and comparative analysis of the models trained. At the end of all of the three phases, the applicability of the BPSO-based feature selection approach will be evident. The proposed methodology is depicted pictorially in Fig. 1.

3.1 BPSO-Based Feature Selection

In the general version (continuous space) of Particle Swarm Optimization (PSO), which is inspired by the optimization behaviour of a flock of birds or a school of fish, there are multiple candidate solutions (called particles) which move around the search space as a group (called swarm) trying to find the global best solution. Each particle has a current position $x_i(t)$ and a current velocity $v_i(t)$. The position of each particle is updated with a velocity update method that also takes into consideration

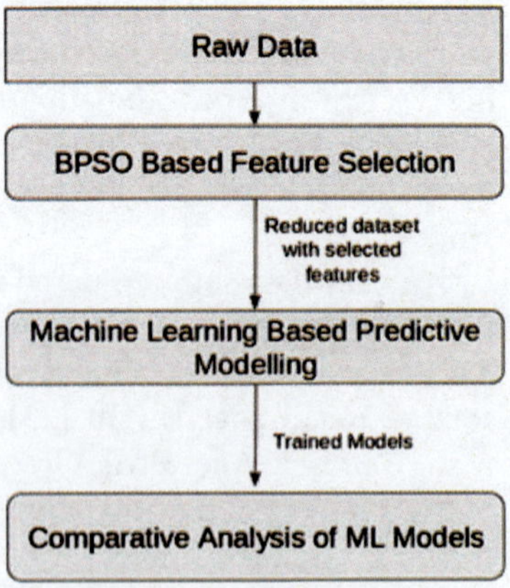

Fig. 1 Proposed methodology for predicting the presence of breast cancer

the best position of the particle so far and the best position found by the entire swarm, thereby performing the exploitation–exploration trade-off (which is done by most optimization approaches in one way or another). The velocity update method is

$$v_{ij}(t+1) = w * v_{ij}(t) + c_1 * r_{ij}(t) * [y_{ij}(t) - x_{ij}(t)] + c_2 * r_{2j} * [\hat{y}_{ij}(t) - x_{ij}(t)]$$

where c_1 and c_2 are called the cognitive and social parameters respectively, that determines how much exploitation is to be done and how much exploration is to be done. Whereas, $y_{ij}(t)$ and $\hat{y}_{ij}(t)$ are the individual best and global best position vectors' jth elements, respectively, w is the inertia parameter, which determines how much will the particular particle be inclined to stick to the direction of its current velocity. With the help of this update method, each particle updates its current position using the following:

$$x_i(t+1) = x_i(t) + v_i(t+1)$$

This update process is repeated for each particle, in each iteration, until convergence or maximum number of iterations. This nature-inspired approach works really well in the continuous search space for optimization of objective functions.

However, in order to use this approach for feature selection, the binary variant of the approach is required. The main idea is that a complete solution for the feature selection problem can be easily represented as a particle in a PSO approach, in terms of a binary position vector, where the 1's represent a selection of a feature and the 0's represent the removal of the feature. Once we have this formulation, then we can make use of a decent classifier like random forest that can find average classification error given any binary vector representing the selection of a subset of features. This works as the objective function that is to be minimized.

For binary PSO, the velocity update to the binary current position vector must result in another binary position vector. In order to achieve this, the approach mentioned in [25] is followed, that suggests that the new position vector $x_{ij}(t+1)$ is instead to be calculated as:

$$X_{ij}(t+1) = \begin{cases} 0 & \text{rand}() \geq S(v_{ij}(t+1)) \\ 1 & \text{rand}() \leq S(v_{ij}(t+1)) \end{cases}$$

where $S(x)$ is the sigmoid function defined as follows:

$$S(x) = \frac{1}{1+e^{-x}}$$

This process essentially makes the velocity updates $v_{ij}(t+1)$ work as probabilities of a bit $x_{ij}(t+1)$ taking on the value of 1 or 0, which enables the adaptation of continuous PSO to its binary variant.

3.2 ML-Based Predictive Modelling

In this phase, six different ML classification algorithms have been considered for testing the prediction system: Random Forest Classifier (RFC), Adaboost Classifier (ABC), Logistic Regression (LR), Decision Tree Classifier (DTC), Support Vector Machines (SVM) and Gradient Boosting Classifier (GBC), which is a mix of some traditional ML algorithms as well as some ensemble methods. 50 different random 80–20 splits of the entire data set are considered, in order to ensure that the average behaviour of all of these approaches is captured. Here the 80% data is used for training purpose and the rest is used for testing.

For performing comparative analysis, these models are evaluated on the basis of multiple evaluation metrics: Prediction Accuracy, Sensitivity, Specificity and Area Under the Curve (AUC) value of Receiver Operating Characteristics (ROC) curve.

4 Dataset Description

Two data sets have been considered in this work to test the applicability of the proposed feature selection approach experimentally.

4.1 Breast Cancer Coimbra Dataset

Breast Cancer Coimbra Dataset (BCCD), created by Patrício et al. [22], contains 116 instances, consisting of 52 healthy controls and 64 breast cancer patients, which is publicly available at https://archive.ics.uci.edu/ml/datasets/Breast+Cancer+Coimbra. Each of the instances are described with the help of nine features and one label that specifies whether it's a patient or healthy control. The nine predictive features are: Age (years), BMI (kg/m^2), Glucose (mg/dL), Insulin (μU/mL), HOMA, Leptin (ng/mL), Adiponectin (μg/mL), Resistin (ng/mL) and MCP-1(pg/dL). As it can be observed, these features consist of some anthropometric data, as well as data that can be gathered from routine blood analysis.

4.2 Breast Cancer Wisconsin (Diagnostic) Dataset

Breast Cancer Wisconsin (Diagnostic) Dataset (BCWDD), which is publicly available through UC Irvine Machine Learning Repository at https://archive.ics.uci.edu/ml/datasets/Breast+Cancer+Wisconsin+(Diagnostic) is utilized in this work, for the purpose of experimentation. This data set consists of 569 number of instances of characteristics pertaining to digitized images of cell nuclei from fine needle aspirate (FNA) of breast masses. The number of benign instances is 357, whereas the number of malignant instances is 212. All of these instances are characterized by 32 features, out of which 30 features are predictive, excluding the patient id feature and the diagnosis feature (which tells if the instances is malignant or not). These 30 features describe various statistics on the characteristics of the FNA digitized image, like radius, texture, perimeter, area, smoothness, compactness, concavity, concave points, symmetry and fractal dimension.

5 Experimental Results and Discussion

A comprehensive experimental study has been carried out in this work for testing the usability of BPSO as a method for feature selection. This section presents the performance of ML-based predictive modelling systems in predicting the presence or absence of breast cancer, both before and after feature selection is performed. The ML techniques used for predictive modelling, consists of a mix of traditional as well as ensemble learners.

The experiments are conducted on a system with the specifications mentioned in Table 1. The implementations are done in Python Version 3.5.2, with the help of Scikit-Learn based implementations of the Machine Learning algorithms applied in this work. The parameter set used while applying these ML algorithms were the default parameter settings of the Scikit-Learn Latest Version 0.22 implementations of the respective algorithms.

In Fig. 2, the performance of various classifiers on BCCD is depicted, before any feature selection is done. It can be observed that GBC achieves the best average prediction accuracy while having decent sensitivity, specificity and AUC values. However, even with GBC an average accuracy of only 76% has been achieved, which is not significant. This is where the prowess of BPSO-based feature selection comes into the picture.

After the initial predictive modelling, BPSO-based feature selection is performed, which resulted in the selection of five out of nine features, to obtain the reduced feature subset: [Age, BMI, Glucose, Resistin, Leptin]. Once the unwanted features are removed from the data set, ML-based predictive modelling is performed again.

Figure 3 depicts the performance of the classifiers on BCCD, after feature selection is performed. Significant improvements across all the classifiers can be observed after feature selection. More notably, in the case of GBC based classification, 4–6%

Table 1 Experimentation environment specifications

Processor	6× Intel(R) core(TM) i5-8500 CPU @ 3.00 GHz
RAM	8 GB
Operating system	Ubuntu 16.04.6 LTS

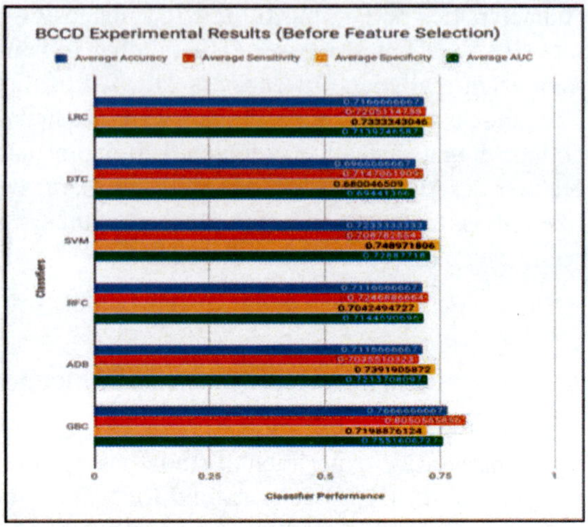

Fig. 2 Performance of various ML classifiers on BCCD, before feature selection

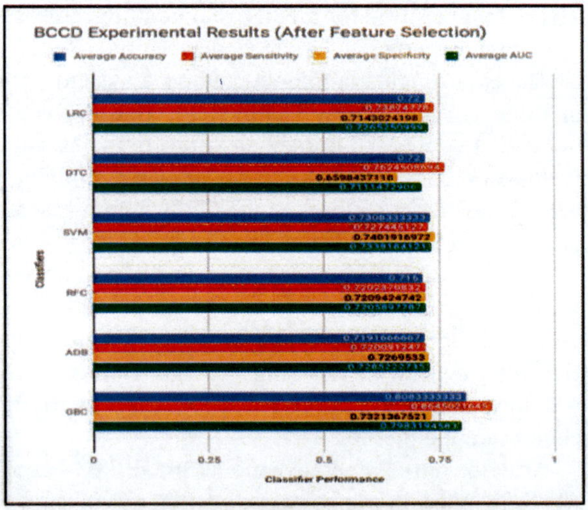

Fig. 3 Performance of various ML classifiers on BCCD, after feature selection

improvement can be observed across different performance evaluation metrics. This demonstrates the utility of the BPSO-based feature selection approach.

Furthermore, a similar process is carried out for BCWDD as well, which is a much higher dimensional data set. In Fig. 4, the performance of various classifiers on BCWDD is shown before feature selection is performed. From the figure, it can be observed that a similar trend is present in the case of BCWDD also. One important thing to note here is that, before even performing any feature selection, the predictive models are already performing very well, achieving average prediction accuracy as high as 97.28%. This means that the room for improvement is quite small in case of BCWDD.

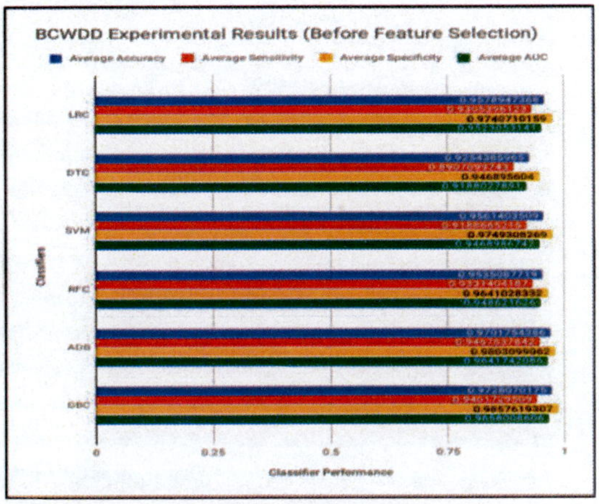

Fig. 4 Performance of various ML classifiers on BCWDD, before feature selection

After these initial experiments, BPSO-based feature selection is performed on BCWDD, which resulted in the selection of 9 out of 30 features, to obtain the reduced feature subset: [Texture mean, Smoothness mean, Concave points mean, Compactness, Texture worst, Area worst, Smoothness worst, Concavity worst, Symmetry worst]. Once the reduced feature subset is obtained, the unwanted features are removed from the data set, so that predictive modelling can be done only using the selected features.

From Fig. 5, it is clear that, even though the room for improvement is very less, in case of BCWDD also, the proposed approach is able to find a better feature subset, due to which improvements are observed across all the classifiers. Post feature selection, GBC based predictive model is able to achieve above 98% accurate results, for all the performance evaluation metrics under consideration.

From the set of experiments in this section, it is clear that BPSO-based feature selection approach can improve the performance of ML classifiers significantly, for the task of breast cancer prediction. The proposed feature selection approach, while combined with GBC, can be very effective at predicting the presence or absence of the disease.

6 Conclusion and Future Work

In this work, the application of BPSO-based feature selection has been studied, to improve the performance of automatic breast cancer prediction systems, by selecting more informative and discriminative features. In order to perform this task, the problem of feature selection is formulated in terms of a discrete optimization problem, on which BPSO can be applied. Two data sets BCCD and BCWDD have been considered for the purpose of experimentation. From the experimental study, after

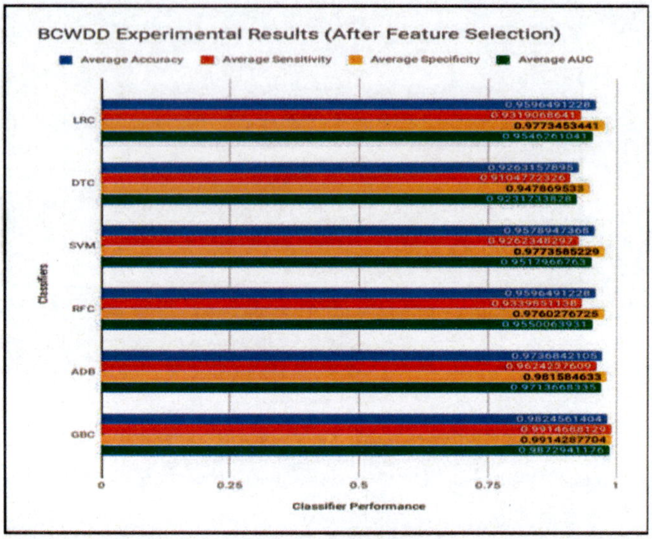

Fig. 5 Performance of various ML classifiers on BCWDD, after feature selection

BPSO-based feature selection, 4–8% and 1–4% improvements have been observed across all the performance evaluation metrics for BCCD and BCWDD, respectively.

The experimental results show that impressive average prediction accuracies of 80.83% and 98.24% can be obtained for BCCD and BCWDD respectively, which further shows that BPSO-based feature selection techniques can be applied to improve automatic breast cancer prediction of clinical decision support systems, in real medical diagnostic scenarios.

Although significant improvements have been observed using the proposed approach, for performing a more detailed evaluation of BPSO-FS, there is a need to consider much higher dimensional data, like Gene Expression Datasets for Breast Cancer. Such kind of experimental study on high dimensional Gene Expression data is part of the future work. Moreover, real-time evaluation of the proposed approach will give even more information regarding its potential utility, which is also a task to be taken up as part of our future work.

Acknowledgements The authors would like to acknowledge TEQIP-III, NIT Silchar for their support.

References

1. WHO breast cancer statistics, World Health Organization breast cancer diagnosis and screening page (2019). https://www.who.int/cancer/prevention/diagnosis-screening/breast-cancer/en/. Last accessed 21 Oct 2019
2. U.T.O.E.D., of group, BC, First results on mortality reduction in the UK trial of early detection of breast cancer. Lancet **332**(8608), 411–416 (1988)
3. K. Kourou, T.P. Exarchos, K.P. Exarchos, M.V. Karamouzis, D.I. Fotiadis, Machine learning applications in cancer prognosis and prediction. Comput. Struct. Biotechnol. J. **13**, 8–17 (2015). https://doi.org/10.1016/j.csbj.2014.11.005
4. D. Delen, G. Walker, A. Kadam, Predicting breast cancer survivability: a comparison of three data mining methods. Artif. Intell. Med. **34**(2), 113–127 (2005). https://doi.org/10.1016/j.artmed.2004.07.002
5. P.J. Lisboa, A.F. Taktak, The use of artificial neural networks in decision support in cancer: a systematic review. Neural Netw. **19**(4), 408–415 (2006). https://doi.org/10.1016/j.neunet.2005.10.007
6. D. West, P. Mangiameli, R. Rampal, V. West, Ensemble strategies for a medical diagnostic decision support system: a breast cancer diagnosis application. Eur. J. Oper. Res. **162**(2), 532–551 (2005). https://doi.org/10.1016/j.ejor.2003.10.013
7. A.M. Abdel-Zaher, A.M. Eldeib, Breast cancer classification using deep belief networks. Expert Syst. Appl. **46**, 139–144 (2016). https://doi.org/10.1016/j.eswa.2015.10.015
8. G. Litjens, C.I. Snchez, N. Timofeeva, M. Hermsen, I. Nagtegaal, I. Kovacs, J. Van Der Laak, Deep learning as a tool for increased accuracy and efficiency of histopathological diagnosis. Sci. Rep. **6**, 26286 (2016). https://doi.org/10.1038/srep26286
9. J. Wang, X. Yang, H. Cai, W. Tan, C. Jin, L. Li, Discrimination of breast cancer with microcalcifications on mammography by deep learning. Sci. Rep. **6**, 27327 (2016). https://doi.org/10.1038/srep27327
10. J. Arevalo, F.A. Gonzlez, R. Ramos-Polln, J.L. Oliveira, M.A.G. Lopez, Representation learning for mammography mass lesion classification with convolutional neural networks. Comput. Methods Programs Biomed. **127**, 248–257 (2016). https://doi.org/10.1109/embc.2015.7318482
11. A.E. Hassanien, T.H. Kim, Breast cancer MRI diagnosis approach using support vector machine and pulse coupled neural networks. J. Appl. Logic **10**(4), 277–284 (2012). https://doi.org/10.1016/j.jal.2012.07.003
12. U.R. Acharya, E.Y.K. Ng, J.H. Tan, S.V. Sree, Thermography based breast cancer detection using texture features and support vector machine. J. Med. Syst. **36**(3), 1503–1510 (2012). https://doi.org/10.1007/s10916-010-9611-z
13. H.L. Chen, B. Yang, G. Wang, S.J. Wang, J. Liu, D.Y. Liu, Support vector machine based diagnostic system for breast cancer using swarm intelligence. J. Med. Syst. **36**(4), 2505–2519 (2012). https://doi.org/10.1007/s10916-011-9723-0
14. R. Ramos-Polln, M.A. Guevara-Lpez, C. Surez-Ortega, G. Daz-Herrero, J.M. Franco-Valiente, M. Rubio-Del-Solar, I. Ramos, Discovering mammography-based machine learning classifiers for breast cancer diagnosis. J. Med. Syst. **36**(4), 2259–2269 (2012). https://doi.org/10.1007/s10916-011-9693-2
15. P. Jiang, J. Peng, G. Zhang, E. Cheng, V. Megalooikonomou, H. Ling, Learning-based automatic breast tumor detection and segmentation in ultrasound images, in *ISBI*, pp. 1587–1590 (2012, May). https://doi.org/10.1109/isbi.2012.6235878
16. C. Nguyen, Y. Wang, H.N. Nguyen, Random forest classifier combined with feature selection for breast cancer diagnosis and prognostic. J. Biomed. Sci. Eng. **6**(05), 551 (2013). https://doi.org/10.4236/jbise.2013.65070
17. J. Dheeba, N.A. Singh, S.T. Selvi, Computer-aided detection of breast cancer on mammograms: a swarm intelligence optimized wavelet neural network approach. J. Biomed. Inform. **49**, 45–52 (2014). https://doi.org/10.1016/j.jbi.2014.01.010

18. A.T. Azar, S.A. El-Said, Performance analysis of support vector machines classifiers in breast cancer mammography recognition. Neural Comput. Appl. **24**(5), 1163–1177 (2014). https://doi.org/10.1007/s00521-012-1324-4
19. A. Addeh, H. Demirel, P. Zarbakhsh, Early detection of breast cancer using optimized ANFIS and features selection, in *2017 9th International Conference on Computational Intelligence and Communication Networks (CICN)*, IEEE, pp. 39–42 (2017, Sept). https://doi.org/10.1109/cicn.2017.8319352
20. C. Baneriee, S. Paul, M. Ghoshal, A comparative study of different ensemble learning techniques using wisconsin breast cancer data set, in *2017 International Conference on Computer, Electrical Communication Engineering (ICCECE)*, IEEE, pp. 1–6 (2017, Dec). https://doi.org/10.1109/iccece.2017.8526215
21. E. Alikovi, A. Subasi, Breast cancer diagnosis using GA feature selection and rotation forest. Neural Comput. Appl. **28**(4), 753–763 (2017). https://doi.org/10.1007/s00521-015-2103-9
22. M. Nilashi, O. Ibrahim, H. Ahmadi, L. Shahmoradi, A knowledge-based system for breast cancer classification using fuzzy logic method. Telematics Inform. **34**(4), 133–144 (2017). https://doi.org/10.1016/j.tele.2017.01.007
23. M. Patrcio, J. Pereira, J. Crisstomo, P. Matafome, M. Gomes, R. Seia, F. Caramelo, Using Resistin, glucose, age and BMI to predict the presence of breast cancer. BMC Cancer **18**(1), 29 (2018). https://doi.org/10.1186/s12885-017-3877-1
24. Y. Li, Z. Chen, Performance evaluation of machine learning methods for breast cancer prediction. Appl. Comput. Math. **7**(4), 212–216 (2018)
25. J. Kennedy, R.C. Eberhart, A discrete binary version of the particle swarm algorithm, in *1997 IEEE International Conference on Systems, Man, and Cybernetics*. Computational Cybernetics and Simulation, vol. 5 (IEEE, pp. 4104–4108) (1997, Oct)

Repulsion-Based Grey Wolf Optimizer

Ankita Wadhwa and Manish Kumar Thakur

Abstract In this paper, a novel algorithm called Repulsion-based Grey Wolf Optimizer (R-GWO) is presented. The proposed algorithm is an improvement of the already existing popular swarm intelligence algorithm, called Grey Wolf Optimizer (GWO). The aim of the proposed work is to enhance the balance between exploitation and exploration in GWO by introducing a repulsion factor. This repulsion factor drifts the population away from the non-promising regions and brings it closer to the optimal solution. To compare the performance of R-GWO and GWO, both the algorithms are run on 16 benchmark fitness functions. The results indicate that R-GWO performs better as compared to the conventional GWO in terms of convergence speed and in discovering the global optimal solution.

Keywords GWO · R-GWO · Repulsion factor · Convergence rate

1 Introduction

From the last two decades, nature-inspired optimization algorithms have been gaining popularity exponentially. All these algorithms are working towards the same goal of finding the best solution and better convergence performance. This goal can be achieved by an algorithm well equipped with exploration and exploitation abilities that ensure the search of the global optimum solution. Few of these algorithms are Particle Swarm Optimization (PSO) [1], Gravitational Search Algorithm (GSA) [2], Ant Colony Optimization (ACO) [3], Grey Wolf Optimizer [4], Bat algorithm (BA) [5], and Salp Swarm Algorithm [6].

A. Wadhwa (✉) · M. K. Thakur
Department of Computer Science Engineering and Information Technology,
Jaypee Institute of Information Technology, Sector 62, Noida, UP, India
e-mail: ankitawadhwa89@gmail.com

M. K. Thakur
e-mail: mthakur.jiit@gmail.com

P. Bansal et al. (eds.), *Proceedings of International Conference on Artificial Intelligence and Applications*, Advances in Intelligent Systems and Computing 1164,
https://doi.org/10.1007/978-981-15-4992-2_36

Global Optimization searches for a vector in a given domain to find the finest (fittest) vector (solution) from all conceivable solution vectors. It is important to attain global optimum in many domains such as routing, graph coloring, wireless networks, etc. Solving the computationally hard optimization problems is a big challenge especially when it is required to deal with nonlinear data having multiple local optima.

Most of the nature-inspired algorithms go through two stages: Exploitation and Exploration. Exploitation is the potentiality to further drill down the function alongside good results to attain convergence and exploration is the ability to wander the search space at different places. In swarm intelligence, exploitation and exploration by the swarm are not easily intelligible as there exists a tradeoff between the duos. Enhancing one capability may affect the other's performance and vice versa.

No free lunch theorem [4], a popular theorem in nature-inspired computing needs a special mention. This theorem states that there is no single nature-inspired algorithm (NIA) which can solve all the optimization problems for global optima. One algorithm may perform better on one application and give poorer results on some other application. Put differently, different NIA may perform contrasting on different applications. This theorem motivates researchers and creates a lot of scope for the development of new as well as modified nature-inspired algorithms. The inclination towards the development of improved and hybridized metaheuristic algorithms is increasing over the years. These variants claim to have high precision, robustness, and speedy convergence [7, 8] as compared to their conventional counterparts.

Grey Wolf Optimizer is a well known metaheuristic algorithm introduced by Mirjalili et al. [4]. This algorithm inspires from the poaching approach of grey wolves that are well known to form and follow a hierarchical grouping. A lot of optimization problems have been solved in the recent past using GWO and its modified and hybridized versions. These domains include and are not limited to scheduling problems [9], feature selection [10], etc.

A number of variants of GWO have been developed. Some of them are parallelized [11] and discrete [10] versions of the algorithm. Other category is hybridization of GWO with some other algorithms like with DE [9], with Genetic Algorithm (GA) [10], with Simplex Method [12] and with Sine Cosine Algorithm [13]. In order to enhance the convergence rate of GWO, authors of [14] proposed a modified Grey Wolf Optimizer (mGWO). To enhance the balance between exploration and exploitation stages with the progressing generations, the authors used an exponentially decaying function. In [15], the author talks about the intelligent gravitational search algorithm, which is an improvement on the already existing GSA algorithm with the use of a repulsion factor. In [16], author experiments on an Intelligent Water Drops Algorithms which is based on swarm optimization inspired from water droplets flowing in the river.

With an extensive background available in modified swarm intelligence algorithms and usage of repulsion factor by authors of [15], in this paper, we introduce a novel enhanced algorithm; called repulsion-based Grey Wolf Algorithm (R-GWO). It focuses on improving the already existing swarm intelligence based Grey Wolf Optimizer Algorithm by incorporating a repulsion factor while updating the position of wolf. Thus, a wolf decides its direction not only based on the best solution, but also considers moving away from the worst solution. It may result in better exploitation and this behavior is explored in this paper.

This paper is organized as follows: Sect. 2 describes Grey Wolf Optimizer Algorithm, followed by Sect. 3 which talks about the repulsion-based improved Grey Wolf Optimizer. Section 4 discusses the results. Finally, Sect. 6 concludes the paper and its findings.

2 Grey Wolf Optimization

Grey Wolf Optimizer is inspired by group behavior of grey wolves. GWO [4] mimics the chasing system and initiates the chain of command of the wolves. In GWO, the wolf group is categorized in following four distinct gatherings, viz. alpha, beta, delta, and omega. These groupings simulate a leadership pecking order (hierarchy) observed in the wolves.

Grey wolves often desire and are seen living in a pack (group). The pioneers of the pack are a female or a male known as alphas (α). Next to pioneers in the progressive system of grey wolves are beta wolves (β). The betas (β) are junior wolves (assistants) that assist the main leaders (alpha) in their leadership-related tasks. The betas (β) should help the alphas (α) in their tasks as secretaries, whereas arranges the remaining lower-level sub-groups among the pack. The β thus fortifies the 'α', arranges all through the pack and offers criticism to pioneers (α).

The third level sub pack in the pecking order of grey wolves is delta (δ). They submit to upper hierarchy and dominate lower hierarchy. They generally include caretakers, senior citizens, etc. It might emerge that the deltas are not given a condemning recognition in the pack; however, it is observed that the whole wolves pack faces inner battles and troubles in absence of the delta. The fourth level grey wolf sub pack is omega (ω). This wolf assumes the job of a fall guy. Fourth level grey wolf dependably need to bow to the various predominant wolf. They are given the least preference in privileges like eating.

Furthermore, three principle ventures of searching for prey, poaching or hunting of prey and assaulting it are executed to catch the prey. To search and encircle the target prey, each member of the pack determines its direction and hence updates its position according to the mathematical Eqs. (1) and (2).

$$\overrightarrow{\text{dir}} = \left|\vec{C}.\overrightarrow{x_p}(t) - \vec{x}(t)\right| \tag{1}$$

$$\vec{x}(t+1) = \vec{x}_p(t) - \vec{A}.\overrightarrow{\text{dir}} \tag{2}$$

where dir is the direction vector, $x(t+1)$ indicates the position of a wolf at $t+1$th iteration and $x_p(t)$ indicates the position of grey wolf's prey. A and C are coefficient vectors. They are mathematically formulated in (3) and (4).

$$\vec{A} = 2\vec{g}.\vec{r_1} - \vec{g} \tag{3}$$

$$\vec{C} = 2.\vec{r_2} \tag{4}$$

The poaching behavior is simulated by taking directions from α, β, and δ wolves which are represented as three best solutions among the entire population. Equations simulating the hunting step are presented from (5) to (11).

$$\overrightarrow{\text{dir}}_\alpha = \left|\vec{C}_1.\vec{x}_\alpha - \vec{x}\right| \tag{5}$$

$$\overrightarrow{\text{dir}}_\beta = \left|\vec{C}_2.\vec{x}_\beta - \vec{x}\right| \tag{6}$$

$$\overrightarrow{\text{dir}}_\delta = \left|\vec{C}_3.\vec{x}_\delta - \vec{x}\right| \tag{7}$$

$$\vec{x}_1 = \vec{x}_\alpha - \vec{A}_1.\left(\overrightarrow{\text{dir}}_\alpha\right) \tag{8}$$

$$\vec{x}_2 = \vec{x}_\beta - \vec{A}_2.\left(\overrightarrow{\text{dir}}_\beta\right) \tag{9}$$

$$\vec{x}_3 = \vec{x}_\delta - \vec{A}_3.\left(\overrightarrow{\text{dir}}_\delta\right) \tag{10}$$

$$\vec{x}(t+1) = \frac{\vec{x}_1 + \vec{x}_2 + \vec{x}_3}{3} \tag{11}$$

G is decreased from 2 to 0 and thus A also decreases over the iterations and lies in the range [−2 to 2 g]. The value $|A| > 1$ results in divergence from the target prey and $|A| < 1$ results in convergence towards the prey.

In nutshell, the optimization process starts with randomly placed wolves in the search space, which update their positions to find the position of prey. Thus update is done based on the position of alpha, beta and delta. Candidate solution vectors diverge and converge from the prey based on value of vector $|A|$. Ultimately the algorithm is stopped after a certain number of iterations or at meeting of convergence criteria. Next section presents a modified version of this algorithm called repulsion-based improved Grey Wolf Optimizer.

3 Repulsion Based Improved Grey Wolf Optimizer

In swarm intelligence approaches, the most important aspect of any algorithm is to preserve a balance between its exploration and exploitation capabilities. A misbalance between these two may result in the stagnation of algorithm in the local optima. To tranquilize this, in this paper we propose a variant of GWO called the Repulsion-based Grey Wolf Optimizer (R-GWO).

Most of the swarm intelligent algorithms update the swarm positions based on the best solution obtained so far. In R-GWO, we use the worst solution (one of the omega wolves) obtained along with the usual three best solutions to update the position of wolf pack. Put differently, each wolf's direction is decided with the help of four wolves namely alpha (best solution), beta (next best solution), delta (third-best solution) and weakest omega wolf (worst solution). Introducing the usage of worst wolf, adds a repulsion factor to the GWO along with conventional attraction factor. This smart behavior of R-GWO controls the balance between explore and exploit the capabilities of GWO.

To implement the above-mentioned idea, we introduced a repulsive factor 'repel' in the position update equation mentioned in (2) and the modified equation with a smart repulsive component is shown in (12) and (13). Repel of any wolf is defined using its own position and position of the prey. Vector difference between the two ensures movement towards the prey. The vector difference between the position of wolf and the worst solution (omega wolf) weakens the movement defined in the numerator, hence creating a repulsive force. Thus the solution drifts away from the non-promising regions of the search space.

$$\text{repel} = \frac{\left(\overrightarrow{x_p}(t) - \vec{x}(t)\right)}{\left|\rho * \vec{x}_\omega(t) - \vec{x}(t)\right|} \tag{12}$$

$$\vec{x}(t+1) = \vec{x}_p(t) - \vec{A}.\overrightarrow{\text{dir}} + r * \text{repel} \tag{13}$$

In the above equations, r is a random number to add randomness in searching behavior and ρ controls the movement away from worst solution and is set to 0.7. Thus Eqs. (8)–(11) in GWO are modified and corresponding equations in R-GWO are shown from (14) to (17). The pseudocode of R-GWO is also presented.

$$\vec{x}_1 = \vec{x}_\alpha - \vec{A}_1.\left(\overrightarrow{\text{dir}_\alpha}\right) + r_1 * \text{repel}_\alpha \tag{14}$$

$$\vec{x}_2 = \vec{x}_\beta - \vec{A}_2.\left(\overrightarrow{\text{dir}_\beta}\right) + r_2 * \text{repel}_\beta \tag{15}$$

$$\vec{x}_3 = \vec{x}_\delta - \vec{A}_3.\left(\overrightarrow{\text{dir}_\delta}\right) + r_3 * \text{repel}_\delta \tag{16}$$

$$\vec{x}(t+1) = \frac{\vec{x}_1 + \vec{x}_2 + \vec{x}_3}{3} \tag{17}$$

Pseudocode of R-GWO

```
1. Initialize the wolf pack,X(set of xi, i = 1 to packsize)
2. Initialize coefficient vectors
3. Calculate the fitness of each wolf in pack
4. xα = the best wolf
```

```
5. xβ = the next best wolf
6. xδ = the third best wolf
7. xω = the worst wolf
8. while (t < Max number of iterations)
a. For each wolf in pack
i. Calculate direction vectors using (6) to (8)
iii. Update current wolf's position using (18)
 end For
b. Update g, A and C vectors
c. Update fitness of all wolves using new positions
d. Update xα, xβ, xδ, xω
e. t = t + 1
 end while
9. return xα
```

4 Results and Discussions

In this section, we investigate the performance of the proposed R-GWO by running it on 16 benchmark functions [17]. These mathematical functions are used to test the ability of R-GWO to reach to global optima and compare its convergence performance with conventional GWO. The considered benchmark functions lie in three-dimension groups viz. 2, 10, and 30. The details of 16 benchmark functions viz. their name, dimensions, and range of parameters are presented in Table 1.

The implementation of GWO and R-GWO is done in GoLang on 8 GB RAM, Intel i5 Processor 5th Generation. The size of wolf pack is taken to be 5. Maximum number of iterations is set to 1000. Both the algorithms were run 30 times on each benchmark function. The best fitness values obtained and other statistical numbers like the worst, median, mean, and standard deviation of fitness values obtained in 30 runs are presented in Table 2 for both the algorithms.

From Table 2, it can be observed that for four fitness functions, the best value obtained in 30 runs is same for GWO as well as R-GWO. For two fitness functions viz. F1 and F2 GWO gives better fitness as compared to R-GWO in terms of best fitness obtained in 30 runs. For majority of the fitness functions i.e. 10 functions, we observe that R-GWO gives better 'best value' in 30 runs. The better values are highlighted in bold in Table 2.

Also, we observe from Table 2, that the mean of the fitness values obtained in 30 runs is same for both the algorithms for only two fitness functions viz. F9 and F14. For two other functions namely F7 and F16, it can be seen that GWO gives better mean fitness values as compared to R-GWO. For all the other 10 fitness functions, we notice that R-GWO is giving better mean fitness values in comparison to GWO. The better mean values among GWO and R-GWO are highlighted in bold.

Convergence curve plots the fitness value attained versus iteration number. It helps to analyze which algorithm is able to attain optima faster. Figures 1 and 2 present the convergence graphs of GWO (blue curve) and repulsion-based improved Grey Wolf Optimizer (R-GWO) (orange curve) for some of the fitness functions.

Table 1 Benchmark functions

S. No.	Function name	Dim	Input range
1	Ackley N.4 function	10	[−35, 35]
2	Shubert 3 function	30	[−10, 10]
3	Schwefel 2.20 function	10	[−100, 100]
4	Qing function	10	[−500, 500]
5	Alpine N.1 function	10	[0, 10]
6	Sum squares function	10	[−10, 10]
7	Rastrigin function	3	[−5.12, 5.12]
8	Quartic function	2	[−1.28, 1.28]
9	Exponential function	2	[−1, 1]
10	Sphere function	10	[−5.12, 5.12]
11	Brown function	10	[−1, 4]
12	Powell sum function	10	[−1, 1]
13	Rosenbrock function	10	[−5, 10]
14	Salomon function	10	[−100, 100]
15	Styblinski-tank function	2	[−5, 5]
16	Zakharov function	30	[−5, 10]

In Fig. 1, we present the convergence curves of five fitness functions where it was observed that R-GWO converges significantly faster than GWO to reach the optimal solution. These fitness functions are F1, F2, F5, F8, and F14. For the other fitness functions F3, F4, F6, F7, F9 to F13, F15 and F16 it was observed that R-GWO converges minutely faster or the same as compared to GWO. Convergence curves for four of these functions viz. F3, F7, F9, and F16 are shown in Fig. 2.

From the above analysis, it can be inferred that R-GWO doesn't converge prematurely and also takes only a rational amount of iterations. It is able to conserve an equilibrium between explore and exploit operations and thus avoid local optima. To sum up, all simulation results assert that the R-GWO algorithm is very helpful in improving the efficiency of the GWO in terms of result quality as well as computational efforts.

5 Conclusion

In this paper, a novel improved version of GWO is proposed. The proposed algorithm is repulsion-based improved variant of GWO (R-GWO). The key highlight of the algorithm is conception of a repulsive factor to drive the solutions away from non-promising zones of search space and hence conserve a balance between exploration and exploitation. Quality of R-GWO is tested and compared with GWO using 16

Table 2 Comparison of GWO and R-GWO statistical fitness values on 16 benchmark functions

	GWO					R-GWO				
	Best	Worst	Median	Mean	SD	Best	Worst	Median	Mean	SD
F1	**−2.59E+1**	−1.53E+1	−2.30E+01	−2.09E+01	3.43E+00	−2.55E+01	−1.32E+1	−2.17E+01	−2.18E+01	3.07E+00
F2	**−1.27E+2**	−9.89E+1	−1.09E+02	−1.04E+02	8.33E+00	−1.23E+02	−8.81E+1	−1.02E+02	**−1.10E+02**	9.10E+00
F3	4.16E−51	1.79E−48	4.31E−49	7.55E−47	6.34E−49	**0.OOE+OO**	4.99E−48	1.58E−49	**5.92E−49**	7.49E−47
F4	1.85E+01	1.01E+02	7.30E+01	2.76E+08	2.95E+01	**0.OOE+OO**	1.65E+07	6.41E+01	**6.22E+01**	2.74E+08
F5	−5.13E+01	−3.49E+1	−4.39E+01	−5.11E+01	5.15E+00	**−5.59E+01**	−4.64E+1	−5.15E+01	**−4.31E+01**	2.81E+00
F6	2.28E−102	1.59E−91	1.95E−96	4.26E−04	4.75E−92	**0.OOE+OO**	9.12E−04	3.47E−72	**1.59E−92**	4.14E−04
F7	**0.OOE+OO**	0.OOE+OO	0.OOE+OO	**0.OOE+OO**	0.OOE+OO	**0.OOE+OO**	3.43E−03	0.OOE+OO	9.75E−04	1.21E−03
F8	8.49E−03	2.11E−02	1.43E−02	2.60E−02	3.29E−03	**0.OOE+OO**	4.66E−02	2.13E−02	**1.46E−02**	1.30E−02
F9	**−1.0E+00**	−1.00E+0	−1.OOE+OO	**−1.OOE+OO**	0.OOE+OO	**−1.OOE+OO**	−1.OE+OO	−1.OOE+OO	**−1.OOE+OO**	3.17E−08
F10	1.58E−94	7.94E−89	7.20E−90	1.80E−04	2.42E−89	**0.OOE+OO**	6.92E−04	2.99E−78	**1.83E−89**	2.31E−04
Fll	6.34E−95	2.28E−89	5.39E−91	6.48E−02	6.63E−90	**0.OOE+OO**	6.39E−02	1.45E−02	**3.33E−90**	4.86E−02
F12	4.97E−195	1.23E−173	2.10E−181	3.20E−01	0.OOE+OO	**0.OOE+OO**	9.17E−01	1.48E−03	**1.25E−174**	3.51E−01
F13	6.41E+00	8.07E+00	7.21E+00	7.63E+00	4.36E−01	**0.OOE+OO**	8.70E+00	7.23E+00	**7.12E+00**	2.48E+00
F14	9.99E−02	9.99E−02	9.99E−02	**9.99E−02**	1.32E−07	**0.OOE+OO**	9.99E−02	9.99E−02	**9.99E−02**	3.16E−02
F15	**−7.83E+1**	−6.42E+1	−7.83E+01	−7.69E+01	4.24E+00	**−7.83E+01**	−7.83E+1	−7.83E+01	**−7.83E+01**	1.19E−02
F16	**0.OOE+OO**	0.OOE+OO	0.OOE+OO	**0.OOE+OO**	0.OOE+OO	**0.OOE+OO**	3.23E−21	0.OOE+OO	1.98E+10	1.89E+10

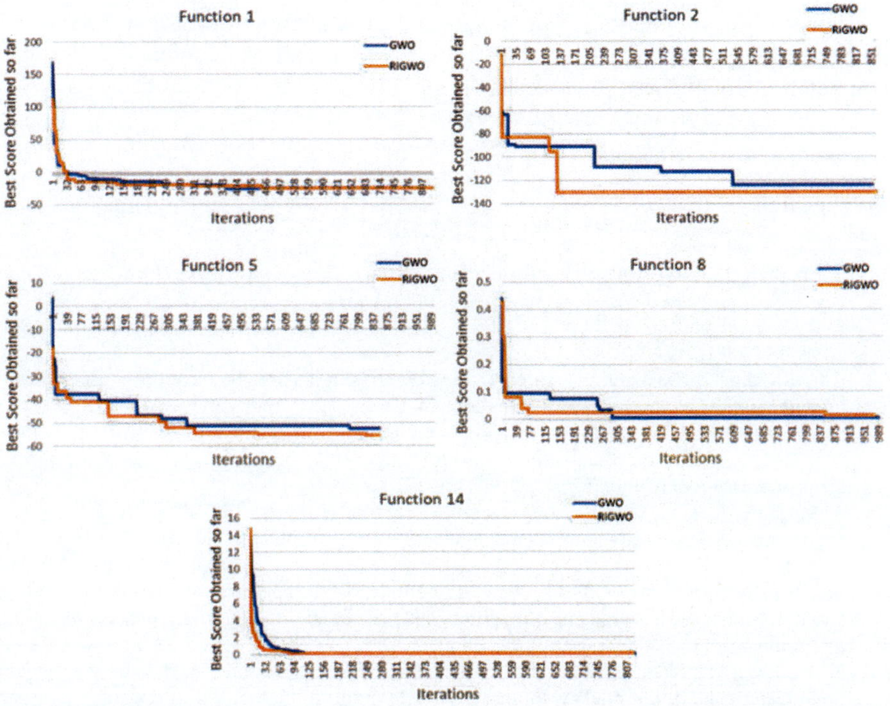

Fig. 1 Convergence curves of five fitness functions representing faster convergence behavior of R-GWO over GWO

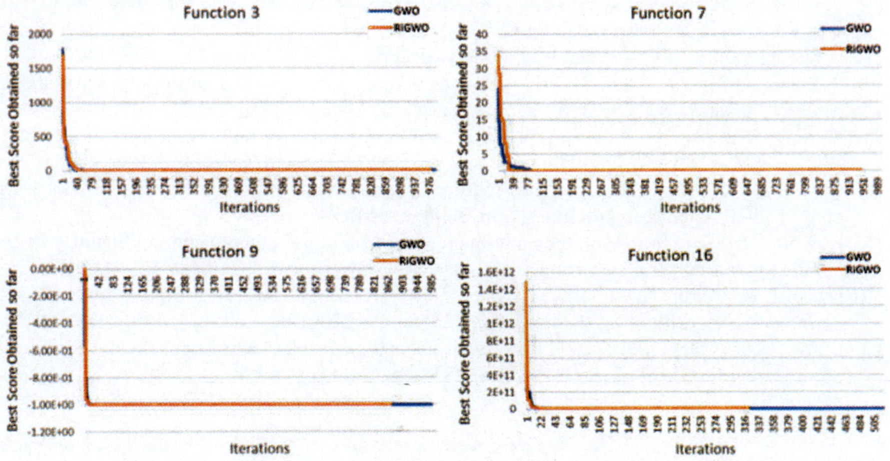

Fig. 2 Convergence curves of four fitness functions representing similar or minutely better convergence behavior R-GWO over GWO

benchmark fitness functions. The results and their analysis prove that R-GWO is more reliable as compared to GWO in terms of fitness value obtained and also computationally efficient as it takes a decent number of iterations to converge.

References

1. R. Eberhart, J. Kennedy, Particle swarm optimization, in *Proceedings of the IEEE International Conference on Neural Networks*, pp. 1942–1948 (1995)
2. E. Rashedi, H. Nezamabadi-Pour, S. Saryazdi, GSA: a gravitational search algorithm. Inf. Sci. 2232–48 (2009)
3. B.C. Mohan, R. Baskaran, A survey: ant colony optimization based recent research and implementation on several engineering domain. Expert Syst. Appl. 4618–27 (2012)
4. S. Mirjalili, S.M. Mirjalili, A. Lewis, Grey wolf optimizer. Adv. Eng. Softw. 46–61 (2014)
5. X.S. Yang, A new metaheuristic bat-inspired algorithm, in *Nature Inspired Cooperative Strategies for Optimization* (Springer, Berlin, 2010). 65–74
6. S. Mirjalili, A.H. Gandomi, S.Z. Mirjalili, S. Saremi, H. Faris, S.M. Mirjalili, Salp swarm algorithm: a bio-inspired optimizer for engineering design problems. Adv. Eng. Softw. 163–91 (2017)
7. S. Yu, Z. Wu, H. Wang, Z. Chen, A hybrid particle swarm optimization algorithm based on space transformation search and a modified velocity model, in *High Performance Computing and Applications* (Springer, Berlin, 2010), pp. 522–527.
8. X. Yu, J. Cao, H. Shan, L. Zhu, J. Guo, An adaptive hybrid algorithm based on particle swarm optimization and differential evolution for global optimization. Sci. World J. (2014)
9. A. Zhu, C. Xu, Z. Li, J. Wu, Z. Liu, Hybridizing grey wolf optimization with differential evolution for global optimization and test scheduling for 3D stacked SoC. J. Syst. Eng. Electr. 317–28 (2015)
10. E. Emary, H.M. Zawbaa, A.E. Hassanien, Binary grey wolf optimization approaches for feature selection. Neurocomputing 371–81 (2016)
11. U.K. Chakraborty, Genetic and evolutionary computing. Inf. Sci. (Ny) **178**, 4419–4420 (2008)
12. S. Zhang, Q. Luo, Y. Zhou, Hybrid grey wolf optimizer using elite opposition-based learning strategy and simplex method. Int. J. Comput. Intell. Appl. (2017)
13. N. Singh, S.B. Singh, A novel hybrid GWO-SCA approach for optimization problems. Eng. Sci. Technol. 1586–601 (2017)
14. N. Mittal, U. Singh, B.S. Sohi, Modified grey wolf optimizer for global engineering optimization. Appl. Comput. Intell. Soft Comput. (2016)
15. H. Mittal, M. Saraswat, An automatic nuclei segmentation method using intelligent gravitational search algorithm based superpixel clustering. Swarm Evol. Comput. 15–32 (2019)
16. H. Shah-Hosseini, The intelligent water drops algorithm: a nature-inspired swarm-based optimization algorithm. Int. J. Bio. Inspired comput. (2009)
17. P. Agarwal, S. Mehta, Empirical analysis of five nature-inspired algorithms on real parameter optimization problems. Artif. Intell. Rev. 383–439 (2018)

LFC of Thermal System with Combination of Renewable Energy Source and Ultra-Capacitor

Arindita Saha, Lalit Chandra Saikia, and Naladi Ram Babu

Abstract An effort is made to explore load frequency control (LFC) of uneven three-area multiple source schemes including thermal, solar thermal power plant (STPP) in addition ultra-capacitor (UC) considering generating rate constraint. An amalgamation of whole number order proportional+integral+derivative (PID) using filter (PIDN) with fractional-order integrator (FOI) controller (PIDN-FOI) is projected as second controller for the scheme. Whale Optimization Algorithm (WOA) finds its usage for concurrent optimization of controller gains along with parameters. The execution of PIDN-FOI is found enhanced compared to other typical controllers regarding less settling time, peak deviation and reduced oscillations. The analysis says that with the integration of STPP in all areas and UC in area-1 the scheme provides enhanced responses. Sensitivity analysis concludes that WOA optimized PIDN-FOI gains/other parameters at basic circumstances are robust to any alterations in scheme condition or factors and further reset is not required.

Keywords Generating rate constraint · PIDN-FOI controller · Solar thermal power plant · Ultra-capacitor · Whale optimization algorithm

1 Introduction

The intention of load frequency control (LFC) is to manage frequency and tie-line power flow during usual and minor disorders. Elgerd et al. [1] were the first to give the awareness of demonstrating a multiple area system for AGC [2, 3]. Most of

A. Saha (✉)
Regent Education and Research Foundation Group of Institutions, Kolkata, WB 700121, India
e-mail: sahaarindita.91@gmail.com

L. C. Saikia · N. R. Babu
National Institute of Technology, Silchar, Assam 788010, India
e-mail: lcsaikia@yahoo.com

N. R. Babu
e-mail: ram1992babu@gmail.com

P. Bansal et al. (eds.), *Proceedings of International Conference on Artificial Intelligence and Applications*, Advances in Intelligent Systems and Computing 1164,
https://doi.org/10.1007/978-981-15-4992-2_37

the past studies are limited to only conventional thermal generations[2, 3]. Due to certain drawbacks like carbon emissions and as conventional resources are reducing day by day it's an urgent need to find out some alternative sources. Energy from the sun is the best easily obtainable abundant source of alternative clean energy. There are very least chance of emission of greenhouse gases while converting solar energy into electrical energy. The use of this particular energy source leads to lesser consumption of conventional energy sources. Das et al. [4] have considered the idea of solar thermal power plant (STPP) meant for a hybrid isolated system. The use of STPP in LFC for multiple area system is done by Sharma et al. [5]. But their learning was restricted to integration of STPP simply in sole control area however further areas included only thermal conventional generation. Hence LFC with STPP for multi-area multisource, system needs further studies.

A fast-acting energy storage device called ultra-capacitor (UC) can make use of intended for checking out oscillations in power system which offers additional storing capability. It is made use of by Mufti et al. [6] aimed at two identical area scheme. The application of UC is found in the autonomous hybrid power system [4]. Surprisingly, no authors have studied the performance of UC for an interconnected conventional system having more than two control areas and different sources in the system incorporating STPP. Thus, this claims for further study.

Recent trend of research in LFC is the proposal of second controller. In most of the historical works, the studies have been done for one or the other multiple area thermal or hydrothermal schemes using various typical controllers alike integral (I), proportional-integral (PI), proportional-integral-derivative (PID). A fractional-order controller in [7] shows better performance than classical controllers. A fresh cascade controller entitled as proportional-integral-derivative-fractional-order integrator (PID-FOI) [7] has been brought into being in additional regions of research whose application is not initiated in LFC. Modern construction of PID contains a derivative filter (N_i) and PID-FOI using N_i can be entitled as PIDN-FOI. Accordingly, the examination of LFC by means of PIDN-FOI and assessment with newly presented controllers is required to be done.

For the proper performance of the controller, its gains/other parameters should be at best. The selection of finest values is done by using typical over and above recent modern meta-heuristic optimization procedures. Typical practices are tiresome and problematic to obtain finest result. Due to growing intricacies and suspicions associated with prevailing techniques authors are examining for fresh and extra advanced approaches. A newly technologically advanced nature motivated meta-heuristic procedure specifically whale optimization Algorithm (WOA) [8] is developed. This imitates the societal behaviour of humpback whales. It preserves stability amid exploitation plus exploration which assistances in judgment of global bests. WOA has a specialty likened to additional algorithms as it accomplishes two stages' exploration and exploitation individually via two diverse calculations. In the main partial part of iterations, WOA demonstrates great exploration everyplace whales are permitted to travel arbitrarily and in succeeding half it displays great exploitation and convergence where they quickly re-locate themselves about near the best-found position so far. So, it illustrates reduced probabilities of getting stuck in limited bests and great convergence. Though WOA has not been applied in the considered typical system of AGC till date accordingly it necessitates widespread studies.

Going through the past works in LFC, it is seen that the controller gains/other parameters of the scheme are optimized at base loading conditions. 1% step load perturbation (SLP) is utilized as disorder. Impression on scheme changing aspects due to change in loading condition need to be studied. In order to verify the healthiness of finest factors attained at base condition, sensitivity investigation is undertaken. Sensitivity analysis for system integrated with UC for multiple source systems using finest gains of PIDN-FOI controller is not accomplished in any past literatures of LFC. Thus, the main points of the current effort are as follows:

(a) Consideration of three-area unequal conventional multiple source thermal-thermal scheme considering GRC and single reheat turbine. Concurrent optimization of controller gains along with parameters using WOA of I, PI, PIDN, and PIDN-FOI when considered individually to relate the dynamic responses to discover the greatest under 1% SLP.
(b) Integration of STPP in the scheme stated in a) in existence of finest controller found and compares the dynamic responses to discover the impact of STPP.
(c) Integration of UC in multisource thermal system with STPP and study impact on system dynamics in existence of best controller.
(d) Sensitivity analysis of the finest controller for scheme mentioned in (c) for widespread change in scheme loading.

2 System Understudy

Three schemes are considered for examination. They are (i) multiple area multiple source thermal schemes (Fig. 2b: block diagram, (ii) multiple area multiple source thermal scheme with STPP, (iii) multiple area multiple source thermal scheme with STPP and UC (Fig. 1): scheme considered in (ii) is equipped with UC in area-1. The ultra-capacitor (UC) is a lately technologically advanced device also familiar as super-capacitor. The alteration in frequency is reflected as error signal to UC. The numeric values of gain (K_{UC}) and time constant (T_{UC}) of UC are mentioned in Appendix. 1% SLP is reflected in area-1 in all the systems to obtain the system dynamics. Area capacities for the systems in (i), (ii), and (iii) are considered the same. Area participating factors (apf's) are reflected at random in every area. They are $apf_{11} = 0.2$, $apf_{12} = 0.4$, $apf_{13} = 0.4$, $apf_{21} = 0.1$, $apf_{22} = 0.45$, $apf_{23} = 0.45$, $apf_{31} = 0.25$, $apf_{32} = 0.375$, $apf_{33} = 0.375$. Numerous second controllers such as I, PI, PID, and PIDN-FOI are in use individually for examination. Controller parameters are simultaneously optimized using WOA considering integral squared error (ISE) as cost function specified by (1).

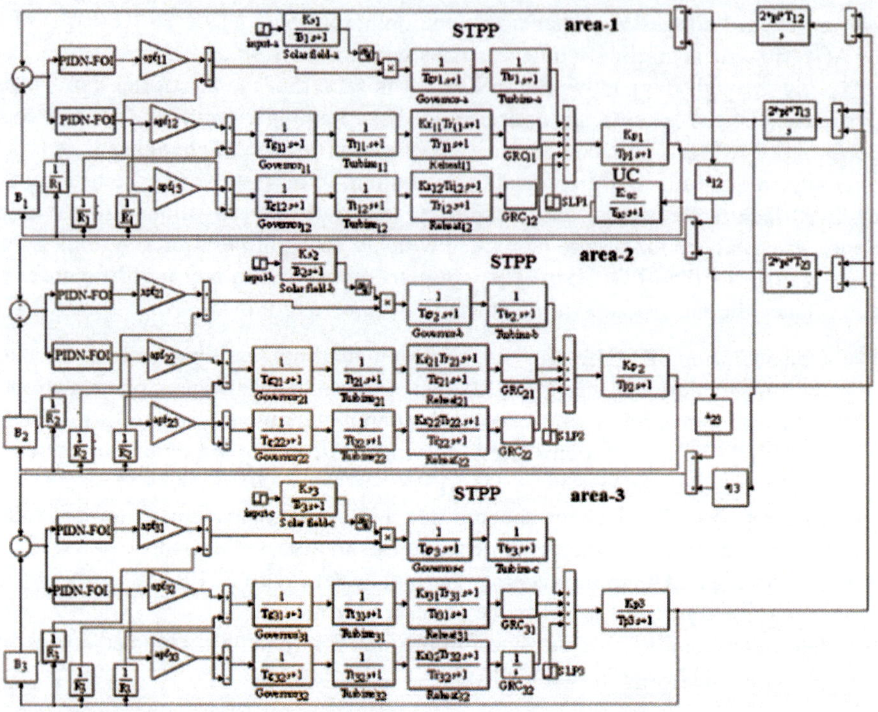

Fig. 1 The investigated system model

$$J_{\mathrm{ISE}} = \int_0^T \left\{ (\Delta f_i)^2 + (\Delta_{\mathrm{Ptie}i-j})^2 \right\} \tag{1}$$

3 PIDN-FOI Controller

A secondary controller named as proportional+integral+derivative with derivative filter-fractional-order integrator (PIDN-FOI) is proposed for the system. The block diagram of this controller is shown in Fig. 1c where the entire controller has been divided into two blocks $C_i(s)$ and $F_i(s)$ where $C_i(s)$ and $F_i(s)$ represent the structure of PIDN and fractional integrator, respectively. $R_i(s)$ and $U_i(s)$ are the input to and output from the complete controller, respectively. For PIDN-FOI controller $C_i(s)$ and $F_i(s)$ are represented by (2) and (3), respectively.

$$C_i(s) = \frac{(K_{Pi} + K_{Di} N_i)s^2 + (K_{Ij} + N_i)s + K_{Ii} N_i}{s(s + N_i)} \tag{2}$$

$$F_i(s) = \frac{1}{s^{\alpha_i}} \quad (3)$$

Thus, PIDN-FOI controller consists of five controlling parameters in comparison to three for PID controller. As stated before that $R_i(s)$ is the controller's input is basically an error signal e(t) = Area control error (ACE), the output $U_i(s)$ is given by (4) and (5).

$$U_i(s) = (C_i(s)F_i(s)) * R_i(s) \quad (4)$$

$$U_i(s) = \left\{ \left(\frac{(K_{Pi} + K_{Di}N_i)s^2 + (K_{Ii} + N_i)s + K_{Ii}N_i}{s(s + N_i)} \right) * \frac{1}{s^{\alpha_i}} \right\} * \mathrm{ACE}_i \quad (5)$$

The gains/other parameters are improved by means of WOA by diminishing J_{ISE} specified by (1) on the basis of settings. The lesser and higher bounds for every factor of PIDN-FOI are1 and 0 respectively other than N_i whose is reflected as 0 and 100. The equations are derived by the knowledge of cascaded controllers mentioned in [7], and other combinations of fractional and integer order controllers in [9, 10].

4 Whale Optimization Algorithm

The WOA method was projected by Mirjalili et al. [8] constructed on the aggressive actions of whales of humpback type. This process sustains evenness amid both exploitation and exploration through the assistance of twofold dissimilar set of equations. Consequently, the benefit of not getting caught in limited targets is experienced. The exhaustive description of WOA is specified in [8]. The input variables of WOA are numeral of search agents, extreme iterations and $\vec{a}$ which decreases from 2 to 0 during iterations. Besides, the output variables are the controller gains along with parameters.

5 Results and Analysis

5.1 Dynamic Responses of Three-Area Multisource Thermal Scheme

Here, the three unequal areas with thermal-thermal (TT) multisource systems have been considered for study (Fig. 2b). The I, PI, PID, and PIDN-FOI controllers are in use individually and in every single instance controller factors are optimized applying WOA taking ISE as performance index. The finest values are specified in Table 1 and the response evaluation regarding finest values of gains of every one controller are

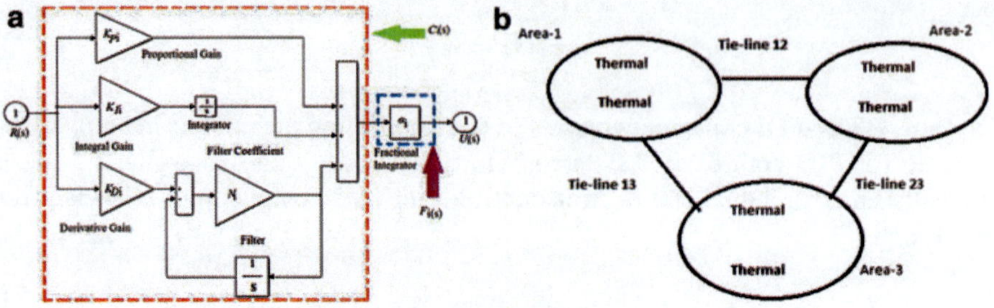

Fig. 2 **a** Controller structure. **b** Block representation of thermal system

Table 1 Finest values of gains and other parameters for controllers for TT system

Controllers	Optimum gains/other parameters
I	$K_{I1}* = 0.99860$ $K_{I2}* = 0.86460$ $K_{I3}* = 0.97530$
PI	$K_{P1}* = 0.00025$ $K_{P2}* = 0.00026$ $K_{P3}* = 0.00015$ $K_{I1}* = 0.69180$ $K_{I2}* = 0.78260$ $K_{I3}* = 0.94630$
PID	$K_{P1}* = 0.00001$ $K_{P2}* = 0.00001$ $K_{P3}* = 0.00012$ $K_{I1}* = 0.59344$ $K_{I2}* = 0.10230$ $K_{I3}* = 0.64965$ $K_{D1}* = 0.00016$ $K_{D2}* = 0.00763$ $K_{D3}* = 0.00634$
PIDN-FOI	$K_{P1}* = 0.69930$ $K_{P2}* = 0.19280$ $K_{P3}* = 0.08390$ $K_{I1}* = 0.99960$ $K_{I2}* = 0.99950$ $K_{I3}* = 0.99940$ $K_{D1}* = 0.26390$ $K_{D2}* = 0.64340$ $K_{D3}* = 0.13490$ $N_1* = 39.70$ $N_2* = 42.27$ $N_3* = 60.47$ $\alpha_1* = 0.0000015$ $\alpha_2* = 0.0000296$ $\alpha_3* = 0.0000198$

related in (Fig. 3). Two responses are given. It is observed oscillations in responses corresponding to PIDN-FOI are less than that of others. Thus, on the basis of degree of oscillations and settling time (ST) it can be concluded that the performance of PIDN-FOI controller is much enhanced than others. To test superiority of responses

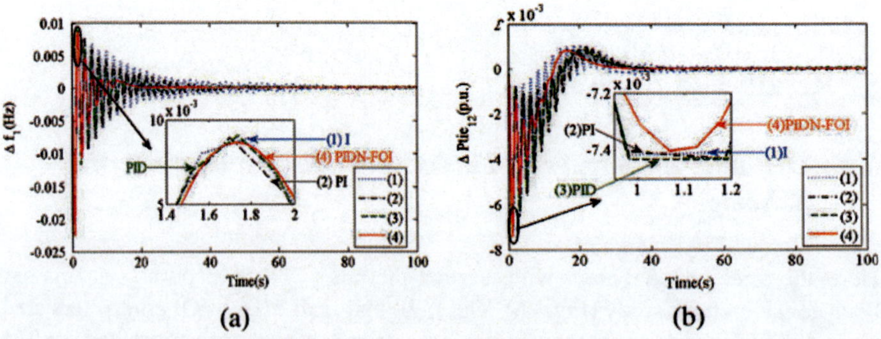

Fig. 3 Response valuation of I, PI, PIDN and PIDN-FOI controllers for **a** Δf_1 versus time, **b** $\Delta P_{\text{tie1}-2}$ versus time

Table 2 Performance comparison (J) for different controllers for TT system

Controllers	Performance index (J)	
	ISE	FOD
I	0.00145	–
PI	0.00134	37,980.09
PID	0.00121	16,943.19
PIDN-FOI	0.00106	10,457.45

corresponding to each controller quantitatively one more performance index called figure of demerit (FOD) is also considered. Here, in FOD it is easy to evaluate the performance in aspect of numerical values. The values of J_{FOD} are found and values of J_{ISE} at the end of iteration are also prominent in Table 2 for each controller. From Table 2 it is evidently realized that PIDN-FOI has lesser value of J_{ISE} and J_{FOD}. Thus, PIDN-FOI controller is better than other.

5.2 *Dynamic Responses of Scheme with PIDN-FOI Controller with and Without STPP*

With the system mentioned in Fig. 2b, an additional solar thermal power plant (STPP) is combined with thermal units in each area (Fig. 1). PIDN-FOI controller is made usage as second controller in all areas. The gains/other factors of PIDN-FOI controllers are optimized by means of WOA and finest values are shown in Table 3. The response evaluation corresponding to finest values are associated with that obtained for the scheme without STTP for PIDN-FOI controller (Fig. 4). From Fig. 4, it infers PIDN-FOI performs much better when STPP is incorporated due to reduced oscillations and ST. The value of J_{ISE} is 0.0010 and J_{FOD} is 8885.85 reveals that incorporation of STPP improves the system dynamics.

Table 3 Finest values of gains along with parameters for PIDN-FOI controller for TT+STPP system

Optimum gains/other parameters
$K_{P11}* = 0.1779$ $K_{P12}* = 0.3709$ $K_{P21}* = 0.3689$ $K_{P22}* = 0.2263$ $K_{P31}* = 0.3709$ $K_{P32}* = 0.4699$
$K_{I11}* = 0.6798$ $K_{I12}* = 0.6059$ $K_{I21}* = 0.6792$ $K_{I22}* = 0.6799$ $K_{I31}* = 0.7099$ $K_{I32}* = 0.6995$
$K_{D11}* = 0.9739$ $K_{D12}* = 0.9233$ $K_{D21}* = 0.9643$ $K_{D22}* = 0.9387$ $K_{D31}* = 0.9800$ $K_{D32}* = 0.9435$ $N_{11}* = 99.00$ $N_{12}* = 99.54$ $N_{21}* = 93.00$ $N_{22}* = 95.00$ $N_{31}* = 96.11$ $N_{32}* = 98.91$ $\alpha_{11}* = 0.00168$ $\alpha_{12}* = 0.00191$ $\alpha_{21}* = 0.00102$ $\alpha_{22}* = 0.000011$ $\alpha_{31}* = 0.00175$ $\alpha_{32}* = 0.00168$

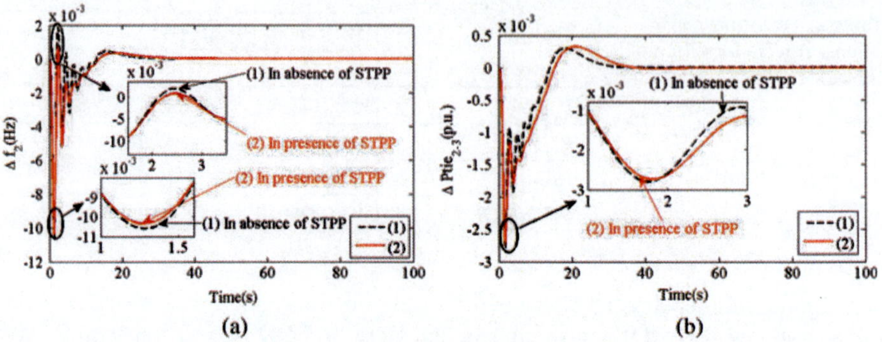

Fig. 4 Response valuation of system with and without STPP for **a** Δf_2 versus time, **b** $\Delta P_{\text{tie}2-3}$ versus time

5.3 *Dynamic Response of Scheme with STPP and UC*

The energy storing device entitled ultra-capacitor (UC) is taken, to study its impact on system dynamics in the scheme with STPP. The dynamic responses of the scheme with STPP and UC in area-1 are compared with that of system with STPP in Fig. 5 using WOA optimized PIDN-FOI controller. The finest values of controller gains and factors are noted (Table 4). The value of J_{ISE} and J_{FOD} are 0.00061 and 6920.3 respectively. Thus, from Fig. 5 and the values J_{ISE} and J_{FOD}, it is noticeably understood that response of the thermal system with STPP and UC provides better dynamics with curtailed peak deviations, intensity of oscillations together with settling time. Therefore, the percentage improvement in the value of J_{ISE} and J_{FOD} for a system with STPP, UC as well as system with STPP with respect to only TT system is reflected in Table 5. This shows that there is a great improvement in the value and hence further work is carried with integration of both STPP and UC in system.

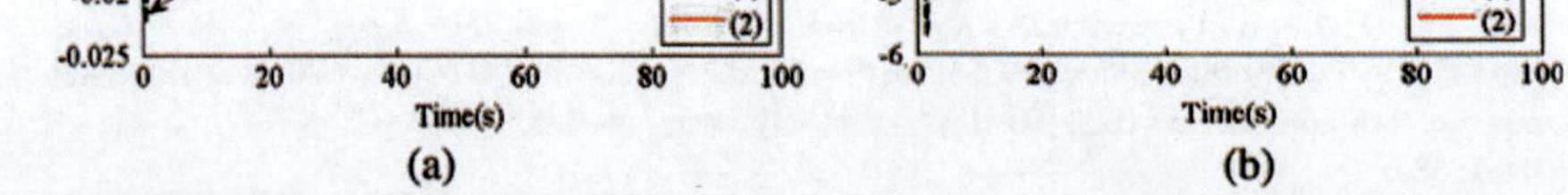

Fig. 5 Response valuation of solar thermal system with and without UC for **a** Δf_1 versus time, **b** Δf_3 versus time

Table 4 Finest values of gains along with parameters for PIDN-FOI controller for TT+STPP+UC system

Optimum gains/other parameters
$K_{P11}* = 0.1187$ $K_{P12}* = 0.9893$ $K_{P21}* = 0.0369$ $K_{P22}* = 0.9893$ $K_{P31}* = 0.1199$ $K_{P32}* = 0.9831$
$K_{I11}* = 0.0203$ $K_{I12}* = 0.7896$ $K_{I21}* = 0.0121$ $K_{I22}* = 0.6895$ $K_{I31}* = 0.0122$ $K_{I32}* = 0.7596$
$K_{D11}* = 0.9787$ $K_{D12}* = 0.9835$ $K_{D21}* = 0.6389$ $K_{D22}* = 0.9989$ $K_{D31}* = 0.9856$ $K_{D32}* = 0.9767$ $N_{11}* = 98.23$ $N_{12}* = 45.21$ $N_{21}* = 94.23$ $N_{22}* = 43.01$ $N_{31}* = 96.32$ $N_{32}* = 42.36$ $\alpha_{11}* = 0.00098$ $\alpha_{12}* = 0.00099$ $\alpha_{21}* = 0.00097$ $\alpha_{22}* = 0.00027$ $\alpha_{31}* = 0.00096$ $\alpha_{32}* = 0.00095$

Table 5 Percentage enhancement in the value of performance index (J) for TT+STPP and TT+STPP+C systems compared to TT system using PIDN-FOI controller

Systems	% improvement J_{ISE}	% improvement J_{FOD}
TT+STPP	3.77	15.03
TT+STPP+UC	42.45	33.82

5.4 Sensitivity Analysis with Changed System Loading

At this juncture, sensitivity investigation is accomplished to see the healthiness of WOA optimized parameters of PIDN-FOI obtained at base conditions and system parameters to extensive deviations in scheme circumstances and factors in scheme comprising of TT, STPP, and UC in area-1. The system loading is changed by ±30%. Each change is considered independently and in each change, PIDN-FOI controller factors are optimized by WOA. The finest values are excluded here. The responses evaluation of scheme at alteration condition is correlated for finest values in regard to transformed condition and base condition (Fig. 6). It is realized that responses

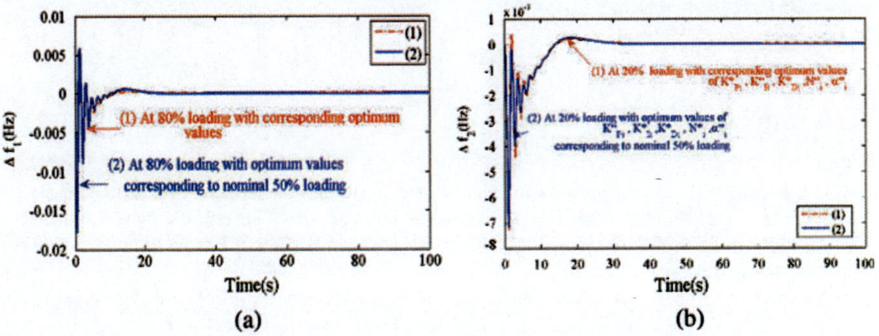

Fig. 6 Response valuation of ultimate system for ±30% of system loading **a** Δf_1 versus time for +30% of system loading, **b** Δf_2 versus time for −30% of system loading

are more or less identical. Thus additional rearranging is not required with extensive fluctuations in scheme condition and factor.

6 Conclusion

The act of a second controller entitled 'PIDN-FOI', cascade amalgamation of whole number order PIDN with fractional-order integrator (FOI) has been assessed for the first time in LFC work of multiple areas multiple source schemes combined with STPP and UC. The controller gains along with parameters are effectively optimized with assistance of simple and fast novel algorithm named as Whale optimization algorithm. This algorithm has the benefit of not getting caught in limited targets. For each and every controller the parameter values are acquired individually through proper optimization. The assessment of scheme dynamics among I, PI, PID, PIDN-FOI controllers in unequal multiple area thermal multiple source scheme discloses health ieraction of the latter. The value of performance index is carefully chosen after numerous optimizations after attaining all the parameter values of controllers. Integration of STPP and UC has a better-quality dynamic action of the scheme. Sensitivity investigation discloses that parameters of PIDN-FOI controller attained at base condition are robust sufficient to extensive alteration in scheme loading in thermal-thermal, STPP and UC in area-1 system. Thus no alteration of controller parameters is essential.

Appendix

$f = 60$ Hz; loading $= 50\%$; $K_{\text{pi}} = 120$ Hz/p.u. MW; $T_{\text{pi}} = 20$ s; $T_{12} = 0.086$ p.u. MW/rad; $H_i = 5$ s; $D_i = 8.33 \times 10^{-3}$ p.u. MW/Hz; $B_i = \beta_i = 0.425$ p.u. MW/Hz; $R_i = 2.4$ Hz/p.u. MW; $T_{\text{gi}} = 0.08$ s, $T_{\text{ti}} = 0.3$ s, $K_{\text{ri}} = 0.5$, $T_{\text{ri}} = 10$ s; $K_{\text{uci}} = -0.7$, $T_{\text{uci}} = 0.9$.

References

1. O.I. Elgerd, *Electric energy systems theory* (McGraw-Hill, New York, 1982), pp. 299–362
2. L.C. Saikia, J. Nanda, S. Mishra, Performance comparison of several classical controllers in AGC for multi-area interconnected thermal system. Int. J. Electr. Power Energy Syst. **33**(3), 394–401 (2011)
3. Y. Wang, R. Zhou, C. Wen, New robust adaptive load-frequency control with system parametric uncertainties, IEE Proc. Gene. Trans. Distrib. **141**(3), 184–190 (1994)
4. D. Das, A.K. Roy, N. Sinha, GA based frequency controller for solar thermal–diesel–wind hybrid energy generation/energy storage system. Int. J. Electr. Power Energy Syst. **43**(1), 262–279 (2012)

5. Y. Sharma, L.C. Saikia, Automatic generation control of a multi-area ST—thermal power system using grey wolf optimizer algorithm based classical controllers. Int. J. Electr. Power Energy Syst. **73**, 853–862 (2015)
6. M.U.D. Mufti, S.A. Lone, S.J. Iqbal, M. Ahmad, M. Ismail, Super-capacitor based energy storage system for improved load frequency control. Electr. Power Syst. Res. **79**(1), 226–233 (2000)
7. B. Maâmar, M. Rachid, IMC-PID-fractional-order-filter controllers design for integer order systems. ISA Trans. **53**(5), 1620–1628 (2014)
8. S. Mirjalili, A. Lewis, The whale optimization algorithm. Adv. Eng. Softw. **95**, 51–67 (2016)
9. A. Saha, L.C. Saikia, Utilization of ultra-capacitor in load frequency control under restructured STPP-thermal power systems using WOA optimized PIDN-FOPD controller. IET Gener. Trans. Distrib. **11**(13), 3318–3331 (2017)
10. A. Saha, L.C. Saikia, Performance analysis of combination of ultra-capacitor and superconducting magnetic energy storage in thermal-gas AGC system with utlilization of whale optimization algorithm optimized cascade controller. J. Renew. Sustain. Energy, **10**, 014103 (2018)

Economic Load Dispatch with Valve Point Loading Effect Using Optimization Techniques

Sachin Prakash, Jyoti Jain, Shahbaz Hasnat, Nikhil Verma, and Sachin

Abstract Economic load dispatch is a method to meet the load demands at minimum cost and losses that may occur at the same time. The practical generation units involve the complex limitations. Among these limitations, valve point loading effect is considered. ELD problem can be solved using various intelligent optimization techniques. In this paper, ELD problem has been solved for IEEE 13-bus system using two intelligent techniques, namely bacterial foraging optimization (BFO) algorithm and particle swarm optimization (PSO). These techniques are inspired by the biological evolution of the living organisms. The results of two algorithms have been compared.

Keywords Economic dispatch problem (EDP) · Valve point loading effect · Evolutionary optimization techniques · Particle swarm optimization (PSO) · Bacterial foraging optimization algorithm (BFOA)

S. Prakash (✉) · J. Jain · S. Hasnat · N. Verma · Sachin
Department of Electrical and Electronics Engineering, Maharaja Surajmal Institute of Technology, New Delhi 110058, India
e-mail: sachin2603@hotmail.com

J. Jain
e-mail: jyotijain_in@yahoo.com

S. Hasnat
e-mail: shahbazhasnat9@gmail.com

N. Verma
e-mail: nikhilverma763@gmail.com

Sachin
e-mail: sachinsingh1825@gmail.com

P. Bansal et al. (eds.), *Proceedings of International Conference on Artificial Intelligence and Applications*, Advances in Intelligent Systems and Computing 1164,
https://doi.org/10.1007/978-981-15-4992-2_38

1 Introduction

Development of infrastructure is always a priority of any country for overall development of the country. Power sector is one of the important infrastructures of a country, which is the prime mover of overall economic development. Power system distributes the electrical load in the network. The load distribution from power generation units is done by complex schemes which should have less fuel cost and maintenance. Economic load dispatch (ELD) [1] is a problem in which the total load demands are divided among the generating units in a power system in such a manner that the total power demands are supplied economically, efficiently and at the same time, satisfying the constraints of power system.

ELD problem has been formulated using equality and inequality constraints. The practical generation units involve the complex limitations. Among these limitations, valve point loading effect is considered [1]. ELD problem can be solved using various intelligent techniques [2, 3], i.e. particle swarm optimization, genetic algorithms, ant colony optimization, grey wolf optimization, etc., and conventional methods. These intelligent techniques are based on some natural phenomena, such as genetics or the actions of some living organism such as bacteria or birds [3, 4].

In this paper, bacterial foraging optimization (BFO) [5, 6] and particle swarm optimization (PSO) [3, 7–10] techniques have been implemented for solving the ELD problem considering the equality and inequality constraints [1]. Both the techniques have been implemented on 13-bus system and their results have been compared.

2 Problem Formulation

The ELD problem can be solved mathematically by minimizing an objective function as given below [1, 6–8]:

Minimize

$$F_T = \sum_{i=1}^{Ng} F_i(P_{Gi}) = \sum_{i=1}^{Ng} (a_i P_{Gi}^2 + b_i P_{Gi} + c_i) \tag{1}$$

Subject to two types of constraints

Equality constraints:

$$\sum_{i=1}^{Ng} P_G = P_{\text{load}} + P_{\text{losses}} \quad (i = 1, 2, 3, \ldots, N_G) \tag{2}$$

Inequality constraints:

$$P_{\min} < P_{Gi} < P_{\max} \tag{3}$$

where

F_T	Generation cost ($/hr)
F_i	Fuel cost of *i*th generator in ($/hr)
P_{Gi}	Load output of ith generator in (MW)
a, b, c	constants of the generator
I	*n*th number of generator
N_g	Total number of generator
P_{load}	Total load demand (MW)
P_{loss}	System transmission losses (MW)
P_{min}	Minimum generation limits
P_{max}	Maximum generation limits.

Large turbo alternators have a set of valves. When power demand increases, these valves are opened in a sequential manner which leads to losses called throttling losses. These losses have a tendency to make the cost-curve nonlinear and non-convex. When the valve is just opened, these losses are very large and considerable as compared to when the valve is fully opened after some time. This process is known as valve point loading effect [1, 6].

In this paper, valve point loading effect has also been considered. The simulation of a power system when this effect is considered requires superimposition of a sinusoidal function along with the quadratic equation described above. Considering this effect, we re-write the equation [1, 6] as:

$$\text{Min } F_T = \sum_{i=1}^{Ng} = F_i(P_{Gi}) = \sum_{i=1}^{Ng} = (a_i P_{Gi}^2 + b_i P_{Gi} + c_i) + |e_i \sin f_i\{P_{\text{min}} - P_{Gi}\}| \tag{4}$$

where e_i and f_i are constants for valve point effect (Fig. 1).

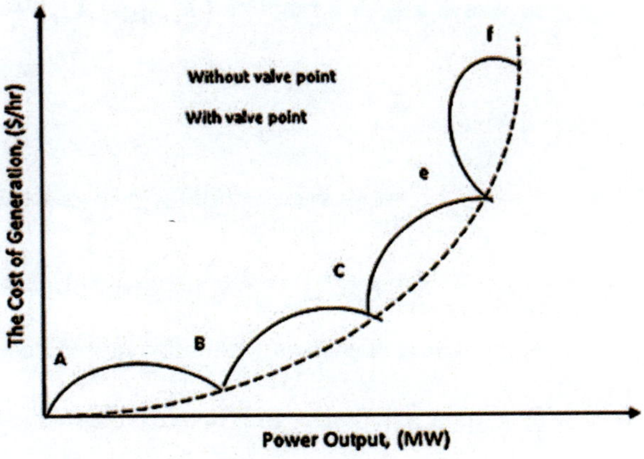

Fig. 1 Load curve with loading effect with constants *a, b, c, e* and *f* [11]

3 Particle Swarm Optimization (PSO)

James Kennedy and Russell C. Eberhart described the particle swarm optimization algorithm in 1995 [3, 10]. It has behaviour of particles such as bird flocking and fish schooling. It utilizes a population-based search procedure [8]. Each particle regulates its flying velocity in adjustment with its adjacent flying velocity. The values of all the particles are given by the fitness function. The given function is then optimized and has velocity, which directs the flying of particles.

Implementation

The initial generations of particles and future updates of particles are generated in such a manner that the sum of the outputs of the particles equals power demand. Hence, the demand is always satisfied, thus only feasible solutions are considered.

Each particle is in given space and represented as

$$X_i = (x_{i1}, x_{i2}, \ldots, x_{\text{id}}) \tag{5}$$

P_{best}: the best solution achieved by that particle of ith particle

$$P_{\text{besti}} = (P_{\text{besti1}}, P_{\text{besti2}}, \ldots, P_{\text{bestid}}) \tag{6}$$

G_{best}: the best value obtained by any particle in the neighbourhood of that particle.

The velocity of particle [7, 8, 10] is

$$V_i = (V_{i1}, V_{i2}, \ldots, V_{\text{id}}) \tag{7}$$

Update P_{bestid} and G_{bestid}

Update velocity and position of each particle calculated from updated P_{bestid} to G_{bestid} as shown [3, 7, 8, 10] in

$$V_{\text{id}}^{j+1} = W * V_{\text{id}}^{j} + C_1 * \text{rand}() * \left(P_{\text{bestid}} - X_{\text{id}}^{j}\right) + C_2 * \text{rand}() * (G_{\text{bestid}} - X_{\text{id}}^{j}) \tag{8}$$

Inertia weight factor [3, 7, 8]

$$W = W_{\max} - ((W_{\max} - W_{\min}) * \text{iter})/\text{iter}_{\max} \tag{9}$$

$$X_{\text{id}}^{j+1} = X_{\text{id}}^{j} + V_{\text{id}}^{t+1} \tag{10}$$

The optimal load of each generator is represented as the optimal value of particles [4] (Fig. 2).

The important parameters in PSO are tabulated in Table 1.

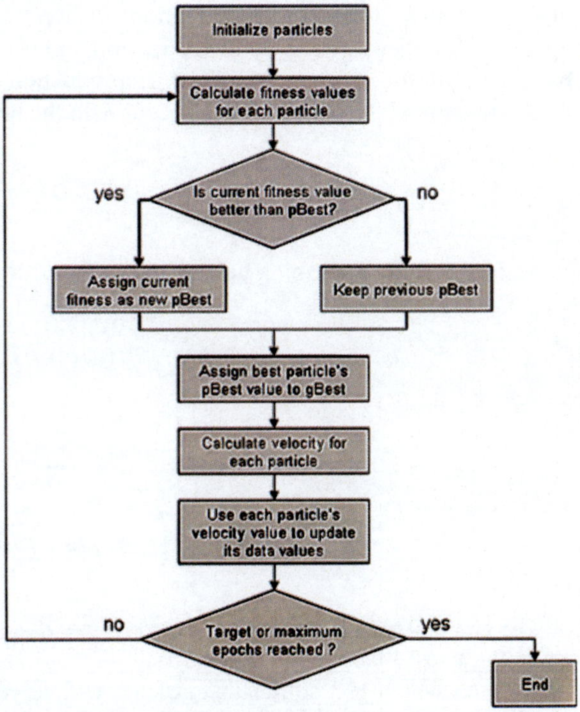

Fig. 2 Flowchart for PSO [12]

Table 1 PSO parameters

Parameters	Size
Population size	30
Maximum iterations	200
W_{min}	0.4
W_{max}	0.9
C_1, C_2	3.1
Vd_{max}	$0.1500 * P_{max}$
Vd_{min}	$-0.1500 * P_{max}$

4 Bacterial Foraging Optimization Algorithm (BFOA)

BFO algorithm process which tends to eliminate the bacteria with weak abilities and choice who having healthy abilities such as *E.coli* and *M.xanthus* [5]. Based on cell–cell interaction, bacteria may repel or ignore each other. The foraging strategy in this optimization is completed by four steps: *Chemotaxis, Swarming, Reproduction, Elimination and Dispersal* [6].

Implementation Chemotaxis process is based on search of food. The chemotaxis behaviour of bacteria is to absorb the nutrients from the environment and move towards or away from specific direction. Bacteria start with tumble, move in all direction, and then swim to defined direction. Let $\theta^i(j, k, l)$ is be the position of ith bacteria at jth (chemotactic step), kth (reproduction step) and the lth (elimination dispersal event). The position of the bacteria in the next step is given [5, 6] by

$$\theta^i(j+1, k, l) = \theta^i(j, k, l) + C(i) * \frac{\Delta(i)}{\sqrt{\Delta^T(i) * \Delta(i)}} \tag{11}$$

Mathematically, swarming behaviour can be written as [5, 6].

$$\begin{aligned} J_{cc}(\theta, p(j,k,l)) &= \sum_{i=1}^{n} J_{cc}\left(\theta, \theta^i(j,k,l)\right) \\ &= \sum_{i=1}^{s}\left[-d_{\text{attract}} \exp\left(-w_{\text{attract}} \sum_{i=1}^{n}\left(\theta_m - \theta_m^i\right)^2\right)\right] \\ &+ \sum_{i=1}^{s}\left[h_{\text{repellent}} \exp\left(-w_{\text{repellent}} \sum_{i=1}^{n}\left(\theta_m - \theta_m^i\right)^2\right)\right] \end{aligned} \tag{12}$$

The coefficients for swarming behaviour are $d_{\text{attract}}= 0.1$, $w_{\text{attract}}=0.2$, $h_{\text{repellant}}= d_{\text{attract}}$, $w_{\text{repellent}}= 10$.

Bacteria which had sufficient nutrients will reproduce and the others will not survive. The healthier of the population will contribute to next generation. This makes the population of bacteria constant. This evolution process will help in reducing the behaviour of stagnation. This creates a new environment of healthy population (Fig. 3).

The important parameters in BFO are tabulated in Table 2.

5 Case Study

The two optimization algorithms such as PSO and BFO are used to solve economic load dispatch with value point loading effects. The results for 13 generating unit test system on load demand of 1800 MW with non-convex fuel cost functions are presented.

The constants a, b and c are for the objective function and also the constants e and f for valve point loading effect. It also gives the inequality constraints $P_{\min}$ and $P_{\max}$ for each of the 13 generating units.

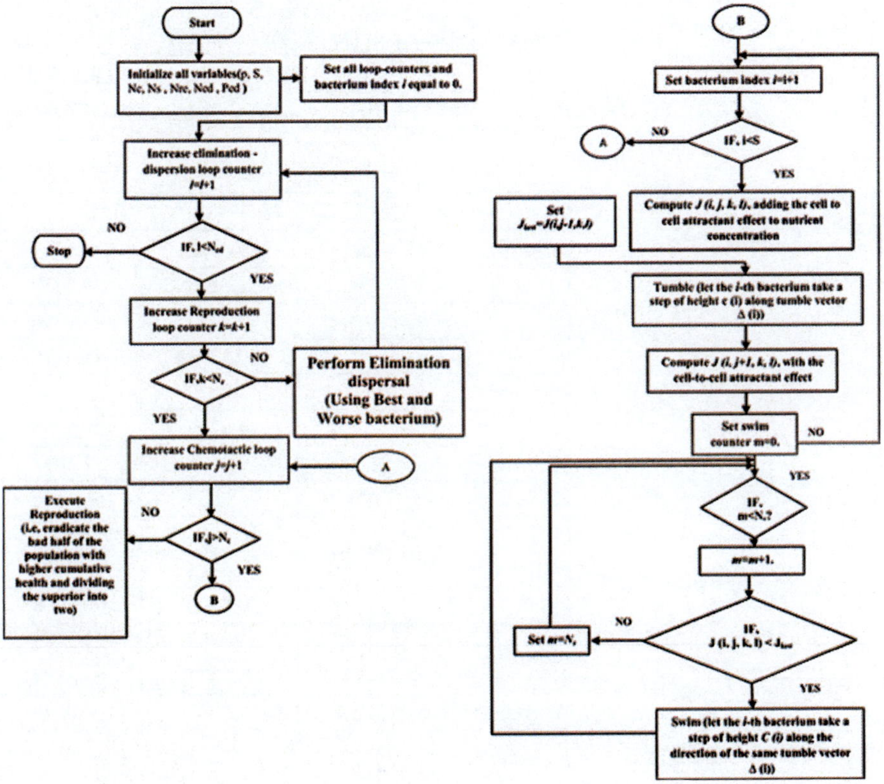

Fig. 3 BFO flowchart [13]

Table 2 BFO parameters

Parameters	Size
Population size	30
Maximum iterations	200
N_r	10
N_c	5
N_s	4
C	0.001
P_{ed}	0.25

6 Results and Discussion

Table 3 represents the final results obtained from the two techniques considered in the paper, namely PSO and BFO. It was observed that the results obtained by BFO are slightly better than that obtained by PSO. The mean generating fuel cost in BFO

Table 3 comparison of results with cost (13 unit system)

Method	Mean cost ($)	Max. cost ($)	Min. cost ($)
BFO	17,978.500	17,986.000	17,971.000
PSO	18,173.107	18,233.974	18,112.240

Table 4 Comparison of results

Generation unit	PSO	BFO
P1	359.0322	654.1417
P2	224.6140	59.9432
P3	152.2342	173.0056
P4	150.3791	98.9275
P5	148.0519	97.3155
P6	94.8369	129.5865
P7	159.3195	77.2891
P8	95.3559	146.1686
P9	106.1354	101.8506
P10	71.5395	54.7101
P11	120.0000	78.5368
P12	55.0988	65.1685
P13	63.4027	63.3584
Total power output (MW)	1800.00	1800.002

was observed to be $17,978.500 and that in PSO is $19,176.500 with a difference of nearly $194.607. The maximum fuel cost in BFO is $17,986.000 and PSO at around $18,233.974, with BFO is lower by $247.974, while the minimum cost in BFO is about $141.24 lower at $17,971.000 as compared to $18,112.240 for PSO. This indicates that BFO is better than PSO as the cost of generation in BFO is lower than that in PSO.

Table 4 represents comparisons of results of PSO and BFO algorithms, implemented on IEEE 13-unit systems for setting of parameters given in Tables 1 and 2.

The graph is plotted of BFO and PSO results is in Fig. 4. These graphs are plotted between number of iterations and generation cost.

7 Conclusion

In this paper, economic load dispatch problem considering valve point loading effect has been solved. The proposed intelligent techniques were performed on MATLAB software. The BFO and PSO algorithms on IEEE 13-bus system with load demand 1800 MW have been implemented. The results obtained indicate that BFO is a better

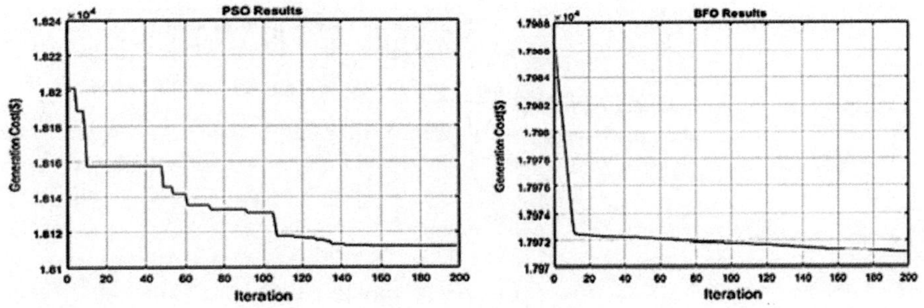

Fig. 4 Results

optimization technique as compared to PSO since the cost of generation comes out be less in BFO. The mean cost in BFO \$194.607 or 1.07% lower than PSO. The maximum costs in BFO \$247.974 or 1.35% lower than PSO whereas the minimum cost in BFO is \$141.24 or about 30.77% lower than that obtained in PSO. The differences obtained are not much significant for the load demand of 18,000 MW and will not have any noticeable effect on consumer tariffs; however, for much larger loads, the difference might become significant enough to have an impact on the tariffs, making BFO preferable over PSO.

The application of BFO and PSO algorithms to ELD problem of IEEE 13-bus system with load demand 1800 MW has been successfully demonstrated.

It is observed that BFO gives lower fuel cost than PSO.

References

1. J. Wood, B.F. Wollenberg, *Power Generation Operation and Control*, 2nd edn (Willey, Hoboken, NJ, 1996), pp. 39–43
2. J. Brownlee, *Clever Algorithms: Nature-Inspired Programming Recipes*, pp. 257–264 (2011)
3. N.K. Jain, U. Nangia, J. Jain, A review of particle swarm optimization. J. Instit. Eng. (India): Series B **99**(4), 407–411 (2016)
4. Y. Liu, K.M. Passino, Biomimicry of social foraging bacteria for distributed optimization: models, principles, and emergent behaviors. J. Optim. Theory Appl. **115**(3), 603–628 (2002)
5. Kevin M. Passino, bacterial foraging optimization. Int. J. Swarm Intell. Res. **1**(1), 1–16 (2010)
6. R. Vijay, Intelligent bacterial foraging optimization technique to economic load dispatch problem. Int. J. Soft Comput. Eng. **2**(2), 55–59 (2012)
7. Z.L. Gaing, Particle swarm optimization to solving the economic dispatch considering the generator constraints. IEEE Trans. Power Syst. **18**(3), 1187–1195 (2003)
8. S. Tiwari, M.P. Dave, B. Dwivedi, Economic load dispatch using particle swarm optimization. Int. J. Appl. Innov. Eng. Manage. **2**(4), 476–485 (2013)
9. N.K. Jain, U. Nangia, J. Jain, An improved PSO based on Initial selection of particle, in *IEEE First International* Conference on Power *Electronics, Intelligent Control and Energy* (Delhi, 2016), pp.1–5
10. N.K. Jain, U. Nangia, J. Jain, An improved PSO based on Initial selection of particles (ISBPSO) for Economic load Dispatch, in *IEEE First International* Conference on Power *Electronics, Intelligent Control and Energy* (Delhi, 2016) pp. 1–5

11. S. Banerjee, D. Maity, C.K. Chanda, Teaching learning based optimization for economic load dispatch problem considering valve point loading effect. Int. J. Electr. Power. Energy Syst. **73** (2015)
12. http://www.mnemstudio.org/particle-swarm-introduction.htm
13. S. Devi, M. Geethanjali, Application of modified bacterial foraging optimization algorithm for optimal placement and sizing of distributed generation. Expert Syst. Appl. **41**(6), 2772–2781 (2014)

Training Multi-Layer Perceptron Using Population-Based Yin-Yang-Pair Optimization

Mragank Shekhar

Abstract This work proposes the Population-Based Yin-Yang-pair Optimization (PYYPO) algorithm and its implementation for successfully training an multi-layer perceptron (MLP) neural network. The proposed PYYPO trainer is then used on five healthcare-related classification datasets: breast cancer, diabetes, liver disorders, Parkinson disease, and vertebral column. The results of the proposed trainer are compared with four other algorithms: Particle swarm optimization, genetic algorithm, cuckoo search, and moth flame optimization algorithms. Lastly, it is shown in the findings that the proposed trainer achieves better scores on most datasets when compared to other algorithms on one performance metrics, namely classification accuracy.

Keywords Metaheuristic · Supervised classification · Multi-layer perceptron · Yin-Yang-pair optimization (YYPO)

1 Introduction

Multi-layer perceptron (MLP) is a system of inter-connected network of nodes such that it mimics the working of neurons present in the brain. MLP network is a type of feedforward neural network (NN) [1] and it is widely used in classification and regression problems [2]. MLP network is often paired with gradient-based algorithms such as backpropagation algorithm [3] to optimize its connection weights and biases. However, the gradient-based algorithms tend to incline more toward exploitation [4] rather than exploring the search space and as the complexity of the problem increases, the dependency on the initial solution becomes more, leading to high possibility of local optimum stagnation [5, 6] and slow convergence [7]. Due to these downsides of using gradient-based algorithms to train MLP, many researchers moved their focus on to stochastics-based approach [8] for training MLP which were reliable alternatives

M. Shekhar (✉)
Bennett University, Greater Noida, Uttar Pradesh, India
e-mail: ms8939@bennett.edu.in

P. Bansal et al. (eds.), *Proceedings of International Conference on Artificial Intelligence and Applications*, Advances in Intelligent Systems and Computing 1164,
https://doi.org/10.1007/978-981-15-4992-2_39

to gradient-based algorithm. Usage of different metaheuristic algorithms for training NN has gained higher recently, due to its stochastic nature leading to avoidance of local optimum.

Metaheuristic algorithms like genetic algorithm (GA) [9], moth flame optimizer (MFO) [10], firefly algorithm [11], particle swarm optimization (PSO) [12], artificial fish swarm [13], gray wolf optimizer [14], cuckoo search (CS) [15], and whale optimization algorithm [16] have been used as a trainer to train MLP networks. According to no free lunch (NFL) theorem [17, 18] no metaheuristic algorithm is suitable for all classes of problems. Therefore, to make the metaheuristic algorithm suitable for a particular class of problems, it is necessary that some modifications should be introduced in it. Hence, the main motivation of this work is to incorporate some modifications to a metaheuristic algorithm for the purpose of training MLP networks.

Yin-Yang-pair Optimization (YYPO) algorithm was proposed by V. Punnathanam and P. Kotecha in 2016 [19]. YYPO algorithm initializes two points known as Yin and Yang and depending on the dimension of the problem, it generates new points around Yin and Yang. To apply YYPO algorithm as a trainer for training MLP network, this paper proposes a modification in the YYPO algorithm by incorporating a population parameter in it. To differentiate the YYPO algorithm from the proposed algorithm, the proposed algorithm is called Population-Based Yin-Yang-pair Optimization (PYYPO).

The rest of the paper is organized as follows:

Section 2 contains a brief overview of YYPO algorithm. Section 3 gives details about development of PYYPO algorithm and PYYPO-based MLP trainer. In Sect. 4, experimental setup is given. Section 5 presents the results and discussions. In Sect. 6, the major conclusions of the work and possible future work are reported.

2 Yin-Yang-Pair Optimization

YYPO uses two points (P_1 and P_2), these points explore the search space by acting as the centers to the hypersphere volumes defined by two radii of r_1 and r_2, respectively, in the space. Point P_1 is designed to focus more on exploitation, whereas Point P_2 is designed to focus more on exploration of the search space. YYPO comprises of two stages, the archive stage and the splitting stage. At every archive stage, the two radii self-adapt in the following way, r_1 tends to decrease periodically, whereas r_2 tends to increase periodically. YYPO begins with randomly initializing the two points (P_1 and P_2) within the domain of 0 and 1. Then, the splitting stage generates new solutions based on the two initialized points by either performing one-way splitting using Eqs. (1) and (2) or D-way splitting using Eq. (3).

$$S_j^j = S^j + \upsilon r,\ j \in 1, 2, \ldots, D \tag{1}$$

$$S^j_{j+D} = P^j - vr,\ j \in 1, 2, \ldots, D \tag{2}$$

$$\left.\begin{array}{l} S^j_k = P^j + v\left(r/\sqrt{2}\right) \text{ if } B^j_k = 1 \\ S^j_k = P^j + v\left(r/\sqrt{2}\right) \text{ otherwise} \end{array}\right\} \begin{array}{c} k \in 1, 2, \ldots, 2D \\ \text{and } j \in 1, 2, \ldots, D \end{array} \tag{3}$$

Here, v denotes random number between 0 and 1, r denotes the radius of the selected point P, B^j_k denotes the (k, j)th element of the randomize binary $2D \times D$ matrix B, and S is a matrix where 2D number of copies of one of the two points are stored. After this, fitness of each point in matrix S is calculated and the point with best fitness is then chosen, this point then appropriately replaces P_1 and P_2. The points P_1 and P_2 are added to an archive after every iteration. Archive stage executes after random intervals. In archive stage, the best points are selected, and the corresponding search radii are modified using Eq. (5) where α is a user defined parameter, and finally, the archive is emptied, and the interval is changed.

$$r_1 = r_1 - (r_1/\alpha) \text{ and } r_2 = r_2 + (r_2/\alpha) \tag{4}$$

A detailed explanation of YYPO is not provided here for the sake of brevity. It can be obtained from literature [19].

3 Development of PYYPO-Based MLP Trainer

This section presents the development details of PYYPO algorithm and PYYPO-based MLP trainer.

3.1 Development of PYYPO

One modification has been introduced in YYPO algorithm for its implementation as a trainer to train an MLP network. A user-defined population parameter PS is introduced in the PYYPO, and PS denotes the total number of solutions that points P_1 and P_2 will create in splitting stage. One can now suitably select the population size of the PYYPO algorithm for different problems with different complexities. In each splitting stage of PYYPO, total $PS/2$ copies are stored in $(PS/2) \times D$ matrix S. In PYYPO algorithm, Eqs. (1), (2) and (3) of YYPO algorithms are replaced by Eqs. (5), (6) and (7), respectively. In these equations, the subscript denotes the row, and the superscript denotes the column of the matrix. In PYYPO algorithm, except for splitting stage, all other stages are similar to that of YYPO algorithm.

$$S^j_i = S^j + vr, \quad j \in 1, 2, \ldots, D \tag{5}$$

$$S^j_{i+PS/4} = P^j - vr, \quad \text{and } i \in 1, 2, \ldots, PS/4 \tag{6}$$

$$\left.\begin{array}{l} S^j_k = P^j + v\left(r/\sqrt{2}\right) \text{ if} B^j_k = 1 \\ S^j_k = P^j - v\left(r/\sqrt{2}\right) \text{ otherwise} \end{array}\right\} \begin{array}{l} k \in 1, 2, \ldots, PS/2 \\ \text{and } j \in 1, 2, \ldots, D \end{array} \tag{7}$$

3.2 PYYPO-Based Trainer

This sub-section presents the process of using PYYPO algorithm to train an MLP network with one hidden layer; the proposed algorithm will be named PYYPO-MLP. There are two important properties which have been considered to design PYYPO-based trainer. First property is related to the encoding scheme of the point in PYYPO algorithm and second property is related to the selection of the objective function. Each point in PYYPO is encoded as a $(1 \times D)$ vector to represent a candidate MLP network. D denotes the size of each vector and it is equal to the total number of parameters present in the MLP network which is set as $((m \times i) + (2 \times i) + i)$, where m denotes the number of input features and i denotes the number of neurons present in the hidden layer. Here, weights and biases of the MLP network are regarded as parameters.

The mean square error (MSE) fitness function is used to measure the fitness value of each point in PYYPO; in MSE, we calculate the difference between the actual target values and predicted values which are generated by the MLP networks for every training sample present in the dataset. MSE is calculated using Eq. (8) where y is the actual target value, y_p is the value predicted by MLP network, and n is number of samples in the training dataset.

$$\text{MSE} = \frac{1}{n}\sum_{i=1}^{n}\left(y - y_p\right)^2 \tag{8}$$

The general steps of the PYYPO algorithm based MLP trainer are as follows:

1. Randomly initialize the two points (candidate MLP networks) P_1 and P_2 in the domain of 0 and 1 and initialize two radii r_1 and r_2 with value of 0.5.
2. Evaluate both MLP networks using the fitness function MSE.
3. If fitness value of P_2 is better than P_1, then interchange P_1 and P_2, and r_1 and r_2.
4. Push both P_1 and P_2 in an archive.
5. Execute the splitting stage for P_1 and P_2 using Eqs. (5)–(7), and evaluate all generated MLP networks and receive the updated P_1 and P_2.
6. Periodically execute the archive stage and update two radii r_1 and r_2 using Eq. (4) and empty the archive.

Table 1 Binary-classification datasets

Dataset	#Training samples	#Test samples	#Attributes	MLP structure
Breast cancer	461	238	8	8-17-1
Diabetes	506	262	8	8-17-1
Liver disorder	227	118	6	6-13-1
Parkinson disease	128	67	22	22-45-1
Vertebral column	204	106	6	6-13-1

7. If the maximum number of iterations is not reached, Go to step 2.
8. Return the point P_1 as the best candidate MLP network and validate it on test dataset.

4 Experimental Setup

Five healthcare-related binary-classification datasets have been used to test the efficiency of the PYYPO-based MLP trainer. These datasets are obtained from the machine learning repository of University of California at Irvine (UCI) [20].

Table 1 gives details about the datasets such as number of features, training samples, and test samples and corresponding MLP structure of each dataset. Each dataset is normalized using Eq. (9) and then partitioned into two parts for training and testing purpose such that 66% of the dataset is used for training and rest 34% is used for testing the MLP network.

$$q_i' = \frac{q_i - \min_q}{\max_q - \min_q} \tag{9}$$

Python 3.7 on a 64-bit Linux operating system has been used to conduct all the experiments presented in this work. To minimize the impact of randomness in the meta-heuristic algorithms all experiments are carried out on 30 separate runs to get statistically meaningful results. For all experiments, parameter settings for every algorithm are borrowed from literature [9, 10, 12, 15, 19] and parameters such as population size and number of iterations are set to 50 and 250 runs, respectively. The number of neurons present in hidden layer is equal to $2 \times m + 1$; here, m is the total number of attributes in the dataset which is referred from literature [10, 14, 16].

5 Results and Discussions

This section presents the benchmark results of the conducted experiments. Classification accuracy is the main parameter which is considered for comparison of

Table 2 Classification accuracy of algorithms for five datasets

Datasets	Metric	PYYPO	GA	PSO	MFO	CS
Breast cancer	AVG	**0.97759**	0.97394	0.96694	0.97437	0.96989
	STD	**0.00497**	0.00577	0.00852	0.00665	0.00732
	BEST	0.98319	0.98319	0.98319	**0.98739**	0.98319
Diabetes	AVG	**0.75687**	0.74605	0.74618	0.74936	0.74020
	STD	**0.00482**	0.00980	0.01522	0.00822	0.01542
	BEST	**0.76717**	**0.76717**	0.75954	0.76335	0.76335
Liver disorders	AVG	0.76299	0.74124	0.74096	**0.76553**	0.71807
	STD	0.01248	0.02266	0.02958	**0.01205**	0.02960
	BEST	0.78813	**0.79661**	0.77966	0.78813	0.78813
Parkinson disease	AVG	**0.87611**	0.82537	0.76766	0.84228	0.82437
	STD	**0.02514**	0.04029	0.04359	0.03670	0.02719
	BEST	**0.91044**	**0.91044**	0.83582	**0.91044**	0.86567
Vertebral column	AVG	**0.87955**	0.86666	0.87106	0.87484	0.84528
	STD	**0.01255**	0.01480	0.01722	0.01464	0.02617
	BEST	0.90566	0.89622	**0.91509**	0.90566	0.89622

the proposed PYYPO-based trainer and four metaheuristic algorithms: GA, PSO, MFO, and CS-based trainers. Three metrics are considered for statistical comparison, namely average value (AVG), standard deviation (STD), and the best (BEST) value of classification accuracy obtained in the last iteration of five metaheuristic algorithms, Table 2 shows the results using these three metrics for every algorithm on the given datasets.

Looking at the results for the breast cancer dataset, it is clear that PYYPO-based trainer outperformed other algorithms by obtaining highest average accuracy of 0.97759 and low standard deviation of 0.00497 for the classification accuracy. GA and MFO came at third and second place and are slightly behind than PYYPO. Best accuracy value was achieved by MFO algorithm with 0.98739 which is slightly better than other algorithms.

PYYPO-based trainer outperforms all algorithms on all three metrics for diabetes dataset by obtaining highest average accuracy of 0.75687 and low standard deviation of 0.00482. PYYPO and GA have the best accuracy of 0.76717.

MFO-based trainer outperformed other algorithms for liver disorders dataset; it achieved the highest average accuracy of 0.76553 and low standard deviation of 0.01205; PYYPO-based trainer came second by obtaining an average accuracy of 0.76299 and a standard deviation of 0.01248. GA obtained best accuracy of 0.79661.

Looking at the results for the Parkinson disease dataset, the PYYPO-based trainer outperformed every algorithm in all three metrics. It obtained an average accuracy value of 0.87611 and a standard deviation of 0.02514. PYYPO, MFO, and GA algorithms have the best accuracy of 0.91044.

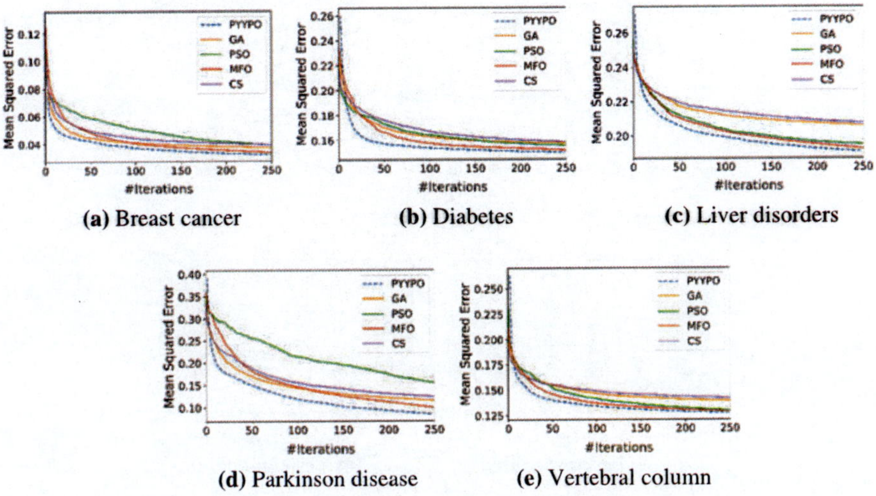

(a) Breast cancer (b) Diabetes (c) Liver disorders

(d) Parkinson disease (e) Vertebral column

Fig. 1 Convergence curves of algorithms over healthcare-related datasets

PYYPO-based trainer got the better results for vertebral column dataset when compared to all other trainers. It obtained an average accuracy of 0.87955. Moreover, PYYPO has low standard deviation value of 0.01255. PSO obtained the best classification accuracy with 0.91509.

Figure 1a–e presents the convergence curves of the PYYPO, GA, MFO, PSO, and CS-based trainers for five classification datasets. These convergence curve shows the convergence of the average MSE over 250 iterations with 30 independent runs. All five convergence curves show that PYYPO-based trainer converges faster than other trainers and achieves best fitness function value (MSE) after 250 iterations for breast cancer, diabetes, liver, Parkinson disease, and vertebral column datasets.

Figure 2a–e presents the boxplots for five classification datasets which were used for evaluating PYYPO, GA, MFO, PSO, and CS-based trainers for 30 different runs. Each boxplot shows 30 MSE values for each trainer. All five boxplots prove that MLP networks trained using PYYPO algorithm have a high level of agreement with one another. Therefore, PYYPO-based trainer achieved more stable results than other algorithm-based trainers.

6 Conclusion

This work proposed the Population-Based Yin-Yang Yin-Yang-pair Optimization (PYYPO) algorithm and its implementation as trainer for training an MLP neural network. The proposed PYYPO trainer is then used on five healthcare-related classification datasets: breast cancer, diabetes, liver disorders, Parkinson disease, and vertebral column where the objective was to minimize the average fitness function

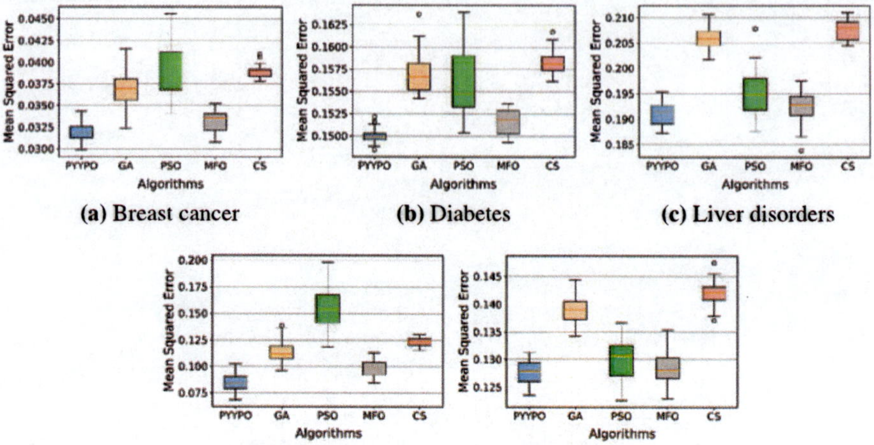

(a) Breast cancer **(b)** Diabetes **(c)** Liver disorders

(d) Parkinson disease **(e)** Vertebral column

Fig. 2 Boxplot charts of algorithms for five datasets

MSE for 30 different runs. The results of the proposed trainer are compared with four other algorithms: Particle swarm optimization, genetic algorithm, cuckoo search, and moth flame optimization.

The findings of the work showed that the proposed trainer outperformed other algorithms on most datasets. It achieved highest average classification accuracy among other algorithms on four out of five datasets. Moreover, the proposed trainer converged faster than other algorithm-based trainers and achieved lowest average MSE value after 250 iterations; this is because, PYYPO algorithm have the ability of maintaining balance between exploration and exploitation of the search space by adaptively controlling the radius of hypersphere volumes, and this helps in finding promising solutions by avoiding local optima stagnation therefore leading to fast convergence.

For future work, performance of the proposed PYYPO-based trainer can be evaluated on other types of neural network architecture. Researchers can also analyze the impact of each user-defined parameters in PYYPO algorithm for training different MLP networks.

References

1. W.S. McCulloch, W. Pitts, A logical calculus of the ideas immanent in nervous activity. Bull. Math. Biophys. **5**, 115–133 (1943)
2. J. Schmidhuber, Deep learning in neural networks: an overview. Neural Netw. **61**, 85–117 (2015)
3. J. Hertz, Introduction to the theory of neural computation, vol. 1 (Basic Books, New York, NY, 1991)

4. J.-R. Zhang, J. Zhang, T.-M. Lok, M.R. Lyu, A hybrid particle swarm optimization–back-propagation algorithm for feedforward neural network training. Appl. Math. Comput. **185**(2), 1026–1037 (2007)
5. M. Gori, A. Tesi, On the problem of local minima in backpropagation. IEEE Trans. Pattern Anal. Mach. Intell. **14**(1), 76–86 (1992)
6. Y. Lee, S.H. Oh, M.W. Kim, An analysis of premature saturation in back propagation learning. Neural Netw. **6**(5), 719–728 (1993)
7. S.E. Fahlman, An empirical study of learning speed in back-propagation networks (1988)
8. V.K. Ojha, A. Abraham, V. Snášel, Metaheuristic design of feedforward neural networks: a review of two decades of research. Eng. Appl. Artif. Intell. **60**, 97–116 (2017)
9. U. Seiffert, Multiple layer perceptron training using genetic algorithms, in *ESANN*, pp. 159–164 (2001)
10. W. Yamany, M. Fawzy, A. Tharwat, A.E. Hassanien, Moth-flame optimization for training multi-layer perceptrons, in *2015 11th International computer engineering Conference* (IEEE, 2015) pp. 267–272
11. I. Brajevic, M. Tuba, Training feed-forward neural networks using firefly algorithm, in *Proceedings of the 12th International Conference on Artificial Intelligence, Knowledge Engineering and Data Bases*, pp. 156–161 (2013)
12. V.G. Gudise, G.K. Venayagamoorthy, Comparison of particle swarm optimization and back-propagation as training algorithms for neural networks. in *Proceedings Swarm Intelligence Symposium, 2003. SIS'03*, pp. 110–117 (2003)
13. S. Hasan, S.Q. Tan, S.M. Shamsuddin, R. Sallehuddin, Artificial neural network learning enhancement using artificial fish swarm algorithm, in *Proceedings of the Third International Conference on Computing and Informatics*, pp. 8–9 (2011)
14. S. Mirjalili, How effective is the grey wolf optimizer in training multi-layer perceptrons. Appl. Intell. **43**(1), 150–161 (2015)
15. A.A.L. Kawam, N. Mansour, Metaheuristic optimization algorithms for training artificial neural networks. Int. J. Comput. Inf. Technol. **1**(2012), 156–161 (2012)
16. I. Aljarah, H. Faris, S. Mirjalili, Optimizing connection weights in neural networks using the whale optimization algorithm. Soft. Comput. **22**, 1–15 (2016)
17. D.H. Wolpert, W.G. Macready, No free lunch theorems for optimization. IEEE Trans. Evol. Comput. **1**(1), 67–82 (1997). https://doi.org/10.1109/4235.585893
18. Y.-C. Ho, D.L. Pepyne, Simple explanation of the no-free-lunch theorem and its implications. J. Optim. Theory Appl. **115**(3), 549–570 (2002)
19. V. Punnathanam, P. Kotecha, Yin-Yang-pair optimization: a novel lightweight optimization algorithm. Eng. Appl. Artif. Intell. **54**, 62–79 (2016)
20. A. Asuncion, D. Newman, UCI machine learning repository (2007)

Maiden Application of Hybrid Crow-Search Algorithm with Particle Swarm Optimization in LFC Studies

Naladi Ram Babu, Lalit Chandra Saikia, Sanjeev Kumar Bhagat, and Arindita Saha

Abstract This article presents the hybridization of crow-search algorithm with particle swarm optimization (hCA-PSO) in load frequency control (LFC) studies. A new ancillary controller entitled as tilt-integral-derivative with filter (TIDN) is proposed and its dynamic performance is found to be better over PID, PIDN. The proposed hCA-PSO technique is used to optimize the parameters of controller and is found to be better over CA and PSO algorithms in terms of setting time. It is also observed that the responses with AC-HVDC tie-line are found to be better over AC tie-line. Moreover, sensitivity analysis reveals that TIDN controller parameters optimized by hCA-PSO are found to be robust at system varied conditions like loading and inertia conditions. Furthermore, the superiority of hCA-PSO is analyzed with system comprising TIDN and thermal-HVDC in terms of dynamic performance and convergence over CA and PSO.

Keywords LFC · AC-HVDC tie-line · hCA-PSO · Sensitivity analysis and TIDN controller

N. R. Babu (✉) · L. C. Saikia · S. K. Bhagat
National Institute of Technology, Silchar, Assam 788010, India
e-mail: rambabu.nits@yahoo.com

L. C. Saikia
e-mail: lcsaikia@yahoo.com

S. K. Bhagat
e-mail: sksanju1070@gmail.com

A. Saha
Regent Education and Research Foundation Group of Institutions, Kolkata, WB 700121, India
e-mail: sahaarindita.91@gmail.com

P. Bansal et al. (eds.), *Proceedings of International Conference on Artificial Intelligence and Applications*, Advances in Intelligent Systems and Computing 1164,
https://doi.org/10.1007/978-981-15-4992-2_40

1 Introduction

LFC is defined as the problem of managing the power mismatch in the power system. It aims to retain the frequency and tie-line power within prescribed values during sudden load perturbations [1–3]. The power exchange occurs through AC interconnected tie-lines. The system stability can be retained by AC tie-lines backing the system during abnormal conditions. To boost up the power dispatch capability and to achieve synchronization among interconnected areas, high voltage direct current (HVDC) tie-lines are utilized [4]. The approximate model of HVDC is developed [5]. Authors in [6] presented the benefits of AC/DC links which provide flexibility in the power system. They have better controllability over AC tie-lines [7–10]. Rakhshani and Rodriguez [11] exhibited the LFC through HVDC links. However, the above studies related to LFC are confirmed to two-area systems only. This provides scope for further investigations.

The presence of nonlinearities like GRC, GDB, and disturbances leads to oscillations in frequency [12, 13]. These oscillations can be minimized by the use of secondary controllers. Numerous secondary controllers like intelligent, integer, fractional, and cascade combinations are available. Controllers like PID [12], PID with filter (N) [14] are of integer type. FO-PI [15] is of fractional type. Fuzzy [16] and neuro fuzzy [16] are of intelligent type. From the above literature, TID [17] with filter (TIDN) is not yet utilized for the L studies which need further investigation.

Secondary controllers outperform well when their values are optimized. Approaches like classical and evolutionary algorithms (EA) are available for optimization to obtain the optimum controller values. The classical approach is laborious and fails to provide optimal values due to sticking at local optima, whereas EA provides optimal search with global optima. Various algorithms like cuckoo search (CS) [12], firefly (FF) [14], biogeography algorithm (BGA) [15], bacterial fogging algorithm (BFA) [18], ant-lion (AL) [19], particle swarm optimization (PSO) [20], hybrid BFA-PSO [21], and gravity search [22] are available in LFC literatures. A new metaheuristic algorithm named by crow-search algorithm CA [23] is accessible. Due to hybridizing property of PSO, CA can be hybridized with PSO. Surprisingly, its application is not found in LFC studies. This provides scope for further investigations.

Rahman et al. [19, 24] presented the controller sturdiness by the deviations in system loading and inertia conditions through the use of BGA and ALO techniques by performing sensitivity analysis. However, the same is to be tested with the hCA-PSO.

From literatures, the objectives are as follows:

(a) Application of hCA-PSO algorithm for optimizing the values of PID, PIDN, and the proposed TIDN controller.
(b) Incorporation of AC-HVDC tie-line in the system (a) by replacing AC tie-lines and to analyze its dynamic responses using finest controller obtained in (a).
(c) Sensitivity analysis of the best controller parameters attained at nominal conditions for checking the sturdiness.
(d) Comparison of system dynamic responses, convergence characteristics for various algorithms like PSO, CA, and the proposed hCA-PSO.

2 The Investigated System

The area system capacity ratio of 1:2:4 is considered. The thermal systems are provided with drop of 4%, governor dead band (GDB) of ±0.036, high voltage direct current (HVDC) transmission, generation rate constraint (GRC) of 3%/min, and their nominal parameters are taken from [3]. The nonlinearities like GRC, GDB, and droop are considered for realistic approach. The power is transmitted through AC and HVDC tie-lines and their respective tie-line power flow equations of AC ($\Delta P_{\text{tie } ij \text{ AC}}$), HVDC ($\Delta P_{\text{tie } ij \text{ HVDC}}$), and parallel AC-HVDC ($\Delta P_{\text{tie } ij}$), are given by (1–2).

$$\Delta P_{\text{tie}\,ij} = \Delta P_{\text{tie}\,ij\,\text{AC}} + \Delta P_{\text{tie}\,ij\,\text{DC}} \tag{1}$$

$$\Delta P_{\text{tie}\,ij} = \left(\frac{2\Pi T_{ij}}{s} + \frac{K_{\text{DC}}}{1 + sT_{\text{DC}}}\right) \cdot (\Delta F_i - \Delta F_j) \tag{2}$$

where i and j are area numbers ($i, j = 1, 2$ and 3; $i \neq j$), T_{DC}, K_{DC}, and F are time constant gain and frequency of HVDC tie-line. The system dynamic responses are acquired with the TIDN controller and hCA-PSO technique by applying 1% SLP in less capacity area with ISE function in (3).

$$\eta_{\text{ISE}} = \int_0^T \{(\Delta F_i)^2 + (\Delta P_{i-j})^2\} \cdot \mathrm{d}t \tag{3}$$

3 The Proposed TIDN Controller

The design of TID [23] is similar to PID but the difference is the presence of proportional constant in PID is replaced by "$(1/s)^n$". The transfer function PID, PIDN, and TIDN are given by (4–6).

$$\text{TF}_{\text{PID}} = K_{Pj} + K_{Ij}/s + K_{Dj} \cdot s \tag{4}$$

$$\text{TF}_{\text{PIDN}} = K_{Pj} + K_{Ij}s + K_{Dj} \cdot s\left(\frac{N_j}{s + N_j}\right) \tag{5}$$

$$\text{TF}_{\text{TIDN}} = K_{Tj} \cdot (1/s)^{n_j} + K_{Ij}s + K_{Dj} \cdot s\left(\frac{N_j}{s + N_j}\right) \tag{6}$$

where n is a real number ($n \neq 0$). K_{Pj}/K_{Tj}, K_{Ij}, and K_{Dj} are proportional, derivative, and integral constants of PID, PIDN, and TID controllers. TID has an advantage of achieving fast disturbance rejection over PID and PIDN controllers. The structure of TIDN is shown in Fig. 2a. The controller values are optimized by hCA-PSO with a set of constraints in (7).

$$\left.\begin{array}{c} 0 \le K_{Tj} \le 0,\ 2 \le n_j \le 8 \\ 0 \le K_{Ij} \le 1,\ 0 \le K_{Dj} \le 1 \text{ and } 0 \le N_j \le 100 \end{array}\right\} \tag{7}$$

4 Hybrid Crow-Search Algorithm with Particle Swam Optimization Technique

4.1 Crow-Search Algorithm (CA)

Askarzadeh [25] proposed CA. It is a population-based algorithm with crow as a search agent. Crow has ability in storing, hiding, and recognizing the food. CA stabilizes the exploitation and exploration phenomenon. The population of search agents of crow-*a* is given in (8) with *d*-dimension.

$$[X^{a,\text{iter}} = X_1^{a,\text{iter}}, X_2^{a,\text{iter}}, \ldots, X_d^{a,\text{iter}}] \tag{8}$$

where iter = 1, 2, 3, …, maxiter and a = 1, 2, 3, …, n (flock length). In every iteration, the crow's optimum position is stored and tracked by all the search agents. Two prospectives are available for crow-*b* to visit the crow-*a*'s hidden place.

Prospective-1 Crow-*a* does not know that crow-*b* is following, and it approaches the hiding place and is given by (9).

$$X^{a,\text{iter}+1} = X^{a,\text{iter}} + r_a \times \text{FL}^{a,\text{iter}} \times (m^{b,\text{iter}} - X^{b,\text{iter}})\ r_b \ge \text{AP}^{b,\text{iter}} \tag{9}$$

where $0 \le (r_a) \le 1$, memory (m) and lower, higher flight length (FL) values tend to local and global optima. Awareness probability (AP) decides the parameters of intensification and diversification. With reduce in AP, CA provides best solution. Lower values of AP provide good intensification results.

Prospective-2 Crow-*a* knows the following of crow-*b*, and crow-*a* tends to a new point given by (10).

$$X^{a,\text{iter}+1} = \text{random position} \tag{10}$$

4.2 Particle Swarm Optimization (PSO)

Eberhart et al. [20] introduced PSO. It is a swarm algorithm and uses birds or fishes as search agent. It comprises of *d*-dimension solution set with particles (a) at different velocities (v) and positions (x). The particle solution (i) in iteration is given by (11).

$$\left[x_i^t = x_{i,1}^t, x_{i,j}^t, \ldots, x_{i,d}^t\right] \quad i = 1, 2, \ldots, a \tag{11}$$

The present solution of each particle is updated with concern of global and local optima and are given by (12, 13), respectively.

$$v_{i,j}^{t+1} = \omega \cdot v_{i,j}^t + C_1 \cdot r_1 |(p_{i,j}^t - x_{i,j}^t)| + |c_2 \cdot r_2 (p_{g\,\text{best}\,j}^t - x_{i,j}^t) \tag{12}$$

$$x_{i,j}^{t+1} = x_{i,j}^t + v_{i,j}^{t+1} \tag{13}$$

where j is an element in dimension (d) with inertia weight (ω) which stimulates convergence speed. The local and global optima ($p_{i,j}^t$, $p_{g\,\text{best}}^t$) represent the present and best solution over a time period (t) with cognitive (C_1, C_2), and random (r_1, r_2).

4.3 Hybrid CA-PSO (HCA-PSO)

Huang et al. [21] proposed a new technique that employs the concepts of both CA and PSO. The initialization process of creating a random population is done in similar to CA and its global optimum solution is obtained from (9) and (10), respectively. In order to obtain the local best solution the obtained global optimum solution is updated by the velocity and particle positions from (12) to (13) in PSO. The present solution is stored and its fitness is calculated. If the present fitness is found to be best over the previous stored value, its value is updated and the procedure gets repeated until the best solution is found. The flowchart of hCA-PSO is in Fig. 2b.

5 Results and Analysis

5.1 Investigations with HCA-PSO Technique

5.1.1 Various Controller Comparisons with HCA-PSO Technique

The system shown in Fig. 1 is provided with various ancillary controllers like PID, PIDN, and the cascaded TIDN controller. The controller values are optimized by hCA-PSO and are listed in Table 1. The corresponding responses are plotted in

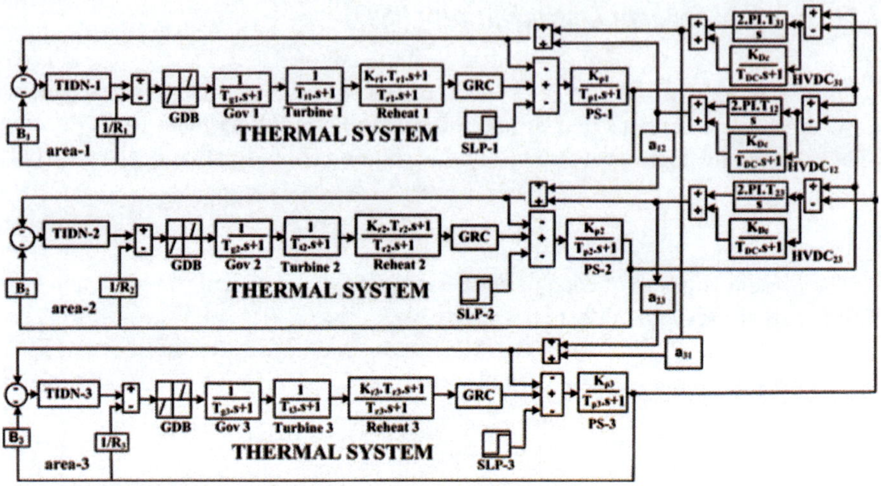

Fig. 1 Transfer function model of the investigated system

Fig. 3a, b. Careful investigations reveal that the response with TIDN controller outperforms over others in terms of settling time and peak undershoot.

5.1.2 Technique System Dynamics with Integration of HVDC Tie-Line and HCA-PSO Technique

The AC tie-line system in Fig. 1 is integrated with AC-HVDC tie-line. It is subjected to TIDN controller and hCA-PSO technique. Table 2 shows the optimum values and its corresponding dynamics are plotted in Fig. 4a, b. Careful observations explore that the integration of HVDC tie-lines in a system gives fewer oscillations and provides better settling time.

5.2 Sensitivity Analysis with the Proposed HCA-PSO Technique and TIDN Controller

A study is conducted to check the sturdiness of the TIDN controller with hCA-PSO technique at system nominal conditions like inertia ($H = 5$ s) and loading (50%). The loading condition is varied by ±20% from 50% by keeping other parameter as fixed. Similarly inertia value (H) is also varied by ±20%. In each varied conditions, hCA-PSO technique optimizes the TIDN controller gains. The responses related to varied and nominal conditions are plotted in Fig. 5a–d. Critical observations of Fig. 5 reveal that the responses obtained at both nominal and varied conditions are nearly same.

Table 1 hCA-PSO optimized gains an parameters of PID, PIDN, and TIDN controller

Controller gains	PID			PIDN			TIDN		
	Area-1	Area-2	Area-3	Area-1	Area-2	Area-3	Area-1	Area-2	Area-3
K_{Tj}/K_{Pj}	0.4822	0.4015	0.5570	0.1526	0.4258	0.9373	0.2363	0.5810	0.7134
n_i	–	–	–	–	–	–	4.5742	7.3535	5.4914
K_{Ij}	0.9547	0.9847	0.3518	0.6127	0.5063	0.8548	0.7841	0.7608	0.6309
K_{Dj}	0.4446	0.5914	0.8930	0.2451	0.5284	0.3677	0.0058	0.9237	0.4522
μ_j	–	–	–	–	–	–	0.3881	0.1351	0.4697
N_j	–	–	–	90.468	95.186	96.306	63.038	49.985	24.847

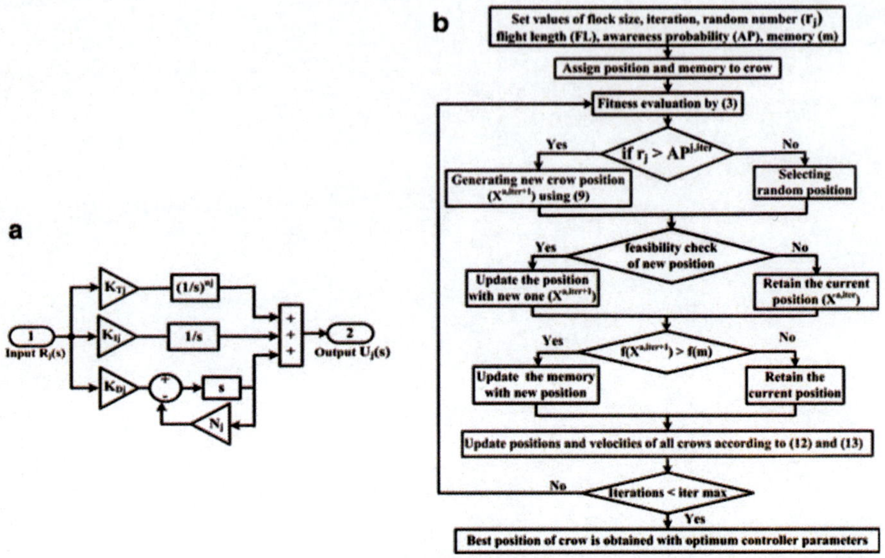

Fig. 2 **a** Block diagram of the proposed cascade TIDN controller. **b** Flowchart of hCA-PSO algorithm

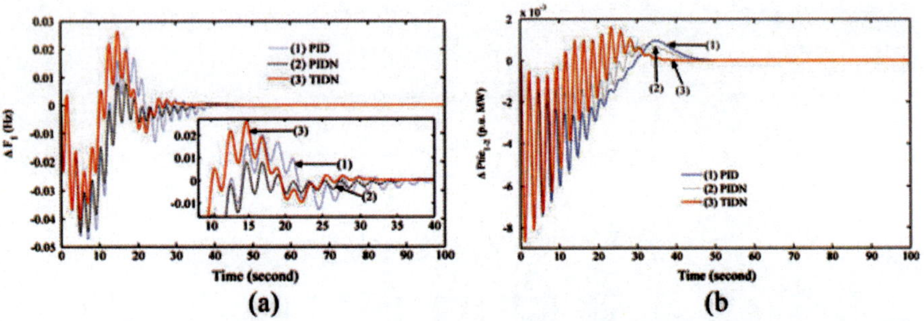

Fig. 3 Dynamic response comparison among PID, PIDN, and the TIDN controller versus time. **a** Frequency deviation in area-1 and **b** Tie-power deviation among area-1 and area-2

Table 2 hCA-PSO optimized gains and parameters of TIDN controller with integration of HVDC tie-line

Gains	K_{Pj}	n_i	K_{Ij}	K_{Dj}	μ_j	N_j
Area-1	0.8191	4.5742	0.7113	0.3245	0.7050	50.960
Area-2	0.9566	7.3535	0.9391	0.4152	0.4508	52.218
Area-3	0.8979	4.0914	0.9762	0.9277	0.4390	15.817

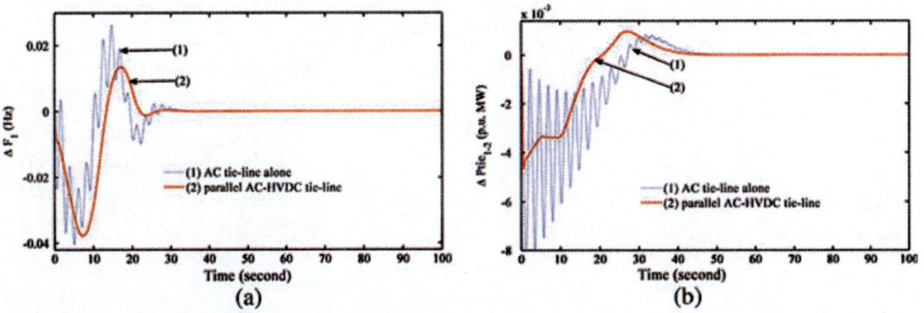

Fig. 4 System dynamics with integration of HVDC tie-line versus time. **a** Frequency deviation in area-1 and **b** Tie-power deviation among area-1 and area-2

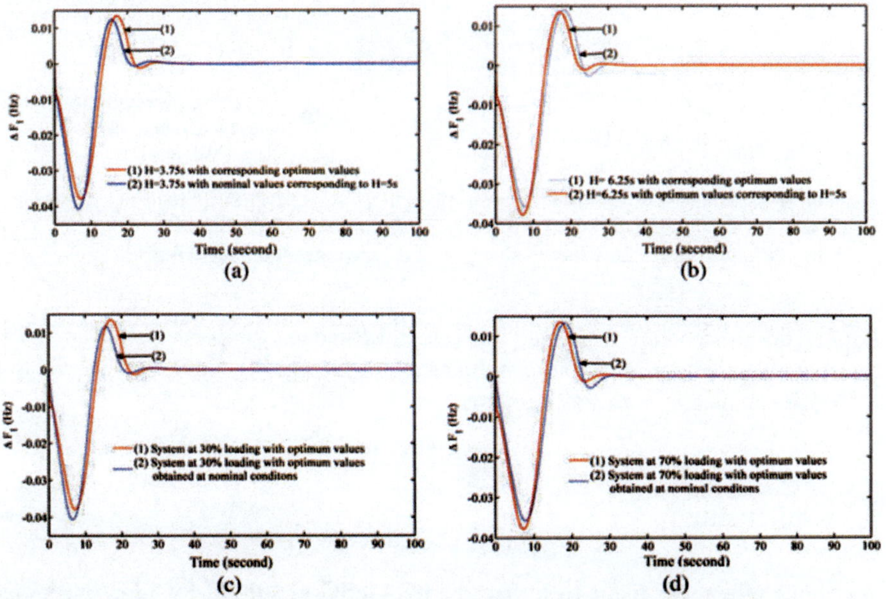

Fig. 5 Comparison of frequency deviation in area-1 versus time. **a** with variation in −20% of inertia, **b** with variation in +20% of inertia, **c** with variation in −20% of system loading, and **d** with variation in +20% of system loading

5.3 Convergence and Dynamic Response Comparison with CA, PSO, and HCA-PSO Techniques

The system in Fig. 1 is considered for investigation. It is subjected with TIDN controller and values are optimized by CA, PSO and hCA-PSO. The obtained responses are compared and illustrated in Fig. 6a–c. It is observed that the responses with hCA-PSO provides less settling time and less oscillation magnitude over CA and

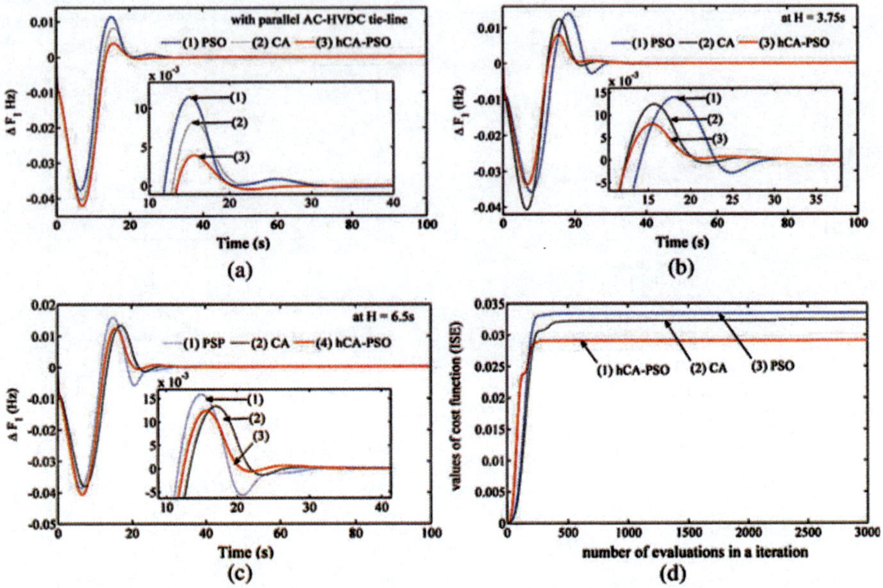

Fig. 6 Comparison among various algorithms like PSO, CA, hCA-PSO. **a** Frequency deviation in area-1 with HVDC tie-line, **b** Area-1 frequency deviation with −20% change in inertia, **c** Area-1 frequency deviation with +20% inertia change and **d** Convergence characteristics

PSO algorithms and are registered in Table 3. Moreover, the convergence curves of the above-studied optimization techniques are plotted in Fig. 6d. It is observed that hCA-PSO converges earlier than others.

6 Conclusions

A maiden effort was made to utilize the hCA-PSO algorithm for LFC multi-area studies. Investigations conclude that hCA-PSO technique optimized parameters lead to better system dynamics over CA and PSO algorithms. Moreover, a new controller named by TIDN is utilized for the LFC studies for the first time along with hCA-PSO technique. Simulations reveal that the responses with TIDN controller outperforms over PID, and PIDN in terms of overshoot and settling time. Further sensitivity analysis suggests that the response obtained at nominal conditions is robust and need not reset again. Furthermore, it is concluded that the integration of HVDC tie-line improves systems dynamics.

Table 3 Various comparisons of the algorithm with generation size of 100 and iteration size of 50

Case study	Run time (s)			Settling time (s)			Magnitude of peak over shoot (mHz)			Cost function (ISE) × (10^{-3})		
	PSO	CA	hCA-PSO	PSO	CA	hCA-PSO	PSO	CA	hCA-PSO	PSO	CA	hCA-PSO
ΔF_1 with HVDC tie-line	610	595	570	29	25	20	13.2	10.2	4.5	33	33	27
ΔF_1 at $H = 6$ s	630	610	579	28	26	23	12	11	8	35.5	34	30.5
ΔF_1 at 70% loading	635	614	561	30	28	22	16	13.5	12	37	34	29

CA crow-search algorithm, *PSO* particle swarm optimization, *hCA-PSO* hybrid crow-search with particle swarm optimization

Appendix

1. Nominal thermal system parameters: $F = 60$ Hz, $K_{pj} = 120$ Hz/MW pu, $T_{jk,\text{AC}} = 0.086$ pu MW/rad, $D_j = 8.33 \times 10^{-3}$ pu MW/Hz, $H_j = 5$ s, $R_j = 2.4$ pu MW/Hz, $B_j = 0.425$ pu MW/Hz, $T_{gj} = 0.08$ s, $T_{tj} = 0.3$ s, $T_{rj} = 10$ s, $K_{rj} = 0.5$ s, $T_{pj} = 20$ s, SLP = 1%, loading = 50%, $P_{r1} = 1000$ MW, $P_{r2} = 2000$ MW, $P_{r3} = 4000$ MW.
2. Crow search algorithm: Flock size = 50, flight length = 0.2, awareness Probability = 0.1, maximum generation = 100.
3. Particle swarm optimization: swarm size = 50, $C_1 = 0.5$, $C_2 = 0.3$, mutation rate = 0.1, maximum generation = 100.

References

1. O.I. Elgerd, *Electric Energy Systems Theory: An Introduction* (McGraw-Hill, New Delhi, 2007)
2. P. Kundur, *Power System Stability and Control* (McGraw-Hill, New Delhi, 1993)
3. W. Tasnin, L.C. Saikia, Performance comparison of several energy storage devices in deregulated AGC of a multi-area system incorporating geothermal power plant. IET Renew. Power Gener. **12**(7), 761–772 (2018)
4. M.J. Chandrashekar, R. Jayapal, Design and comparison of I, PI, PID and fuzzy logic controller on AGC deregulated power system with HVDC link, in *International conference on Circuits, Controls and Communications* (Bengaluru, 2013), pp. 1–6
5. K.K. Challa, P.S.N. Rao, Analysis and design of controller for two area thermal-hydro-gas AGC system, in *International Conference on Power Electronics, Drives and Energy Systems & Power India* (New Delhi, 2010), pp. 1–4
6. T.A. Kumar, V.S. Krishna, N.V. Ramana, Improvement of dynamic performance of three area thermal system under deregulated environment using AC tieline parallel with HVDC link, in *International Conference on Power and Energy Systems* (Chennai, 2011), pp. 1–6
7. Y. Arya, N. Kumar, AGC of a multi-area multi-source hydrothermal power system interconnected via AC/DC parallel links under deregulated environment. Electr. Power Energy Syst. **75**, 127–138 (2016)
8. Y. Arya, Impact of hydrogen aqua electrolyzer-fuel cell units on automatic generation control of power systems with a new optimal fuzzy TIDF-II controller. Renew. Energy **139**, 468–482 (2019)
9. Y. Arya., N. Kumar, AGC of a two-area multi-source power system interconnected via AC/DC parallel links under restructured power environment. Optimal Control Appl. Meth. **37**(4), 1–18 (2015)
10. O.D. Adeuyi, M. Cheah-Mane, J. Liang et al., Frequency support from modular multilevel converter based multi-terminal HVDC schemes. Power Energy Soc. Gen. Meet. 1–5 (2015)
11. E. Rakhshani, P. Rodriguez, Inertia emulation in AC/DC interconnected power systems using derivative technique considering frequency measurement effects. IEEE Trans. Power Syst. **32**(5), 3338–3351 (2017)
12. G. Sharma, I. Nasiruddin, K.R. Niazi, Robust automatic generation control regulators for a two-area power system interconnected via AC/DC tie-lines considering new structures of matrix Q. IET Gener. Trans. Distrib. **10**(14), 3570–3579 (2016)
13. J. Morsali, K. Zare, M.T. Hagh, AGC of interconnected multi-source power system with considering GDB and GRC nonlinearity effects, in *6th Conference on Thermal Power Plants* (Tehran, 2016), pp. 12–17

14. K. Jagatheesan, B. Anand, S. Samanta, N. Dey, A.S. Ashour, V.E. Balas, Design of a proportional-integral-derivative controller for an automatic generation control of multi-area power thermal systems using firefly algorithm. IEEE/CAA J. Automatica Sinica **6**(2), 503–515 (2019)
15. I. Pan, S. Das, Fractional order AGC for distributed energy resources using robust optimization. IEEE Trans. Smart Grid **7**(5), 2175–2186 (2016)
16. G. Sharma, K.R. Niazi, R.C. Bansal, Adaptive fuzzy critic based control design for AGC of power system connected via AC/DC tie-lines. IET Gener. Trans. Distrib. **11**(2), 560–569 (2017)
17. P.N. Topno, S. Chanana, Tilt integral derivative control for two-area load frequency control problem, in *2nd International Conference on Recent Advances in Engineering & Computational Sciences,* pp. 1–6 (2015)
18. J. Nanda, S. Mishra, L.C. Saikia, Maiden application of bacterial foraging-based optimization technique in multiarea automatic generation control. IEEE Trans. Power Syst. **24**(2), 602–609 (2009)
19. M. Raju, L.C. Saikia, N. Sinha, D. Saha, Application of antlion optimizer technique in restructured automatic generation control of two-area hydro-thermal system considering governor dead band, in *Innovations in Power and Advanced Computing Technologies* (Vellore, 2017), pp. 1–6
20. K.-W. Huang, Z.-X. Wu, CPO a crow particle optimization algorithm. Int. J. Comput. Intell. Syst. **12**(1), 426–435 (2018)
21. R.K. Sahu et al., Optimal gravitational search algorithm for automatic generation control of interconnected power system. Ains Shams Eng. J. (5), 721–733 (2014)
22. N. Rambabu, L.C. Saikia, Automatic generation control of a solar thermal and dish-stirling solar thermal system integrated multi-area system incorporating accurate HVDC link model using crow search algorithm optimised FOPI Minus FODF controller. IET Renew. Power Gener. **13**(2), 2221–2231 (2019)
23. D. Guha, P.K. Roy, S. Banerjee, Maiden application of SSA-optimised CC-TID controller for load frequency control of power systems. IET Gener. Trans. Distrib. **13**(7), 1110–1120 (2019)
24. A. Rahman, L.C. Saikia, N. Sinha, Load Frequency control of a hydro-thermal system under deregulated environment using biogeography-based optimised three-degree-of freedom integralderivative controller. IET Gener. Trans. Distrib. **9**(15), 2284–2293 (2015)
25. A. Askarzadeh, A novel metaheuristic method for solving constrained engineering optimization problems: crow search algorithm. J. Comput. Struct. **169**, 1–12 (2016). https://doi.org/10.1016/j.compstruc

Recent Trends in Computational Intelligence and Data Science

Hybrid KFCM-PSO Clustering Technique for Image Segmentation

Jyoti Arora and Meena Tushir

Abstract In this paper, an image segmentation algorithm is proposed which is Kernel-based Fuzzy c-means using Particle Swarm Optimization (KFCM-PSO). This algorithm uses kernel function as it is a generalized distance metric that maps data points into high-dimensional plane where the data points are more clearly separable. Kernel-based Fuzzy c-means are integrated with Particle Swarm Optimization because the traditional Fuzzy c-means algorithm falls into local optima problem whereas PSO is a global optimization algorithm. Comparison of the proposed algorithm with existing FCM, KFCM, and FCM-PSO algorithms is done and the results show that the proposed algorithm gives better results than others.

Keywords Fuzzy C-Means · Particle Swarm Optimization · Kernel metrics · Image segmentation

1 Introduction

Clustering is a task in which objects are classified into homogeneous groups such that elements of the same group are more similar than the elements of different groups. Clustering involves the partitioning of data sets into subsets. Clustering is performed in various applications like taxonomy, image processing, pattern recognition, etc. Clustering is mainly divided into hard, soft, probabilistic, and possibilistic partitions. One type of soft clustering is Fuzzy c-means (FCM) proposed by Zadeh [1], which is very effective as it is efficient, straightforward, and easy to implement,

J. Arora (✉)
Department of Information Technology, Maharaja Surajmal Institute of Technology, GGSIPU, New Delhi 110058, India
e-mail: joy.arora@gmail.com

M. Tushir
Department of Electrical and Electronics Engineering, Maharaja Surajmal Institute of Technology, GGSIPU, New Delhi 110058, India
e-mail: meenatushir@yahoo.com

P. Bansal et al. (eds.), *Proceedings of International Conference on Artificial Intelligence and Applications*, Advances in Intelligent Systems and Computing 1164,
https://doi.org/10.1007/978-981-15-4992-2_41

but, it is very sensitive to initial values because of which result varies. Initial cluster center is chosen randomly due to which results are affected. Due to this random selection of centroids, iterative process of FCM falls into local optimal solution. To solve this problem, we are going to introduce a new clustering approach using nature-inspiring algorithm. Nature-inspiring algorithms consider different processes from nature as an inspiration. These algorithms are very effective in optimizing real-world complex engineering problems. With an increase in the development of industrialization, engineering problems are becoming complex. This happens because of the increasing time complexity, space complexity, variables, dimensions, etc. So to overcome these problems, optimization is required for which nature-inspiring algorithms are designed. Nature-inspiring algorithms are mainly divided into swarm intelligence based algorithms and evolutionary algorithms. Charles Darwin's theory [2] became an inspiration for evolutionary algorithms that are based on the evolutionary behavior of natural processes. Swarm intelligence based algorithms (also known as swarm optimization techniques) mimics the behavior of natural swarms and hence optimizes certain problems. J. Kennedy and R. Ederhart introduced Particle Swarm Optimization (PSO) in 1965 [3] which is based on the birds' flocking behavior. In population, particles cooperate and compete to generate swarm intelligence for optimizing search. The basic concept of PSO lies in optimizing a problem iteratively to improve the solution by accelerating the particles towards its personal best and global best position. Since PSO algorithm is a global optimization algorithm that is why it is incorporated with FCM algorithm and hence overcome the problems of FCM. It leads to a new clustering approach which is FCM-PSO [4]. To map the data points into high-dimensional space, kernel matrix [5] is used. Kernel matrix clusters data points that are not linearly separable by changing the linear algorithm into non-linear algorithm. Because of this advantage of kernel matrix, we incorporate kernel function into FCM-PSO and introduced a new clustering approach with the benefits of kernel metrics and proposed kernel-based FCM-PSO (KFCM-PSO).

This paper is organized as follows: Sect. 2 reviews the Fuzzy C-Means algorithm, kernel matrix, and Particle Swarm Optimization algorithm. Section 3 presents a new hybrid clustering approach which is kernel Fuzzy C-Means using Particle Swarm Optimization algorithm. Section 4 illustrates the experiment and results followed by the drawn conclusion in Sect. 5.

2 Preliminaries

2.1 Fuzzy C-Means (FCM)

Fuzzy c-means (FCM) algorithm is based on fuzzy set theory [6]. It is an effective method of clustering in which a data element can belong to two or more clusters. FCM is proposed by Dunn [7], which is further improved by Bezdek [8, 9] and is generally used in pattern recognition. In FCM, the elements are clustered based on the membership degree assigned to each element. In this technique, soft boundaries of data set among the clusters are considered.

The standard objective function is used for the partitioning of the data $\{x_b\}_{b=1}^{P}$ into C clusters is defined as [8]:

$$O_{\mathrm{FCM}}(M,Y)=\sum_{a=1}^{C}\sum_{b=1}^{s}m_{ab}^{z}\|x_b-y_a\|^2 \tag{1}$$

where array $M=\{m_{ab}\}$ is the partition matrix, $Y=\{y_a\}_{a=1}^{C}$ is the prototype of cluster,$\|x_b-y_a\|^2=d_{ab}=d(x_b,y_a)$ denotes the distance between the cluster center y_a and the pixel x_b.

Using this distance, partition matrix and cluster centroid are calculated as [8]:

$$m_{ab(\mathrm{IFS})}=\frac{1}{\sum_{k=1}^{C}\left(\frac{E_{\mathrm{IFS}}(x_b,y_a)}{E_{\mathrm{IFS}}(x_b,y_k)}\right)^{2/z-1}} \tag{2}$$

$$y_a=\frac{\sum_{k=1}^{S}m_{ab(\mathrm{IFS})}^{z}m(x_k)}{\sum_{k=1}^{S}m_{ab(\mathrm{IFS})}^{z}} \tag{3}$$

High membership value is assigned to a pixel that is close to centroid and low membership is assigned to a pixel that is far from the centroid.

The major drawbacks of FCM are: (1) efficiency of this algorithm is very low as it does not consider information about its neighborhood due to which it is not able to identify outliers and noise points, (2) it randomly initializes the centroid due to which the whole algorithm may stuck into local optimum solution.

2.2 Kernel Metrix

A kernel function is a generalized distance metric. It calculates the distance between the data points that are mapped into a high-dimensional plane where the data could be more easily separated and less price has to be paid for performing certain operations in the high-dimensional plane [5]. The kernel function is employed using a kernel trick which uses a dot product operation and transforms linear algorithm into a non-linear one [10]. Some of the kernel functions are linear kernel, Polynomial kernel, Gaussian kernel, Hyperbolic tangent kernel, etc.

When the knowledge about the data is not known prior then a general-purpose kernel is used which is known as Gaussian kernel. In this paper, we have used Gaussian kernel function [11, 12].

$$\mathrm{GK}(u_i,u_j)=\exp\left(-\frac{\|u_i-u_j\|}{2\rho^2}\right) \tag{4}$$

where GK is a Gaussian kernel function and $\|u_i - u_j\|$ is the square of distance between u_i(data points) and u_j(cluster center), ρ is kernel width which is a positive constant.

2.3 Particle Swarm Optimization (PSO)

Particle Swarm Optimization is a nature-inspiring global optimization algorithm inspired by the behavior of bird flocking and fish schooling. In the year 1995, Kennedy and Eberhart [13, 14] proposed the PSO algorithm. PSO is an iterative method used to optimize a problem by repeatedly iterating to find the best possible solution to it. In PSO, there are several particles that move around a search space and each particle has its own position and velocity. The particles have memory due to which they keep a count of their previous best position and their personal best position. Each group of particles has its own best position which is known as the global best position. Each particle has its own fitness value. The fitness value is calculated using a fitness function. In each iteration, the position, velocity and the fitness value of the particles are updated. The position is updated using the personal best position and the global best position. Each particle strives to attain their personal best position and the global best position.

The following equations are used to update the position and velocity of a particle [13]:

$$v(t+1) = pv(t) + qe_1(p(t) - x(t)) + re_2(g(t) - x(t)) \tag{5}$$

$$x(t+1) = x(t) + v(t+1) \tag{6}$$

where v: velocity of the particle, x: position of the particle, p: inertia weight, q, r: acceleration coefficients, pb: personal best position, gb: global best position, e_1, e_2: positive random values in the range of [0, 1], av(t): Inertia term, $\text{qe}_1(\text{pb}(t) - x(t))$: cognitive component, $re_2(\text{gb}(t) - x(t))$: social component.

In Eq. (5), velocity of the particle is updated by adding inertia term, cognitive term and social component and this updated velocity are added to the current position to obtain the new position of the particle as shown in Eq. (6).

PSO was integrated with FCM and hence, a new clustering approach was discovered. This new approach is known as FCM-PSO [15].

3 Proposed Algorithm

3.1 Kernel-Based Fuzzy C-Means Particle Swarm Optimization (KFCM-PSO)

In the FCM-PSO algorithm, Euclidean distance metric was used. The Euclidean distance is sensitive to outliers and perturbations. This problem is solved by using kernel functions. A kernel function is a generalized distance metric that calculates the distance between the data points that are mapped into high-dimensional plane. In the high-dimensional plane the data points are more clearly separable. The advantage of using kernel function is that it can be used to change the linear algorithm into the non-linear algorithm by using a dot product. Therefore, we proposed a new hybrid clustering algorithm named kernel Fuzzy C-Means Particle Swarm Optimization (KFCM-PSO).

KFCM-PSO Algorithm

Kernel Fuzzy C-Means are used for image segmentation to perform clustering. We optimized this approach by integrating it with Particle Swarm Optimization.

Following are the steps of our proposed algorithm:

Step 1: The algorithm starts by choosing P particles and C clusters for a given dataset.

Step 2: Calculate the cluster centroid using FCM algorithm with Kernel distance.

Step 3: Initialize the PSO parameters, i.e., inertia weight (p), acceleration coefficients (q, r).

Step 4: Set up PSO data structure in which swarm velocity, swarm position and swarm local best fitness are initialized.

Step 5: Initialize swarm position with the cluster centroid which was calculated in Step 2.

Step 6: For iterations = 1 to max-iteration do

(i) For particle = 1 to P

(a) Calculate kernel distances to all centroids using Eq. (4).

(b) Data vector d_v is assigned to cluster C to satisfy the following criteria:

$$d(d_v, s_{pk}) = \min \forall_{i=1,\ldots,C_i} \{d(d_v, s_{\text{pc}})\} \quad (7)$$

where s_{pk} denotes the kth cluster centroid of pth particle in cluster R_{pk}.

(c) Fitness is calculated using

$$O_f = \frac{\sum_{k=1}^{C_i} \left[\sum_{\forall d_v \in R_{pk}} d(d_v, s_k)/|R_{pk}| \right]}{C_i} \quad (8)$$

(ii) Update the global best fitness and global best position value using local fitness.
(iii) Update swarm position using Eq. (5) and swarm velocity using Eq. (6)

Step 7: Optimized cluster centroid is calculated using swarm position
Step 8: Partition matrix is calculated using the cluster centroid obtained in Step 7.

4 Experiments and Results

In this section, we have compared the proposed technique KFCM-PSO with the widely used FCM, KFCM, and FCM-PSO on the UCI machine learning datasets and further performed the segmentation of some synthetic and natural images.

4.1 UCI Datasets

The UCI datasets include Zoo, Glass, and Ecoli. The detailed information of these datasets are listed in Table 1.

To show the effectiveness of our proposed algorithm, we used the following validity measures such as misclassification, accuracy and error rate. The error rate and misclassification of the algorithms are shown in Table 2. According to the comparison in Table 2, our proposed algorithm KFCM-PSO exhibits a minimum error rate and misclassification among all four algorithms for all three datasets.

The accuracy gained by all the four algorithms for the three datasets zoo, glass and Ecoli depicted in Table 3. It can be seen that KFCM has slightly better accuracy than. KFCM-PSO has better accuracy than FCM-PSO. Also, it can be observed that in zoo and glass datasets the accuracy is similar for KFCM and FCM-PSO whereas for Ecoli dataset FCM-PSO has slightly better accuracy than KFCM. Hence it has been observed that KFCM-PSO obtains the maximum accuracy.

Table 1 The detailed specifications of UCI datasets

Dataset	Data points	Features	Number of clusters
Zoo	101	16	7
Glass	214	9	7
Ecoli	336	7	8

Table 2 Clustering results of UCI datasets in terms of error rate and misclassification

Dataset	Algorithm	Error rate (%)	Misclassification
Zoo	FCM	20.792	21
	KFCM	19.802	20
	FCM-PSO	19.802	20
	KFCM-PSO	**15.842**	**16**
Glass	FCM	38.318	82
	KFCM	36.916	79
	FCM-PSO	36.916	79
	KFCM-PSO	**35.981**	**77**
Ecoli	FCM	21.131	71
	KFCM	21.131	71
	FCM-PSO	17.857	60
	KFCM-PSO	**15.179**	**51**

Table 3 Clustering results of UCI datasets in terms of accuracy

Dataset	Algorithm	Accuracy (%)
Zoo	FCM	79.208
	KFCM	80.198
	FCM-PSO	80.198
	KFCM-PSO	**84.158**
Glass	FCM	61.682
	KFCM	63.084
	FCM-PSO	63.084
	KFCM-PSO	**64.019**
Ecoli	FCM	78.869
	KFCM	78.869
	FCM-PSO	82.143
	KFCM-PSO	**84.821**

4.2 Segmentation of Images

A grayscale image of coins is taken with two segments. The image is further corrupted with 0.01 level of Gaussian noise. Figure 1a represents the image corrupted with noise. Figure 1b represents the segments achieved after FCM with the presence of noise. Figure 1c represents the results of KFCM with a better definition of segments in the presence of noise. Figure 1d represents the results of FCM-PSO with the presence of noise. Figure 1e represents the results of the proposed technique KFCM-PSO with better clarity of segments as compared to FCM, KFCM, and FCM-PSO.

Figure 2a represents a synthetic image created in Matlab with the dimension of (100 × 100). Two segments are created with black and white pixels. The image

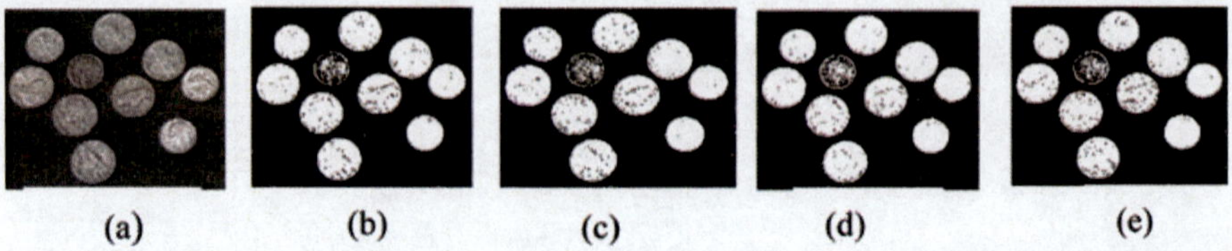

Fig. 1 **a** Image of coins corrupted with 0.01 gaussian noise, **b** segmentation achieved after FCM, **c** segmentation achieved after KFCM, **d** segmentation achieved after FCM-PSO, **e** segmentation achieved after proposed KFCM-PSO

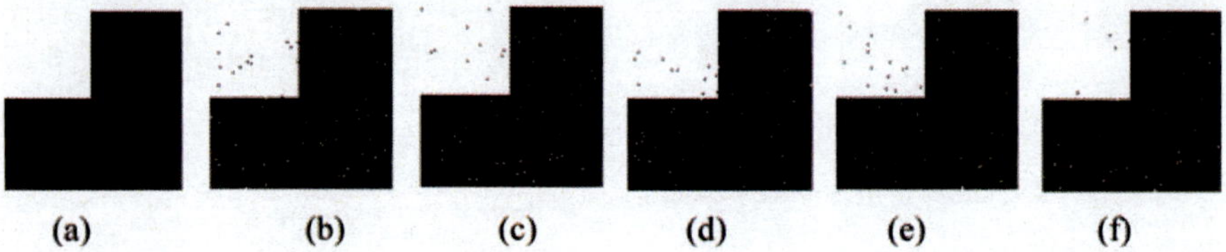

Fig. 2 **a** Synthetic image of with two segments, **b** image corrupted with 0.01 salt and pepper noise, **c** segmentation achieved after FCM, **d** segmentation achieved after KFCM, **e** segmentation achieved after FCM-PSO, **f** segmentation achieved after proposed KFCM-PSO

is further corrupted with 0.01 level of salt and pepper noise. Figure 2b represents the corrupted image. Figure 2c represents of segmentation achieved after FCM. The result shows the presence of noise. Figure 2d represents the image segmentation achieved after KFCM. The result shows better as compared to segmentation achieved after FCM. Figure 2e shows the result of FCM-PSO with the presence of noise. Figure 2f shows the result of the proposed technique showing better robustness against noise as compared to FCM, KFCM, and FCM-PSO.

5 Conclusion

In this paper, we have presented a new hybrid KFCM-PSO algorithm. The drawback of FCM is that it is sensitive to initial values and outliers hence can be trapped in local minima. In order to achieve global minima, PSO was integrated with FCM. Further, we have integrated FCM-PSO with kernel function which maps the data points into high-dimensional plane where the data points are more clearly separable. We have compared our proposed algorithm with other algorithms, i.e., FCM, KFCM, and FCM-PSO on various data sets, i.e., Zoo, Glass, and Ecoli using validity measures like accuracy, error rate, and misclassification. It can be observed that the proposed algorithm has achieved maximum accuracy amongst all the other algorithms. Further, experiments were carried on some synthetic images. The results proved the robustness of the proposed technique.

References

1. L.A. Zadeh, Fuzzy sets, Inf. Control **8**(3) (1965)
2. A. Parashar, K.K. Swankar, Genetic algorithm using to the solution of unit commitment. Int. J. Eng. Trends Technol. **4**(7), 2986–2990 (2013)
3. J. Kennedy, R. Eberhart, Particle swarm optimization, in *IEEE International Conference on Neural Networks*, pp. 1942–1948 (1995)
4. R. Kakkar, A. Singh, Image segmentation using hybrid PSO-FCM. IJISET Int. J. Innov. Sci. Eng. Technol. **2**(7) (2015)
5. T. Hofmann, B. Scholkopf, A.J. Smola, Kernel methods in machine learning. Ann. Stat. 1171–1220 (2008)
6. J.C. Bezdek, Measuring fuzzy uncertainty. Fuzzy Syst. **2** (1994)
7. J. Dunn, A fuzzy relative of the ISODATA process and its use in detecting compact well separated clusters. J. Cybern. **3**(3), 32–57 (1974)
8. J.C. Bezdek, *Pattern Recognition with Fuzzy Objective Function Algorithms* (Plenum, New York, 1981)
9. J.C. Bezdek, R. Ehrlich, W. Full, FCM: the fuzzy c-means clustering algorithm. Comput. Geosci. (1984)
10. K.-R. Muller, S. Mika, G. Ratsch, K. Tsuda, B. Scholkopf, An introduction to kernel-based learning algorithms. IEEE Trans. Neural Netw. **12**, 181–201 (2001)
11. X. Yang, G. Zhang, J. Lu, J. Ma, A kernel fuzzy c-means clustering-based fuzzy support vector machine algorithm for classification problems with outliers or noises. IEEE Trans. Fuzzy Syst. **19**, 105–115 (2011)
12. D.M. Vargas, B.E. Carvajal-Gamez, G. Ochoa, J.J. Rubio, Robust gaussian-base radial kernel fuzzy clustering algorithm for image segmentation. Electron. Lett. **55**(15), 835–837 (2019)
13. J. Kennedy, R. Eberhart, *Swarm Intelligence* (Morgan Kaufmann Publishers, San Francisco, CA, 2001)
14. J. Kennedy, R.C. Eberhart, A discreet binary version of the particle swarm algorithm. (1997)
15. H. Izakian, A. Abraham, V. Snášel, Fuzzy clustering using hybrid fuzzy c-means and fuzzy particle swarm optimization, in *World Congress on Nature & Biologically Inspired Computing* (2009)

Performance Analysis of Different Kernel Functions for MRI Image Segmentation

Jyoti Arora and Meena Tushir

Abstract Nowadays, Kernel-based clustering methods have gained popularity over the widely used Euclidean distance-based methods in the area of image segmentation. Kernel-based methods generally replace the inner product with the positive definite function and perform a non-linear mapping of input data into a high dimensional feature space. The use of these non-linear kernel methods makes the process of clustering more robust and accurate. Different types of kernel methods have been used for the process of clustering. In this paper, we are going to explore the performance of different kernel functions applied with fuzzy c-means for the process of image segmentation. We have incorporated Gaussian kernel function, hyper-tangent kernel function, log-based kernel function and Cauchy kernel function. Experiments are performed on various synthetic and real MRI brain images. The performance is measured on the basis of segmentation accuracy, false-negative ratio and false-positive ratio.

Keywords Fuzzy c-means · Kernel functions · MRI image segmentation

1 Introduction

Clustering is an unsupervised process of segmenting an image into different groups or partitions. Segmentation basically subdivides an image on the basis of the measure of similarity between the pixel properties of an image. It is used in extensive applications such as segmentation of medical images, satellite images, pattern recognition, feature

---

J. Arora (✉)
Department of Information Technology, Maharaja Surajmal Institute of Information Technology, GGSIPU, New Delhi 110058, India
e-mail: joy.arora@gmail.com

M. Tushir
Department of Electrical and Electronics Engineering, Maharaja Surajmal Institute of Technology, GGSIPU, New Delhi 110058, India
e-mail: meenatushir@yahoo.com

P. Bansal et al. (eds.), *Proceedings of International Conference on Artificial Intelligence and Applications*, Advances in Intelligent Systems and Computing 1164,
https://doi.org/10.1007/978-981-15-4992-2_42

extraction, etc. Over the past decade, many fuzzy-based clustering techniques have been used for this process. Among the fuzzy-based clustering techniques, fuzzy c-means (FCM) [1] algorithm is the most widely used clustering technique.

The main advantage of FCM algorithm is that it allows pixels to belong to more than one cluster with some degrees of membership grades. FCM has been proved to retain information more efficiently as compared to crisp methods of segmentation. But the similarity metrics of FCM use Euclidean distance. It does not perform well with data that has been affected by noise. It only detects clusters with spherical shape [2]. In the past few years, kernel methods have been used with fuzzy clustering. The kernel-based clustering has been proven to perform well by preserving the inherent structure of the segments in the input space [3]. In literature, several studies have proven the effectiveness of kernel-based clustering methods [4–6]. The performance of kernel-based clustering methods highly depends on the choice of the kernel function and there are several types of kernel functions available. The selection of kernel function is data specific and the problem of the discussion.

In this paper, fuzzy c-means is embedded with different kernel functions and the performance of kernel-based fuzzy clustering is evaluated on different synthetic and real medical images. The paper is presented as follows: Sect. 2 gives the review of literature of kernel-based clustering; Sect. 3 describes the fuzzy c-means clustering incorporated with different kernel functions; Experiments and discussions are performed in Sects. 4 and 5 concludes the paper.

2 Review of Kernel-Based Clustering

Kernel-based clustering algorithm is defined where the Euclidean-based similarity metrics are substituted by a kernel function. Kernel methods are used to partition nonlinear datasets. These methods are used by many researchers and several clustering methods have been modified with the use of different kernel functions.

Wu and Xie [3] developed kernel-based fuzzy clustering method incorporated with FCM. Kannan [7] used a robust kernel-based clustering method to deal with noisy data. Song and Zhang [8] incorporated spatial terms with kernel-based FCM for the process of image segmentation. Chaira [9] has incorporated hyper-tangent based kernel method with an intuitionistic clustering method for the process of MRI image segmentation. Gong et al. [10] used a radial basis kernel function for the process of segmentation of image along with local information. Tushir and Srivastava [11] proposed hybrid kernel-based clustering (KPFCM) incorporated with probabilistic c-means.

Kannan et al. [12] proposed an effective robust kernel-based fuzzy clustering method for analyzing cancer database. They have incorporated Laplacian kernel-induced distance and implemented it with high dimensional cancer database.

Vargas et al. [13] proposed robust Gaussian-based radial kernel fuzzy clustering algorithm for the process of image segmentation. They have worked on kernel functions in order to improve the typical information tolerance. Gaussian kernel function is a commonly used kernel function. Very few researchers have worked on different types of kernel functions for the process of image segmentation. Tushir et al. [14] have explored different types of kernel functions with hybrid c-means and tested the results on different synthetic and real datasets.

In this paper, we have explored the effectiveness of different kernel functions such as Gaussian, log-based, hyper-tangent and Cauchy kernel functions in the area of image segmentation.

3 Kernel-Based Fuzzy C-Means with Different Kernel Function

In order to study the functioning of kernel methods with fuzzy c-means, a KFCM clustering with Gaussian function (KFCM-G) is defined in this section. Further, subsection defines modification made by incorporating different kernel functions with KFCM.

The objective function of FCM is modified as KFCM by embedding kernel function

$$J_{\text{KFCM}}(\mu, v) = \sum_{K=1}^{N} \sum_{i=1}^{c} \mu_{ik}^{m} \|\phi(x_k) - \phi(v_i)\|^2 \tag{1}$$

where

$$\|\phi(x_k) - \phi(v_i)\|^2 = K(x_k, x_k) - 2K(x_k, v_i) + K(v_i, v_i)$$

Using Kernel Function,

$$\|\phi(x_k) - \phi(v_i)\|^2 = 2(1 - K(x_k, v_i))$$

By employing the transformation of kernel metrics into objective function:

Step 1: Eq. (1) can be written as

$$J_{\text{KFCM-G}}(\mu, v) = 2 \sum_{k=1}^{N} \sum_{i=1}^{c} \mu_{ik}^{m} (1 - K(x_k, v_i)) \tag{2}$$

where $K(x_k, v_i) = \exp\left(-\frac{\|x_k - v_i\|^2}{2\sigma^2}\right)$ is a Gaussian kernel function. After choosing the kernel function and the objective function is minimized using Lagrange's multiplier using zero gradient condition and values of centroids and membership matrix is calculated as.

Step 2: Compute the new centroids as

$$v_i = \frac{\sum_{k=1}^{N} \mu_{ik}^{m} K(x_k, v_i) x_k}{\sum_{k=1}^{N} \mu_{ik}^{m} K(x_k, v_i)} \tag{3}$$

Step 3: Compute new membership as

$$\mu_{ik} = \frac{\left({}^{1}\!/_{(1 - K(x_k, v_i))} \right)^{{}^{1}\!/_{m-1}}}{\sum_{i=1}^{c} \left({}^{1}\!/_{(1 - K(x_k, v_i))} \right)^{{}^{1}\!/_{m-1}}} \tag{4}$$

Repeat steps 1–3 until the following termination criteria is satisfied

$$\left\| J_{\text{KFCM-G}}^{\text{present}} - J_{\text{KFCM-G}}^{\text{previous}} \right\| \prec \varepsilon$$

Here the initial parameters such as centroids v_i and membership matrix μ_{ik} are set from FCM and termination criteria ε is set to 0.0001. The $m = 2$ is a measure of degree of fuzziness.

3.1 Kernel-Based Fuzzy C-Means Using Hyper-Tangent Function (KFCM-H)

Hyper-tangent kernel function is defined as

$$K(x_k, v_i) = 1 - \tanh\left(\frac{\|x_k - v_i\|^2}{2\sigma^2} \right), K(x, x) = 1$$

Using hyper-tangent kernel function, the objective function of KFCM-H clustering is derived from Eq. (2)

$$J_{\text{KFCM-H}}(\mu, v) = 2 \sum_{k=1}^{N} \sum_{i=1}^{c} \mu_{ik}^{m} \left[\tanh\left(\frac{\|x_k - v_i\|^2}{\sigma^2} \right) \right] \tag{5}$$

The calculations of centroids $v_i =$ for $i = 1, 2, \ldots, c$ with hyper-tangent kernel function proceeding as follows:

$$v_i = \frac{\sum_{k=1}^{N} \mu_{ik}^m K(x_k, v_i)\left(1 + \tanh\left(\frac{\|x_k - v_i\|^2}{\sigma^2}\right)\right) x_k}{\sum_{k=1}^{N} \mu_{ik}^m K(x_k, v_i)\left(1 + \tanh\left(\frac{\|x_k - v_i\|^2}{\sigma^2}\right)\right)} \tag{6}$$

The expression for membership matrix μ_{ik} is derived as

$$\mu_{ik} = \frac{\left(1/(1 - K(x_k, v_i))\right)^{1/m-1}}{\sum_{i=1}^{c} \left(1/(1 - K(x_k, v_i))\right)^{1/m-1}} \tag{7}$$

3.2 Kernel-Based Fuzzy C-Means Using Log Function (KFCM-L)

The objective function of the Kernel-based Fuzzy c-means using log function is define

$$J_{\text{KFCM-L}}(\mu, v) = 2 \sum_{k=1}^{N} \sum_{i=1}^{c} \left(\mu_{ik}^m\right) \log\left(1 + \beta \|x_k - v_i\|^2\right) \tag{8}$$

The calculations of modified membership matrix μ_{ik} and new centroids v_i are as

$$v_i = \frac{\sum_{k=1}^{N} \mu_{ik}^m \left(1/1 + \beta \|x_k - v_i\|^2\right) x_k}{\sum_{k=1}^{N} \mu_{ik}^m \left(1/1 + \beta \|x_k - v_i\|^2\right)} \tag{9}$$

$$\mu_{ik} = \frac{\log\left(1 + \beta \|x_k - v_i\|^2\right)^{-1/m-1}}{\sum_{i=1}^{c} \log\left(1 + \beta \|x_k - v_i\|^2\right)^{-1/m-1}} \tag{10}$$

3.3 Kernel-Based Fuzzy C-Means Using Cauchy Function (KFCM-C)

The objective function of the Kernel-based Fuzzy c-means using Cauchy function (KFCM-C) is defined as

$$J_{\text{KFCM-C}}(\mu, v) = \sum_{k=1}^{N}\sum_{i=1}^{c}\mu_{ik}^{m}\left[1 - \left[\frac{1}{1+\beta\|x_k - v_i\|^2}\right]\right] \tag{11}$$

The values of centroid v_i and membership matrix u_{ik} with Cauchy kernel function are defined as:

$$v_i = \frac{\sum_{k=1}^{N}\mu_{ik}^{m}\left({}^{1}/{1+\beta\|x_k - v_i\|^2}\right)x_k}{\sum_{k=1}^{N}\mu_{ik}^{m}\left({}^{1}/{1+\beta\|x_k - v_i\|^2}\right)} \tag{12}$$

$$\mu_{ik} = \frac{\left({}^{1}/{(1 - K(x_k, v_i))}\right)^{{}^{1}/{m-1}}}{\sum_{i=1}^{c}\left({}^{1}/{(1 - K(x_k, v_i))}\right)^{{}^{1}/{m-1}}} \tag{13}$$

4 Experimental Results and Discussions

In this section, the performance of all the kernel-based techniques is evaluated on the synthetic image of the phantom and further on slices of MRI image of the human brain acquired from Internet Brain Segmentation Repository [15]. The evaluation and analysis were done both qualitatively and quantitatively. The segmentation performance of KFCM-G, KFCM-H, KFCM-L and KFCM-C is done based on the segmentation accuracy, false-negative ratio and false-positive ratio. All the algorithms are implemented in MATLAB (R2015a).

The qualitative result presents the pictorial representation of the different segments of an image achieved after the clustering. The smoothness in the visualization represents the robustness of the algorithm against noise. To analyze the segmentation performance quantitatively, different measures such as false-positive ratio, false-negative ratio and overall segmentation accuracy [16] have been used. The false-positive ratio fp_r for a segment represents the error due to extra pixels and false-negative ratio fn_r for a segment represents the error due to lost pixels. The segmentation accuracy SA defines the overall accurately classified pixels by the total number of pixels.

4.1 Segmentation of Synthetic MRI Image

A synthetic image of head phantom is generated using MATLAB R2015 to analyze the accuracy of different kernel-based clustering algorithms. The generated image is a grayscale intensity image that consists of one large ellipse (representing the

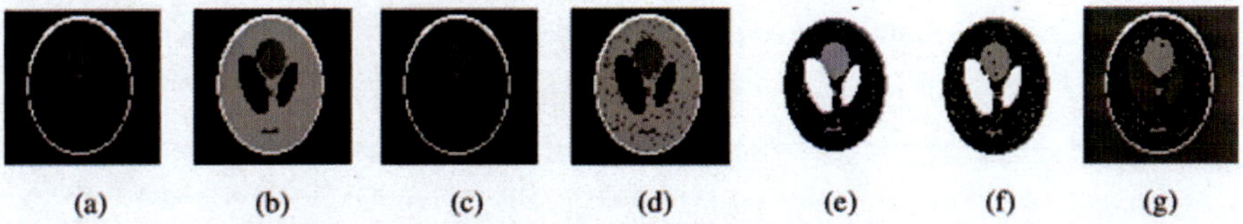

Fig. 1 **a** Original image of synthetic brain image. **b** Ground truth image. **c** Image corrupted with 0.01 gaussian noise. **d** Segmentation achieved after KFCM-G. **e** Segmentation achieved after KFCM-H. **f** Segmentation achieved after KFCM-C. **g** Segmentation achieved after KFCM-L

brain) which contains several smaller ellipses (representing features in the brain). The image is corrupted with 0.01% of Gaussian noise. Figure 1a shows the original image. Figure 1b shows the ground truth image of the brain used to evaluate the performance of the segmentation performed. Figure 1c shows the image corrupted with 0.01 level of Gaussian noise. Figure 1d shows the results of overall segmentation achieved after KFCM-G. The result represents better robustness against noise as compared to KFCM-C and KFCM-L. The segmentation performed by KFCM-H also shows better results as compared to KFCM-C and KFCM-L. Table 1 shows the result of segmentation accuracy (SA), false-negative ratio and false-positive ratio of every segment of synthetic MRI image. The overall segmentation results are also calculated to analyze the performance. The Table shows that KFCM-H performs better as compared to all other kernel-based clustering techniques. KFCM-L performs better as compared to KFCM-G and KFCM-C. KFCM-C shows the poorest results as compared to other kernel-based techniques in robustness and segmentation accuracy.

4.2 Segmentation of Real MRI Image

The segmentation of magnetic resonance image (MRI) of the human brain is segmented into three main segments as gray matter (GM), white matter (WM) and cerebrospinal fluid (CSF). The background is considered as pathological region is segmented as fourth segment. The experiments are conducted on the T-1 weighted axial slice (slice no. 110) of real brain image [15]. Figure 2 gives the visualization of all the four segments using different kernel-based fuzzy c-means techniques. Figure 2a represents the original image, Fig. 2b represents the ground truth image of four segments. Figure 2c, d show the better visualization of all the segments as compared to Fig. 2e, f. This shows that KFCM-G and KFCM-H are more robust to noise as compared to KFCM-C and KFCM-L. Table 2 gives the quantitative results of segmentation accuracy obtained per segment. The overall segmentation results are also calculated. KFCM-H performs well with 100% accuracy, KFCM-G performs with 98.6% accuracy, KFCM-C and KFCM-L perform with 98.2% accuracy.

Table 1 Segmentation results of synthetic brain image

Technique	Segment	False-negative	False-positive	Segmentation accuracy
KFCM-G	1	0	0.0013	0.999
	2	0.0056	0.576	0.781
	3	0.0479	0.008	0.96
	4	0	0	1
	Overall segmentation	0.026	0.146	**0.935**
KFCM-H	1	0	0	1.00
	2	0.04	0.36	0.827
	3	0	0.009	0.999
	4	0.05	0.005	0.979
	Overall segmentation	0.022	0.093	**0.951**
KFCM-C	1	0.071	0.03	0.961
	2	0	0	1
	3	0.022	0.53	0.779
	4	0	0.0004	0.999
	Overall segmentation	0.0232	0.140	**0.934**
KFCM-L	1	0.0602	0.003	0.967
	2	0	0.002	0.999
	3	0.022	0.429	0.812
	4	0	0	1
	Overall segmentation	0.0205	0.108	**0.944**

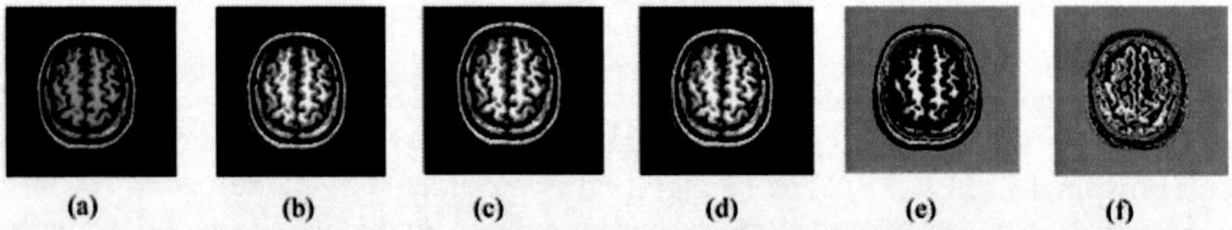

Fig. 2 **a** Original MRI image of all the four segments. **b** Ground truth image. **c** Segmentation achieved after KFCM-G. **d** Segmentation achieved after KFCM-H. **e** Segmentation achieved after KFCM-C **f** Segmentation achieved after KFCM-L

5 Conclusion

This paper presents the performance analysis of different kernel functions embedded with fuzzy c-means in the application of medical image segmentation. In literature, Gaussian kernel function is widely used for the purpose of image segmentation. In this paper, we have analyzed Gaussian, hyper-tangent, Cauchy and log-based kernel function. KFCM-H performs well on synthetic and real MRI images. KFCM-G is

Table 2 Segmentation results of real T-1 weighted MRI image

Technique	Segment	False-negative ratio	False-positive ratio	Segmentation accuracy
KFCM-G	1	0	0	1.00
	2	0	0	1.00
	3	0.0427	0	0.978
	4	0	0.649	0.968
	Overall segmentation	0.0108	0.162	**0.986**
KFCM-H	1	0	0	1.00
	2	0	0	1.00
	3	0	0	1.00
	4	0	0	1.00
	Overall segmentation	0	0	**1.00**
KFCM-C	1	0	0.065	0.968
	2	0	0.030	0.985
	3	0.042	0	0.978
	4	0.005	0	0.997
	Overall segmentation	0.012	0.024	**0.982**
KFCM-L	1	0	0.005	0.997
	2	0	0.03	0.985
	3	0.042	0.00	0.978
	4	0	0.0648	0.968
	Overall segmentation	0.0105	0.0297	**0.982**

robust to noise and performs well on noisy images. KFCM-L and KFCM-C are also another class of kernel functions that can be used for the segmentation of images and can be explored further.

References

1. J.C. Bezdek, *Pattern Recognition with Fuzzy Objective Function Algorithms* (Plenum, New York, 1981)
2. W. Cai, S. Chen, D. Zhang, Fast and robust fuzzy C-means clustering algorithm incorporating local information for image segmentation. Pattern Recognit. **40**(3), 825–838 (2007)
3. Z.D. Wu, W.X. Xie, Fuzzy c-means clustering algorithm based on kernel method, in *Proceedings of the Fifth International Conference on Computational Intelligence and Multimedia Applications*, pp. 47–54 (2003)

4. D.Q. Zhang, S.C. Chen, Fuzzy clustering using kernel methods, in *Proceedings of this Conference on Control and Automation* (China, 2002), pp. 123–128
5. D.Q. Zhang, S.C. Chen, Clustering incomplete data using kernel-based fuzzy c-means algorithm. Neural Process. Lett. **18**(3), 155–162 (2003)
6. S. Zhou, J. Gan, Mercer kernel fuzzy c-means algorithm and prototypes of clusters, in *Proceedings of Conference on Data Engineering and Automated Learning*, pp. 613–618 (2004)
7. S.R. Kannan, Robust kernel FCM in segmentation of breast medical images. Intl. J. Expert Syst. Appl. **38**(4), 4382–4389 (2011)
8. J. Song, Z. Zhang, A modified robust FCM model with spatial constraints for brain MR image segmentation. Information **10**(74), 1–15 (2019)
9. M. Gong, Y. Liang Y, J. Shi J, W. Ma, J. Ma, Fuzzy c-means clustering with local information and kernel metric for image segmentation. IEEE Trans. Image Proc. **22**(2), 573–584 (2013)
10. T. Chaira, A. Panwar, An atanassov's intuitionistic fuzzy kernel clustering for medical image segmentation. Intl. J. Comput. Intll. Syst. 1–13 (2011)
11. M. Tushir, S. Srivastava, A new kernelized hybrid c-means clustering model with optimized parameters. J. Appl. Soft Comput. **10**(2), 381–389 (2010)
12. S.R. Kannan, M. Siva, R. Ramathilagam, R. Davi, Effective kernel-based fuzzy clustering systems in analyzing cancer database. Data Enabled Discov. Appl. 2–5 (2018)
13. D.M. Vargas, B.E. Carvajal-Gamez, G. Ochoa, J.J. Rubio, Robust gaussian-base radial kernel fuzzy clustering algorithm for image segmentation. Electron. Lett. **55**(15), 835–837 (2019)
14. M. Tushir, S. Srivastava, Exploring different kernel functions for kernel-based clustering. Int. J. Artific. Intell. Soft Comput. **5**(3) (2016)
15. Internet Brain Segmentation Repository (IBSR), Available: http://www.cma.mgh.harvard.edu/ibsr/
16. A.P. Zijdenbos, B.M. Dawant, Brain segmentation and white matter lesion detection in MR images. Criti. Rev. Biomed. Eng. **22**(5), 401–465 (1994)

A Novel Approach for Predicting Popularity of User Created Content Using Geographic-Economic and Attention Period Features

Divya, Vikram Singh, and Naveen Dahiya

Abstract Today, the rapid growth of the internet has led to the rapid dissemination of the User Generated Content, which is visible in the exponential growth of websites like Twitter, YouTube, Facebook and Instagram. With this rapid development, identification of the content which is going to be popular has posed some interesting paradigm. Since the application domain of the topic include a big set including network dimensioning, server load balancing, marketing strategic decisions, recommendation systems, etc. this topic poses an interesting field. In this paper, we propose a new framework for the popularity prediction of a User Created Content by exploiting the features that include the attention period of any viewer and the economic factors of the publisher network.

Keywords Popularity prediction · Strategic decision making · User created content sites · Attention period

1 Introduction

Today the information age is characterized by the rapid advancements in the proliferation of User Created Content sites such as Facebook, YouTube, Twitter and Instagram. All these platforms have become an important source for the dissemination of multimedia information worldwide, with YouTube topping the list having more than 1 billion viewers watching hundreds of millions of hours and generating over a billion views [1]. When talking about social networking sites, Facebook is

Divya (✉) · V. Singh
Chaudhary Devi Lal University, Sirsa, Haryana, India
e-mail: divyajatain@msit.in

V. Singh
e-mail: vikramsinghkuk@yahoo.com

Divya · N. Dahiya
Maharaja Surajmal Institute of Technology, New Delhi, India
e-mail: naveendahiya@msit.in

P. Bansal et al. (eds.), *Proceedings of International Conference on Artificial Intelligence and Applications*, Advances in Intelligent Systems and Computing 1164,
https://doi.org/10.1007/978-981-15-4992-2_43

most popular with over 1.5 billion registered users [2], out of which around 85% are actually engaged in the content creation process [3], videos being the most popular content. Next comes Instagram having over 400 million users [4] and Twitter having a network of around 300 million users tweeting over 500 million Tweets per day [5].

Owing to the large scale, heterogeneous content format and the decentralization of the contents and for the effective maintenance and growth of these UGC sites, identifying the contents that will gain popularity is very important. Predicting the view count of a video may be purposeful in various domains including video caching, video recommendation and video procurement, to load balancing the throughput of the servers and data centers, predictive forecasting, disaster management and reputation management to name a few. It can also help the media houses to adjust marketing strategies by enabling the funds allocation for digital marketing to be deployed wisely and efficiently.

In recent years, a lot of research has been carried out in the field of popularity prediction due to its intensive applications in a lot of fields. Today we are living in an information age where predictions of content popularity can serve the application areas which are enumerated as but not limited to following:

Popularity prediction can make the filtering of a huge amount of information easy and efficient. This serves as a cause of study for recommendation systems. To reach a greater number of people, the content generators and suppliers can better organize their content and develop more efficient delivery platforms. The possibility to predict popular items create scenarios of additional profits for advertising media houses (Fig. 1).

Advertising networks can generate more refined and fruitful advertising strategies to reach the desired goal of number of users to provide a better user experience as well as to have the monetary gains. It can be very helpful for content optimization. For instance, news websites select and organize articles from a very lively and organically changing content pool. Accurate popularity predictions can boost a site's traffic and publishers can increase the traffic through social media optimizations. Predicting the content which is likely to be popular can also help when the network is congested by diverting the necessary load or by offloading the mobile data.

In the past few years, a lot of research has been done on addressing the content popularity prediction paradigm [6–11] be it in the form of textual data like news articles [8], Tweets [12], images [10, 13] or videos [7, 9, 14, 15]. Visual information is the most prominent of the humans' senses to seek any information, so videos are inherently more popular. In our paper, we are proposing an approach to improve the success rate of popularity prediction of video content on the basis of inclusion of geographic-economic and attention period features.

The remaining paper is organized as follows: Section 2 deals with an overview of the related state-of-art. Section 3 discusses the proposed approach. Section 4 deals with the conclusion and scope of future work in the area.

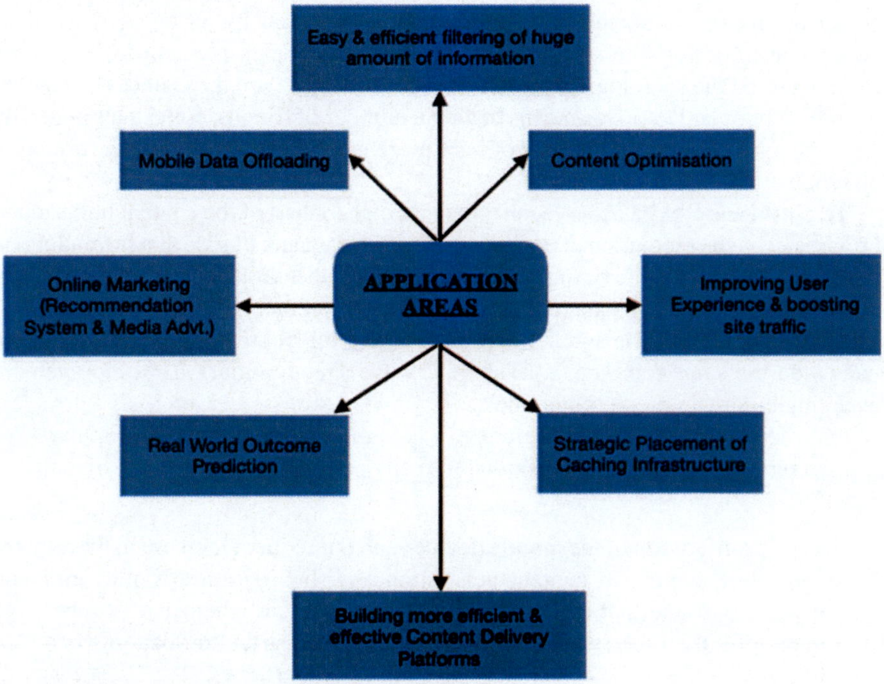

Fig. 1 Application areas of popularity prediction

2 Related Work

In the age of attention mechanism, the content that is likely to get more attention is of prime focus due to its implications in the different domains as discussed earlier. The content that can be discussed may be either text, image or video. A lot of work has been done that focuses on web access patterns [16] specifically in the video contents [17].

Tatar et al. observed that after the publication of a video, the view count can be considered a measure to identify whether the content is popular or not, and it is found to exhibit a heavy-tailed distribution [18].

Szabo and Huberman, however, established a log-linear relationship between the early publication stages and at later times in case of YouTube videos [6]. Most of these prediction methods consider early distribution patterns [6, 7, 9].

The Social-Forecast method given by Xu et al. [11] takes into account the social structure of the network of the publisher by incorporating the follower count. However it shows significant improvement in a tradeoff between prediction accuracy and prediction timing, still the limitation is that it is quite difficult to find out the detailed structure of the social network. the size and structure of the publisher's network is also a relevant factor.

According to the sociological weak ties theory, used by Wang et al. in their paper, a user/viewer's behaviour is affected by their social interactions. The more loosely bound the sociological interactions are, the more are the chances of content becoming popular [19]. This means having a number of friends, colleagues or family with a spread out span of the network are more critical for content and information dissemination.

Trzcinski and Rokita observed that the external context of the content holds much importance in the prediction of popularity [20]. This means a video that is trending on the other social is likely to be more popular than any other video which is nontrending.

Using multi-platform data for content popularity prediction used by Castillo et al. and Roy et al. [21, 22] is another approach where the popular Tweets are linked to videos on the same topic on YouTube and have shown almost 70% improvement over single-domain prediction models.

On a detailed survey of the literature, it is found that predicting the popularity of a video remains a challenging task due to many reasons. Some of them are enlisted as:

1. Due to proliferation of the mobile devices and consequent increase in the number of people engaging in content generation and distribution activities in social media, it is found that external context of the content whether it is popular or trending on the other platforms is pivotal in the distribution patterns of video [20].
2. Implicit popularity which is identified by the number of views reflects the viewer's attention, still everything that gets attention is not necessarily appreciated. Moreover, in order to have some fruitful outcomes to predict popularity, the user attention should be engaged for at least 6–8 min [23, 26].
3. It is also found that the early view patterns of popular and unpopular videos are similar. So the early view pattern-based classification has its own limitations. This method also limits the use of multi-source data for popularity prediction [24].
4. The prediction is more precise for the submissions for which the attention decays quickly like Digg stories, but in case of videos like those uploaded at YouTube, the contents are available for a long time and have a spread span of user attention, thus making is more prone to inaccurate predictions [6].
5. The characteristics such as the duration of a video, gender of the publisher, the geographical location as well as the structure of the economic background of the publisher also play a vital role in making content more popular or less popular one [25]. It also includes the activities and interactions between various viewers. This calls for an inquisitive study of the corresponding domain of popularity prediction.

3 Proposed Approach

Regression is a statistical way to design a model to show the interrelationship between a dependent variable and independent variable. The Univariate Linear Regression models the relationship using linear associations between dependent and independent variables of the model whereas the Multivariate Linear Regression considers the relationship between more than one independent variable and a dependent variable.

As mentioned in [6, 9] predicting popularity can be defined as a regression task in which the aim is an accurate prediction of the view count V at time T_t, while the initial T_d number of days after publication ($T_d < T_t$) is given.

We first discuss the state-of-art methods namely Univariate Linear Regression (ULR), Multivariate Linear Regression(MLR) and Multivariate Radial Basis Function(MRBF) Regression and then describe our proposed method called geographic-economic SVR to predict the popularity.

For Univariate Linear Regression (ULR) Szabo et al. [6] mentioned that there is a high correlation between log-transformed early and late popularity counts. So, mathematically, the future popularity of a content video C_v can be predicted at the time T_t by

$$\tilde{n}(C_v, T_d, T_t) = \exp(\alpha(T_d, T_t).lnn(C_v, T_d)) \quad (1)$$

where exp is the exponential function, $\tilde{n}(C_v, T_d, T_t)$ represents the view count at time T_t for the given video C_v when the prediction is done at time T_d. The weight learnt from the training videos $C_v \in T$ is given as $\alpha(T_d, T_t)$ and it is computed using ordinary least squares method. $n(C_v, T_d)$ is the number of views gathered at time T_d.

For Multivariate Linear Regression(MLR) the univariate model can be extended by including views accrued by the video before T_r. Pinto et al. [9] worked on a Multivariate Linear Regression model which uses feature values' linear combination to predict popularity of video C_v at time T_d. Increasing the dimension of input feature vector, the number of views received in the timeline between T_0 when the video is published and time T_d, which is the reference is sampled, and the view count gathered in the duration is used to form a feature vector.

$$\tilde{n}(C_v, T_d, T_t) = \sum_{j=1}^{r} \alpha\left(T_j, T_t\right).\left(n\left(C_v, T_j\right) - n\left(C_v, T_{j-1}\right)\right) \quad (2)$$

where $\alpha\left(T_j, T_t\right)_{j=1}^{r}$ are the model learning parameters and the term $\left(n\left(C_v, T_j\right) - n\left(C_v, T_{j-1}\right)\right)$ refers to the view deltas in ith sampling interval.

The major limitation observed in this model is that the innate variations of the evolution patterns cannot be captured by the weights achieved from the training data. Extending the MLR model [9] proposed to introduce the concept of correlation between the videos on the basis of their evolution patterns.

The Multivariate Radial Basis Function regression uses Radial Basis Function and predicts the number of views using distances to a set of pre-selected training videos $C_V \in C$ and view increments.

$$\tilde{n}(C_v, T_d, T_t) = \sum_{j=1}^{r} \alpha\big(T_j, T_t\big) . \big(n\big(C_v, T_j\big) - n\big(C_v, T_{j-1}\big)\big) + \Sigma W_{C_v} \Phi(V, C_v) \quad (3)$$

In MRBF, Gaussian Radial Basis Function is used as an alternate for measuring likeness between the evolution patterns of a given video and a set of representative videos $C_v \in C$ from a data set used for training. Here we select the representative samples from a uniformly distributed random set of values. The Gaussian RBF function computes the distance (to be used in the above equation) between the video input and the sample set and evaluates RBF features' term of the formula.

Popularity-SVR as mentioned in [20] uses Support Vector Regression (SVR) [26] with the feature vectors mapped into a non-linear space using Gaussian Radial Basis kernel function for popularity prediction. Using this method results in easy pattern capture and a powerful prediction method. In this method, the prediction of popularity of a video V can be done as

$$\tilde{n}(C_v, T_d, T_t) = \sum_{j=1}^{r} \alpha_j . \Phi(X(C_v, T_d), X(j, T_d)) + c \quad (4)$$

where $\Phi(a, b) = \exp\Big(-\frac{||a-b||^2}{2\sigma^2}\Big)$ is a Gaussian Radial Basis Function with σ parameter, $X(C_v, T_d)$ is a feature vector for video C_v available at time T_d and $X(k, T_r)_{k=1}^{K}$ is a set representing support vectors values given by the Support Vector Regression algorithm along with the coefficients $\alpha_{k\,k=1}^{k=K}$ and intercept c. Normally, we use the feature vector as proposed in [9].

$$X(C_v, T_d) = \Big\{\tilde{n}\big(C_v, T_j\big) - \tilde{n}\big(C_v, T_j - 1\big)_{j=1}^{d}\Big\} \quad (5)$$

All the above-mentioned methods have their own set of pros and cons. However, none of them is found to include the Geographic-economic conditions of the area from which the data is being created or uploaded. This is found to have a significant impact on the popularity of a video. For instance, a video that is being uploaded from some developed country says, any European nation is likely to have a higher popularity rate in comparison to that being uploaded from some third world country. Moreover, for any video to be viral, the duration for which the user's attention is retained is also vital. As mentioned by Bielski et al. [23] the role of Click Through Rate (i.e. the count of clicks on titles and thumbnails per number of times the titles and thumbnails are shown) and the duration of viewer engagement (approximately 6–8 min) should be factored in while making a decision about how popular any video can become. Any content which is clicked might not be appreciated, and any content that is watched for just a few seconds or minutes is also less probably liked, which is evident from the trends of viewers moving from one video to another after a few moments' lapses. So we define the term attention period—which is measured by the time duration for which content engages a user's attention in a particular video. By making a framework that incorporates geographic-economic features and the attention period feature in the existing state-of-art methods as mentioned above, we are likely to get a better model for predictive forecasting of popularity of the content.

4 Conclusion and Future Scope

In our paper, we discussed the state-of-art techniques for predictive analysis of popularity of video content. Though we have conducted a detailed literature survey from the sources such as IEEE Transactions, Springer, ACM, etc. but there might be a possibility that the literature review is not exhaustive. Keeping in view the study of the literature carried out in this paper, we identified that the factors like geographic-economic conditions and attention period is not considered in any of the techniques considered so far. To understand the social dynamics of the user network in a better way and the impact of the above-mentioned factors in the prediction of the popularity of a user created video, we propose to enhance the cues in the traditional methods and design a new framework that can outperform the existing techniques by providing better results.

References

1. YouTube, Press statistics, 2019, Accessed on: July 10, 2019. (Online) Available: https://www.youtube.com/yt/press/statistics.html
2. Facebook, Company info, 2019, Accessed on: July 10, 2019 (Online). Available: http://newsroom.fb.com/company-info/
3. Adage.com, Facebook 85 users creating content, 2015, Accessed on: Oct. 6, 2015, (Online). Available: http://adage.com/article/ digital/facebook-85-users-creating-content/236358/
4. Instagram, Our Story. 2019, Accessed on: July 10, 2019. (Online). Available: https://instagram-press.com/our-story/
5. Twitter, Company info. 2019, Accessed on: Oct. 6, 2019. (Online). Available: https://about.twitter.com/company
6. G. Szabo, B.A. Huberman, Predicting the popularity of online content. Commun. ACM **53**(8), 80–88 (2010)
7. Y. Borghol et al., Characterizing and modelling popularity of user—generated videos. Performance Eval. **68**(11), 1037–1055 (2011)
8. R. Bandari, S. Asur, B. A. Huberman, The pulse of news in social media: forecasting popularity. CoRR, 2012. (Online). Available: http://arxiv.org/abs/1202.0332
9. H. Pinto, J.M. Almeida, M.A. Gonc¸alves, Using early view patterns to predict the popularity of YouTube videos. in *Proceedings of the sixth ACM international conference on Web search and data mining*, 2013, pp. 365–374
10. A. Khosla, A.D. Sarma, and R. Hamid, What makes an image popular? in *Proceedings of the 23rd international conference on World wide web*, 2014, pp. 867–876

11. J. Xu, M. van der Schaar, J. Liu, H. Li, Forecasting popularity of videos using social media. CoRR, 2014. (Online). Available: http://arxiv.org/abs/1403.5603
12. M. Osborne, V. Lavrenko, Rt to win! predicting message propagation in twitter. in *Fifth International AAAI Conference on Weblogs and Social Media*, 2011, pp. 586– 589
13. F. Gelli, T. Uricchio, M. Bertini, A.D. Bimbo, S.-F. Chang, Image popularity prediction in social media using sentiment and context features. in *Proceedings of the 23rd ACM international conference on Multimedia*, 2015, pp. 907–910
14. B. Wu, W.-H. Cheng, Y. Zhang, T. Mei, Time matters: Multi-scale temporalization of social media popularity. in *Proceedings of the 24th ACM international conference on Multimedia*, 2016, pp. 1336–1344
15. B. Wu, T. Mei, W.-H.C. Y. Zhang, Unfolding temporal dynamics: Predicting social media popularity using multi-scale temporal decomposition. in *Proceedings of 30th AAAI Conference on Artificial Intelligence*, 2016, pp. 272–278
16. V. Almeida, A. Bestavros, M. Crovella, A. de Oliveira, Characterizing reference locality in the WWW. Proc. Conf. Parallel Distrib. Inform. Syst., 1996, pp. 92–103
17. M. Chesire, A. Wolman, G.M. Voelker, H.M. Levy, Measurement and analysis of a streaming-media workload. Pro. USENIX Symp. Internet Technol. Syst., 2001, pp. 1–12
18. A. Tatar, M.D. de Amorim, S. Fdida, P. Antoniadis, A survey on predicting the popularity of web content. J. Internet Serv. Appl. **5**, 1–20 (2014)
19. X. Wang, B. Fang, H. Zhang, S. Su, Predicting the popularity of online content based on weak ties theory. in *IEEE Conference on Data Science in Cyberspace (DSC)*, Gaungzhou, 2018, pp. 386–391
20. T. Trzcinski, P. Rokita, Predicting popularity of online videos using support vector regression. IEEE Trans. Multimedia **19**(11), 2561–2570 (2017)
21. C. Castillo, M. El-Haddad, J. Pfeffer, M. Stempeck, Characterizing the life cycle of online news stories using social media reactions. in *Proceeding of the ACM Conference Computer Supported Cooperative Work Social Computing*, 2014, pp. 211–223
22. S.D. Roy, T. Mei, W. Zeng, S. Li, Towards cross-domain learning for social video popularity prediction. IEEE Trans. Multimedia **15**(6), 1255–1267 (2013)
23. A. Bielski, T. Trzcinski, Pay Attention to virality: understanding popularity of social media videos with the attention mechanism. in *2018 IEEE/CVF Conference on Computer Vision and Pattern Recognition Workshops* pp. 2398–2400, 2018
24. Z. Tan, Y. Zhang, Predicting the Top-N popular videos via a cross-domain hybrid model. IEEE Trans. Multimedia (2018)
25. A.L. Jia, S. Shen, D. Li, S. Chen, Predicting the implicit and explicit video popularity in a user generated content site with enhanced social features. Comput. Netw. **140**(20), 112–125 (2018)
26. H. Drucker, C.J.C. Burges, L. Kaufman, A.J. Smola, V. Vapnik, Support vector regression machines. Proc. Neural Inform. Process. Syst. 1996, pp. 155–161

Medical Assistance Using Drones for Remote Areas

Vanita Jain and Nalin Luthra

Abstract In this paper, we have proposed a system of drones that can transfer blood and other medical supplies from a designated medical facility to remote areas/villages where road travel can be expensive and time taking. The novel system of drones works on autonomous navigation using PX4 flight stack on Pixhawk and uses GPS to be aware of its location in x-y axis whereas sharp sensors along with ultrasonic sensors are used to understand the environment. Proposed system has been developed on Robot Operating System (ROS) using its modular approach. Analyzed environment is used to control altitude, speed and direction of a drone when an obstruction such as trees, building or any other object is detected using sharp sensors.

Keywords ROS · Computer vision · Drone

1 Introduction

Drone usage in different fields such as photography, mapping, policing [1], survey [2], and monitoring agriculture [3], forest or land and water resources [4]. It saves farmers the time, yield better land management and therefore helps them monitor their land [4]. Drones have also been used to capture footage of illegal hunting or poaching in restricted areas that are not easy to cover. Assessing traffic congestion and police supervisory for routine security sweeps are also wide applications witnessed [5]. Relatively new emerging use of drones in construction [6].

Drones in commercial industries have been emerging at a fast pace [7]. Commercial drones are designed and built by individuals, organizations, and businesses to be used for commercial purposes. Most of the drones built for commercial industries are built using cheap material and easily available components which are suitable for payload for 4 lb [8]. The emergence has mostly of drones has been mostly supported

---

V. Jain (✉) · N. Luthra
Bharati Vidyapeeth's College of Engineering, New Delhi, India
e-mail: vanita.jain@bharatividyapeeth.edu

P. Bansal et al. (eds.), *Proceedings of International Conference on Artificial Intelligence and Applications*, Advances in Intelligent Systems and Computing 1164,
https://doi.org/10.1007/978-981-15-4992-2_44

by individuals and businesses for profit without being subjected to scrutiny, usually due to low awareness and no public interest.

Population and limited land resources have changed our way of living from big houses to multi-story skyscrapers or bullock carts to sports cars. We are looking at huge traffic jams and long waiting queues for basic necessities such as medical services. Spending at least 40 days on road per year stuck in a traffic jam and we are bound to contemplate that these figures will increasingly proliferate [9].

Emergency medical requirement of medicine in locations which lack the basic infrastructure such as road makes it practically impossible for the government to provide medical assistance and medicines. So, in such a scenario drone specifically quadcopter [10] can be used for medicines disposal.

So, the leading companies in the field of health care often ponder over the fact to utilize UAVs in order to minimize the distance for providing aid to expand their coverage of support and thereby study other different strategies [11]. Hence, there is a need for an idea that can tackle these circumstances with ease.

Drones privacy [12, 13] is also a major concern in different nations regarding photography. Usage of camera on drones is causing nations to make protocol to restrict its free usage [14]. Our proposed idea does not use camera for path planning hence privacy concerns in the proposed idea have been dealt with easily.

When a distress signal is received for medical assistance, a drone (quadcopter) is loaded with the supplies such as blood, medicines or vaccines and then is set off for its journey. In the case of location management and to make sure that the drone reaches the correct destination, the use of Global Positioning System (GPS) is done to ensure efficient navigation. The drone can be handled using software or mobile app for navigation and tracking.

This paper is organized in 5 sections. Section 2 explains the system architecture and design. Section 3 describes the functioning of the system with a brief explanation of path planning, server and path analysis. Sections 4 and 5 concludes while discussing future scope of the proposed system.

2 System Architecure

System Architecture is divided into four different modules consists of ground station, server, client and vehicle as shown in Fig. 1. Ground station is further divided into ROS enabled computer, mavlink data transmitter module and data/coordinate receiver. Server is attached to ground station and acts as a bridge between all three modules for data transmission which uses mavlink protocol along with TCP/IP. Client's user interface consists of web app/android. Finally, vehicle consists of onboard computer along with companion computer and different sensors.

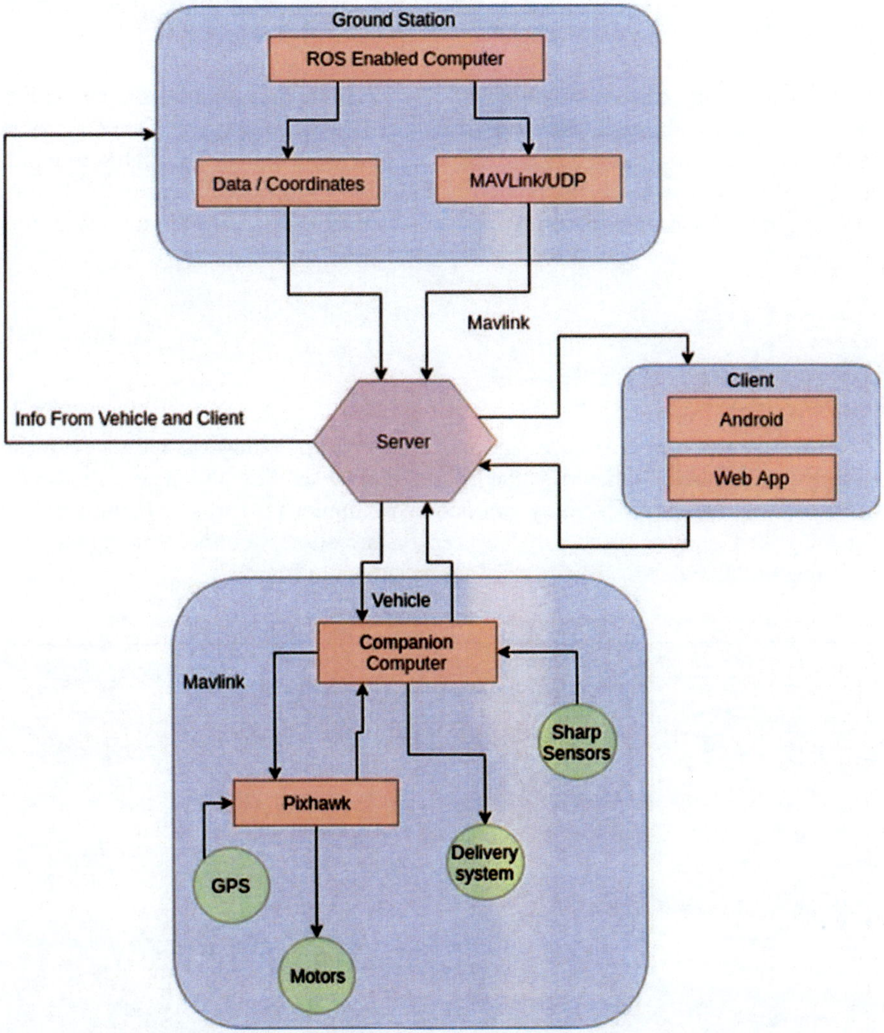

Fig. 1 System architecture

2.1 Onboard Computer

Onboard companion computer is used for communication with Pixhawk to send commands via mavlink protocol for controlling the drone's movements and receive feedback from GPS of current coordinates for path planning. All computations are performed on companion computers such as path planning, sending feedback of location, coordinates for dropping packages and dynamic interacting with base station for live location of the drone.

2.2 Sensors

We are using sharp sensors and ultrasonic sensors both, depending upon the ambient light conditions with respect to time. During the day, sunlight prevails hence sharp sensors operating with infrared rays technology measure distance but at times of peak sunlight the sun rays make sharp sensors unusable so ultrasonic sensors will play a vital role. These sensors will analyze the environment and fetch it to the companion computer for dynamic path planning.

2.3 Setup

We are using raspberry pi as a companion computer with Pixhawk as a flight controller and PX4 flight stack. Pixhawk is attached to GPS for sending global coordinates and these coordinates are received by companion computer for further planning of path and integrated analysis. Whereas sharp sensors are directly connected to companion computers to give details of surroundings as shown in Fig. 2.

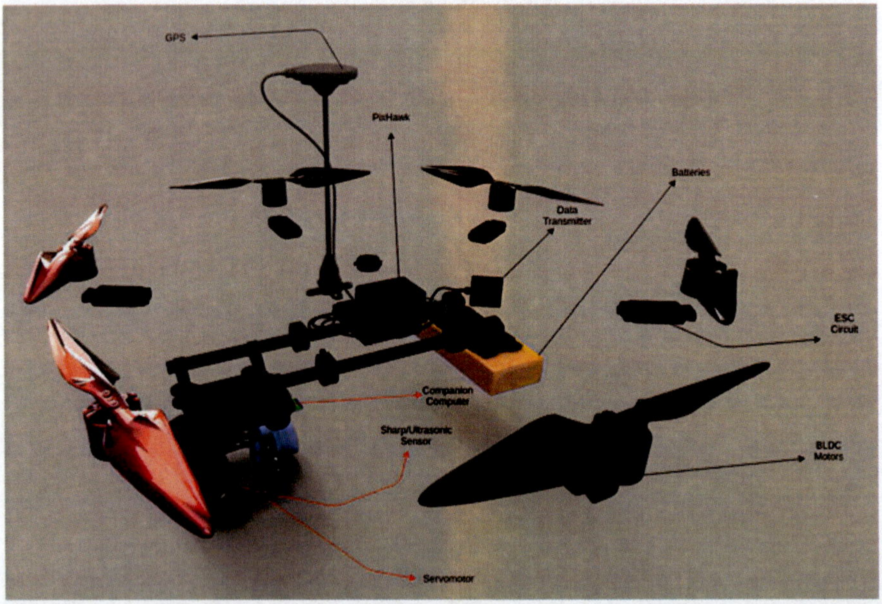

Fig. 2 Drone intersection

3 Functioning of the System

The medical practitioner is provided with a mobile application of our system. Whenever he/she is in need of certain medical equipment/medicines immediately in an area that is not easily accessible then he can send a request for drone to the remote server via the mobile application. The application will send the geographic coordinates and the list of requirements to the server which notifies the operator who loads the required supplies in the container of the drone.

The destination address is then fed on the onboard companion computer. The drone then takes off and heads towards the destination address. It scans its immediate surrounding using sharp sensors/ultrasonic sensors and plan its path as described further in this section. The drone repeatedly sends its location, sensor data and all the required onboard computer logs to the remote server. The location and estimated remaining time are also sent to the mobile application being used by the medical practitioner. When the drone reaches its destination the mobile application notifies the medical practitioner who can then collect the medical supplies.

After this, the drone notifies the remote server that the return journey has started. The in-flight logs saved on the server are then analyzed to retrieve the necessary information which can be used in the return journey or else when the same path is to be traveled again by the drone. After this, the completion of the task is notified to the remote server. In case of any accident, this data can be used to analyze the conditions in which the accident took place so as to give back the vital information required to make the system safer.

3.1 Path Planning

Shortest path from point A to B is directly flying in a straight line without taking any turns, we are using a greedy algorithm to plan a path in our system. According to greedy algorithm locally optimal solution is selected for global optimal solution, the same approach is used to our path planning system. As shown in Fig. 3 direct path is selected for traveling from A to B, red color line is a general traveling path by any automobile whereas a drone can directly travel in a straight line from A to B which is represented in green.

The main problem with straight line traveling is related to an obstacle which is encountered by drone during its journey. Many buildings or trees act as an obstacle for drones to travel from point A to point B. To avoid these obstacles sharp/ultrasonic sensors plays an important role these sensors are attached to servo motor as shown in Fig. 4. These servo motors are controlled by companion computers connected directly to sense the environment.

As shown in Fig. 5, the following path is planned, green path represents the ideal case, whereas red path represents current path planned by drone. Path shown in orange will be dynamically generated by a drone when it reaches a certain range.

Fig. 3 Map tracing

Fig. 4 Sharp sensor attached with servo

3.2 *Application and Remote Server*

Complete system consists of two endpoints, server, and client. Client is handled by a doctor whereas the server is completely autonomous. Server handles requests sent by a client which consists of medical supplies such as medicines, blood or injections as per requirement.

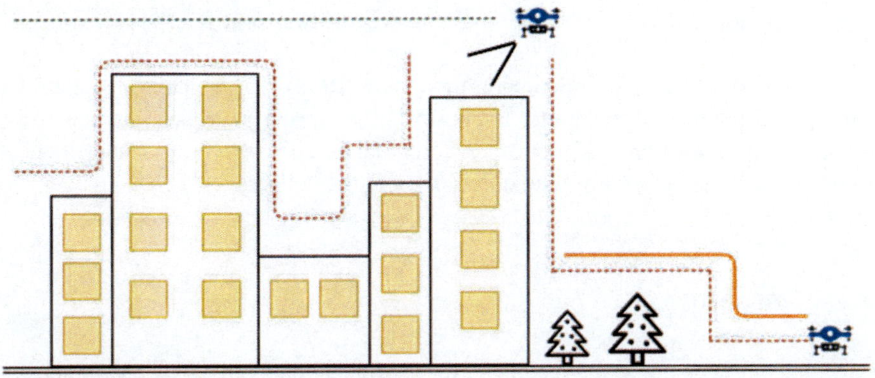

Fig. 5 Path analysis

3.2.1 Client Application

Client application is controlled by a doctor or medical person. He/she place an order on the application which directly sent to remote server. Remote server informs the concerned person about the placed order which is loaded to drone for transmission. Figure 6 shows graphic user interface of the client-side application with all the features.

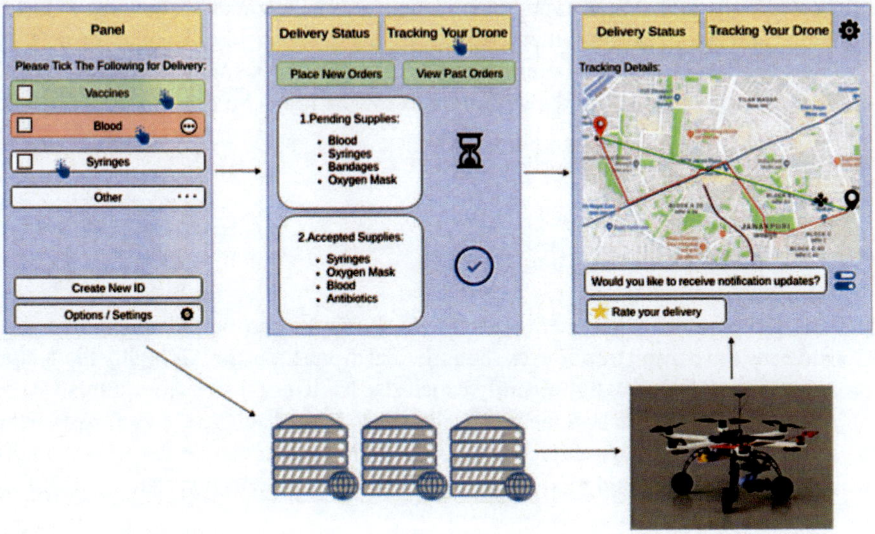

Fig. 6 Client GUI architecture

3.2.2 Remote Server

Remote server is based robotic operating system (ROS) [15] which is also the main software in companion computer to control drone during path planning and delivery. Different nodes on ROS control different tasks from inventory management to informing drones about dropping coordinates of the delivery.

3.3 Path Analysis

Path analysis depends on the reading of the sensors, as data is received by the sensors it is transmitted to the companion computer. As shown in Fig. 5, the red path is how path is planned for first time and green path is the most ideal condition. During multiple journeys to the same location, system will generate bulk load of data which will allow the system to memorize the path which later will be used to find the most ideal path.

4 Conclusions

The proposed interconnected system in this paper is capable of responding to remote medical supply or medical emergency using a single medical application is faster than any other conventional method in place right now without violating laws of privacy. The novel proposed architecture of the system is simple to design, while computation power used by the system is as low as it can be directly implemented on raspberry pi.

5 Future Scope

In future geological parameters such as no-fly zones can be taken into account. Considering the current trends in the commercial drone industry it is highly likely that drone traffic will increase significantly which also has to be taken into account to avoid object collision. This can be done in accordance with local aviation laws. Considering these factors a protean, multifaceted and versatile system can be developed which would be more robust and modular than the currently proposed system.

References

1. L. Marin, K. Krajčíková, Deploying drones in policing southern European borders: constraints and challenges for data protection and human rights, *Drones and Unmanned Aerial Systems* (Springer, Cham, 2016), pp. 101–127
2. L.P. Koh, S.A. Wich, Dawn of drone ecology: low-cost autonomous aerial vehicles for conservation. Trop. Conserv. Sci. **5**(2), 121–132 (2012)
3. P. Tripicchio, M. Satler, G. Dabisias, E. Ruffaldi, C.A. Avizzano, Towards smart farming and sustainable agriculture with drones. in *2015 International Conference on Intelligent Environments*, 2015, pp. 140–143. IEEE
4. A. Abdelgader, *Drones and its Application in Agriculture*, Drones in agriculture (2015). https://doi.org/10.13140/rg.2.2.20397.90088
5. G, Salvoa, L. Carusoa, A. Scordoa, Urban traffic analysis through an UAV, University of Palermo—Department DICAM, Viale Delle Scienze ed 8, Palermo, 90128, Italy
6. I. Zaychenko, A. Smirnova, A. Borremans, Digital transformation: the case of the application of drones in construction. in *MATEC Web of Conferences* vol. 193. EDP Sciences (2018)
7. B. Rao, A.G. Gopi, R. Maione, The societal impact of commercial drones. Technol. Soc. **45**, 83–90 (2016)
8. R. Austin, Unmanned aircraft systems: UAVS design, development and deployment vol. 54. (Wiley, 2011)
9. M.S. Wara, S.K. Thakur, S.I. kar, Department of EIE, Application of Quadcopter For Emergency Medical Assistance 2018. In. J. PureAppl. Math. (2018)
10. A. Choi-Fitzpatrick, D. Chavarria, E. Cychosz, J.P. Dingens, M. Duffey, K. Koebel, S. Siriphanh, M. Yurika Tulen, H. Watanabe, T. Juskauskas, J. Holland, L. Almquist, *Up in the AIR: A Global Estimate of Non-Violent Drone Use 2009–2015*, ed by B. Joan, Kroc School of Peace Studies at Digital@USanDiego, University of SanDiego, 2016
11. K.R. Dayananda, R. Gomes, J. Straub, An interconnected architecture for an emergency medical response unmanned aerial system. in *2017 IEEE/AIAA 36th Digital Avionics Systems Conference (DASC)* (2017), pp. 1–6. IEEE
12. V. Chang, P. Chundury, M. Chetty, Spiders in the sky: user perceptions of drones, privacy, and security. in *Proceedings of the 2017 CHI Conference on Human Factors in Computing Systems* pp. 6765–6776 (2017)
13. P. Finn, Domestic use of aerial drones by law enforcement likely to prompt privacy debate. Washington Post, 22 (2011)
14. R. Luppicini, A. So, A technoethical review of commercial drone use in the context of governance, ethics, and privacy. Technol. Soc. **46**, 109–119 (2016)
15. M. Quigley, B. Gerkeyy, K. Conleyy, J. Fausty, T. Footey, J. Leibsz, E. Bergery, R. Wheelery, A. Ng, ROS: an open-source Robot Operating System. in *International Conference on Robotics and Automation, ser. Open-Source Software workshop*, 2009

The Curious Case of Modified Merge Sort

Harsh Sagar Garg, Vanita Jain, and Gopal Chaudhary

Abstract We propose an O(N logN) sorting method by Modified Merge Sort Method, which shows huge potential to sort big data sets, and even better potential to sort big data sets with a smaller range. Six sorting algorithms: Selection Sort, Insertion sort, Merge sort, Quick sort, Bubble Sort and Modified Merge Sort were implemented in C++ programming languages and tested for the random sequence input of length 100, 1000, 4000, 9000 on random numbers and the IPL Dataset.

Keywords Sorting · Merge sort · IPL · Bubble sort · Random numbers · Fast sorting · Radix sort · Bit-masking

1 Introduction

Sorting, as a key operation on data, has attracted intense interests from the beginning of computing [1]. Lots of excellent algorithms have been composed and applied, such as Bubble Sort, Selection Sort, Insertion Sort, etc. However, it's been proven that sorting algorithms based on comparison have a basic requirement of (N log N) comparisons [2], which means the time complexity is at least O(N log N) [3]. The non-comparison sorting algorithms, such as Bucket Sort, Counting Sort and Radix Sort are not limited by the (N log N) boundary and can reach O(N) complexity but these algorithms have very limited uses. Most of the state-of-art sorting algorithms use parallel computing to handle big datasets and have achieved exceptional achievements [4] For example in 2015, FuxiSort [5] developed by Alibaba Group, is a very interesting sorting algorithm. FuxiSort can complete the 100 TB Daytona GraySort benchmark in 377 s on a random non-skewed dataset and 510 s on a skewed dataset, and Indy GraySort benchmark in 329 s. These algorithms are still restricted by the lower boundary complexity of comparison sorting algorithm and time-consuming networking. Radix Sort also works well for big data sets and data sets with limited

H. S. Garg (✉) · V. Jain · G. Chaudhary
Bharati Vidyapeeth's College of Engineering, New Delhi, India
e-mail: harshgarg.work@gmail.com

P. Bansal et al. (eds.), *Proceedings of International Conference on Artificial Intelligence and Applications*, Advances in Intelligent Systems and Computing 1164,
https://doi.org/10.1007/978-981-15-4992-2_45

frequency range, but it slows down if the data has a higher number of bits, which is not the case with Modified Merge Sort.

The idea is to develop a sorting algorithm that uses bitwise operators to improve the complexity of sorting. This will particularly have huge applications in the field of Healthcare and Sports.

2 Related Works

In the literature, there are numerous research works on comparing the performance of sorting algorithms. In [6], a comparative study of some parallel sort algorithms such as merge sort, alternative quick sort, map sort and STL sort was carried out. The aim was to introduce the merge sort algorithm and its ability to sort an n-sized array of elements in O(N log N) time complexity. The approach adopted in developing this algorithm was a divide-and-conquer method. The empirical and theoretical analyses of the merge sort algorithm were presented in addition to its pseudo-codes. Algorithms such as bubble, insertion and selection sort algorithms were tested against the Merge Sort algorithm. After multiple iterations, it was concluded that the other algorithms have quadratic $O(N^2)$ time.

According to [7], a large number of research works have discussed issues related to dedicated and parallel machines [4]; but only a little research has been carried out on performance evaluations of parallel sorting algorithms on the clustered station. Hence, the goal of [7] is to compare some parallel sorting algorithms with overall execution time as a performance parameter. A comparative study of some parallel sort algorithms such as merge sort, map sort, alternative quick sort and STL sort was carried out, the goal was to introduce the merge sort algorithm and its ability to sort an n-sized array of elements in O(N log N) time complexity. The divide-and-conquer approach was used to develop the method. In addition to the pseudo-codes, the empirical and theoretical analyses of the merge sort algorithm were also presented.

The sorting algorithm used in C++ STL is a hybrid of Heap Sort and Randomized Quicksort. Heapsort is a comparison-based sorting algorithm. Heapsort is somewhat based on Selection Sort sorting algorithm, and it uses Heap Data Structure to sort the elements, Heapsort is an in-place algorithm which means that it doesn't use considerable extra space [8].

3 Modified Merge Sort Algorithm

In digital computer programming, a bitwise operation operates on one or more bit patterns or binary numerals at the level of their bits. It is a fast and simple action, directly supported by the process, and is used to manipulate values for comparisons and calculations. Several times faster than multiplication and division operators, whereas they are even faster than the addition operator in most cases. A lot of modern processors perform multiplication and addition operations almost as fast as bitwise operations due to their longer instruction pipeline, bitwise operations use lesser resources and power.

Another source of motivation is the brilliant performance of Randomized Quick Sort in practical usage because of shuffling. Similarly, the use of bitwise operators will improve the complexity of the algorithm in practical usage more than just theoretical usage. To implement the sorting algorithm, we will have to implement a hybrid concept that draws inspiration from three main concepts:

1. Bitwise Operators
2. Breaking array into smaller sub-arrays
3. Merging the sorted sub-arrays.

3.1 Bitwise Operations

The algorithm starts by using a very simple concept that any number when OR with any other number results in a number greater than or equal to the existing number. Similarly if we OR all the array elements, we can say that we have generated the maximum possible element that could have been present in the array, that no individual element in the array will be bigger than this element. We can say this to be the maximum possible element of the array. Once this is done, the next step would be to calculate the rightmost set bit in the maximum possible element, that can simply be done by taking the log of maximum possible element at the base 2. This is the position of the biggest set bit, which will be our deciding factor to split the array into smaller sub-array. The crux of the whole algorithm lies in this part itself. Like in merge sort, we have a fixed point which is the center of the array. In quick sort, we choose pivot element, similarly, in this case, we are using bitwise operators from the start of the algorithm itself.

```
// a[]: array
maxPossibleElement  = 0;
for(inti = 0;i < 1000;i ++){
      maxPossibleElement  =  maxPossibleElement|a[i];
            }
position =  log10(maxPossibleElement)/log10(2);
```

3.2 Merge Function

```
void merge(x[],y[], xcnt, ycnt, a[]){
        i = 0, j = 0;
        while(i < ycnt){
        a[i] = y[i];
        i ++;
        }
        while(j < xcnt){
        a[i] = x[j];
        j ++;i ++;
        }
}
```

3.3 Breaking into Sub-Problems

This is an interesting step, which draws its inspiration from Induction in Mathematics. We have set up a base case, which is like the starting point in mathematics, then we have written a recursive function for the bigger array considering it will surely work for smaller sub-arrays. The computation stops when the size of array becomes zero or one or when the position of rightmost set bit becomes negative, that is when it goes out of bounds, which is the base case. A function Modified Merge Sort is made, which takes the array, the number of elements and the position of rightmost set bit as arguments. Then it checks if the position is less than zero, then it returns else it checks if the number of elements equal to zero or one, then also it returns. Then it creates a variable mask, which is nothing but a number which has all bits zero apart from the bit at the position, now based on this we will break the bigger array into two parts. The first part will contain elements whose bit at position p is set, the second half will contain elements whose bit at position p is unset.

```
modifiedMergeSort(a[], n, place){
         if(place < 0){
         return;
         }
         if(n ==0){
         return;
         }
         mask  =  1 < <place;
         x[10000],y[10000];
         xcnt = 0;
         ycnt = 0;
         for(inti = 0;i < n;i ++){
         if(a[i]&mask){
                x[xcnt] = a[i];
                xcnt ++;
        }
        else{
```

```
                y[ycnt] = a[i];
                ycnt ++;
        }
        }
        modifiedMergeSort(x,xcnt,place-1);
        modifiedMergeSort(y,ycnt,place-1);
        merge(x,y,xcnt,ycnt,a);
        }
```

4 Comparative Study and Discussion

All the six sorting algorithms (Selection Sort, Insertion sort, Merge sort, Quick sort, Bubble Sort and Modified Merge Sort) were implemented in C++ programming languages and tested for the random sequence input of length 100, 1000, 4000, 9000. All the six sorting algorithms were executed on machine Operating System having Intel(R) Core(™) i5-4210U CPU @ 1.70 GHz and installed memory (RAM) 4 Gb. Time is in (Clocks per second)*10^6 (Fig. 1).

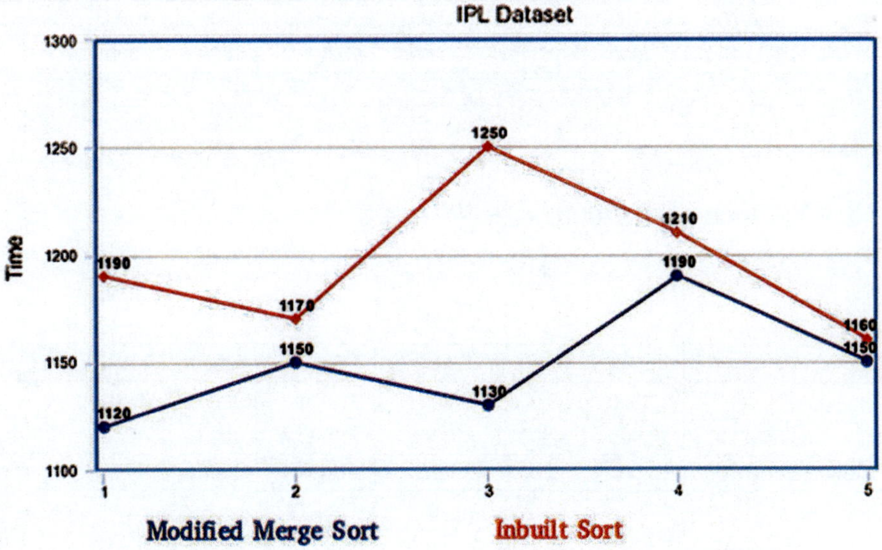

Fig. 1 Comparison chart

4.1 Number of Elements = 100

Input	Bubble	Insertion	Merge	Selection	Inbuilt	Modified merge
Random	5000	100	2000	4000	100	2000
Random mod 10	2000	100	2000	4000	100	1000
Random mod 100	2000	200	1000	3000	100	1000

4.2 Number of Elements = 1000

Input	Bubble	Insertion	Merge	Selection	Inbuilt	Modified merge
Random	13,000	3000	3000	9000	1000	8000
Random mod 10	12,000	2000	3000	9000	1000	1000
Random mod 100	15,000	2000	4000	9000	1000	1000

4.3 Number of Elements = 4000

Input	Bubble	Insertion	Merge	Selection	Inbuilt	Modified merge
Random	13,000	53,000	8000	22,000	2000	10,000
Random mod 10	12,000	42,000	7000	21,000	2000	2000
Random mod 100	15,000	2000	7000	21,000	2000	2000

4.4 Number of Elements = 9000

Input	Bubble	Insertion	Merge	Selection	Inbuilt	Modified merge
Random	$>10^6$	203,000	17,000	$>10^6$	7000	18,000
Random mod 10	$>10^6$	209,000	17,000	$>10^6$	5000	5000
Random mod 100	$>10^6$	202,000	17,000	$>10^6$	4000	3000

4.5 IPL Dataset

The IPL Dataset contains all Indian Premier League Cricket matches between 2008 and 2016. This is the ball-by-ball data of all the IPL cricket matches till season nine. The dataset contains two files: deliveries.csv and matches.csv. matches.csv contains details related to the match such as location, contesting teams, umpires, results, etc. deliveries.csv is the ball-by-ball data of all the IPL matches including data of the batting team, batsman, bowler, non-striker, runs scored, etc. [9].

References

1. P. Adhikari, A.K.K. Suleiman, I.M. AlTurani, I.A. Mahmoud, I. AlTurani, N.I. Zanoon, Review on sorting algorithms, "a comparative study on two sorting algorithm", Mississppi state university. Int. J. Comput. Sci. Secur. (IJCSS), **7**(3), 125 (2013)
2. M. Goodrich, R. Tamassia, *Data Structures and Algorithms in Java*, 4th edn, (Wiley 2010), pp. 241–243
3. T. Cormen, C. Leiserson, R. Rivest, C. Stein, *Introduction to Algorithms*, (McGraw Hill, New York, 2001), pp. 320–330
4. J. Kennedy, R.C. Eberhart, Particle swarm optimization. in *IEEE International Conference on Neural Networks*, (Piscataway, NJ, 1995), pp. 942–1948
5. http://corewar.co.uk/assembly/insertion.html
6. R. Sedgewick, K. Wayne, *Algorithms, Pearson Education*, 4th edn, (2011), pp. 248–249
7. H. Deitel, P. Deitel, *C++ How to Program*, (Prentice Hall, 2001), pp. 150–170
8. T. Cormen, C. Leiserson, R. Rivest, C. Stein, *Introduction to algorithms*, Third edn, (McGraw-Hill, New York, 2009), pp. 15–17
9. https://www.kaggle.com/manasgarg/ipl

Effect of Activation Functions on Deep Learning Algorithms Performance for IMDB Movie Review Analysis

Achin Jain and Vanita Jain

Abstract Huge amount of data is generated every moment over the Internet on various platforms such as social networking sites, blogs, customer reviews on various sites where individuals express their views or thoughts about different subjects. Users' sentiments expressed over the Web influence the readers, product vendors, and politicians greatly. This unstructured form of data needs to be analyzed and converted into a well-structured form and for this purpose, we require Sentiment Analysis. Sentiment Analysis is the process of contextual mining of text that is used to identify and extract the expressed mindset or feelings in different manners such as negative, positive, favorable, unfavorable, thumbs up, thumbs down, etc. In this paper, we have used Convolutional Neural Network (CNN), Long Short-Term Memory (LSTM) and a hybrid approach of CNN and LSTM to perform sentiment classification of IMDB Movie Review dataset. We have applied the trained model on the dataset using various activation functions and compared the accuracy achieved. Maximum accuracy (88.35%) is achieved with CNN with ReLU Activation Function whereas minimum accuracy (48.19%) is achieved with LSTM when used with Linear Activation Function.

Keywords Sentiment analysis · Deep learning · Activation function · Long Short-Term memory (LSTM) · Convolutional neural network (CNN)

A. Jain
University School of Information, Communication and Technology, GGSIPU, Sector 16 C, Dwarka, Delhi, India
e-mail: achin.mails@gmail.com

V. Jain (✉)
Bharati Vidyapeeth's College of Engineering, Paschim Vihar, New Delhi, India
e-mail: vanita.jain@bharatividyapeeth.edu

P. Bansal et al. (eds.), *Proceedings of International Conference on Artificial Intelligence and Applications*, Advances in Intelligent Systems and Computing 1164,
https://doi.org/10.1007/978-981-15-4992-2_46

1 Introduction

Sentiment Analysis is the process of analyzing thoughts, feelings, expressions, and opinions about a sentence [1]. It is widely used in many application areas such as building brand perceptions, movie box office predictions, general elections, news analysis and many more [2]. In the modern world, everything is moving towards the Internet and companies are trying very hard to understand the sentiment of people about various products. Major e-commerce players like Amazon and Flipkart are also using Sentiment Analysis to study people's likes and dislikes for presenting them customized products. The sentiment for textual data can be broadly classified into positive, negative or neutral. There is a surge in the use of Sentiment Analysis mainly due to the increasing demand for analysis of hidden information coming from social media platforms like Facebook and Twitter [3, 4].

In modern times, social networking platforms such as Facebook, Twitter, etc. play a crucial role in everyone's life. They allow users to express their opinions about any topic without any fear or hesitation. Sentiment Analysis is built entirely upon the features extracted from textual data. Sentiment Analysis involves many fields such as retrieval of information, Natural Language Processing (NLP), Computational Linguistics, Machine Learning and Artificial Intelligence (AI) [5]. There are different types of features such as Uni-Gram, Bi-Gram, Tri-Gram, etc. Classification of Sentiment Analysis can be done on three levels [6]:

- **Aspect Level**: This refers to the features of the expressions or sub-expressions within a sentence [7].
- **Document Level**: This mainly deals with the analysis of sentiments obtained from a complete document or a paragraph [8].
- **Sentence Level**: This aims to obtain the sentiments of a single sentence [9].

Feature Extraction is one of the biggest challenges in Machine Learning. Sentiment Analysis also involves the use of various text features such as POS Tagging [10], Negation Handling [11], etc. Deep Learning is one such technique that is used to overcome this problem. Deep Learning first came into the picture in the year 2006 when G R Hinton proposed this approach [12]. It is a part of the Machine Learning family and involves the use of Neural Networks whose architecture is inspired by the neurons present in the human brain. In this approach, multiple inputs are provided to layers of neurons for processing where the summation of the product of inputs and their associated weights takes place. Once the summation is done, the activation function is used to limit the output value within a threshold limit. In the final step, the difference between actual output and obtained output is calculated and weights are updated accordingly till the desired output is not obtained.

The idea behind the use of Deep Learning for Sentiment Analysis is to train a network to make predictions on unseen data [14]. The main advantage of using this approach is that it can lead to better performance on big data sets. For time series data, this approach is the most beneficial.

The steps to follow for this are:

(a) Initiate the analysis with a labeled dataset.
(b) The documents should be brought to a standard numeric shape:
    (i) Hot encodings and embeddings
    (ii) Zero Padding and truncation
(c) The network structure is defined. Commonly used layers for text are embedding and LSTM layers.
(d) Train and apply the network.

Deep learning is very significant for both unsupervised and supervised learning. Embedding layers are used as it gives clear meaning to each encoding whereas LSTM [13] is mostly used for time series analysis and to predict the next outcome from the previous output.

Rest of the paper is structured as follows. Section 2 presents related work. Section 3 explains the research methodology used along with the CNN and various activation functions used. Section 4 shows the experimental results obtained and Sect. 5 concludes the work done.

2 Related Work

In [15], the authors have used CNN to visualize Sentiments of Images collected from twitter and the results show that Googlenet performs better than AlexNet. The authors in [16] perform the message and phrase-level Sentiment Analysis task on Semeval-2015 dataset using CNN. The experiment includes initialization of parameters weight so that the addition of new features can be avoided. In [17] detailed research is carried out on sentiment analysis related to microblogs. The authors extracted the user's opinions and attitudes about a topic with the help of CNN. The results show that by using CNN need to extract explicit features can be eliminated. In paper [18], the authors represented seven-layer framework to perform sentiment analysis of sentences. The framework depends on CNN and Word2Vec for vector representation. The model was tested on the Movie Review dataset collected from rottentomatoes.com. The experiment results in 45.5% accuracy.

Many researchers have combined approach of CNN and LSTM to achieve better results in Sentiment Analysis. The researchers in [19] combined DCNN and LSTM to perform classification task on Thai Twitter Data and the results show that the proposed approach achieved better accuracy than SVM and Naïve Bayes. The authors in [20] combined Probabilistic Neural Network (PNN) and two-layered Restricted Boltzmann (RBM) to perform classification on Reviews Dataset. The paper shows significant improvement in accuracy in five different datasets.

3 Research Methodology

In this paper, we present Sentiment Analysis using CNN, LSTM and Hybrid CNN+LSTM using various activation functions. Convolutional Neural Network is one of the most popular and widely used Deep Learning Techniques. CNN comprises of stack of neurons arranged in various layers to process input dataset and calculate the output. CNN is like a Deep Neural Network but also consists of additional layers such as Convolution Layer, Pooling Layer and Fully Connected (FC) Layer. Long Short-Term Memory (LSTM) is a form of Artificial Neural Network which is recurrent in nature. LSTM is very commonly used in Deep Learning with feedback connections.

Activation functions are used on CNN and are also known as transfer functions. The purpose of using activation function is to transform the non-linearity of the input dataset. The accuracy of any classification problem is greatly affected by the choice of activation function. Each activation function comes with both advantages and disadvantages as mentioned by the authors in [21]. One such example is the use of ReLU activation function which solves the vanishing gradient problem of the sigmoid activation function. But ReLU also suffers from one problem which is dying ReLU problem and this shortcoming is overcome by LeakyReLUfunction [22, 23]. Table 1 presents the activation functions used in this paper for experimental purposes.

Table 1 Activation function used in deep learning

S. No.	Activation function	Formula	Figure
1.	Linear	$y = ax$	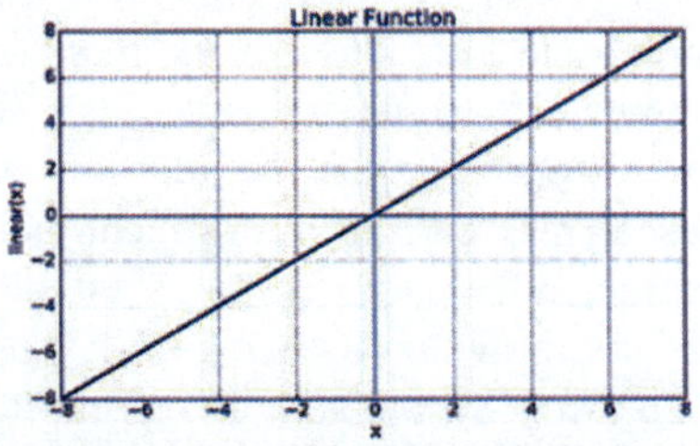
2.	Tanh (hyerbolic tangent)	$f(x) = \tanh(x) = 2/(1 + e^{-2x}) - 1$ OR $\tanh(x) = 2 * \text{sigmoid}(2x) - 1$	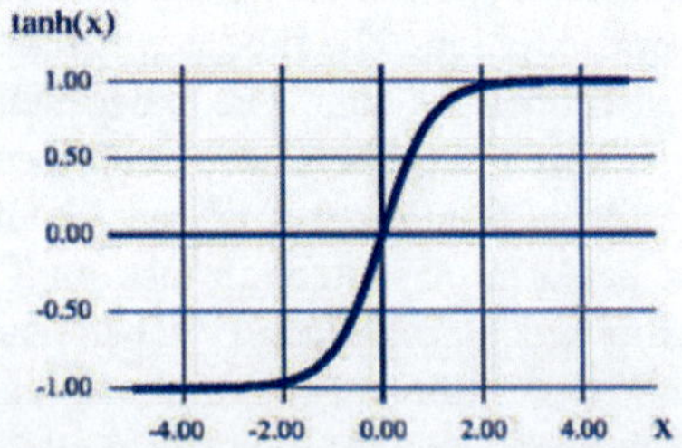

(continued)

Table 1 (continued)

S. No.	Activation function	Formula	Figure
3.	Exponential linear unit (ELU)	$R(x) = \{x\ x > 0\}$ $\{\alpha(e^x - 1)\ x \leq 0\}$	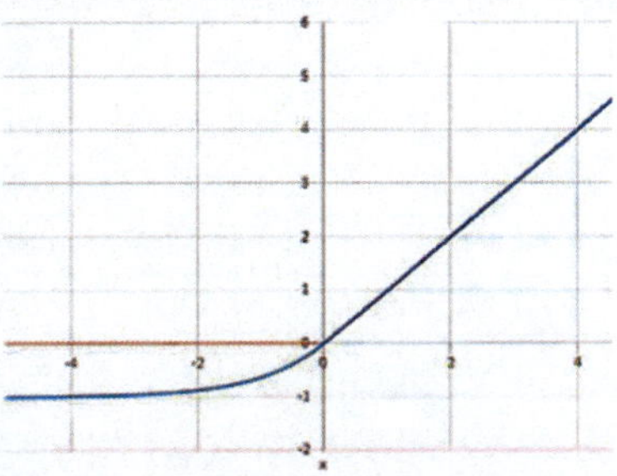
4.	Rectified linear unit (ReLU)	$y = \max(0, z)$	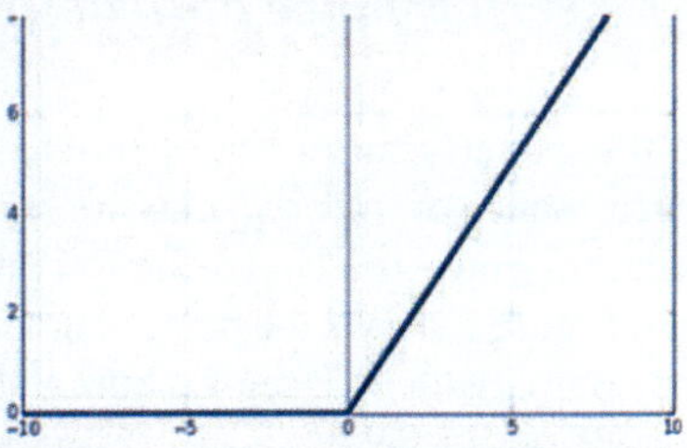
5.	Exponential	$\exp(x) = e^x$	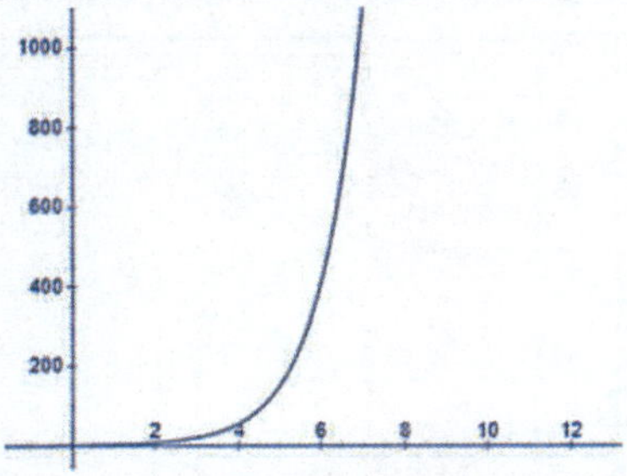

4 Experimental Results

For the experiment, we used Google Collab to run Deep Learning Algorithms on IMDB Movie Review dataset. We used Keras API of Python to implement the experimental work. IMDB Movie Review dataset [24] is present in Keras dataset library and it contains 50,000 rows with equal distribution for training and testing samples. In this paper, three different Deep Learning methods are used with various activation functions to calculate the accuracy of the sentiment classification. Table 2 shows the results obtained in three cases (i) CNN with various activation functions, (ii) LSTM with various activation functions and (iii) Hybrid CNN+LSTM with various activation functions.

Table 2 Accuracy comparison for various activation functions

S. No.	Activation function	CNN		LSTM		CNN+LSTM	
		Accuracy (%)	AUC	Accuracy (%)	AUC	Accuracy (%)	AUC
1	ReLU	88.35	0.95	59.04	0.85	83.22	0.991
2	Tanh	87.81	0.952	60.43	0.856	82.33	0.901
3	ELU	88.12	0.9513	68.51	0.865	81.16	0.903
4	Exponential	87.40	0.9454	50.14	0.8652	82.40	0.9016
5	Linear	88.29	0.954	48.19	0.8597	82.38	0.9057

4.1 Performance Parameters

In this paper, sentiments of movie reviews are computed using classification. The performance of different classifiers with various activation function is analyzed with accuracy and Area Under the Receiver Operating Characteristics curve (AUC) metrics. Figure 1 shows the best accuracy values achieved in three cases and Fig. 2 shows the comparison of Deep Learning algorithm using different activation functions.

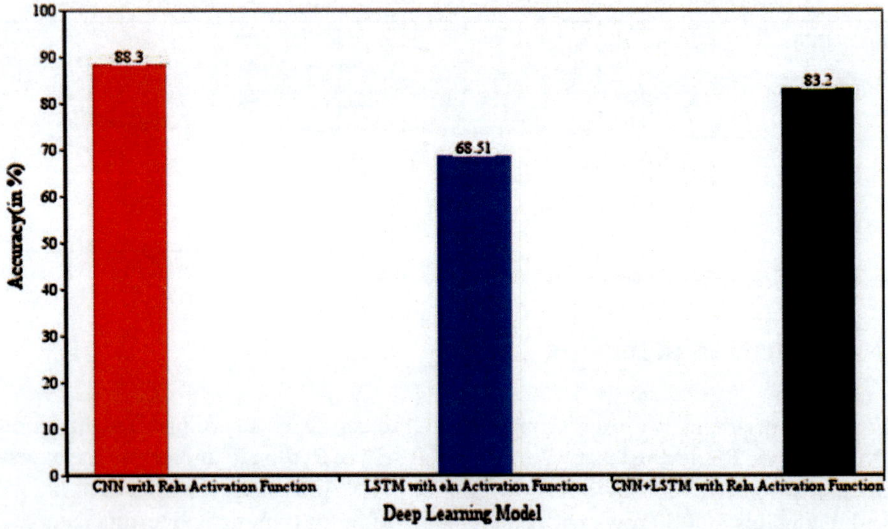

Fig. 1 Accuracy comparison graph

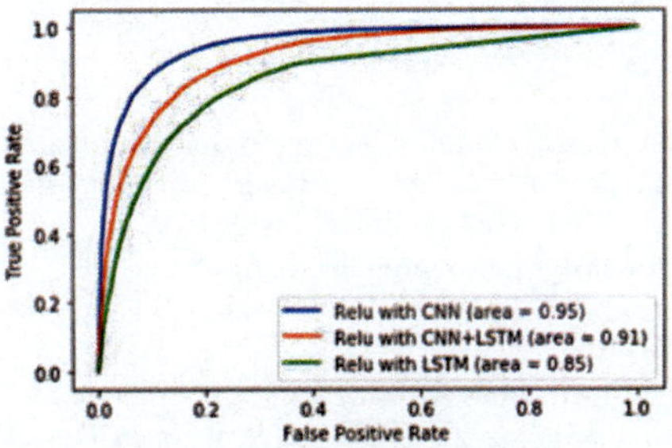

ROC – ReLU Activation Function

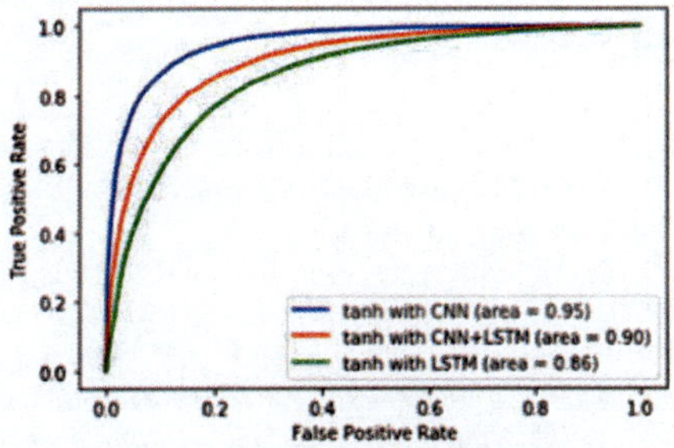

ROC – Tanh Activation Function

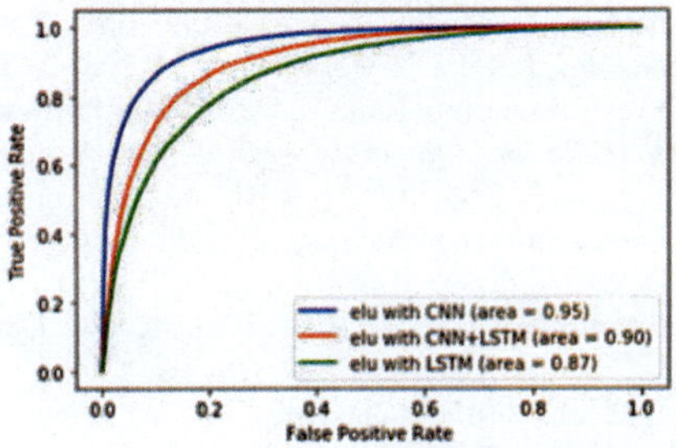

ROC – ELU Activation Function

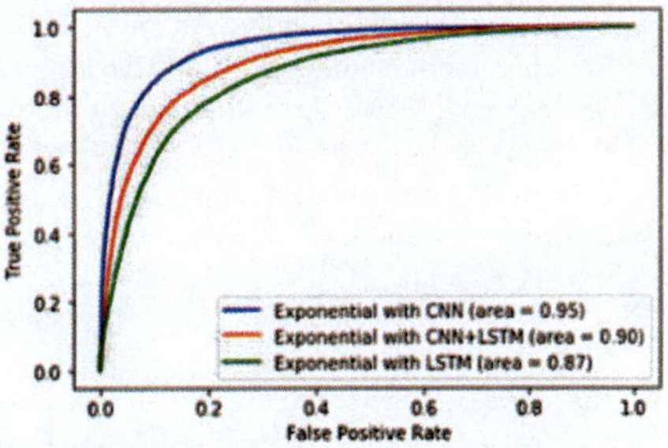

ROC – Exponential Activation Function

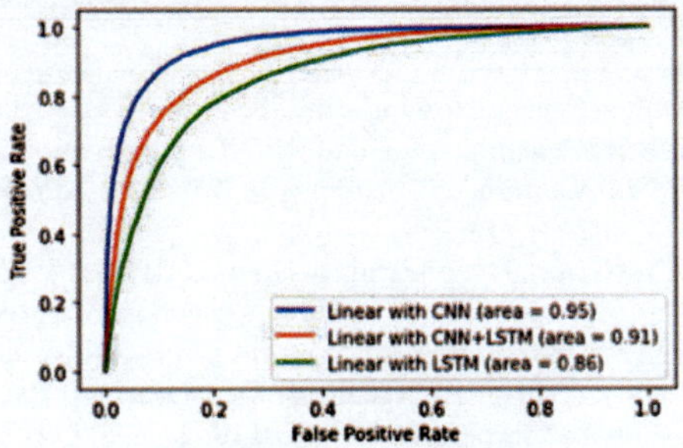

ROC – Linear Activation Function

Fig. 2 ROC comparison for various activation function

5 Conclusion

We carried out the task of sentiment classification on IMDB Movie Review dataset using various Deep Learning algorithms with a variety of activation functions. The dataset is evaluated with three different Deep Learning Models namely CNN, LSTM and hybrid approach of CNN+LSTM. The accuracy of each model is calculated using five different activation function and performance is compared in terms of accuracy and area under the ROC curve. From the results, it is found that CNN and CNN+LSTM provided best accuracy values with ReLU Activation Function and LSTM works best with ELU Activation Function.

References

1. A.V. Yeole, P.V. Chavan, M.C. Nikose, Opinion mining for emotions determination. in *2015 IEEE International Conference on Innovations in Information, Embedded and Communication Systems (ICIIECS)*, (2015)
2. B. Heredia, T.M. Khoshgoftaar, J. Prusa, M. Crawford, Cross domain sentiment analysis: an empirical investigation, in *2016 IEEE 17th International Conference Information Reuse and Integration*, 2016, pp. 160–165
3. F. Luo, C. Li, Z. Cao, Affective-feature-based sentiment analysis using SVM classifier, in *2016 IEEE 20th International Conference on Computer Supported Cooperative Work in Design*, 2016, pp. 276–281
4. M. Haenlein, A.M. Kaplan, An empirical analysis of attitudinal and behavioral reactions toward the abandonment of unprofitable customer relationships. J. Relatsh. Mark. **9**(4), 200–228 (2010)
5. E. Aydogan, M.A. Akcayol, A comprehensive survey for sentiment analysis tasks using machine learning techniques, in *2016 International Symposium on Innovations in Intelligent Systems and Applications*, (2016), p. 17
6. J. Singh, G. Singh, R. Singh, *A review of sentiment analysis techniques for opinionated web text* (CSI Trans, ICT, 2016)
7. K. Schouten, F. Frasincar, Survey on aspect-level sentiment analysis. IEEE Trans. Knowl. Data Eng. **28**(3), 813–830 (2015)
8. R. Moraes, J.F. Valiati, W.P.G. Neto, Document-level sentiment classification: An empirical comparison between SVM and ANN. Expert Syst. Appl. **40**(2), 621–633 (2013)
9. D.M.E.D.M. Hussein, A survey on sentiment analysis challenges. J. King Saud Univ. Eng. Sci. **30**(4), 330–338 (2018)
10. Y. Wang, K. Kim, B. Lee, H.Y. Youn, Word clustering based on POS feature for efficient twitter sentiment analysis. Human-Centric Comput. Inf. Sci. **8**(1), 17 (2018)
11. S. Pandey, S. Sagnika, B.S.P Mishra, A technique to handle negation in sentiment analysis on movie reviews. in *2018 International Conference on Communication and Signal Processing (ICCSP)* 2018, (pp. 0737–0743). IEEE
12. Y. LeCun, Y. Bengio, G. Hinton, Deep learning. Nature **521**(7553), 436–444 (2015)
13. S. Wen, H. Wei, Y. Yang, Z. Guo, Z. Zeng, T. Huang, Y. Chen. Memristive LSTM network for sentiment analysis. in *IEEE Transactions on Systems, Man, and Cybernetics: Systems* (2019)
14. H.H. Do, P.W.C. Prasad, A. Maag, A. Alsadoon, Deep learning for aspect-based sentiment analysis: a comparative review. Expert Syst. Appl. **118**, 272–299 (2019)
15. J. Islam, Y. Zhang, Visual sentiment analysis for social images using transfer learning approach, in *2016 IEEE International Conference on Innovations in Information, Embedded and Communication Systems* (BDCloud), Soc. Comput. Netw. (SocialCom), Sustain. Comput. Commun., 2016 (pp. 124–130)
16. A. Severyn, A. Moschitti, Twitter sentiment analysis with deep convolutional neural networks. in *Proceedings 38th International ACM SIGIR Conference Research Development, Information Retrieval SIGIR*, vol. 15, pp. 959–962 (2015)
17. L. Yanmei, C. Yuda, Research on Chinese micro-blog sentiment analysis based on deep learning, in *2015 8th International Symposium on Computational Intelligence and Design*, 2015 (pp. 358–361)
18. X. Ouyang, P. Zhou, C.H. Li, L. Liu, Sentiment analysis using convolutional neural network, Comput. in *2015 IEEE International Conference on Computer and Information Technology; Ubiquitous Computing and Communications; Dependable, Autonomic and Secure Computing; Pervasive Intelligence and Computing (CIT/IUCC/DASC/PICOM)*, 2015, (pp. 2359–2364)
19. P. Vateekul, T. Koomsubha, A study of sentiment analysis using deep learning techniques on thai twitter data, 2016
20. R. Ghosh, K. Ravi, V. Ravi, A novel deep learning architecture for sentiment classification, in *3rd IEEE International Conference on Recent Advances in Information Technology* (2016) pp. 511–516

21. C. Nwankpa, W. Ijomah, A. Gachagan, S. Marshall, Activation functions: comparison of trends in practice and research for deep learning. arXiv preprint arXiv:1811.03378, pp. 1–20 (Nov 2018)
22. B. Xu, N. Wang, T. Chen, M. Li, Empirical evaluation of rectified activations in convolutional network. arXiv preprint arXiv:1505.00853, pp. 1–5 (May 2015)
23. K. Alrawashdeh, C. Purdy, Fast activation function approach for deep learning based online anomaly intrusion detection. in *International Conference on Big Data Security on Cloud, Omaha*, pp. 5–13 (May 2018)
24. A.L. Maas, R.E. Daly, P.T. Pham, D. Huang, A.Y. Ng, C. Potts, Learning word vectors for sentiment analysis. in *Proceedings of the 49th annual meeting of the association for computational linguistics: human language technologies* vol. 1, Association for Computational Linguistics, 2011 (pp. 142–150)

Human Activity Recognition Using Tri-Axial Angular Velocity

Surinder Kaur, Javalkar Dinesh Kumar, and Gopal

Abstract Human Activity recognition is one of the prime applications of machine learning algorithms nowadays. It is being used in the field of healthcare, game development, smart environments, etc. Data from the sensors connected to a person can be employed to train supervised machine learning models in order to foresee the movement being done by the individual. In this paper, we will utilize data available at UCI machine learning Repository. It contains data generated from the accelerometer, gyroscope and other different sensors of Smartphone to train supervised predictive models using machine learning strategies like KNN, Logistic Regression, etc. to generate a model which can be used to predict the kind of movement being carried out by the individual which is branched into six activities like walking, walking upstairs, walking downstairs, sitting, standing and laying. We will be comparing the accuracy of our model using KNN and Logistic Regression.

Keywords Human activity recognition · KNN · Logistic regression · Accelerometer · Gyroscope

1 Introduction

In the past decade, there has been an exponential increment in the number of smart phone which are highly efficient, easy to use and easily accessible for the users. In the time period of 2012–2014, the smart phone users had increased by more than double from the existing 156 million users to 364 million. With the increased number of advanced mobile phone, the amount of information being developed by the PDA

S. Kaur (✉) · Gopal
Department of Information Technology, Bharati Vidyapeeth's College of Engineering, New Delhi, India
e-mail: pihu2004@gmail.com

J. D. Kumar
Department of Electronics and Communication Engineering, Lingaya's Vidyapeeth, Haryana, India

P. Bansal et al. (eds.), *Proceedings of International Conference on Artificial Intelligence and Applications*, Advances in Intelligent Systems and Computing 1164,
https://doi.org/10.1007/978-981-15-4992-2_47

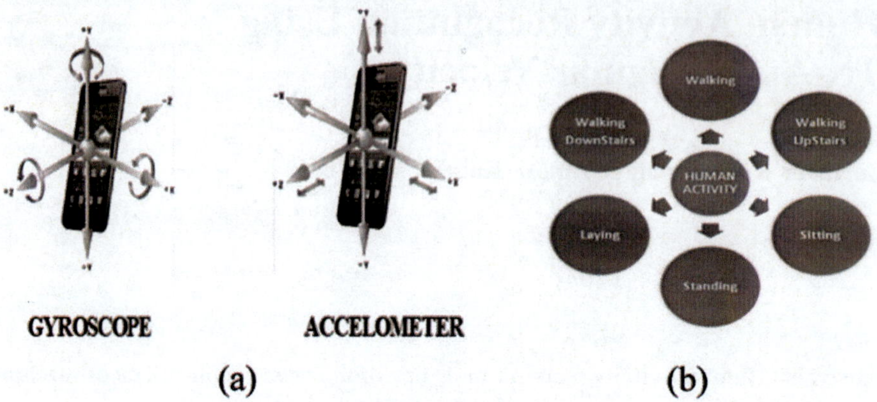

Fig. 1 **a** Gyroscope and accelerometer **b** classification of activities

is becoming more and more useful. The smart phone these days come equipped with sensors such as gyroscope, accelerometer, etc. The readings attained from these sensors of the smartphone depend on the movement of the smartphone [1]. This enables us to track the activity of the person carrying the smart phone using the data of the sensors of their smart phone. Human activity recognition model uses this data to train models such as KNN, Logistic Regression, etc. and then predict the activity being carried out by an individual [2].

The data will be processed through supervised machine learning algorithms to produce predictive classification models results of which will be effectively used to determine the physical activities of an individual under consideration into six categories namely:—"Sitting, Standing, Laying, Walking, Walking upstairs, Walking-Downstairs" [3]. Human activity recognition [6, 7, 8, 9] is becoming a very popular area of research solely with the reason for predicting human activities in a given set of observations along with the adjoining environment. A noteworthy part of this promising application is related closely with the healthcare industry [4]. With the successful implementation of this process, it will be very easy for the doctors to track down the movement of their patients in a non-invasive manner by using easily available devices such as smart phones (Figs. 1) [10, 11, 12, 13, 14].

2 Related Works

A lot of work has been performed in the Human Activity recognition area using the three methods: view-based, wearable sensor-based and smart phone based. The wearable sensor-based method allows users to wear devices containing sensors like accelerometer, gyroscope, GPS, etc. that track the movement and activities of the user. However, here the wearable devices may pose a problem for the user making him/her uncomfortable [5]. On the other hand, the smart phone based approach where user will place the smartphone in his/her pocket and based on the activities performed by the user, the sensors will extract the data and record the activities. The research performed by us uses the smartphone-based method.

Many researchers have performed the work by recognizing various activities. They used various classification algorithms like KNN, Logistic Regression, Decision trees, Support Vector machines, and Neural Networks. However, they achieved accuracy by using an accelerometer and gyroscope. We have considered only tri-axial acceleration and using the values, an accuracy of 92% was achieved in the case of KNN and 96% in the case of Logistic Regression.

Our work turns out to be different from others as apart from removing the redundant and irrelevant features, irrelevant instances have also been eradicated as a result of which our system achieves better accuracy and gains recognition.

3 Dataset and Preprocessing

The dataset used in this paper comes from the UCI machine learning repository. The original dataset was derived by carrying experiments on thirty different individual volunteers within the age of 19–40 years. In the experiments, each volunteer was made to perform the six activities which are decided to be predicted by the machine learning models. During the data acquiring process the volunteer was made to wear a smartphone consisting of sensors on their waist. Using the Accelerometer and Gyroscope present in the smartphone device we acquire and store the values of tri-axial linear acceleration and the tri-axial angular velocity of the movements of the individual under observation. The data acquired from different activities are shown in Figs. 2, 3, 4, 5, 6 and 7.

The dataset thus obtained was reduced into 563 columns and 6106 rows (approx. 60%). Each row of the dataset also contained the Id number between from 1 to 30 (Id number is used to represent the 30 individuals). Also, each row consisted of the cell named activity which was given the name of the activity being carried out by the volunteer. This cell was required for the running of supervised ML models. The dataset thus obtained was stored in csv format for easy reproduction and for convenience in importing while training models in programming language python. The various features present in the data set included—"tBodyAcc-XYZ, tGravityAcc-XYZ, tBodyAccJerk-XYZ, tBodyGyro-XYZ, tBodyGyroJerk-XYZ, tBodyAccMag, tGravityAccMag, tBodyAccJerkMag, tBodyGyroMagetc".

4 Results and Discussion

To assess the achievement of clustered KNN classification for each and every activity. The confusion matrix of clustered KNN is displayed in Table 1. Confusion matrix is a measurement tool normally utilized in administered learning procedures. Every column of the confusion matrix shows the precedents in a forecasted class, while each

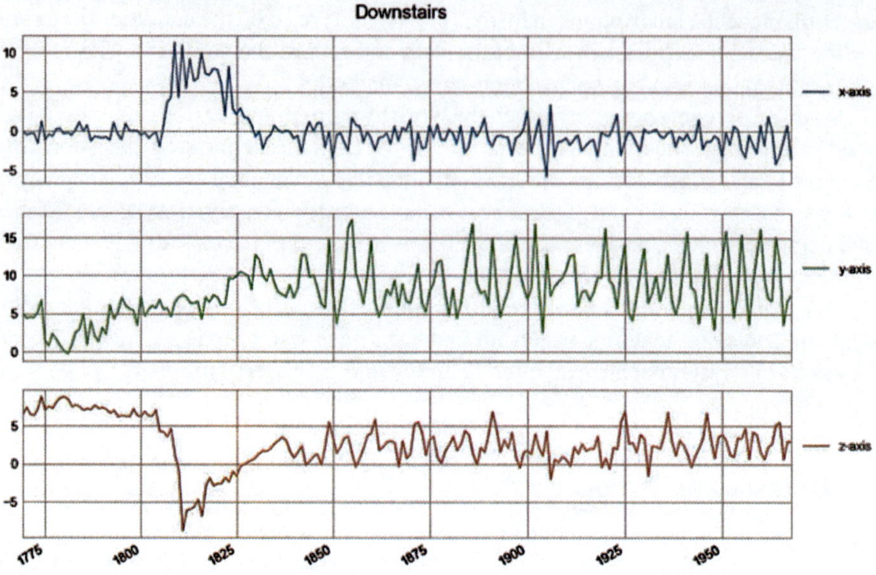

Fig. 2 Tri-axial angular velocity recorded in downstairs activity

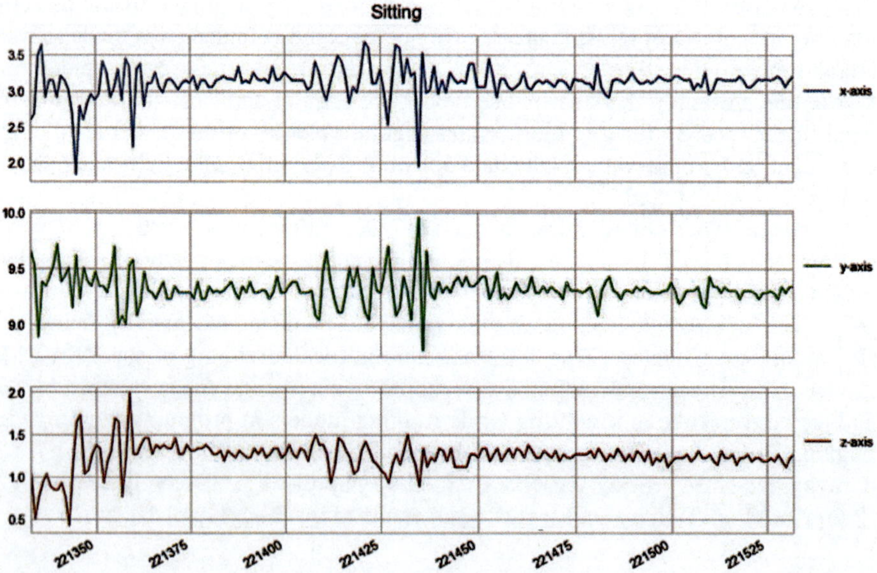

Fig. 3 Tri-axial angular velocity recorded in sitting activity

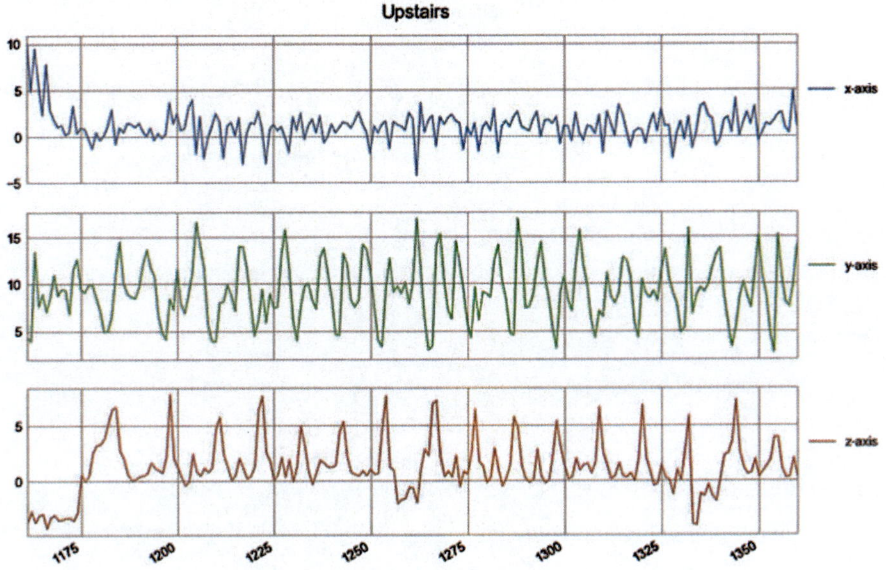

Fig. 4 Tri-axial angular velocity recorded in upstairs activity

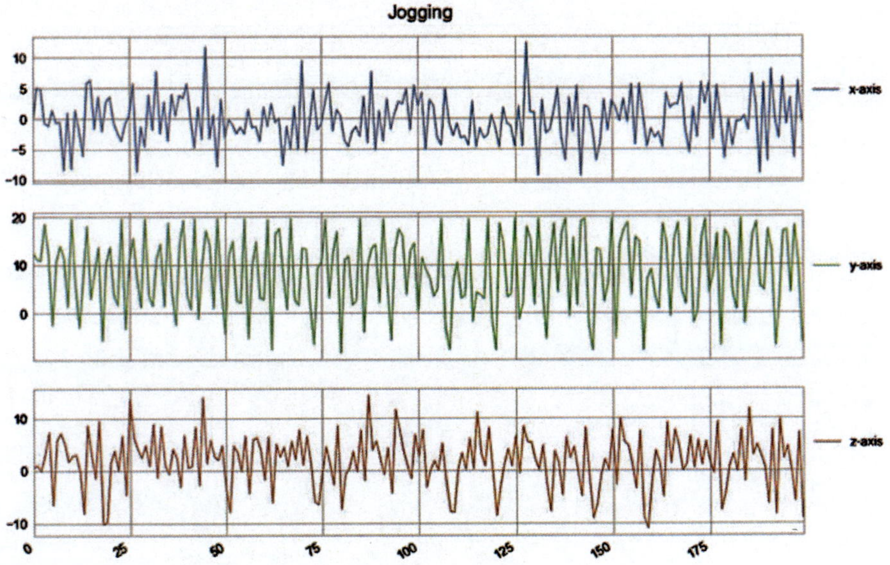

Fig. 5 Tri-axial angular velocity recorded in jogging activity

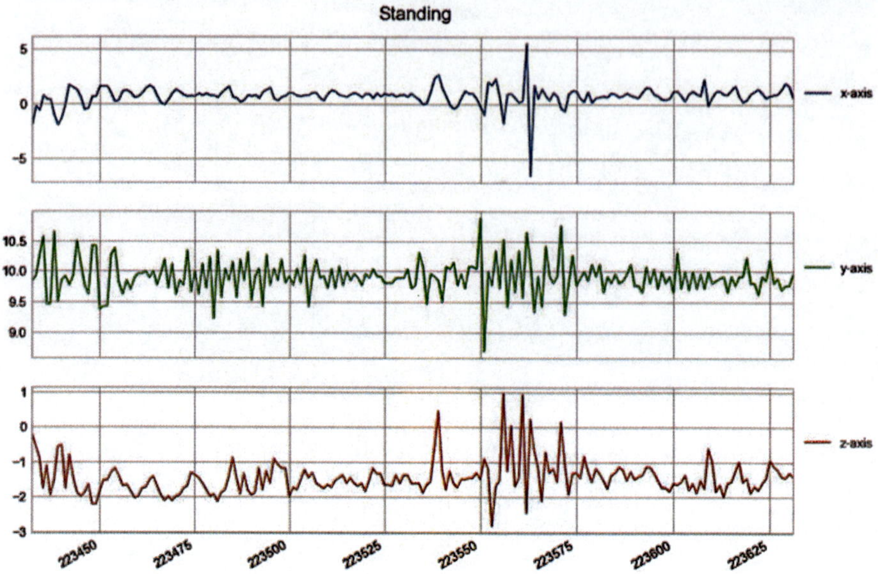

Fig. 6 Tri-axial angular velocity recorded in standing activity

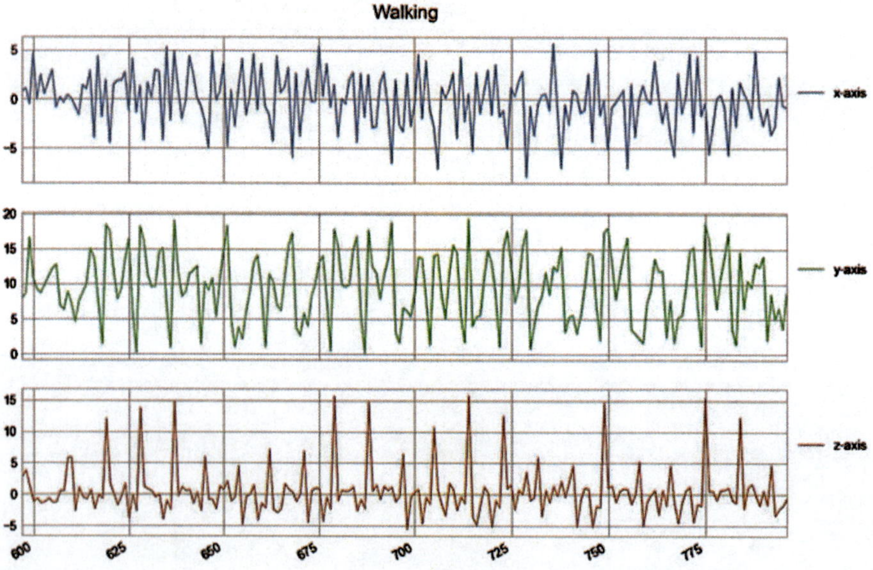

Fig. 7 Tri-axial angular velocity recorded in walking activity

Table 1 Classification Accuracy

Technique used	Accuracy (%)
KNN classifier	91.24
Logistic regression	96

row personifies the precedents in an absolute class. Compared to the performance of activities like running, lying down, standing and sitting, the KNN classifier presents slightly worse performance for walking, where this activity is typically classified as running or standing. The accuracy obtained is shown in Table 1.

Confusion matrix obtained on test data set as shown in Figs. 8 and 9 where rows depict actual action while column represent predicted action. The classification was performed using 6-fold cross-validation utilizing 70% data for training and 30% percent data for testing. In this confusion matrix produced by KNN - walking was predicted with the best accuracy and standing was predicted with lowest accuracy.

In this confusion matrix produced by Logistic Regression—walking was predicted with best accuracy and sitting was predicted with lowest accuracy.

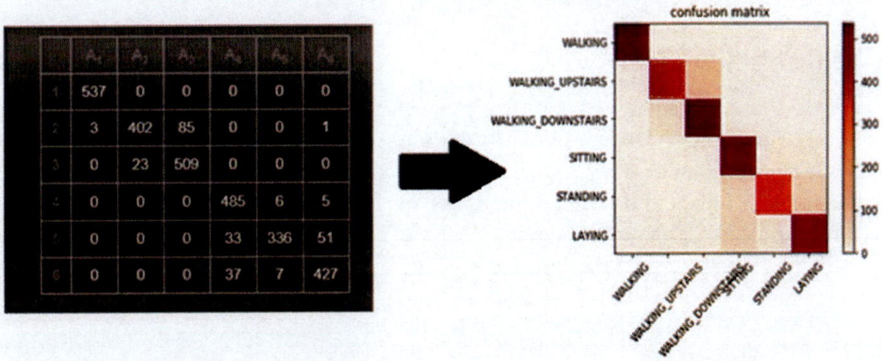

1	537	0	0	0	0	0
2	3	402	85	0	0	1
3	0	23	509	0	0	0
4	0	0	0	485	6	5
5	0	0	0	33	336	51
6	0	0	0	37	7	427

Fig. 8 Confusion matrix of KNN showing classification accuracy evaluated on HAR data set

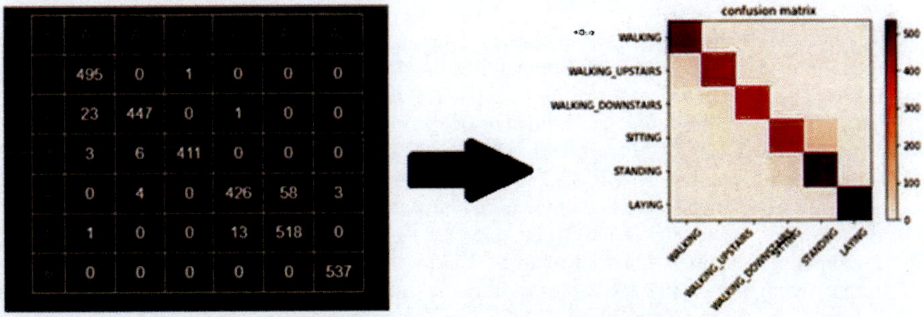

495	0	1	0	0	0
23	447	0	1	0	0
3	6	411	0	0	0
0	4	0	426	58	3
1	0	0	13	518	0
0	0	0	0	0	537

Fig. 9 Confusion matrix of logistic regression showing classification accuracy evaluated on data set

5 Conclusion

This paper studies the best in class in human activity recognition depending on smartphone sensors. The state-of-the-art has been surveyed in this paper on human activity recognition using accelerometers. Human activity recognition has wider applications in therapeutic research and human study framework. The important constituents of every HAR system, i.e. fundamentals of machine learning have also been included in our study. The system generated 31 features in both time and frequency field and diminished the feature dimensionality in order to enhance the performance and accomplishment. The activity data is then prepared and analyzed using machine learning techniques.

6 Future Scope

In future, we will try to predict information about human behavior from human activities which can be very useful in health, advertisements industry, etc. We will implement LSTM neural network using tensor flow and train the data. Trained data will be exported on an android app.

References

1. J.L.R. Ortiz, *Smartphone-based human activity recognition*. Springer, Berlin (2015)
2. H. Vellampalli, Physical human activity recognition using machine learning algorithms. **43**, 3605–3620 (2017)
3. A. Bayat, M. Pomplun, D.A. Tran, A study on human activity recognition using accelerometer data from smartphones. Proced. Comput. Sci. **34**, 450–457 (2014)
4. D. Anguita, A. Ghio, L. Oneto, X. Parra, J.L. Reyes-Ortiz, Human activity recognition on smartphones using a multiclass hardware-friendly support vector machine. in *International Workshop of Ambient Assisted Living (IWAAL 2012),* Vitoria-Gasteiz, Spain. Dec 2012
5. K. VanLaerhoven, K.A. Aidoo, S. Lowette, Real-time analysis of data from many sensors with neural networks. in *Proceedings of the Fifth International Symposium on Wearable Computers, Zurich, Switzerland,* 8–9 Oct 2001, pp. 115–122
6. D. Anguita, A. Ghio, L. Oneto, X. Parra, J.L. Reyes-Ortiz. Energy efficient smartphone-based activity recognition using fixed-point arithmetic. J. Univ. Comput. Sci. Special Issue Ambient Assist. Living Home Care. **19**(9) (2013)
7. D. Anguita, A. Ghio, L. Oneto, X. Parra, J.L. Reyes-Ortiz. Human activity recognition on smartphones using a multiclass hardware-friendly support vector machine. in *4th International Workshop of Ambient Assited Living, IWAAL 2012 Proceedings Lecture Notes in Computer Science*, (Vitoria-Gasteiz, Spain, Dec 3–5, 2012), pp 216–223
8. J.L. Reyes-Ortiz, A. Ghio, X. Parra-Llanas, D. Anguita, J. Cabestany, A. Català, Human activity and motion disorder recognition: towards smarter interactive cognitive environments. in *21th European Symposium on Artificial Neural Networks, Computational Intelligence and Machine Learning, ESANN 2013*. Bruges, Belgium 24–26 Apr 2013

9. K.Y.Chow, Z.Z. Htike, S. Egerton, Real-time human activity recognition using external and internal spatial features. in *2010 Sixth International Conference on Intelligent Environments*, pp. 25–28, Jul (2010)
10. J.R. Kwapisz, G.M. Weiss, S.A. Moore, Activity recognition using cell phone accelerometers. Hum. Factors **12**(2), 74–82 (2010)
11. L. Mo, F. Li, Y. Zhu, A. Huang, Human physical activity recognition based on computer vision with deep learning model. in *IEEE International Instrumentation and Measurement Technology Conference Proceedings*, Taipei (2016)
12. L. Sun, D. Zhang, N. Li, Physical activity monitoring with mobile phones, pp. 104–111 (2011)
13. J.-Y. Yang, J.-S. Wang, Y.-P. Chen, Using acceleration measurements for activity recognition: an effective learning algorithm for constructing neural classifiers. Pattern Recog. Lett. **29**, 2213–2220 (2008)
14. D.M. Karantonis, M.R. Narayanan, M. Mathie, N.H. Lovell, B.G. Celler, Implementation of a real-time human movement classifier using a triaxial accelerometer for ambulatory monitoring. IEEE Trans. Inf Technol. Biomed. **10**, 156–167 (2006)

DCNN-Based Facial Expression Recognition Using Transfer Learning

Puneet Singh Lamba and Deepali Virmani

Abstract Human facial expression recognition has been a lively area of research over the last few years. Traditional approaches perform soundly on datasets prepared under controlled environment but when it comes to unconventional datasets having image variations or frames with partial faces, the performance has declined profusely. With deep learning specifically convolutional neural network (CNN), a lot of models have been proposed to improve the performance. In this paper, transfer learning is employed to get learned weights and biases to upsurge the performance. Trailing layers of ResNet-50 (a state-of-the-art CNN architecture) trained on more than a million images from ImageNet Dataset are used as a part of transfer learning. Further, an additional convolutional layer is suffixed to the first 48 layers of Resnet-50. The entire model is then trained separately using five activation functions (ELU, RELU, Sigmoid, SELU, Exponential) and a comparative analysis is performed. Maximum accuracy (86.80%) with the proposed model is achieved when using SELU as an activation function. The exponential activation function has shown the minimum accuracy among the five activation function used.

Keywords Transfer learning · SELU · ResNet-50 · ImageNet · CNN

P. S. Lamba (✉)
University School of Information, Communication and Technology, GGSIPU, Sector 16 C, Dwarka, Delhi, India
e-mail: singhs.puneet@gmail.com

Bharati Vidyapeeth's College of Engineering, Paschim Vihar, New Delhi, India

D. Virmani
Department of Computer Science, Bhagwan Parshuram Institute of Technology, New Delhi, Delhi, India
e-mail: deepalivirmani@gmail.com

P. Bansal et al. (eds.), *Proceedings of International Conference on Artificial Intelligence and Applications*, Advances in Intelligent Systems and Computing 1164,
https://doi.org/10.1007/978-981-15-4992-2_48

1 Introduction

Human expressions act as an exceedingly important ingredient in social communication. Generally, communication can be verbal or nonverbal. Verbal communication doesn't really require facial human expression recognition. But when it comes to nonverbal communication, emotions are expressed through facial expressions. Facial expressions act as subtle gestures for a longer dialogue. Face is a multi-communication system rather than multi-gesture system transmitting emotion, feel position, age, quality, intelligence, attractiveness, and almost certainly other substances as well [1]. Eye also plays an imperative role in nonverbal communication providing a mixture of ideas [2, 3]. This paper identifies the following human facial expressions: neutral, anger, contempt, disgust, fear, happy, sadness, surprise. A happy human face is expressed by a curved shape eye. Sad emotion is expressed as mounted skewed eyebrows and frown. An angry face is correlated with unfriendly and irksome conditions which are expressed with squeezed eyebrows and stressed eyelids. Disgust faces are expressed with pull-down eyebrows and creased nose. An unpredicted event leads to surprise which is expressed with widened eye and mouth and can be easily recognized. The fear expression is correlated with a surprise that is conveyed with growing skewed eyebrows.

Human facial expression recognition is one of the most prevailing and challenging task in nonverbal communication. A lot of advancement has already been achieved in the past decade [4–6] resulting in a principal framework marking the facial movements—FACS [7] (Facial Action Coding System). Expressions/emotions being an inevitable part of any communication can be articulated in various forms which may or may not be perceived with an open eye. Hence using technical aspects of computer vision, any manifestation before or after is under observation for recognition. Human expressions recognition is expedient in various fields including, but not limited to, human-computer interaction [8], animatronics [9], medical [10, 11] and security [12, 13]. Human expression recognition can be implemented using various human attributes namely face [9, 14, 15], dialogue [12, 16], EEG [17], and even text [18]. Among the mentioned attributes, facial expressions are one of the most popular as required features can be extracted at ease [9, 19, 20].

Facial Expression Recognition is performed in two stages: feature extraction and classification. Feature extraction can be categorized as geometric based and appearance-based. In geometrically based feature extraction focus is concentrated on eye, mouth, nose, eyebrow, and eyelids regions while the appearance-based feature extraction concentrates on a specific section of the face. During classification facial expressions are categorized using the features extracted in the first stage. This paper mainly focuses on extracting expressions from the face using a convolutional neural network (CNN) employing various activation functions (sigmoid, ReLU, ELU, SELU, and exponential) resulting in a comparative analysis of the same. Rest of the paper is structured as follows. Section 2 elaborates on the related study. Section 3 explains the methodology used along with the dataset being processed and the experimental setup. Section 4 analyses the experimental results obtained. Section 5 concludes the study.

2 Related Study

Formerly facial expression recognition (FER) requires a customary two step process, where in the first step, features are extracted from the subject frame resulting in a feature map, and in the later step, classifiers (decision tree, support vector machines, KNN, etc.) are selected to categorize the expression. Local binary pattern [21], histogram of oriented gradients [22], Gabor filters [23] are among some of the traditional features which are used to recognize the expressions. Based on the features extracted, classifiers (according to their working) will then process the features set and dispense the expression/emotion to the image frame. However, these traditional approaches have their limitations, considering the types of complex datasets that are originating. The answer to the limitations is deep learning. With deep learning, specifically, CNN (used for image cataloging and computer vision glitches [24–31]), numerous deep learning-based models have already been developed for FER. Using a zero-bias CNN, high accuracy is attained on Extended-Cohn-Kanade and the Toronto Face Dataset [32]. A model [9] for facial expressions for stylized animated characters is also developed. A neural network [14] with two convolution layers, one max-pooling layer, and four "inception" layers, i.e. sub-networks are also proposed. With the aim to improve the recognition rate of impulsive facial expressions, an incremental boosting CNN (IB-CNN) (enhancing the discriminative neurons) [33] has been proposed. Meng in [34] proposed an identity-aware CNN (IA-CNN), reducing the variations in learning identity and expression-related information. In a nutshell, facial expression recognition using deep learning (CNN) has outperformed the traditional approaches achieving momentous improvement in terms of accuracy, precision, and other performance parameters. In this work, we have used the concept of transfer learning and proposed a framework, which is focusing on salient face regions.

3 Methodology

This section explains the methodology used to recognize various facial expressions. Section 3.1 explains the CNN architecture and activation function used to predict facial expressions. ResNet-50 architecture and transfer learning concepts are described in detail in Sects. 3.2 and 3.3. Dataset used, performance parameters and the experimental setup (proposed framework) are explained in the following sections.

3.1 CNN

CNN is known to be used in various computer vision applications, including facial expression recognition. When it comes to unseen facial/pose deviations, CNN (being robust to facial location changes) performs better than multilayer perceptron [35]. Three types of heterogeneous layers exist in CNN namely convolutional layers, pooling layers, and fully connected layers. The first layer (convolutional) is used to produce numerous forms of activation feature maps, employing a set of learnable filters that convolve through the entire image frame. This layer assists in local connectivity, weight sharing and shift-invariance. The convolutional layer is followed by pooling layer which is employed to reduce the spatial size of the feature maps. For classification, fully connected layers are added to ensure connection with the penultimate layer (every neuron in layer x to every other neuron in the layer $x + 1$). Table 1 explains the various activation function used in the network.

3.2 ResNet-50

ResNet-50 is a standard state-of-the-art CNN architecture trained on more than a million images (ImageNet dataset) which is identical to VGG-16 but with some additional identity mapping capability. The network consists of 50 deep layers and has the capability to catalogue subjects into 1000 categories. Thus, the network has learned rich feature representations for a variety of images having an input size of 224-by-224. In this paper, leading 48 out of 50 layers of ResNet-50 are used in the feature extraction process.

3.3 Transfer Learning

A pre-trained model is one that has been trained beforehand on some database and contains the weights/bias representing unique features of subjects. These unique learned features are usually transferable. Taking the learned weights and biases of a pre-trained model (trained on ImageNet) and applying the same fixing some layers and reequipping the residual layers constitutes transfer learning. In this paper, we apply transfer learning by taking the learned weights from the ImageNet dataset, and retraining a few, later layers on the Cohn-Kanade dataset.

Table 1 Activation function used in CNN

Activation function	Formulae	Figure
Sigmoid	$S(z) = \frac{1}{1+e^{-z}}$	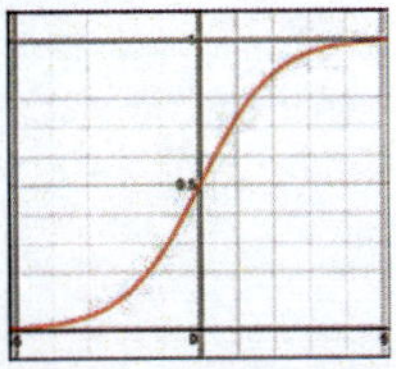
ReLU	$y = \max(0, z)$	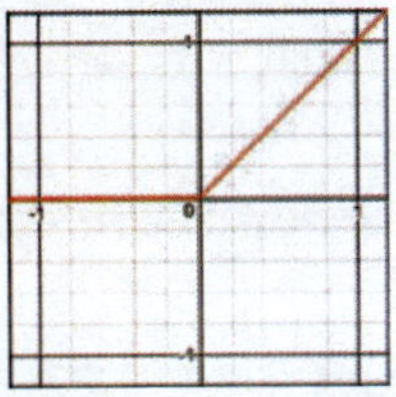
ELU	$R(z) = \begin{Bmatrix} z & z > 0 \\ \alpha(e^{z} - 1) & z \leq 0 \end{Bmatrix}$	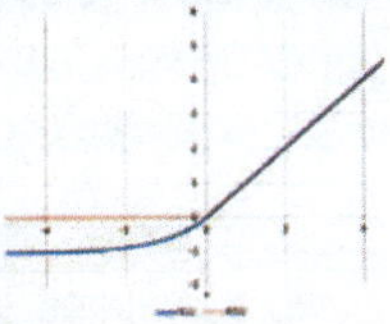
SELU	$\text{selu}(x) = \lambda \begin{cases} x & \text{if } x > 0 \\ \alpha e^{x} - \alpha & \text{if } x \leq 0 \end{cases}$	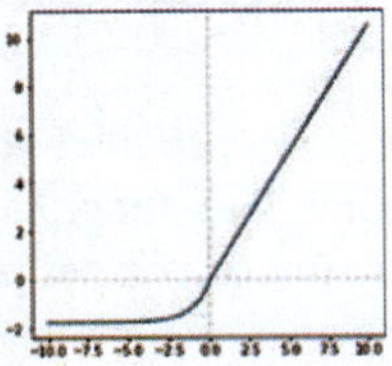
EXP	$\exp(x) = e^{x}$	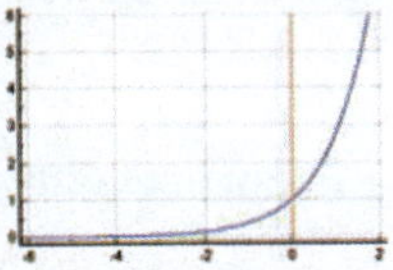

3.4 Dataset

We evaluate one standard database Cohn-Kanade which is widely used to evaluate facial expressions and face recognition as summarized in Table 2.

Table 2 Summary of cohn-kanade dataset

Properties	Descriptions
# of images	486
Single/multiple faces	Single
Gray/Color	Gray
Resolution	640*490
View	Frontal-view only
Facial expression	23 facial displays with single action units or grouping of action units
Ground truth	AU label for final frame in each image sequence

3.5 Performance Parameters

In this paper, facial expressions are recognized using classification. The performance of classification (CNN) with different activation functions are analyzed by measuring the accuracy, precision, recall, receiver operating characteristic curve, F1 score and area under the curve score.

3.6 Experimental Setup

This section provides the experimental setup for the proposed framework required to predict the human facial expressions. Preprocessing is considered as an essential step for image-related processing. Section 3.6.1 explains the preprocessing step that is applied to the raw dataset images. Feature extraction using transfer learning for the proposed framework is elaborated in Sect. 3.6.2.

3.6.1 Preprocessing

To recuperate the performance of the FER system, preprocessing is performed afore feature extraction. Preprocessing comprises of image clarity, scaling, contrast adjustment, and additional enhancement processes. In this work, faces are aligned according to the eye position to reduce the variance in the data as a part of preprocessing step. For locating eye position accurately, MTCNN (Multitask Cascaded Neural Network) has been used. The entire process of preprocessing (pseudocode) is explained in Fig. 1 and the transformation is shown in Fig. 2.

```
Algorithm 1 Face Alignment

procedure FaceAligner(img)
   left_eye_center, right_eye_center ← MTCNN(img)
   angle ←Determine angle b/w right and left eye centers
   desired_right_eye_x = 1 - 0.35
   dist ← sqrt(dx ** 2 + dy ** 2)
   desired_dist = desired_right_eye_x - 0.35
   desired_dist *= 256
   scale ← desired_dist / dist
   eyes_center ← Determine center of line joining both the eyes
   M ← Calculate an affine matrix for 2D rotation using eyes_center, scale and angle
   tx = 0.5 * 256
   ty = 0.35 * 256
   M[0, 2] += (tx - eyes_centerx)
   M[1, 2] += (tx - eyes_centery)
   result ← Apply an affine transformation to the img using M and desired face width and height
   return result
end procedure
```

Fig. 1 Pseudocode for preprocessing

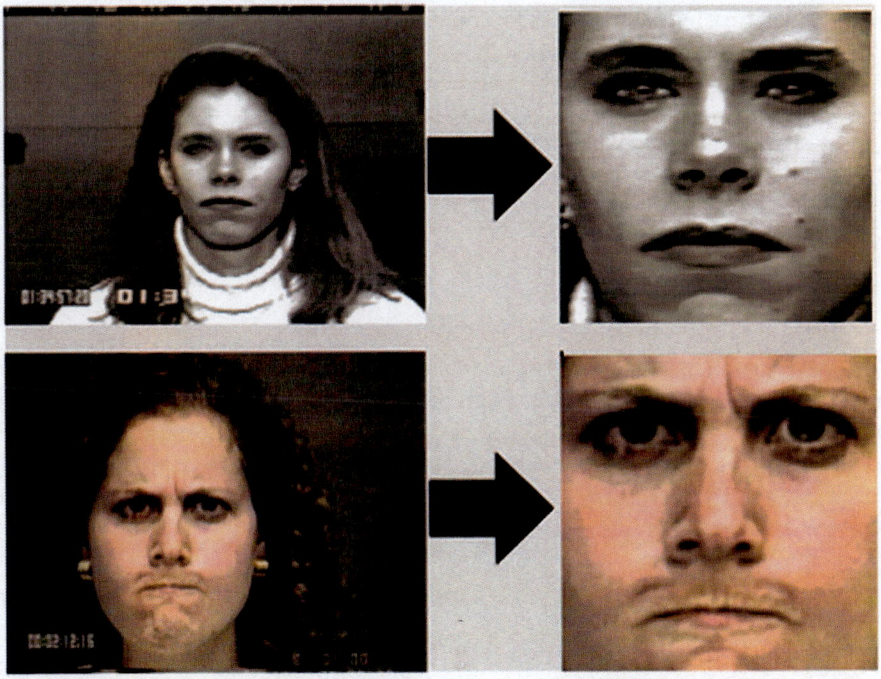

Fig. 2 Preprocessing done to align the faces

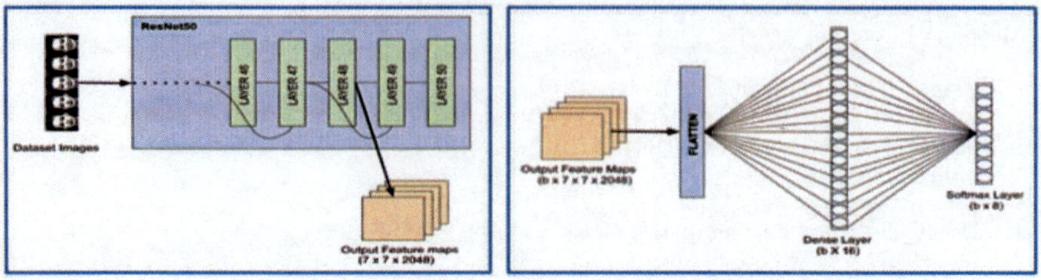

Fig. 3 Proposed model for feature extraction (left) and classification (right) for facial recognition

3.6.2 Feature Extraction Using Transfer Learning

Weights of Resnet-50 trained on ImageNet dataset are used for feature extraction. For our model, an additional convolutional layer (of $k \times k$ kernel and stride s along with a softmax layer of unit 8) is suffixed to first 48 layers of ResNet-50. Last two layers are removed as they play a trifling role in detecting emotions. The loss used for training is categorical cross-entropy with adam (Adaptive moments) optimizer. Preprocessed image dataset is fed into the ResNet-50 model and the output is extracted at layer 48 which acts as input to our model. The proposed framework is shown in Fig. 3. The output of layer 48 is a 2048 feature map of size 7 * 7. These feature maps are extracted for each of the image of the dataset and are flattened for our model. Our model architecture consists of a dense layer with 16 relu units along with a softmax layer of eight units as shown in second half of Fig. 3.

4 Experimental Results

A standard publicly available dataset (Cohn-Kanade) is used to analyze the performance of the model explained in Sect. 3.6. The model is implemented separately using five activation functions and is ranked according to performance parameters explained in Sect. 3.5. The performance is investigated in terms of precision, recall, accuracy, and F1 score. Additionally, ROC plots are utilized to outwardly analyze the execution of the proposed system in verification.

4.1 Result Description and Analysis

Table 3 shows the comparative analysis of the activation function used. It can be clearly observed that SELU activation function delivers the maximum accuracy whereas exponential activation function stood last in the race. Same is confirmed from precision, recall and F1 score of the respective activation functions. Comparative receiver operating characteristics (ROC) curve along with area under the curve

Table 3 Comparative analysis of activation function used

S. No.	Activation function	Precision		Recall		F1-score		Accuracy
		Micro avg	Macro avg	Micro avg	Macro avg	Micro avg	Macro avg	
1.	Elu	0.83	0.75	0.83	0.56	0.83	0.59	0.8274
2.	Relu	0.80	0.87	0.80	0.56	0.80	0.64	0.7980
3.	Sigmoid	0.80	0.75	0.80	0.51	0.80	0.55	0.7970
4.	Selu	0.87	0.90	0.87	0.67	0.87	0.72	0.8680
5.	Exponential	0.72	0.44	0.72	0.38	0.72	0.38	0.7157

for each expression are shown in Fig. 4. In each of the ROC curve, SELU has the highest AUC.

5 Conclusion

Human facial expressions express the emotional state of a subject to the observers. This paper proposes a framework to detect the expressions using CNN and transfer learning. Learned weights/bias of Resnet-50 (trained on ImageNet dataset) are used for feature extraction. Further, an additional convolutional layer is suffixed to first 48 layers of Resnet-50. The framework is evaluated on Cohn-Kanade dataset. As a part of preprocessing, faces are aligned according to the eye position to reduce the variance in the data. The model is evaluated using five activation functions and the performance is compared in terms of accuracy, precision, recall, F1-score, ROC, and area under the curve. Among the activation function used, SELU (with an accuracy of 86.80%) has outperformed its counterparts by a margin.

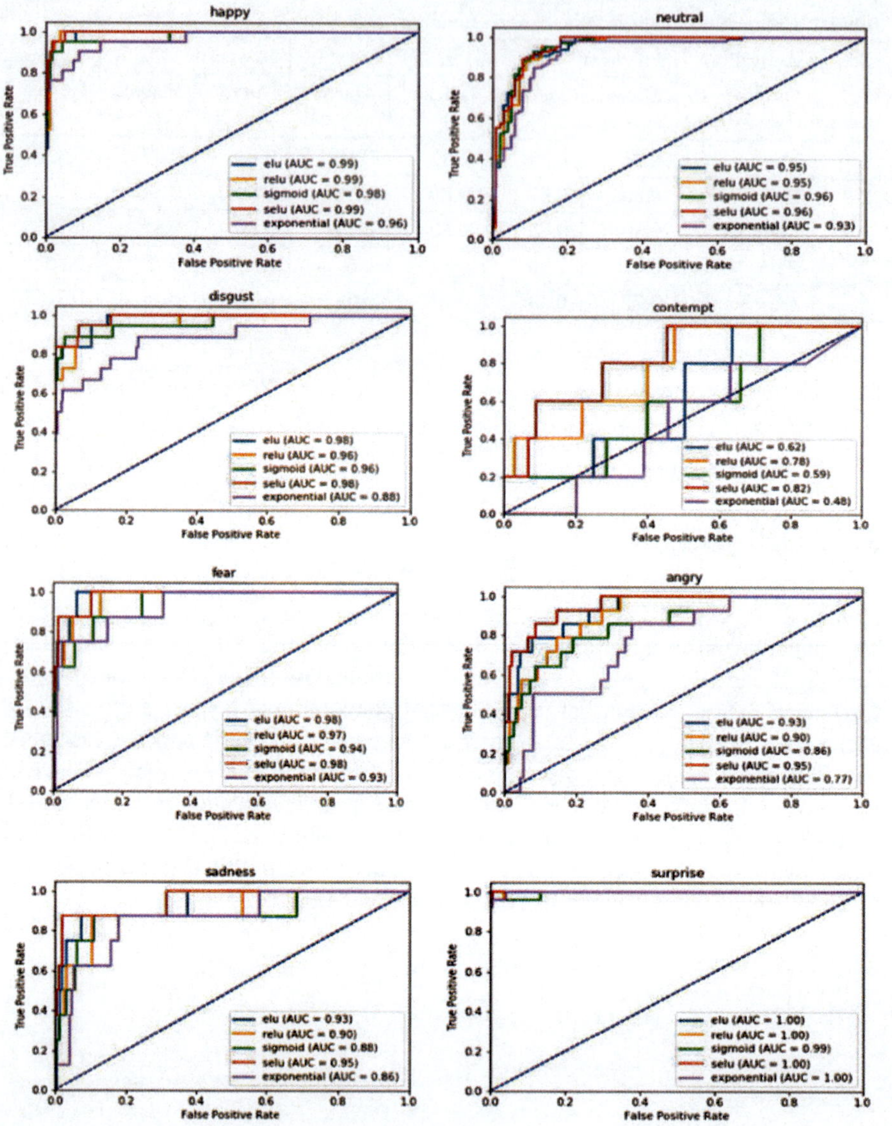

Fig. 4 Comparative ROC curve for each expression

References

1. P. Ekman, W. Friesen, Unmasking the face: a guide to recognizing emotions from facial clues. (2003)
2. P.S. Lamba, D. Virmani, Information retrieval from emotions and eye blinks with help of sensor nodes. Int. J. Electr. Comput. Eng. (IJECE) **8**(4), 2433–2441 (2018)

3. P.S. Lamba, D. Virmani, Reckoning number of eye blinks using eye facet correlation for exigency detection. J Intell Fuzzy Syst. **35**(5), 5279–5286 (2018)
4. Y. Tian, T. Kanade, J. Cohn, Recognizing action units for facial expression analysis. IEEE Trans. Pattern Anal. Mach. Intell **23**(2), (2001)
5. M.S. Bartlett, G. Littlewort, M. Frank, C. Lain-scsek, I. Fasel, J. Movellan, Fully automatic facial action recognition in spontaneous behavior. in *Proceedings of the IEEE Conference on Automatic Facial and Gesture Recognition*, 2006
6. M. Pantic, J.M. Rothkrantz, Facial action recognition for facial expression analysis from static face images. IEEE Trans. Syst. Man Cybern. **34**(3) 2004
7. P. Ekman, W. Friesen, Facial action coding system: a technique for the measurement of facial movement. Consulting Psychologists Press, 1978
8. R. Cowie, E. Douglas-Cowie, N. Tsapatsoulis, G. Votsis, S. Kollias, W. Fellenz, J.G. Taylor, Emotion recognition in human-computer interaction. IEEE Signal Process. Mag. **18**(1), 32–80 (2001)
9. D. Aneja, A. Colburn, G. Faigin, L. Shapiro, B. Mones, Modeling stylized character expressions via deep learning. in *Asian Conference on Computer Vision*, (Springer, Cham, pp. 136–153) 2016
10. J. Edwards, H.J. Jackson, P.E. Pattison, Emotion recognition via facial expression and affective prosody in schizophrenia: a methodological review. Clin. Psychol. Rev. **22**(6), 789–832 (2002)
11. H.-C. Chu, W.W.J. Tsai, M.-J. Liao, Y.M. Chen, Facial emotion recognition with transition detection for students with high-functioning autism in adaptive e-learning. Soft. Comput. **22,** 2973–2999 (2018)
12. C. Clavel, I. Vasilescu, L. Devillers, G. Richard, T. Ehrette, Fear-type emotion recognition for future audio-based surveillance systems. Speech Commun. **50**(6), 487–503 (2008)
13. S.T. Saste, S.M. Jagdale, Emotion recognition from speech using MFCC and DWT for security system. in *2017 International conference of Electronics, Communication and Aerospace Technology* (ICECA), vol. 1, pp. 701–704. IEEE, 2017
14. A. Mollahosseini, D. Chan, M.H. Mahoor, Going deeper in facial expression recognition using deep neural networks. in *2016 IEEE Winter Conference onApplications of Computer Vision (WACV)*, 2016. IEEE
15. P. Liu, S. Han, Z. Meng, Y. Tong, Facial expression recognition via a boosted deep belief network. in *Proceedings of the IEEE Conference on Computer Vision and Pattern Recognition*, pp. 1805–1812, 2014
16. K. Han, D. Yu, I. Tashev, Speech emotion recognition using deep neural network and extreme learning machine. in *Fifteenth Annual Conference of the International Speech Communication Association*, 2014
17. Panagiotis C. Petrantonakis, Leontios J. Hadjileontiadis, Emotion recognition from EEG using higher order crossings. IEEE Trans. Inf Technol. Biomed. **14**(2), 186–197 (2010)
18. C.-H.Wu, Z.-J. Chuang, Y-C. Lin, Emotion recognition from text using semantic labels and separable mixture models. in *ACM Transactions on Asian Language Information Processing (TALIP)*, vol. 5, no. 2, pp. 165–183 (2006)
19. M.J. Lyons, S. Akamatsu, M. Kamachi, J. Gyoba, J. Budynek, The Japanese female facial expression (JAFFE) database. in Third International Conference on Automatic Face and Gesture Recognition, pp. 14–16, 1998
20. P.L. Carrier, A. Courville, I.J Goodfellow, M. Mirza, Y. Bengio, FER-2013 face database. Universit de Montreal, 2013
21. C. Shan, S. Gong, P.W. McOwan, Facial expression recognition based on local binary patterns: a comprehensive study. Image Vis. Comput. **27**(6), 803–816 (2009)
22. C. Junkai, Z. Chen, Z. Chi, H. Fu, Facial expression recognition based on facial components detection and hog features. In International workshops on electrical and computer engineering subfields, pp. 884–888, (2014)
23. M.S. Bartlett, G. Littlewort, M. Frank, C. Lainscsek, I. Fasel, J. Movellan, Recognizing facial expression: machine learning and application to spontaneous behavior. in *2005 IEEE Computer Society Conference on Computer Vision and Pattern Recognition* (*CVPR*), vol. 2 (2005), pp. 568–573. IEEE

24. A. Krizhevsky, I. Sutskever, G.E. Hinton, Imagenet classification with deep convolutional neural networks. Adv. Neural. Inf. Process. Syst. **25**(2), 1097–1105 (2012)
25. K. He, Z. Xiangyu, S. Ren, J. Sun, Deep residual learning for image recognition. in *Proceedings of the IEEE conference on computer vision and pattern recognition*, pp. 770–778, 2016
26. J. Long, E. Shelhamer, T. Darrell, Fully convolutional networks for semantic segmentation. in *Proceedings of the IEEE Conference on Computer Vision and Pattern Recognition*, 2015
27. K. He, G. Gkioxari, P. Dollr, R. Girshick, Mask r-cnn. in *International Conference on Computer Vision*, 2017 (pp. 2980–2988). IEEE
28. S. Minaee, A. Abdolrashidiy, Y. Wang, An experimental study of deep convolutional features for iris recognition. in *Signal Processing in Medicine and Biology Symposium*, 2016. IEEE
29. S. Minaee, I. Bouazizi, P. Kolan, H. Najafzadeh, Ad-Net: audio-visual convolutional neural network for advertisement detection in videos. ArXiv, abs/1806.08612 (2018)
30. I. Goodfellow, J. Pouget-Abadie, M. Mirza, B. Xu, D. Warde-Farley, S. Ozair, A. Courville, Y. Bengio, Generative adversarial nets. in *Advances in Neural Information Processing Systems*, pp. 2672–2680 (2014)
31. S. Minaee, Y. Wang, A. Aygar, S. Chung, X. Wang, Y.W. Lui, E. Fieremans, S. Flanagan, J. Rath, MTBI Identification from diffusion MR images using bag of adversarial visual features. IEEE Trans. Med. Imaging **38**(11), 2545–2555 (2019)
32. P. Khorrami, T. Paine, T. Huang, Do deep neural networks learn facial action units when doing expression recognition? in *Proceedings of the IEEE International Conference on Computer Vision Workshops*, 2015
33. S. Han, Z. Meng, A.-S. Khan, Y. Tong, Incremental boosting convolutional neural network for facial action unit recognition. in *Advances in Neural Information Processing Systems*, pp. 109–117 (2016)
34. Z. Meng, P. Liu, J. Cai, S. Han, Y. Tong, Identity-aware convolutional neural network for facial expression recognition. in *IEEE International Conference on Automatic Face & Gesture Recognition*, 2017 pp. 558–565. IEEE
35. B. Fasel, Robust face analysis using convolutional neural networks. in *Proceedings of the 16th International Conference on Pattern Recognition*, vol. 2 (2002), pp. 40–43. IEEE

Mobile-Based Prediction Framework for Disease Detection Using Hybrid Data Mining Approach

Megha Rathi and Ayush Gupta

Abstract Early detection of life-threatening diseases is a major healthcare problem and early detection an immediate response is required to control diseases. Now a day's mobile technology has become an important medium for doctors, patients and healthcare practitioners for seeking information related to health. In this study, we build a mobile-based disease detection system using hybrid data mining approach. The algorithms used are Sequential Minimal Optimization and Naïve Bayes and these classifiers were combined by using the voting ensemble technique. The proposed Framework enables us to enter the symptoms and then prediction model detects disease based on the symptom entered by the user. Proposed model is able to detect various diseases online. The proposed framework is evaluated on different disease datasets collected from UCI ML Repository. An experiment scheme was designed for checking the efficiency of the proposed hybrid model and the results depict that the proposed model is able to detect disease with average accuracy of 93.305 for different diseases.

Keywords Disease detection · Web framework · Support vector machine · Naïve bayes · Hybrid approach

1 Introduction

In the current lifestyle where all persons are living under stress, health-related problems are increasing day by day. With the busy schedule and heavy work pressure everyone is suffering from minor to severe health problems. Data Mining and Machine Learning are significantly gaining importance in the domain of healthcare. Manual detection of any disease is very much time consuming and a person will go

M. Rathi · A. Gupta (✉)
Computer Science Department, Jaypee Institute of Information Technology, Noida, India
e-mail: ayushgupta2959@gmail.com

M. Rathi
e-mail: megha.rathi@jiit.ac.in

P. Bansal et al. (eds.), *Proceedings of International Conference on Artificial Intelligence and Applications*, Advances in Intelligent Systems and Computing 1164,
https://doi.org/10.1007/978-981-15-4992-2_49

under several tests in order to find out the outcome of any disease. Automation of disease detection has proven to be very beneficial in healthcare and if any such tool exists online anyone can access it and know about the disease type he is suffering from.

The proposed study aims to create an easy way to analyze one's health status in terms of ailments. The project features a simple, yet elegant disease detection system which can be used extensively in the real world. It also suggests health habits and facts details. The android app component provides a mobile platform which one can use on the go as per the requirement. This application renders emergency features like SMS notification alerts to user listed contacts. The project has two subparts which are android and the web platform which have been developed which work in coordination. The website has been deployed on a cloud server for remote access. Classifiers that were used in final implementation were SMO (Sequential Minimal Optimization), an SVM classifier, and, Naïve Bayes Classifier. Naïve Bayes works on the Bayes Theorem, i.e., the classifier uses a probabilistic approach in carrying out predictions. It's easy to understand and also to incorporate. The other classifier that was used is SMO, which was developed initially by John C. Platt under Microsoft Research [1]. Training the SVM classifier requires the optimization of a large quadratic programming (QP) problem. But with SMO, this huge QP problem is divided into smallest attainable chunks of QP problems. The solving of these small QP problems is carried out analytically, rather than the numerical approach to the optimization of QP problems which would have been in an inner loop when implemented. This causes the time complexity of training an SVM classifier to lie between linear and cubic time. Matrix operation such as matrix multiplication is avoided by using SMO and the time complexity ranges between linear and quadratic time. It can be seen clearly that SMO outperforms SVM classifier in the matters of both time and space complexity. It is seen that SMO is several folds better than linear and non-linear SVMs.

The hybridization of the above classifiers, SMO and Naïve Bayes was done using ensemble techniques, to be precise Voting ensemble. Training and testing code were written in Java using the Weka API. For demonstrations of the same, a J2EE web application was developed along with an android application to ensure access for web as well as mobile users.

2 Literature Review

Deepika et al. [2] proposed a hybrid classification methodology to combine J48 (decision tree) and IBk (Instance-based KNN). Individual classifiers were trained at initial level and then they were combined using meta-learners. Their results show an accuracy of 100% when J48 tree and IBk are combined using vote ensemble and an accuracy of 65% when they combined the same classifiers using Stack ensemble. Their paper lists several other classifiers along with their accuracies and also shows pictorially the system architecture involved in training of the hybrid classifiers. This paper was the motivation for our proposed hybrid classification model.

Vijayarani et al. [3] put presents a method for liver disease prediction. In their approach, they have used SVM and Naïve Bayes classifiers individually. They carried out their work on Matlab 2013 tool and have concluded that performance of SVM is better than Naïve Bayes. They have evaluated the performance on the basis of accuracy as well as execution time.

Ahmed et al. [4] have worked on training C4.5 (decision tree, also J48 in weka). They clearly explained the working of decision tree and stated how classification techniques can be used to predict risk parameters of diabetes. They have taken the breast cancer dataset from UCI repository.

In the research [5], Decision Support System for the detection of Breast Cancer is created. The same author combined Fuzzy Logic and Fuzzy Decision Tree to improve classifier performance by introducing new factor, i.e., FDT-stable which is simple to understand and apply. Proposed approach applied on two different datasets of breast cancer both taken from UCI named Breast Cancer and Breast Cancer Wisconsin and it is found that the error rate produced by the system is 0.3661 and 0.1414 which is very low.

In the study [6] neural network is used for diagnosing the patient suffering from breast cancer. In this study author local linear wavelet for detecting breast cancer and trained the model using Recursive Least Square method for improving the performance of the classifier. Accuracy achieved by the model is 97.2%.

Ravi et al. [7] proposed an automated human disease detection system. In their approach, they used data mining technique and a truth discovery mechanism. In the proposed method, first the data set and the user query are preprocessed. Then using Brown Clustering Algorithm similar symptoms are grouped. These similar symptoms are then used to identify diseases using the mapping table. The truth discovery method using the doctor expert model algorithm is used to find the more relevant disease from the set of identified diseases. The proposed method achieved an average accuracy of 85%.

Pal et al. [8] have proposed a model that uses both data mining and machine learning techniques for the prediction of most common diseases. In this model a specific person first enters his/her current health status, these are then extracted using data mining methods and fed into different machine learning models. They have used Naïve Bayes, decision tree, and random forest as the machine learning models and also provided a detailed comparison between these methods. Their model also suggests remedies for the disease predicted.

Babu et al. [9] proposed an approach to predict heart disease based on certain attributes. By using past labeled dataset and new patients' attributes, K-means clustering is applied to these patients' attributes to find similar heart diseases. Then MAFIA algorithm is used to mine the most frequent itemsets from the database. Then decision tree algorithm is used to classify the diseases based on the attributes obtained from MAFIA algorithm.

Dahiwade et al. [10] proposed a general disease prediction method using the symptoms of the patients. There method uses the disease dataset from UCI machine learning repository. They have used Convolutional Neural Network (CNN) and K-Nearest Neighbor (KNN) algorithms to classify a set of symptoms to the disease. They have compared the results obtained by these algorithms and found that CNN performed better than with an accuracy of 84.5% and also the memory requirements of CNN were found to be less than KNN.

Baiju et al. [11] have presented a diabetic prediction approach based on Disease Influence Measure (DIM). In their approach for diabetic prediction, they first preprocess the dataset and then calculate the DIM based on the resultant features. Based on this DIM the diabetic prediction is done using a threshold DIM. They compared the results obtained by their method with different classification algorithms like SVM, ANN, etc. and their method performed better than all the other methods.

From these literature reviews, it can be said hybrid approach performs better than a single classification method. Also, ensemble techniques can improve the overall accuracy of the model.

3 Methodology

Classification is one of the key tasks in data mining. The classification algorithm plays an important role in analyzing and predicting clinical data. There were several algorithms that were used in obtaining the desired results. In the proposed study we have combined SMO (Sequential Minimal Optimization) which is used for training support vector machine [12] and Naïve Bayes algorithms. The rule for the combination was average of probabilities. Classifier Subset Evaluator clubbed with the best first search was also used. Given below is the description of datasets used in the study along with the architecture of the proposed hybrid model for mobile-based disease detection.

3.1 Dataset Description

Datasets were taken from the UCI ML Repository [13]. A number of datasets were taken into consideration so as to ensure generalization of the predictor to some extent. The datasets were split in 80 to 20 ratio for train and test set. We have used Breast Cancer, Hepatitis, Heart disease, HIV, Diabetes and Dermatology datasets in our work. Each dataset consists of attributes describing different parameters in that disease and corresponding values and a label of whether a patient has that disease or not.

3.2 Data Preprocessing and Feature Selection

In order to provide quality results in terms of accuracy, datasets required to be preprocessed as it had missing data, null values, and nan (not a number) values. So, to remove all such anomalies from the original data we use Classifier Subset Evaluator combined with Best First Search for selecting relevant attributes. The following tables show the datasets after feature selection was applied

3.3 Proposed Framework for Disease Detection

The main objective of this study is the development of mobile-based disease detection system which is able to detect all kind of diseases. In this section, we explain the overall architecture of the proposed model. We present the ensemble model architecture along with the training model and web-based architecture.

1. Hybrid Model Architecture

For disease detection, wrong results may produce results which in turn led to the death of any person so it is required in medical domain we must create a model whose error rate is very low. In order to improve the classification accuracy, we create an ensemble technique. For the same, we have combined SMO with Naïve Bayes [8]. Figure 1 depicts the flow diagram of the proposed hybrid model.

2. Proposed Model Training Architecture

For training the proposed model we have used the framework as shown in Fig. 2. The datasets are first integrated and preprocessed. Then an optimized classifier is learned by the model using the training dataset. In each iteration of training, the

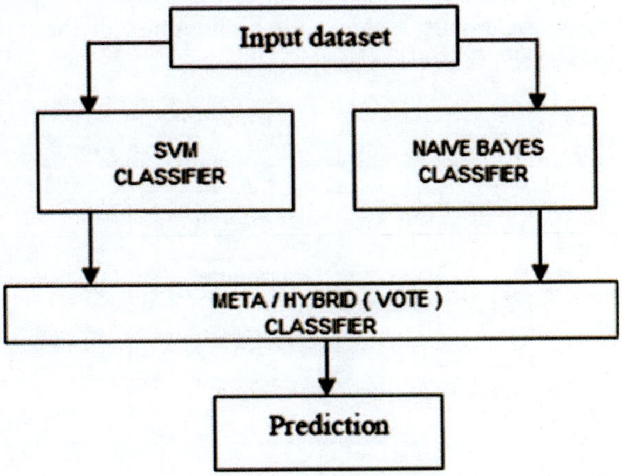

Fig. 1 Proposed hybrid model architecture

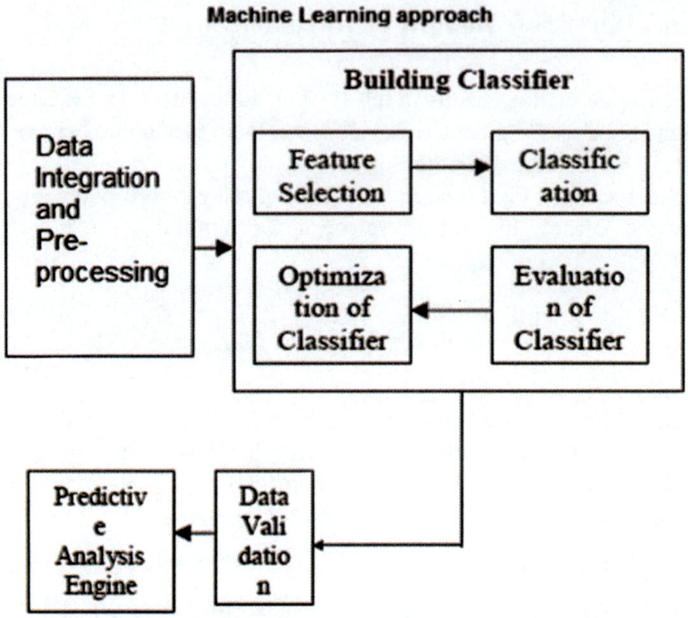

Fig. 2 Proposed hybrid model training architecture

model is evaluated and optimized. Finally, the trained model is tested and its results are described in the next section.

3. Web-Based Architecture

For the web interface, we have used a three-tier architecture. In the client tire, the client interacts with the application using the browser. In the middle tier resides the servers which handle the incoming requests from the clients and generate results based on the request. The middle tier interacts with the EIS tier which stores the database. Figure 3 shows the architecture of the web application that utilized this model to predict diseases.

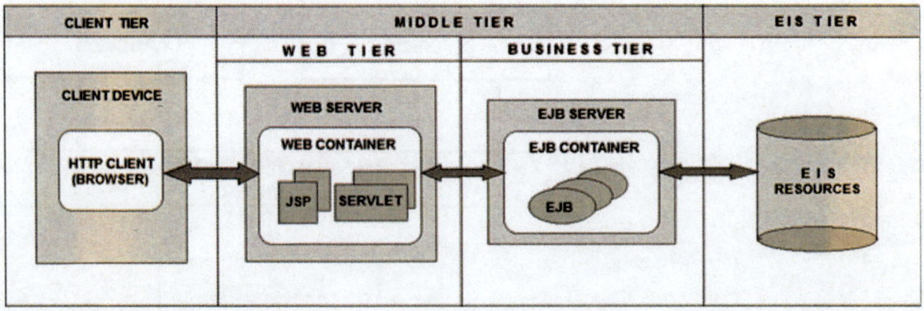

Fig. 3 Web-based architecture

4 Experimental Results

In the development of the proposed hybrid model for disease detection, we have used and tested our approach on 6 different datasets mentioned in Tables 1, 2, 3, 4. We have used accuracy as the metric to evaluate the model performance. First, we apply data preprocessing on all the datasets so that we have quality data after that we apply and tested our approach on all datasets. Table 1 shows the result when a single approach is applied to all datasets. The classification was performed using several algorithms such as Naïve Bayes, SMO, IBk, J48, Random Forest, as individual classifiers.

Further, ensemble techniques were used to compare performance based on accuracy of predictions on test datasets. Tables 2 and 3 present the result of other ensemble techniques which we have tested but result outcome is not appreciable then we tested our proposed approach in which we are getting better results so we selected the same hybrid techniques for creation of the model. Table 4 presents the result of our proposed approach. From all the results shown it has been found that ensemble technique Sequential Minimal Optimization and Naïve Bayes, when combined with voting ensemble produce better results.

Table 1 Classification results after testing with a single data mining approach (in %)

Single classifier	Breast cancer	Diabetes	Heart disease	HIV	Hepatitis	Dermatology
NB	97.86	78.87	88.52	93.85	80	95.89
SMO	97.86	81.17	88.52	94.46	83.33	95.89
IBk	80.71	66.23	62.29	92	63.33	71.23
J48	92.86	76.62	81.97	91.38	83.33	95.89
Random forest	95	79.87	80.33	93.54	80	87.67

Table 2 Ensembling of DM algorithm with bootstrap bagging (in %)

Bootstrap bagging	Breast cancer	Diabetes	Heart disease	HIV	Hepatitis	Dermatology
NB	96.43	77.27	86.88	93.54	86.67	94.52
SMO	96.43	81.82	85.25	95.69	83.33	98.63
J48	96.43	79.22	80.33	92.62	83.33	90.41

Table 3 Ensembling with boosting—adaboost (in %)

Boosting-adaboost	Breast cancer	Diabetes	Heart disease	HIV	Hepatitis	Dermatology
SMO	97.86	81.17	83.61	92.62	80	95.89
NB	97.86	74.67	83.61	90.77	86.67	97

Table 4 Proposed approach results (in %)

Vote	Breast cancer	Diabetes	Heart disease	HIV	Hepatitis	Dermatology
J48 + IBk	80.7	66.23	63.93	94	63.33	75.34
SMO + NB + IBk	95.71	79.87	65.57	95.3	76.67	83.56
SMO + NB	97.8	81.17	88.5	95.69	96.67	100

Our main objective is the development of mobile-based hybrid framework for disease detection. So, after testing classifier performance, we build mobile-based model. Shown below are the snapshots of mobile-based App for disease detection. The main advantage of this mobile-based framework is that all process is online and just after inputting certain values you can detect any disease type. This can ultimately reduce the time, cost and most important mortality rate due to late detection of disease. Figures 4, 5 and 6 presents the snapshot of mobile applications for the detection of different diseases.

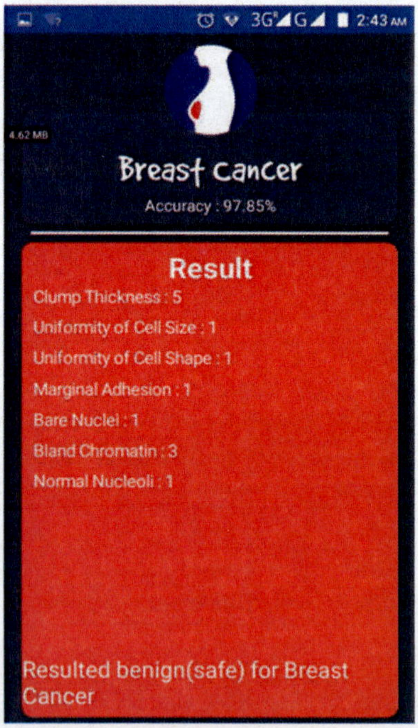

Fig. 4 Breast cancer diagnosis

Fig. 5 HIV diagnosis

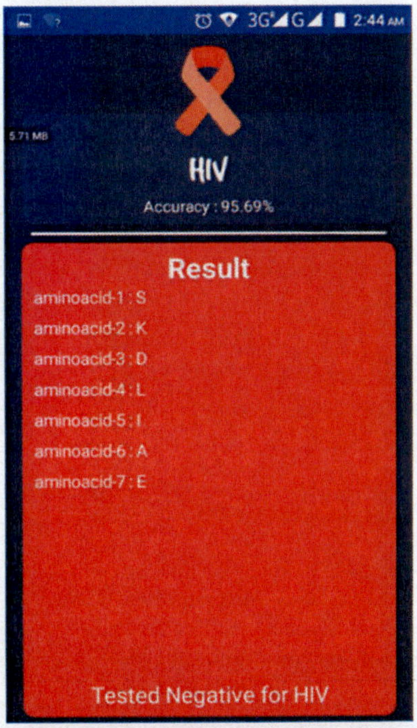

Fig. 6 Diabetes diagnosis

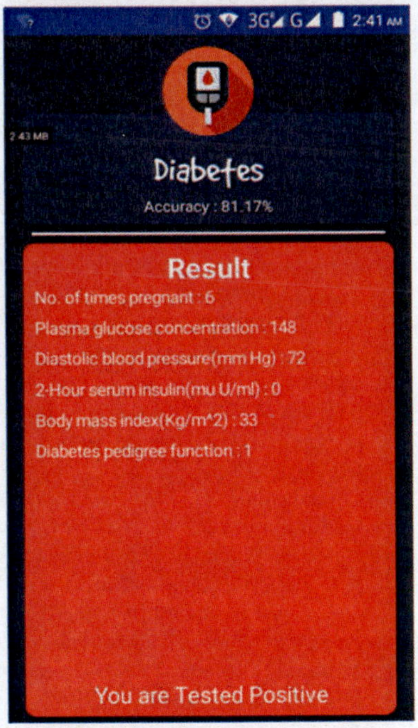

5 Conclusion

This research has introduced mobile-based framework for disease diagnosis. The proposed framework merges two data mining algorithm for the creation of ensemble model for disease detection. The results above clearly show that we got better accuracies using our proposed hybrid classification model, i.e., Voting ensemble of SMO and Naïve Bayes classifiers using average of probabilities as combination rule. The web application is optimized to be used in different browsers and the android app promises a snappy user interface. The proposed technique will be able to show better results in other different applications as well. Further works will focus on parameter tuning during model generation and dataset tuning to get even better results.

References

1. J.C. Platt, Sequential minimal optimization, a fast algorithm for training support vector machines. Microsoft Res. (1998)
2. N. Deepika, S. Poonkuzhali, Design of hybrid classifier for prediction of diabetes through feature relevance analysis. in *International Journal of Innovative Science, Engineering & Technology*, vol. 2 no. 10, Oct (2015)
3. S. Vijayarani, S. Dhayanad, Liver Disease Prediction using SVM and Naivee Bayes algorithm. Int. J. Sci. Eng. Technol. Res. (IJSETR) **4**(4), (2015)
4. A.I. Ahmed, M.M. Hasan, A hybrid approach for decision making to detect breast cancer using data mining and autonomous agent based on human agent teamwork. in *17th International Conference on Computer and Information Technology (ICCIT)*, (2014)
5. V. Levashenko, E. Zaitseva, Fuzzy Decission trees in medical decision making support system. in *Proceedings of the Federated Conference on Computer Science and Information Systems*, pp. 213–219, (2012)
6. M.R. Senapati, A.K. Mohanty, S. Dash, P.K. Dash, Local linear wavelet neural network for breast cancer recognition. Neural Comput. Appl. **22**(1), 125–131 (2013)
7. R.K. Ravi, T.K. Ratheesh, An efficient human disease detection using data mining techniques. in *International Conference on Intelligent Data Communication Technologies and Internet of Things (ICICI)* 2018. vol. 26, pp. 96–966, (2019)
8. A.K. Pal, P. Rawal, R. Ruwala, V. Patel, Generic disease prediction using symptoms with supervised machine learning. Int. J Sci. Res. Comput. Sci. Eng. Inf. Technol. **5**(2), 1082–1086 (2019)
9. S. Babu et al., Heart disease diagnosis using data mining technique. in *2017 International conference of Electronics, Communication and Aerospace Technology (ICECA)*, (Coimbatore, 2017), pp. 750–753
10. D. Dahiwade, G. Patle, E. Meshram, Designing disease prediction model using machine learning approach. in *2019 3rd International Conference on Computing Methodologies and Communication (ICCMC)*, (Erode, India, 2019), pp. 1211–1215
11. B.V. Baiju, D.J. Aravindhar, Disease influence measure based diabetic prediction with medical data set using data mining. in *2019 1st International Conference on Innovations in Information and Communication Technology (ICIICT)*, Chennai, India, 2019, pp. 1–6
12. V. Vapnik, *Statistical Learning theory* (Wiley, New York, 1998)
13. UCI repository of machine learning databases, University of California at Irvine, Department of Computer Science Wisconsin Breast Cancer Database Available: http://archive.ics.uci.edu/ml/machine-learning-databases

Computational Science and its Applications

Nested Sparse Classification Method for Hierarchical Information Extraction

Gargi Mishra and Virendra P. Vishwakarma

Abstract Use of sparse representation method for classification of visual data has proved its efficiency for challenges encountered in unconstrained face identification. The main identified constrains which heavily affect classification accuracy are variation in expression, pose and lighting. In this paper, a novel classification method is developed as nested sparse classification (NSC) method which incorporates the advantages of sparse representation manifold. In NSC method, sparse representation-based classification is implemented in a nested manner which allows the extraction of hierarchical relationship information between test and training samples. The hierarchical relationship information helps the classifier to be discriminative to inter-personal face changes while robust to intra-personal variations. The implementation not only improves the classification performance but also makes the system capable of being scaled. The improved accuracy of proposed NSC method is demonstrated by experiments carried out on two standard databases (ORL database and YALE database). The performance is analysed with an improvement of more than 2% in terms of classification accuracy.

Keywords Nested sparse · Sparse representation · Face recognition · Linear classifier · Hierarchical information extraction

---

G. Mishra · V. P. Vishwakarma (✉)
University School of Information Communication and Technology, Guru Gobind Singh Indraprastha University, New Delhi 110078, India
e-mail: virendravishwa@rediffmail.com

G. Mishra
e-mail: gargi.mishra1983@gmail.com

P. Bansal et al. (eds.), *Proceedings of International Conference on Artificial Intelligence and Applications*, Advances in Intelligent Systems and Computing 1164,
https://doi.org/10.1007/978-981-15-4992-2_50

1 Introduction

Face recognition (FR) has attracted many researchers due to its applications in access control, Internet communication, law enforcement, security, computer entertainment and surveillance [1]. The efficiency of recognition algorithms is limited due to various unconstrained conditions present in real-life facial images. The major constraints affecting the performance of FR algorithms are illumination changes, pose variation, occlusion and expression changes [2]. Thus, research focus is still required in this field. Many FR algorithms have already been proposed with significantly good classification accuracy. During past decade, sparse representation approach has been used in many FR algorithms and is proved to be very efficient and robust. In sparse representation approach, test image is represented as sparse linear combination of training images. Further, the test image is classified to the class of training images having least reconstruction error. Significant number of recently proposed classification approaches based on sparse representation is available in the literature. Wagner et al. [3] proposed face recognition method using sparse representation which addresses illumination changes and image misalignment. Extended SRC proposed by Deng et al. [4] identifies variation between training and testing data by constructing a dictionary for intra-class variants. Wang et al. [5] proposed adaptive sparse representation-based classification (ASRC), in which sparsity and correlation both are considered jointly. In [6, 7], structured sparsity is proposed, which is able to approximate the image deterioration due to illumination, misalignment and occlusion. Superposed SRC proposed by Deng et al. [8] uses sample-to-centroid differences and class centroids for constructing dictionary. Discriminative low-rank representation (DLRR) method proposed by Chen and Yi [9] is further extension of sparse representation method which rectifies the corrupted test images by extracting a low-rank projection matrix. Label consistent K-SVD (LC-KSVD) proposed by Jiang et al. [10] prepares a discriminative dictionary with label information. In [11], a method called LGE-KSVD, proposed by Ptucha and Savakis, focuses on optimizing the dictionary learning through linear extension of graph embedding. Xu et al. [12] proposed a method of FR using sparse representation method in two different phases. Xu and Zhu [13] proposed FR using simple sparse representation method and reported 2–10% performance improvement over nearest neighbour classification method [14].

In this paper, a robust face image recognition method, NSC is proposed, which utilizes sparse representation-based classification method in a nested manner. The use of sparse representation at different levels not only helps to increase the sparsity but also incorporates the advantages of sparse representation manifold. The implementation of NSC method in nested manner also allows the extraction of hierarchical relationship information between training and testing images. The fundamental inspiration of this work is to extract the hierarchical relationship information which provides precise discrimination between different class face images and intra-class variations. The use of hierarchical relationship information provides better classification accuracy along with increasing the system scalability. Proposed method is evaluated for face image recognition on two standard datasets. Extensive simulations are carried out on ORL [15] and YALE [16] datasets to prove the potential of NSC method. Results are compared against the simple sparse representation method for all possible sets. Comparative performance is presented in terms of mean classification accuracy.

Rest of the paper is organized as follows. Section 2 elaborates the details of the proposed method. Section 3 focuses on the simulation results of proposed method on two standard databases and result analysis. Finally, conclusions are discussed in Sect. 4.

2 Proposed Method

For implementation of proposed method, it is assumed that every dataset has total 'C' classes, with each class having 'D' face images $I_1, I_2, \ldots, I_D$. Each dataset is split into two mutually exclusive image sets, training image set and testing image set. Training image set has 'T' images from each class, while testing image set contains remaining $(D - T)$ images per class. Therefore, training image set contains total $(T \times C)$ images whereas testing image set has $(D - T) \times C$ images. Class label for training images is denoted by 'L', i.e. a particular training image may belong to Lth class, where $L = 1, 2, \ldots, C$. In the proposed method, all the dataset images are resized to 40×40 pixels and converted into unit column vectors. For conversion into unit column vector, input image is first converted to column vector and then normalized, Eq. (1).

$$I_j = \frac{I_j(:)}{\|I_j\|_2} \tag{1}$$

where $j = 1, 2, \ldots, (C \times D)$.

In NSC method, sparse approximation-based classification is used at two different levels to improve the classification performance of simple sparse method. At the first level, one nearest training image from each class is identified using intra-class sparse approximation (Intra-CSA). And at the second level, test image is approximated as linear sparse combination of identified nearest training images (Here, number of identified nearest training images is equal to the number of total classes). Further, the nearest training image having highest contribution in representing test image is identified using inter-class sparse approximation (inter-CSA). Finally, the nearest training image having highest contribution value in representing test image is considered to be the finest match for the test image.

Level I: Estimation of C nearest training images using intra-CSA In this level, C nearest training images are identified using intra-CSA which is explained in following points:

1. For each class, test image is approximated as a linear sparse combination of all training images from that particular class. Therefore, the following equation should be satisfied.

$$q = \alpha_1^L t_1^L + \alpha_2^L t_2^L + \cdots + \alpha_T^L t_T^L \quad (2)$$

In Eq. 2, $t_i^L (i = 1, 2, \ldots, T)(L = 1, 2, \ldots, C)$ are T training images from Lth class and $\alpha_i^L (i = 1, 2, \ldots, T)(L = 1, 2, \ldots, C)$ are the corresponding coefficients needed for intra-CSA. All the training images from the Lth class $t_1, t_2, \ldots, t_T$ are represented using a matrix $\mathrm{TI}^L = [t_1, t_2, \ldots, t_T]$. In other words,

$$q = \sum_{i=1}^{T} \alpha_i^L \mathrm{TI}(:, i)^L \quad (3)$$

In matrix form, Eq. (3) can be rewritten as

$$q = \mathrm{TI}^L * \alpha^L \quad (4)$$

where $\mathrm{TI}^L = \left[t_1^L, t_2^L, \ldots, t_T^L\right]$, $\alpha^L = \left[\alpha_1^L, \alpha_2^L, \ldots, \alpha_T^L\right]$ and $L = (1, 2, \ldots, C)$.

Also, values of α^L are restricted real numbers, i.e. value lies between -1 to $+1$ and $\alpha_1^L + \alpha_2^L + \cdots + \alpha_T^L = 1$.

Further, Eq. (4) is solved to estimate the values of α^L. To solve Eq. (4), singularity test is performed on $\left[\mathrm{TI}^L\right]^T\left[\mathrm{TI}^L\right]$. If $\left[\mathrm{TI}^L\right]^T\left[\mathrm{TI}^L\right]$ is not singular, solution is obtained using least squares method, Eq. (5).

$$\alpha^L = \left(\left[\mathrm{TI}^L\right]^T\left[\mathrm{TI}^L\right]\right)^{-1}\left[\mathrm{TI}^L\right]^T q \quad (5)$$

If $\left(\left[\mathrm{TI}^L\right]^T\left[\mathrm{TI}^L\right]\right)$ is nearly singular, values of α^L is estimated using Eq. (6).

$$\alpha^L = \left(\left[\mathrm{TI}^L\right]^T\left[\mathrm{TI}^L\right] + \mu I\right)^{-1}\left[\mathrm{TI}^L\right]^T q \quad (6)$$

Here, value of μ is positive real number and I is identity matrix. In our implementation, value of μ is set to 0.01.

2. Coefficient values (α^L) estimated above are used for calculation of representation contribution of each training image present in $\left[\mathrm{TI}^L\right]$. It can be understood from approximation equation Eq. (1) that each $\left[\mathrm{TI}^L\right]$ makes some contribution towards the representation of test image q. The representation contribution made by ith $\left[\mathrm{TI}^L\right]$ is $\alpha_i^L\left[\mathrm{TI}^L\right]_i$, where $i = (1, 2, \ldots, T)$ and $L = (1, 2, \ldots, C)$. The difference between representation contribution RC of ith training image and test image q is calculated using squared norm_2 distance given by Eq. (7).

$$D_i^L = \left\|q - \mathrm{RC}_i^L\right\|^2 \quad (7)$$

From the above explanation, it is understood that smaller value of D_i^L provides higher representation contribution of ith training image with test image.

Accordingly, one nearest training image having highest representation contribution is selected from each class to estimate total C nearest training images of test image q. Finally, all the C nearest training images $n_1, n_2, \ldots, n_C$ are represented using a matrix $\text{NTI} = [n_1, n_2, \ldots, n_C]$.

Level II: Classification using inter-CSA The implementation is explained in the following points:

3. In this step, test image q is approximated as linear sparse combination of selected C nearest training images. Here, it is assumed that the following equation is perfectly satisfied.

$$q = \alpha_1 n_1 + \alpha_2 n_2 + \cdots + \alpha_C n_C \tag{8}$$

Here, $n_i (i = 1, 2, \ldots, C)$ are C—nearest training images of test image q and $\alpha_i (i = 1, 2, \ldots, C)$ are respective coefficients. Alternatively,

$$q = \sum_{i=1}^{C} \alpha_i \text{NTI}(:, i) \tag{9}$$

Equation (9) can be represented in matrix form as

$$q = \text{NTI} * \alpha \tag{10}$$

Here, $\text{NTI} = [n_1, n_2, \ldots, n_C]$ and $\alpha = [\alpha_1, \alpha_2, \ldots, \alpha_C]^T$. Here, α values are restricted real numbers with values between -1 to $+1$ and also satisfy the equation $\alpha_1 + \alpha_2 + \cdots + \alpha_C = 1$. The solution of Eq. (10) is obtained in similar as in Level I using Eqs. (5) and (6).

4. In this step, representation contribution is calculated for final classification task. For ith nearest training image, representation contribution calculation is done using Eq. (11).

$$\text{RC}_i = \alpha_i * \text{NTI}(:, i) \tag{11}$$

$$D_i = \|q - \text{RC}_i\|^2 \tag{12}$$

Finally, the test image q is classified into the class of nearest training image having smallest value of difference (D), i.e. highest representation contribution Eq. (12). For easier understanding, NSC method is also explained in Algorithm 1.

Algorithm 1: NSC Method

Input *Grey scale/colour image*

For each input image, **do**

Resize input image to 40×40 pixels.

Convert input image into column vector.

Normalize using L_2 norm.

EndFor

Divide the complete image dataset into mutually exclusive training and testing image set.

For each test image, q, **do**

Estimate C nearest training images using intra-CSA

Compute α for intra-CSA, equations (4), (5) and (6).

Calculate representation contribution for each nearest training image.

Find C nearest training images having minimum deviation, equation (7).

Final classification using inter-CSA

Compute α for intra-CSA, equations (5), (6) and (10).

Calculate RC for each nearest training image, equation (11).

Find nearest training image with smallest deviation, equation (12).

EndFor

Output: *Test image, q, classified to the class of nearest training image having smallest deviation.*

3 Results and Analysis

The proposed method is evaluated using two classification experiments conducted with MATLAB on Intel Core i5, 2.71 GHZ and RAM of installed capacity 8 GB. Experiments are performed on two standard face datasets, ORL [15] and YALE [16]. Each dataset is first divided into two mutually exclusive image sets having training and testing images. For each dataset, T images out of D images per class are picked up as training image set and remaining $(D - T)$ images per class are testing image set. Thus, for a particular dataset, the total number of training image sets is $\frac{D!}{T!(D-T)!}$. To showcase the superiority of NSC method over simple sparse method, analysis is done in terms of mean percentage classification accuracy and improvement in classification accuracy on ORL and YALE datasets.

3.1 *Experiment 1: ORL Dataset*

ORL face image dataset has images of 40 different subjects with 10 images of each subject. The image dimension is 92 × 112 pixels with 256 grey levels. Experiments are performed for all possible training image sets and more that 3 training images per class (TIPC). Table 1 shows mean classification accuracy (%) of proposed NSC method and simple sparse method on ORL dataset. The change in mean classification accuracy (%) for different numbers of TIPC of ORL dataset is plotted in Fig. 1. Plot shows two curves of different colours; lower curve in red color depicts mean

Table 1 Mean classification accuracy (%) of proposed NSC method and simple sparse method on ORL dataset

Number of TIPC	Number of training sets	Proposed NSC method	Simple sparse method [13]	Improvement in classification accuracy (%)
4	210	93.50	91.52	2.16
5	252	95.51	93.57	2.07
6	210	96.65	94.76	1.99
7	120	97.45	95.48	2.06
8	45	97.99	96.00	2.07
9	10	98.21	96.25	2.03

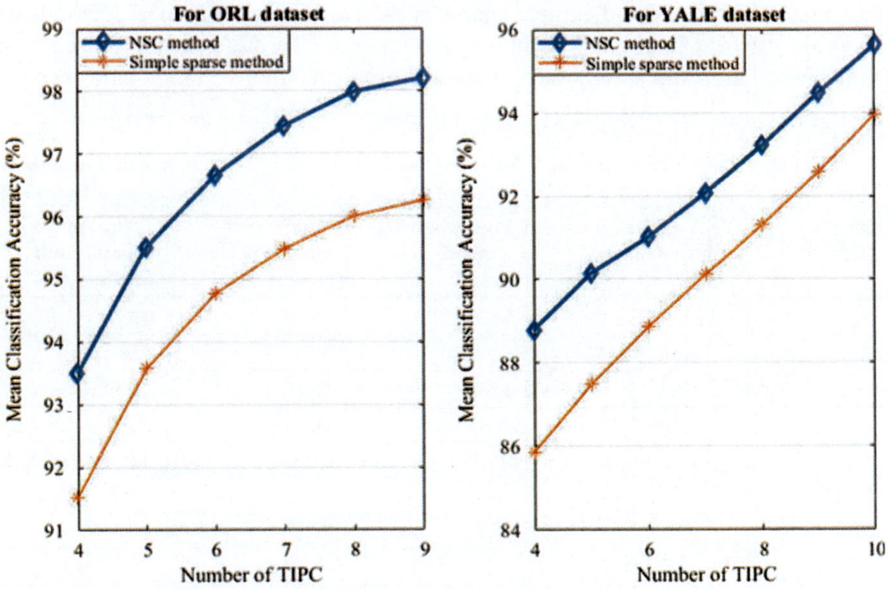

Fig. 1 Mean classification accuracy (%) using NSC method

classification accuracy (%) of simple sparse method, whereas upper curve in blue color represents proposed NSC method. It is obvious from the plot that classification accuracy varies with the change in number of TIPC used in the experiment. Classification accuracy greatly improves with the increase in number of TIPC.

Since increase in the number of TIPC gives rise to memory and computational requirements, therefore these two factors must be handled carefully while designing an FR system.

3.2 *Experiment 2: YALE Dataset*

YALE face image dataset has images of 15 different subjects with 11 images of each subject. The image dimension is 220 × 175 pixels with 256 grey levels. Experiments are performed for all possible training image sets and more that 3 training images per class (TIPC).

Table 2 shows mean classification accuracy (%) of proposed NSC method and simple sparse method on YALE dataset. The change in mean classification accuracy (%) for different number of TIPC of YALE dataset is plotted in Fig. 1. The maximum improvement in classification accuracy is observed as 2.16% for ORL dataset at 4 TIPC and 3.41% for YALE dataset at 4 TIPC. The potential limitations of this research are also discussed here. Larger size of training dataset results into higher number of computations per test with the increased accuracy. This is the fundamental restriction of simple sparse method which also applies to proposed NSC method. In the proposed NSC method, simple sparse is used to evaluate the linear relationship between test face image and training face images. Therefore, in the presence of nonlinear relationship among face images, the classification accuracy may degrade.

Table 2 Mean classification accuracy (%) of proposed NSC method and simple sparse method on YALE dataset

Number of TIPC	Number of training sets	Proposed NSC method	Simple sparse method [13]	Improvement in classification accuracy (%)
4	330	88.79	85.86	3.41
5	462	90.16	87.50	3.04
6	462	91.02	88.86	2.43
7	330	92.07	90.11	2.17
8	165	93.21	91.30	2.09
9	55	94.47	92.55	2.07
10	11	95.62	93.94	1.78

4 Conclusion

This paper presents a novel classification technique called NSC method which is based on simple sparse representation. The use of simple sparse representation in nested manner allows the extraction of hierarchical relationship information among training and test face images which provides precise discrimination between different face images and intra-personal variations. The hierarchical relation information not only contributes to increased classification accuracy but also makes the system scalable. The classification performance of NSC method is evaluated on two standard datasets, ORL and YALE, for more than three training images per class. The improvement in classification accuracy is observed to be more than 2% for all tests conducted. Proposed NSC method can be extended by adding a greater number of nesting levels of simple sparse for improved classification accuracy.

References

1. W. Zhao, R. Chellappa, P.J. Phillips, A. Rosenfeld, Face recognition: a literature survey. ACM Comput. Surv. **35**(4), 399–458 (2003)
2. P.J. Phillips, J.R. Beveridge, B.A. Draper, G. Givens, A.J. O'Toole, D.S. Bolme, S. Weimer, An introduction to the good, the bad, & the ugly face recognition challenge problem, in IEEE International Conference on Automatic Face & Gesture Recognition and Workshops (FG 2011) (2011), pp. 346–353
3. A. Wagner, J. Wright, A. Ganesh, Z. Zhou, H. Mobahi, Y. Ma, Toward a practical face recognition system: robust alignment and illumination by sparse representation. IEEE Trans. Pattern Anal. Mach. Intell. **34**(2), 372–386 (2012)
4. W. Deng, J. Hu, J. Guo, Extended SRC: undersampled face recognition via intraclass variant dictionary. IEEE Trans. Pattern Anal. Mach. Intell. **34**(9), 1864–1870 (2012)
5. J. Wang, C. Lu, M. Wang, P. Li, S. Yan, X. Hu, Robust face recognition via adaptive sparse representation. IEEE Trans. Cybern. **44**(12), 2368–2378 (2014)
6. K. Jia, T.H. Chan, Y. Ma, Robust and practical face recognition via structured sparsity, in Computer Vision–ECCV 2012 (Springer, 2012), pp. 331–344
7. X. Wei, C.T. Li, Y. Hu, Robust face recognition under varying illumination and occlusion considering structured sparsity, in 2012 IEEE International Conference on Digital Image Computing Techniques and Applications (DICTA) (IEEE, 2012), pp. 1–7
8. W. Deng, J. Hu, J. Guo, In defense of sparsity based face recognition, in 2013 IEEE Conference on Computer Vision and Pattern Recognition (CVPR) (IEEE, 2013), pp. 399–406
9. J. Chen, Z. Yi, Sparse representation for face recognition by discriminative low-rank matrix recovery. J. Vis. Commun. Image Represent. **25**(5), 763–773 (2014)
10. Z. Jiang, Z. Lin, L.S. Davis, Label consistent K-SVD: learning a discriminative dictionary for recognition. IEEE Trans. Pattern Anal. Mach. Intell. **35**(11), 2651–2664 (2013)
11. R. Ptucha, A. Savakis, LGE-KSVD: flexible dictionary learning for optimized sparse representation classification, in 2013 IEEE Conference on Computer Vision and Pattern Recognition Workshops (CVPRW). (IEEE, 2013), pp. 854–861
12. Y. Xu, D. Zhang, J. Yang, J.Y. Yang, A two-phase test sample sparse representation method for use with face recognition. IEEE Trans. Circ. Syst. Video Tech. **21**(9), 1255–1262 (2011)
13. Y. Xu, Q. Zhu, A simple and fast representation-based face recognition method. Neural Comput. Appl. **22**(7–8), 1543–1549 (2013)

14. L. Nanni, A. Lumini, Particle swarm optimization for ensembling generation for evidential k-nearest-neighbour classifier. Neural Comput. Appl. **18**(2), 105–108 (2009)
15. F. Samaria, A. Harter, Parameterisation of a stochastic model for human face identification, in: Proceedings of IEEE Workshop Applications of Computer Vision. (1994), pp. 138–142. Available: http://www.cl.cam.ac.uk/research/dtg/attarchive/facedatabase.html
16. P. Belhumeur, J. Hespanha, D. Kriegman, Eigenfaces vs. fisherfaces: recognition using class specific linear projection. IEEE Trans. Pattern Anal. Mach. Intell. **19**(7), 711–720 (1997). Available: http://cvc.yale.edu/projects/yalefaces/yalefaces.html

A Robust Surf-Based Online Human Tracking Algorithm Using Adaptive Object Model

Anshul Pareek, Vsudha Arora, and Nidhi Arora

Abstract Vision-based tracking is a requisite to a large number of applications in computer vision. Tracking algorithms are expected to deal with problems like full occlusion/partial occlusion, luminance variation, scale, pose, and camera motion. Initially, color and contour-based human tracking was a trend but it had inherent shortcomings that were overcome by the interest (descriptor) point-based tracking. But these tracking methods had its own limitations like the number of matching points obtained vary axiomatically from one frame to another and may subside over time, and the high computational complexity in a running video. This paper comes up with an adaptive model to describe the object model which under non-stationary conditions will track a human. This algorithm uses a foreground model of target obviates the need of a classifier, also accounts for human motion and its pose change, and SURF as feature detector descriptor is used along with a simple auto-regressive (AR) predictor to pledge with the cases of partial or full occlusion. The efficacy of the algorithm is proven by the simulation results obtained.

Keywords Human tracking · Object detection · Image matching · Motion model · ORB · SIFT · SURF · AR predictor

---

A. Pareek (✉)
ECE Department, Maharaja Surajmal Institute of Technology, New Delhi, India
e-mail: er.anshulpareek@gmail.com

V. Arora · N. Arora
CSE Department, G.D. Goenka University, Haryana, India
e-mail: vasudha.arora@gdgoenka.ac.in

N. Arora
e-mail: nidhi.arora1@gdgoenka.ac.in

P. Bansal et al. (eds.), *Proceedings of International Conference on Artificial Intelligence and Applications*, Advances in Intelligent Systems and Computing 1164,
https://doi.org/10.1007/978-981-15-4992-2_51

1 Introduction

What is computer vision? In simple words, it is mimicking of human vision by eyes. Computer vision interprets the images using static or mobile cameras to understand their content. Computer vision duties are equivalent to those of human vision-like object recognition/classification, object detection, and object tracking. Tracking algorithms task is to estimate framework of the dynamic system, e.g., feature point positions, object position, their matching, its estimation. The source of information is the frames obtained from video captured by single or multiple cameras which may be static or moving. The application areas are broadly classified as surveillance [1], augmented reality, automatic object detection, human—computer interaction [2], video compression, and assistive robotics [3, 4]. There has been an unparalleled growth witnessed in the last two decades in this field because of low computing power and low-cost setup [5, 6].

Our paper focuses on tracking by detection framework based on interest point detection techniques [7]. From the evaluation of various feature detector–descriptor algorithms [8], ORB SURF and SIFT are the decent choices to overcome tracking challenges in real-time systems. All three feature detection description techniques are applied to the motion model. First preference was given to ORB because it is the fastest one [8], but it did not turn up with sufficient amount of keypoints to continue tracking. So, SURF and SIFT are applied. SIFT is very robust and shows good performance under scale and rotation variances [9]. Affine shifts are a little tricky, but moderate [8]. The advantages of SURF over SIFT are rotational invariancy, blurring, and warp transformation. When scaled images are under consideration, then SIFT is used. The computational complexity os SURF is one-third that of SIFT because it uses integral image and box filter. When illumination variation is encountered, then both SIFT and SURF are efficient [8]. Choosing feature description along with color feature for describing object model, then using only feature description technique would result in more robust performance [10, 11].

We explicitly deal with the problem of tracking a human in a pre-recorded video. This can be applied for tracking/following any human by a mobile robot. Initially, first frame of the video is captured and object to be tracked is selected manually. Descriptors are obtained in the selected region to create an object model. In the next step, the next frame target is detected by matching the features. The target region on new frame is detected by the polygon which contains the keypoints obtained by the correspondences. But in the long run, obtained matching points fade in number. So, to define target polygon, an effective number of good matches is required, and there is a dire need to update the object model from time to time. An auto-regressive predictor AR (p) is used where $p = 10$. This helps in predicting the location of the target when occlusion is encountered because of unavailability of matching points. Gradient–descent algorithm provides the predictor coefficients.

So, here a tracking algorithm detects the target in all the coming frames using object model. The descriptor points in the model evolve over time. These evolved points are obtained from a template pool where stable feature points from each frame are stored. This obviates the need for a pre-trained classifier simultaneously resolving the stability plasticity dilemma problem of visual tracking [12–14].

The coming sections include problem definition. Section 3 gives an overview of the object model, and experimental setup is explained in Sect. 4. The tracking algorithm is explained in Sect. 5 followed by experimental results discussed in Sect. 6. Section 7 states the conclusion.

2 Problem Definition

Target is located in first frame, and this is to be tracked in all frames. Let us assume a set of frames F_i, where $i = 0, 1, 2, 3 \ldots N$. The human is identified by selecting contour; this is achieved manually by pointing human silhouette. Later, a polygon A_o is drawn over these points. Let $S(I) = \{(x_1, s_1), (x_2, s_2), \ldots, (x_n, s_n)\}$ be the set of feature detector keypoints of an image F, where x_i is the two-dimensional keypoint location of the 64-dimensional SIFT or SURF descriptor s_i. The correspondences are the best matching keypoints between two images. A tracking polygon for each frame is calculated. Let A_t be the tracking polygon on image frame F_t.

$$A_t = (b_t, c_t) \tag{1}$$

where b_t is the set of keypoints locations enclosed by a polygon and is defined as:

$$b = \{x_i | (x_i, s_i) \in S_0 \wedge x_i \in A_0\} \tag{2}$$

and

$$c_t = \left(c_x, c_y\right) \text{ is the center of the polygon} \tag{3}$$

Given $F_0, A_0(b_0, c_0)$ and S_0, the task to be accomplished is the computation of $A_t(b_t, c_t)$ for all the frames $t = 1, 2, 3\ldots N$.

3 Object Model

The object model is created by using three variants of sets $k_1(t)$, $k_2(t)$ and $k_3(t)$ where $k_1(t)$ holds the keypoints obtained from matching current frame F_t from previous frame F_{t-1} The second set $k_2(t)$ consists of correspondences between first frame and current frame. The keypoints associated with these correspondences within a polygon constitute $k_2(t)$. The third set is obtained by the correspondence between current frame and descriptors of template pool. The template pool consists of correspondence which evolves over time. This set basically defines a convex hull of the object under tracking, which is to be followed in the captured frame. So the frequently matching keypoints appear in template pool for long as compared to points which are insignificant and appear very less during matching; mostly, they are the false matches. The object model derives its stable feature points from this template pool. The points within the bounded regions are the ones only retained. All the background descriptors are discarded. Hence, the object model is now

$$M_{\mathrm{m}}(t) = \{k_1(t) \cup k_2(t) \cup k_3(t)\} \tag{4}$$

It is clear from the above object model that it is basically a function of the current frame that makes it dynamic in nature [15].

4 Test Bed

For this evaluation, C++ is the software framework. Open CV 3.2.0 is used for implementation of algorithm. Simulation of algorithm is done on Linux-based Apple Macbook Air with core i5, 1.8 GHz. The brute-force matcher is used for both algorithms for matching; it uses nearest neighbor distance ratio (NNDR). The threshold ratio kept here is 0.7.

A manual selection of contour is done using a mouse. For feature detection and extraction SIFT and $\mathrm{SURF}_{(64)}$ are used. Outlier rejection is must in order to obtain dependable object model while forming these above-mentioned sets. The first set $k_1(t)$ uses homography-based RANSAC [15] for outlier rejection with 2000 iterations and 99.5% confidence. The same technique is not applicable to all three because there is significant displacement in the object from first frame to current frame. Hence, for the set $k_2(t)$, KNN [16] and scaling is used to remove outliers. In case of third set $k_3(t)$, outliers are rejected based on the user-defined threshold on the likelihood present between feature points of template pool and current image matching points. The number of descriptors in template pools is user-defined limited to 5000 in this case. The size of template is maintained by removing oldest descriptor on arrival of new descriptor and size remains the same.

5 Tracking Algorithm

Initially, as shown in Fig. 1, target initializations are done where the target is manually selected and framed by polygon A_0. On the second step, object recognition and target model update $k_1(t)$, $k_2(t)$, and $k_3(t)$ sets are obtained as explained before. At last, if number of matching points between current frame and the object model is less than the user-defined threshold, an occlusion is predicted. In such condition, with the help of autoregression predictor, the center of polygon is predicted. Matching points are looked for within rectangular window around this center obtained. If still

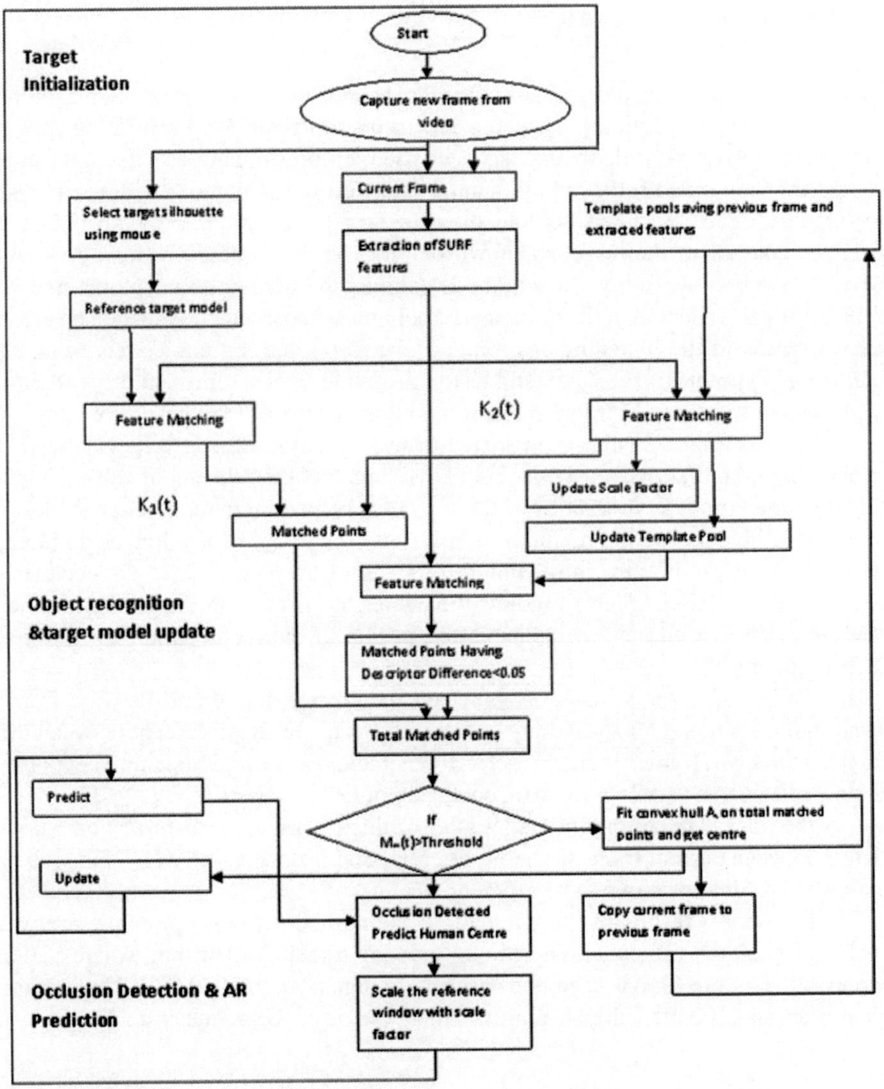

Fig. 1 Block diagram of the tracking algorithm

the matching points are less than the threshold, this confirms occlusion. Tracking is reinitialized once matching points above threshold are obtained.

6 Experimental Result

The video obtained with the tracking algorithm (for different video) is available online for viewing and inspection [17]. A few snapshots of the tracking results are shown in Fig. 2 for which recorded video has been used. Figure 2a manifests the first frame captured of the video followed by manual selection of the contour with the mouse in Fig. 2b. Figure 2c shows contour over the target. In Fig. 2d, polygon is drawn over the contour of the target as shown below. The descriptors obtained by using SURF are presented in Fig. 2e. Figure 2f exhibits the tracker polygon obtained by the algorithm consisting of the three sets of polygons discussed previously. The green circles represent the matching keypoints of set $K_1(t)$, the red circles represent the matching keypoints of set $K_2(t)$, and lastly, the white circles represent the matching keypoints of set $K_3(t)$. Figure 2g–l shows tracking of target via tracker polygon.

Partial and full occlusion is presented in Fig. 2j and k, respectively. It is visible in Fig. 2o and p that there are matching keypoints less than threshold value. In this case, template pool keypoints from the $K_3(t)$ set helps in attaining better tracking. These matching keypoints maintain the size of convex hull for a longer duration, so after occlusion, it helps in regaining the track of the target. Once the occlusion is encountered, AR predictor predicts the center of polygon in the coming frames leading to the search window as exhibited in Fig. 2l, where tracking is recovered after occlusion.

Figure 2n–p presents matching between the current frame and the first frame identified from the video. Matching points reduces as the displacement between the targets in the two frames increases, clearly visible in Fig. 2n–p. Matching keypoints almost reduce to a very low value on occlusion of target as seen in Fig. 2o. Figure 2p shows the tracker regaining the target after occlusion. Figure 2m shows correspondence between current frames. The outlier rejection is done using KNN and scaling as RANSAC did not prove to be efficient.

Table 1 shows results of the simulation; the number of descriptors are average making it memory efficient. Whereas for existing tracking algorithm, where SURF along with color or blob is used, the descriptor range is up to 1000 [18]. The average time is around 400 ms which is almost half of the other techniques used [18].

7 Conclusion

The proposed algorithm states a dynamic object model-based tracking which detects target in all the consecutive frames. This model contains a set of SURF feature points which are updated over the time and again. The algorithm has several advantages over conventional and existing methods; firstly, it overcame the limitation of interest point-based tracking. Secondly, no need of training a classifier since the implemented algorithm uses a model for the foreground only. This reduces the computational

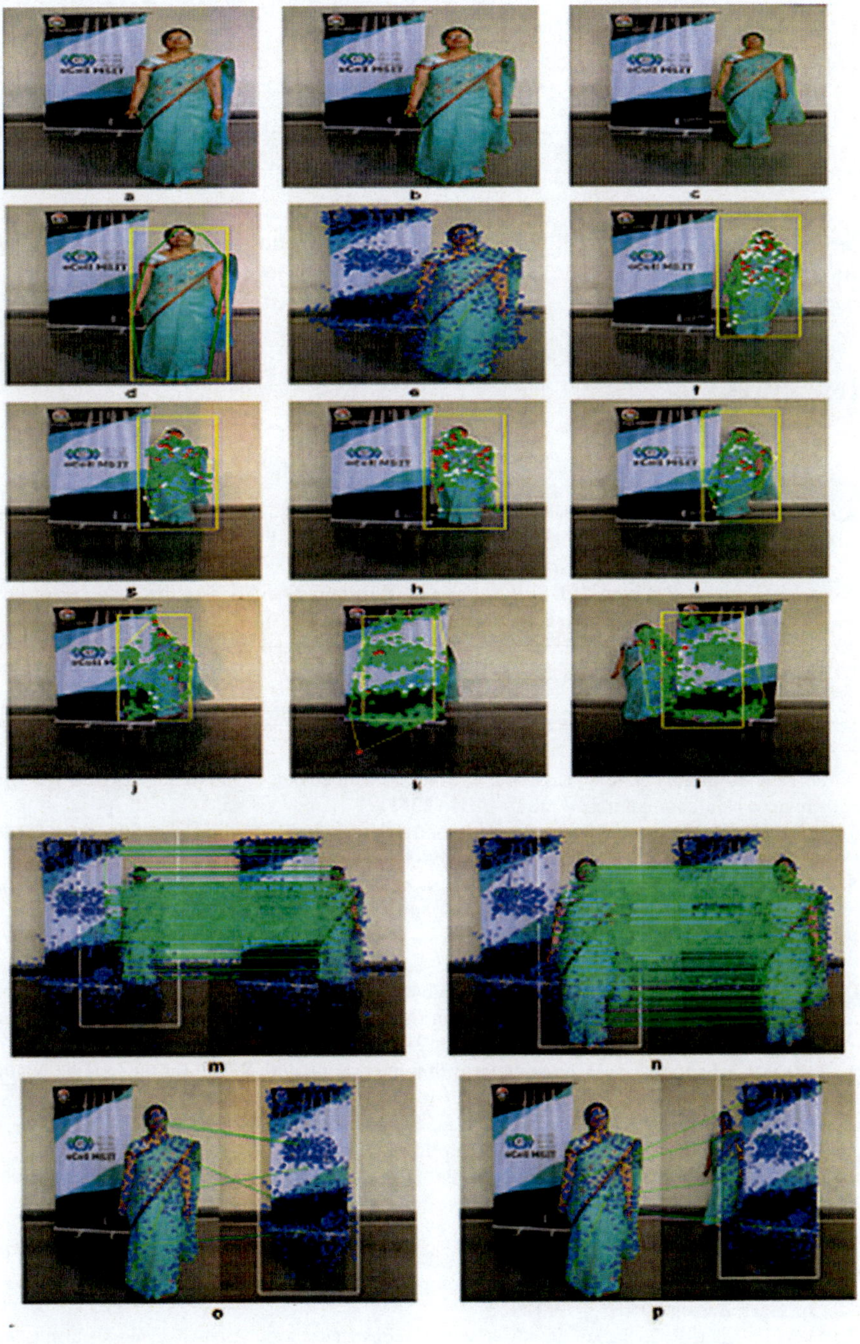

Fig. 2 Tracking results

Table 1 Details of microenvironment of algorithm trial

Dataset	Total number of frames	Occlusion	Camera motion and pose change	Number of descriptors	Average time (ms)
D-1	290	Yes	Smooth, No	100–400	480

complexity. Last but not least, the object model description reports pose variation of humans, and therefore, can be used for long-term tracking.

References

1. N.V. Puri, P.R. Devale, Development of human tracking in video surveillance system for activity analysis. IOSR J. Comput. Eng. **4**(2), 26–30 (2012)
2. J.M. Rincón, D. Makris, C. Urunuela, J.C. Nebel, Tracking human position and lower body parts using Kalman and particle filters constrained by human biomechanics. IEEE Trans. Syst. Man Cybern. Part B. **41**(1), 26–37 (2011)
3. M.J. Mataric, J. Eriksson, D.J. Feil-Seifer, C.J. Winstein, Socially assistive robotics for post-stroke rehabilitation. J. Neuro Eng. Rehabil. **4** (2007)
4. F.M. Campos, L. Correia, J.M.F. Calado, Robot visual localization through local feature fusion: an evaluation of multiple classifiers combination approaches. J. Intell. Rob. Syst. **77**(2), 377–390 (2015)
5. A. Yilmaz, O. Javed, M. Shah, Object tracking: a survey. ACM Comput. Surv. (CSUR) **38**(4) (2006)
6. H. Yang, L. Shao, F. Zheng, L. Wang, Z. Song, Recent advances and trends in visual tracking: a review. Neurocomputing **74**, 3823–3831 (2011)
7. W. Kloihofer, M. Kampel, Interest point based tracking, in International Conference on Pattern Recognition (ICPR) (ACM, 2010), pp. 3549–3552
8. A. Pareek, N. Arora, Evaluation of feature detector-descriptor using RANSAC for visual tracking (23 Feb 2019). Available at SSRN: https://ssrn.com/abstract=3354470
9. D.G. Lowe, Distinctive image features from scale-invariant keypoints. Int. J. Comput. Vis. **60**(2), 91–110 (2004)
10. S. Haner, I.Y. Gu, Combining foreground/background feature points and anisotropic mean shift for enhanced visual object tracking, in International Conference on Pattern Recognition (ICPR), Istanbul (IEEE, 2010), pp. 3488–3491
11. J. Zhang, J. Fang, J. Lu, Mean-shift algorithm integrating with SURF for tracking, in Natural Computation (ICNC), Shanghai. (IEEE, 2011), pp. 960–963
12. W. He, T. Yamashita, L. Hongtao, S. Lao, SURF tracking, in International Conference on Computer Vision, Kyoto (IEEE, 2009), pp. 1586–1592
13. D.-N. Ta, W.-C. Chen, N. Gelfand, K. Pulli, SURFTrac: efficient tracking and continuous object recognition using local feature descriptors, in IEEE Conference on Computer Vision and Pattern Recognition (CVPR), Miami, FL (IEEE, 2009), pp. 2937–2944
14. S. Gu, Y. Zheng, C. Tomasi, Efficient visual object tracking with online nearest neighbor classifier, in 10th Asian Conference on Computer Vision (ACCV), Queenstown, New Zealand (Springer, Berlin, 2010), pp. 271–282
15. M.A. Fischler, R.C. Bolles, Random sample consensus: a paradigm for model fitting with applications to image analysis and automated cartography. Commun. ACM **24**(6), 381–395 (1981)

16. M. Gupta, S. Garg, S. Kumar, L. Behera, An on-line visual human tracking algorithm using SURF-based dynamic object model, in International Conference on Image Processing (ICIP) (IEEE, 2013), pp. 3875–3879
17. A. Pareek, Human tracking using surf-based dynamic model. https://www.youtube.com/watch?v=sT2cr-SfJrU
18. H. Kandil, H. Kandil, A. Atwan, A comparative study between sift-particle and surf-particle video tracking algorithms. Int. J. Signal Process. Image Process. Pattern Recogn. **5**(3), pp. 111–122 (2012)

Emotion-Based Hindi Music Classification

Deepti Chaudhary, Niraj Pratap Singh, and Sachin Singh

Abstract Music emotion detection is becoming a vast and challenging field of research with the increase in digital music clips available online. Emotion can be considered as energy that brings the person in positive or negative motion. Music emotion recognition (MER) is an emerging field of research in various areas such as engineering, medical, psychology, musicology. In this work, the music signals are divided into four categories—excited, calm, sad and frustrated. Feature extraction is carried out by using MIR toolbox. The classification is done by using K-nearest neighbor (K-NN) and support vector machine (SVM). The feature-wise accuracy of both the approaches is compared by considering mean and standard deviation of feature vectors. Results reveal that spectral features provide the maximum accuracy among all the features and SVM outperforms K-NN. Maximum accuracy achieved by SVM is 72.5% and K-NN is 71.8% for spectral features. If all the features are considered, then the accuracy achieved by SVM is 75% and by K-NN is 73.8%.

Keywords Music emotion recognition · Support vector machine · K-nearest neighbor · Human–computer interaction · MIR toolbox

---

D. Chaudhary (✉) · N. P. Singh
Department of Electronics and Communication Engineering, National Institute of Technology Kurukshetra, Haryana 136119, India
e-mail: deeptinitk@gmail.com

N. P. Singh
e-mail: nirajatnitkkr@gmail.com

D. Chaudhary
Department of Electronics and Communication Engineering, UIET, Kurukshetra University, Kurukshetra, Haryana 136119, India

S. Singh
Department of Electrical and Electronics Engineering, National Institute of Technology, Delhi 110040, India
e-mail: sachinsingh.iitr@gmail.com

P. Bansal et al. (eds.), *Proceedings of International Conference on Artificial Intelligence and Applications*, Advances in Intelligent Systems and Computing 1164,
https://doi.org/10.1007/978-981-15-4992-2_52

1 Introduction

Music is the best way to convey the emotional message. Emotion is the energy that brings a person in motion, and music is the energy to induce emotion in person. Thus, music is the energy to control the human motion, or in other words, music acts as the driving force for human beings depending on the emotion induced by the songs. Various types of emotions like excited, peace, calm, frustrated, surprise, affection, sad and lonely can be sensed from music. According to C. C. Pratt, music is considered as mode of expression for emotions [1]. The humans naturally judge the music on the basis of emotions induced by them. The automatic MER system based on human–computer interaction is used to automatically detect the emotion of musical clips [2]. The automatic system used for emotion recognition helps the people to find the songs of their interest from largely available online database. People search the music manually in the playlist and select the required song in earlier music systems. This manual process of searching the required was very strenuous. MER is interdisciplinary research area, and researchers working in the field of psychology, naturopathy and cognitive science can share their knowledge and combine it to design efficient MER. Researchers in music psychology use the MER systems to study instant changes in human mind and emotion perceived by human being by listening. Naturopathy is the study to treat, heal and cure the body itself by internal energy of persons. Such treatment processes include meditation, exercise, relax and getting in touch with nature. Thus, to provide proper cure and healing to human beings, music is to be classified in various categories and MER system is required for efficient treatment. Cognitive science is the study of process of reaction of nervous system to different conditions. MER system helps the researchers working in this field to detect the reaction of nervous system to different types of songs.

The automatic MER system follows the five main procedures—dataset collection, preprocessing, annotation, feature extraction and classification [3]. These fives procedures are adopted in this work, and accuracy result of all the features is compared. The various types of standard datasets such as MediaEval 2017 [4], DEAP [5], CAL 500 [6] are available online for research in MER. In preprocessing, the dataset is converted to precise format of 44,100 Hz sampling frequency and 16 bits precision. Annotation process consists of categorizing the songs in different classes. The emotions are conceptualized by Hevner's model for categorical approach [7] and Russel's and Thayer's model for two-dimensional approach [8, 9]. Various types of features such as energy, rhythm, timbre and tonality can be extracted by various tools such as Psysound [10], Marsyas [11], MIR toolbox [12]. The classification process can be conducted by various classifiers such as support vector machines (SVM), Gaussian mixture model (GMM), K-nearest neighbor (K-NN) and combination of artificial neural network (ANN). Finally, accuracy of the system is determined in terms of fulfilling the listener's desire according to their emotional choice for song [13].

In this research work, the dataset considered belongs to Hindi songs and collected from freely available online. The features such as dynamics, rhythm, timbre, spectral and tonality are calculated by using MIR toolbox. The music signals are divided into four categories. The accuracy is calculated by all the features individually by the use of SVM and K-NN. In the proposed work, the feature-wise accuracy by considering mean and standard deviation of feature vectors is calculated and compared for SVM and K-NN classifiers. The most efficient feature is identified for MER.

2 Related Work

Automatic emotion recognition has been growing interest among researchers of various fields such as musicology, engineering, medical and psychology. The proposed work aims to identify the most efficient feature for MER. Thus, the previous work related to feature extraction is represented in this section. In 2002, G. Tzanetakis and P. Cook proposed an algorithm to classify the music signals in various categories on the basis of their genres [14]. Authors proposed three feature sets—timbre, rhythm and pitch. In 2003, Feng et al. analyzed the concept of detecting the mood by retrieving information from music [15]. Authors classified the mood as happiness, sadness, anger and fear, respectively, by considering tempo and articulation. Muyuan et al. made use of adaptive scheme to detect the emotion of music in 2004 [16]. Authors make use of statistical and perceptual features to conduct their research. Lu et al. presented hierarchical framework for automatic detection and tracking of mood of songs by extracting intensity, rhythm and timbre features [17]. 84% recall and 81% precision are achieved experimentally in this work. Yang et al. focused on the issues related to detection of emotion in music signals and proposed a regression approach for predicting AV values on the basis of which the songs are categorized in different classes [2]. Psysound and Marsyas are used by authors for feature extraction, and R^2 statistics achieved 58.3% for arousal and 28.1% for valence. Saari et al. proposed wrapper selection feature reduction approach for improving classification process [18]. Features such as fluctuation, timbre, tonality, rhythm are extracted by using MIR toolbox. Authors reported 56.5% classification rate with only four features. Wang et al. proposed maximum a posteriori approach to analyze personalized music emotion by using acoustic emotion Gaussian's model [19]. Authors extracted the features by using MIR toolbox and validated the effectiveness of their proposed approach. Ahsan et al. analyzed the annotation as multilabel classification by extracting timbre and rhythm [20]. Xiao Hu and Yi-Hsuan Yang proposed regression-based models for mood detection using fifteen audio features and classified songs in five mood categories by extracting loudness, timbre, pitch, rhythm and harmony [21].

Experimental results of this work prove that the performance of regression system is affected by size of training database and accuracy of annotation. Shasha Mo and Jianwei

Niu proposed a technique for analyzing music signal for detecting the emotions [22]. The technique makes use of orthogonal matching pursuit, Wigner function for distribution and Gabor functions, and ATF features are extracted. Experiments were carried out by using four datasets, and mean accuracy of 69.5% was achieved by this technique. Patra et al. made use of LibSVM and neural networks to develop music emotion detection system based on lyrics of Hindi and western music songs and extracted audio and lyrics features [23]. The annotation process is subjective in this work. Authors achieved F-measure of 0.751 and 0.83 for above two techniques. In the related work mentioned above, the features used by different authors are discussed. In this proposed work, the accuracy of all the features is compared and most efficient feature is detected among all.

The main points considered are summarized in the proposed work.

(1) Hindi songs are considered for this research.
(2) The model considered for categorization of emotion classes in the proposed work is 2-d, and four classes are considered for both the approaches.
(3) SVM and K-NN are used as classifier, and the results in terms of accuracy are compared.

This paper is organized as described further. The literature review of previous papers is described in Sect. 2. Section 3 describes the methodology for proposed approach. Results and discussions are presented in Sect. 4. Conclusions are mentioned in Sect. 5.

3 Methodology

In this section, methodology of proposed work is discussed. The basic MER system is described by the block diagram shown in Fig. 1. It consists of five basic steps—data collection, preprocessing, annotation, feature extraction and classification. The above-mentioned five steps are discussed in this section.

3.1 Database Collection

Three hundred fifty Hindi songs of various genres such as pop, classical, patriotic, and jazz and ghazals are collected from freely available online resources [1, 2]. Seventy percentage of database is considered for training, and 30% is treated as test data. The songs are collected from the online source at the sampling rate of 44,100 Hz. The emotion perceived from a song is not stable for its entire duration. The complete

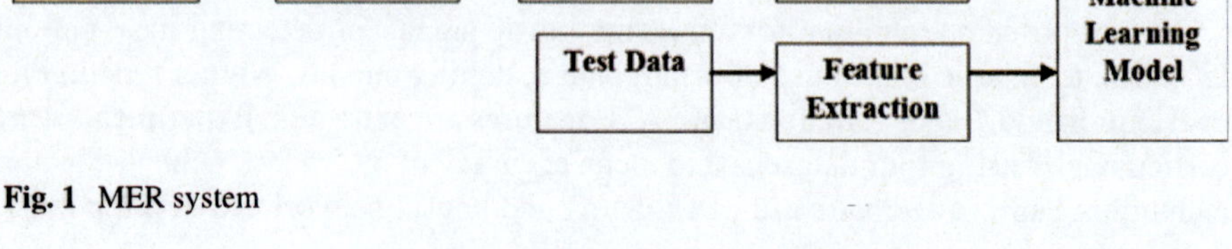

Fig. 1 MER system

Table 1 Sample songs of different categories

Song ID	Song description	Category
1	(i) Rimjhim rimjhim, rumjhum rumjhum (ii) Kyu naye lag rahe hain ye dharti pawan	Excited
2	(i) Aaj purani rahon se (ii) Jab koi baat bigad jaye	Calm
3	(i)Tera gam agar na hota (ii) Mere khwabon ka hare k naksh mitade koi	Sad
4	(i) Ek tha dil, ik thi dhadkan (ii) Judda hoke bhi tu mujhme kahi baki hai	Frustrated

music can contain sections of different emotions. Thus, 30 s segment representing most effective emotion of songs is considered in this work. The sample dataset is given in Table 1.

The excited song belongs to positive valence and positive arousal. The calm song belongs to positive valence and negative arousal. The sad song belongs to negative valence and negative arousal. The frustrated song belongs to negative valence and positive arousal. The term arousal deals with the energy related to song, and valence is related to the type of emotion belonging to a song.

3.2 Preprocessing

Preprocessing mainly consists of two steps; they are windowing and framing. Windowing is directly in co-operated with the Fourier transform function. Hamming window is used to preprocess the signal. The sound signals are non-stationary; thus, the analysis of sound signals is carried out by considering short time signals. The process of transforming the sound signal in short time signals is framing. Frame length of 25 ms is considered in the proposed work.

3.3 Annotation

Annotation is the process of categorizing the songs in different classes. In this work, the songs are distributed to four groups of eight people each. These groups are asked to categorize the songs in four categories, Excited, Calm, Depress and Angry, according to Thayer's two-dimensional model as shown in Fig. 2 [3]. The average of the decision of all the groups is considered as final class of the songs.

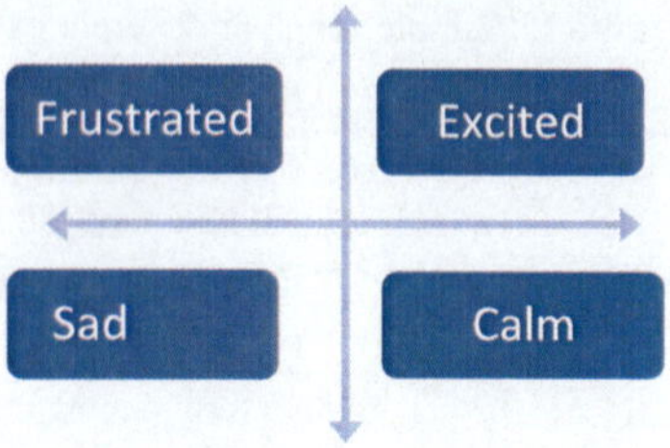

Fig. 2 Two-dimensional representation of emotions

Table 2 Detail of features [12]

Type	Description
Dynamics	RMS energy
Rhythm	Tempo, attack time, attack slope
Timbre	Zero cross, low energy, spectrum flux
Spectral	Centroid, brightness, spread, skewness, kurtosis, spectral roll-off 95% and spectral roll-off 85%, entropy, flatness, roughness, irregularity, MFCC
Tonality	Chromagram (peak, centroid), mode, key clarity

3.4 Feature Extraction

Feature extraction is mainly used to extract various features that are associated with the audio signal. In this research, the features are evaluated by using MIR toolbox [12]. Basically, five types of features, dynamics, rhythm, timbre, pitch and tonality, can be evaluated using MIR toolbox as mentioned in Table 2.

3.5 Classification

In this research work, the classification is carried out by using SVM and K-NN. SVM classifiers are based on supervised learning techniques and algorithms. The dataset divided is divided into training and testing part. The training data is used to train the SVM by marking the particular category. Further, the test data is analyzed to check their category by determining the hyperplane that maximizes the distance between the classes [24, 25]. K-NN classifier is used to store the data for all the categories, and new classes are classified based on distance functions. The test data is classified based on the majority vote of its neighbors.

4 Result and Discussion

The proposed work is implemented by using MATLAB. The features are extracted by using MIR toolbox 1.6.1 in MATLAB. The features explained in Table 2 are extracted. The mean and standard deviation of the feature vectors resulted by MIR toolbox is considered. The accuracy of all the features is calculated individually and thereafter compared with each other. Accuracy is considered as performance evaluation parameter. The accuracy achieved by considering mean of all the features individually is shown in Table 3, and accuracy achieved by considering standard deviation is shown in Table 4 for SVM.

The summarized results for SVM mentioned in Tables 3 and 4 are described below.

(1) Accuracy achieved by SVM is 60.8% for dynamics if mean of feature is considered and achieves 59.6% accuracy if standard deviation of feature vector is considered.
(2) Accuracy achieved by SVM is 62.5% for rhythm if mean of feature is considered and achieves 62.2% accuracy if standard deviation of feature vector is considered.
(3) Accuracy achieved by SVM is 62.7% for timbre if mean of feature is considered and achieves 61.3% accuracy if standard deviation of feature vector is considered.
(4) Accuracy achieved by SVM is 68.6% for tonality if mean of feature is considered and achieves 68.2% accuracy if standard deviation of feature vector is considered.

Table 3 Accuracy comparison of mean of feature vectors by SVM

Feature	Accuracy by SVM (%)
Dynamics	60.8
Rhythm	62.5
Timbre	62.7
Tonality	68.6
Spectral	72.5
All features	75

Table 4 Accuracy comparison of standard deviation of feature vectors by SVM

Features	Accuracy by SVM (%)
Dynamics	59.6
Rhythm	62.2
Timbre	61.3
Tonality	68.2
Spectral	71.9
All features	72.8

Table 5 Accuracy comparison of mean of feature vectors by K-NN

Feature	Accuracy by K-NN (%)
Dynamics	60
Rhythm	61.5
Timbre	61.6
Tonality	68.4
Spectral	71.8
All features	73.8

Table 6 Accuracy comparison of standard deviation of feature vectors by K-NN

Features	Accuracy by K-NN (%)
Dynamics	58.8
Rhythm	61.1
Timbre	60.8
Tonality	69.8
Spectral	71.2
All features	70.6

(5) Accuracy achieved by SVM is 72.5% for spectral if mean of feature is considered and achieves 71.9% accuracy if standard deviation of feature vector is considered.

(6) The accuracy achieved by SVM if all the features are considered is 75% if mean is taken into consideration and 72.8% if standard deviation of feature vectors is considered.

The authors also determine the accuracy of the system by using K-NN for the same set of data and features that are considered for SVM. The results are shown in Tables 5 and 6.

The summarized results for K-NN mentioned in Tables 5 and 6 are described below.

(1) Accuracy achieved by K-NN is 60% for dynamics if mean of feature is considered and achieves 58.8% accuracy if standard deviation of feature vector is considered.

(2) Accuracy achieved by K-NN is 61.5% for rhythm if mean of feature is considered and achieves 61.1% accuracy if standard deviation of feature vector is considered.

(3) Accuracy achieved by K-NN is 61.6% for timbre if mean of feature is considered and achieves 60.8% accuracy if standard deviation of feature vector is considered.

(4) Accuracy achieved by K-NN is 68.4% for tonality if mean of feature is considered and achieves 69.8% accuracy if standard deviation of feature vector is considered.

(5) Accuracy achieved by K-NN is 71.8% for spectral if mean of feature is considered and achieves 71.2% accuracy if standard deviation of feature vector is considered.
(6) The accuracy achieved by K-NN if all the features are considered is 73.8% if mean is taken into consideration and 70.6% if standard deviation of feature vectors is considered.

The above-mentioned points for SVM and K-NN reveal that the spectral features results in best accuracy among all the other features considered. It is also concluded that the results of feature vectors by considering mean gives better performance than by considering standard deviation of features.

Graphical representation of accuracies achieved by considering mean and standard of features by using SVM and K-NN is also shown graphically in Figs. 3 and 4.

The results mentioned above reveal that the spectral feature provide the best accuracy among all features and mean of feature vectors provide better accuracy

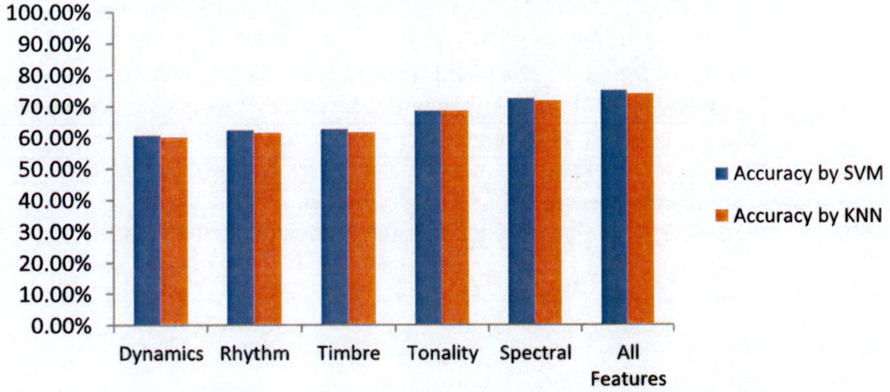

Fig. 3 Accuracy comparison by using mean of feature vectors in SVM and K-NN

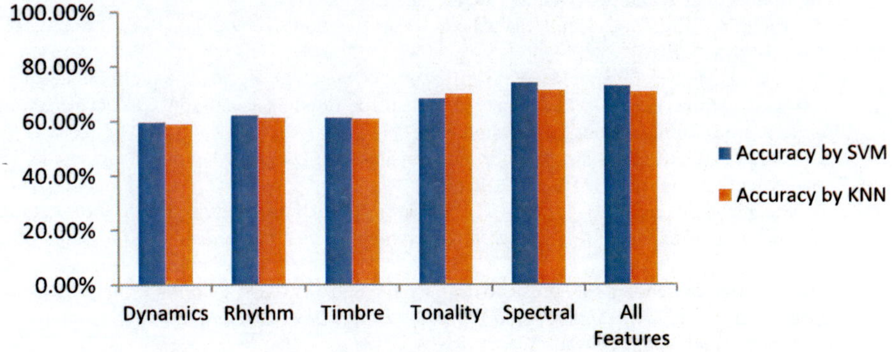

Fig. 4 Accuracy comparison by using standard deviation of feature vectors in SVM and K-NN

than standard deviation of feature vectors. It is also proved that SVM provides better result than K-NN.

5 Conclusion

In this work, the mean and standard deviation of features is compared to determine the accuracy. SVM and K-NN are used as classifiers for both the cases. Accuracy of SVM and K-NN is also compared for both mean and standard deviation of feature vectors. The results show that the accuracy achieved by considering mean is higher than that achieved by consideration of standard deviation of features. Results reveal that spectral features achieve the maximum accuracy among all the features. The accuracy achieved by considering spectral features individually is 72.5% for SVM and 71.8% for K-NN if mean of feature vectors is considered. The accuracy achieved by considering spectral features individually is 71.9% for SVM and 71.2% for K-NN if standard deviation of feature vectors is considered. It has also been shown that accuracy achieved by using SVM is 75% and by K-NN is 73.8% if mean of features is considered for all the combined features and accuracy achieved by using SVM is 72.8% and by K-NN is 70.6% is standard deviation of features is considered. Therefore, results prove that SVM outperforms K-NN for both the cases. In further approach related to this work, more number of features can be considered and the same technique may also be applied by using other datasets and other classification techniques such as Gaussian mixture models and artificial neural networks.

References

1. C.C. Prat, Music as the language of emotion. Lecture delivered in Whittall Pavilion, The Library of Congress 21 Dec 1952
2. Y.-H. Yang, Y.-C. Lin, Y.-F. Su, H. H. Chen, A regression approach to music emotion recognition. IEEE Trans. AUDIO, SPEECH, Lang. Process. **16**(2), 448–457 (2008)
3. Y.-H. Yang, Y.-F. Su, Y.-C. Lin, H. H. Chen, *Music Emotion Recognition* (Boca Raton, CRC Press, 2 February 2011)
4. B. Bischke, P. Helber, C. Schulze, V. Srinivasan, A. Dengel, D. Borth, The multimedia satellite task at mediaeval 2017: emergency response for flooding events, in *CEUR Workshop Proceedings* (Dublin, Ireland, 13–15 September 2017)
5. S. Koelstra et al., DEAP: A database for emotion analysis; Using physiological signals. IEEE Trans. Affect. Comput. **3**(1), 18–31 (2012)
6. S.-Y. Wang. J.-C. Wang. Y.-H. Y. Wang, H.-M. Wang, Towards time-varying music auto-tagging based on CAL500 expansion, in *International Conference on Multimedia and Expo* (Chengdu, China, 14–18 July 2014)
7. D. Su, P. Fung, These words are music to my ears: recognizing music emotion from lyrics using AdaBoost, in *Asia-Pacific Signal and Information Processing Association Annual Summit and Conference* (Kaohsiung, Taiwan, 29 Oct–1 Nov 2013)
8. A. Shakya, B. Gurung, M. S. Thapa, M. Rai, Music classification based on genre and mood, in *International Conference on Computational Intelligence, Communications and Bussiness*

Analytics, CICBA 2017. Communications in Computer and Information Science, vol. 776, (Springer, Singapore), pp. 168–183
9. J. Grekow, Audio features dedicated to the detection of arousal and valence in music recordings, in *IEEE International Conference on Innovations in Intelligent SysTems and Applications* (Gdynia, Poland, 3–5 July 2017)
10. D. Cabrera, S. Ferguson, E. Schubert, *Psysound3: Software for Acoustical and Psychoacoustical Analysis of Sound Recordings* ('PSYSOUND3': software for acoustical and Empirical Musicology Group, University of New South Wales, NSW 2052, Australia, May 2014, 2002)
11. G. Tzanetakis, P. Cook, MARSYAS: a framework for audio analysis. Organised Sound **4**(3), 169–175 (2000)
12. P.T. Olivier Lartillot, A Matlab toolbox for musical feature extraction from audio, in *International Conference on Digital Audio Effects* (Bordeaux, France, 10–15 Sept 2007)
13. X. Hu, A framework for evaluating multimodal music mood classification. J. Assoc. Inf. Sci. Technol. **68**(2), 273–285 (2017)
14. G. Tzanetakis, P. Cook, Musical genre classification of audio signals. IEEE Trans. Speech Audio Process. **10**(5), 293–302 (2002)
15. Y. Feng, Y. Zhuang, and Y. Pan, Popular music retrieval by detecting mood, in *Proceedings of the 26th Annual International ACM SIGIR Conference on Research and Development in Information Retrieval* (Toronto, Canada, 28 July–1 Aug 2003), pp. 375–376
16. W. Muyuan, Z. Naiyao, Z. Hancheng, M. Wang, N. Zhang, H. Zhu, User-adaptive music emotion recognition, in *Proceedings of the 7th International Conference on Signal Processing, 2004* (Beijing, China, 31 Aug–4 Sept 2004)
17. L. Lu, D. Liu, H. J. Zhang, Automatic mood detection and tracking of music audio signals. IEEE Trans. Audio, Speech Lang. Process. **14**(1), 5–18 (2006)
18. P. Saari, T. Eerola, O. Lartillot, Generalizability and simplicity as criteria in feature selection: application to mood classification in music. IEEE Trans. Audio, Speech Lang. Process. **19**(6), 1802–1812 (2011)
19. J. Wang, Y. Yang, H. Wang, S. Jeng, Personalized music emotion recognition via model adaptation, in *Asia Pacific Signal and Information Processing Association Annual Summit and Conference* (Hollywood, CA, USA, 3–6 Dec 2012)
20. H. Ahsan, V. Kumar, C.V. Jawahar, Multi-label annotation of music, in *8th International Conference on Advances in Pattern Recognition* (Kolkata, India, 4–7 Jan 2015)
21. X. Hu, Y.H. Yang, Cross-dataset and cross-cultural music mood prediction: a case on western and chinese pop songs. IEEE Trans. Affect. Comput. **8**(2), 228–240 (2017)
22. S. Mo, J. Niu, A novel method based on OMPGW method for feature extraction in automatic music mood classification. IEEE Trans. Affect. Comput. **3045** (2017)
23. B.G. Patra, D. Das, S. Bandyopadhyay, Multimodal mood classification of Hindi and Western songs. J. Intell. Inf. Syst. 1–18 (2018)
24. C.W. Hsu, C.C. Chang, C.J. Lin, A practical guide to support vector classification. BJU Int. **101**(1), 396–400 (2008)
25. C. Cortes, C. Cortes, V. Vapnik, V. Vapnik, Support vector networks. Mach. Learn. **20**(3), 273–297 (1995)

Analysis of Offset Quadrature Amplitude Modulation in FBMC for 5G Mobile Communication

Ayush Kumar Agrawal and Manisha Bharti

Abstract Next-generation mobile communication is required to allocate the resources to all users smoothly and efficiently. But the allocation of resources is difficult in present 4G mobile communication systems using OFDM and hence in this manuscript, a new technique based on Filter Bank Multicarrier (FBMC) modulation and demodulation are designed and employed. The proposed scheme will help devices to communicate automatically employing machine learning techniques including higher data rates as per the need of users with less bit error rate. Offset quadrature amplitude modulation (OQAM) technique for FBMC is discussed and analyzed to achieve a lower bit error rate for the latest generation, 5G wireless communication

Keywords OQAM · FBMC · BER · OFDM · MIMO · 5G mobile communication

1 Introduction

Due to many more benefits over different types of technologies such as carrier and symbol synchronization, low ICI, total set of narrow bands over a limited available bandwidth and many, OFDM has been widely adopted in many broadband wired and wireless communication systems for three decades. The time-frequency resources cannot be allocated easily and resulting in bad efficiency and spatial behavior of large data rates [1]. Filter bank multicarrier (FBMC) with OQAM offers a solution for low spectral efficiency by using both time and frequency pulses found. Many of FBMC's properties are identical to OFDM, and one of the key issues is its high PAPR value [2].

A. K. Agrawal (✉) · M. Bharti
National Institute of Technology Delhi, Delhi 110 040, India
e-mail: 182220002@nitdelhi.ac.in; ayush6295@gmail.com

M. Bharti
e-mail: manishabharti@nitdelhi.ac.in

P. Bansal et al. (eds.), *Proceedings of International Conference on Artificial Intelligence and Applications*, Advances in Intelligent Systems and Computing 1164,
https://doi.org/10.1007/978-981-15-4992-2_53

FBMC inclusive of OQAM offers many benefits than that of conventional OFDM systems such as (i) No requirement of the cyclic prefix (ii) Its PSD side lobe is very low [3]. Many works on reducing PAPR for FBMC systems have been proposed in recent years; in [4] Rahim et al. In OFDM or OQAM Systems the reduction system which is being used is the chipping system in his research. The Visual Networking Index (VNI) of Cisco projected that by the end of 2019, the total number of devices linked to the cellular network will be smart devices [5]. Each smartphone consumer is expected to download approximately 1 TB of data annually by the end of 2020 [6] on average. The tremendous use of smart devices has resulted in a huge increase in mobile user traffic. Current researchers, however, were exposed to new emerging technologies such as IoT, device to device communications, M 2 M communications, vehicle contact Healthcare, and the large number of devices utilizing tactile internet. Some of the 5 G systems' basic requirements are [7]:

- 100% Connectivity for any site and everywhere
- Long life of the battery
- Could manage a large number of devices at the same time
- Low delay
- Large Gbps ($\geq$1 Gbps) network speed, meaning the time needed to download the movie may be less than a second.

Present wireless technology cannot handle the severe and ever-increasing data rate and provide connectivity to the enormous number of users at the same time. Due to the out-of-band emission and high PAPR, LTE systems are unable to handle this type of rapid increase in the rate of the data including the connectivity of more devices.

Compared to existing systems, it is taken into consideration that the next cellular communication systems will meet the massive and tactile internet-connected devices [8]. FBMC attracts wireless communication to the next generation of researchers and industry. Currently, FBMC with OQAM is an alternative waveform contender for the orthogonal frequency division multiplexing for 5 G mobile communication to achieve the basic requirements of high data rates and better spectral properties [9, 10]. Compared to conventional methods, the FBMC method has several advantages. There are a few significant advantages I (i) More radiation must be transmitted outside the required bandwidth (ii) signal without cyclic prefix (iii) high environmental robustness. Because of the use of pulse shaping filters, FBMC has lower spectral sidelobes compared to traditional OFDM systems, which means that the FBMC is capable of variable parameters and asynchronous transmission [11–13]. Only in the real domain can the filters used in OQAM-FBMC satisfy the orthogonality property. In OQAM, half of the symbol cycle [14, 15] delays the real and imaginary part of the symbols. In 5 G communication systems, OQAM-FBMC systems are the opportunity for physical layer engineering. Bouhadda et al. [16], investigated the performance of OQAM-FBMC investigated in terms of bit error rate by considering phase error and intrinsic interference expressed in terms of the distribution of probability in the discrete domain. This BER study was carried out with 4-OQAM and the channel noise was considered an additive white Gaussian noise. Zakaria et al. [17] have given the symbol error rate expression by considering the probability density of Gaussian one, for 2 m pulse amplitude modulated data mapping function and constant phase error only for the channel has additive white Gaussian noise. HanenBouhadda et al. [18] conducted a high-power amplifier analytical analysis and a nonlinear distortion effect on BER on both multicarrier modulation techniques: FBMC/OQAM and OFDM. In this paper, the performance of bit error rates in FBMC waveforms with OQAM was significantly improved. We provided simulation results between signal-to-noise ratio (SNR) and the bit error rate (BER) for various subcarrier values compared to theoretical results as well as traditional methods.

2 Multicarrier Modulation

Information is transmitted through pulses in multicarrier systems that generally have an intersection in frequency and time. Time and frequency overlap. Its major benefit is that these vibrations typically only take up a tiny bandwidth in order to transform frequency-selective broadband channels into multiple, essentially flat sub-channels (subcarriers) with negligible interference. It allows a very easy uni-tap equalizer to be implemented, which leads to the detection of maximum likelihood symbols of G noise. In addition, the estimation of the channel is made easy in many situations, adaptive modulation and coding techniques are applied, and Multiple Input Multiple Output can be used directly [19].

In the following chapters, the various techniques are discussed in more detail.

2.1 Cyclic Prefix—OFDM

Cyclic Prefix—OFDM is the commonly utilized multi-career scheme and is used in Non-Wired LAN and LTE, for example. Cyclic Prefix—OFDM uses rectangular transmission and generates pulses that significantly decrease the complexity of the computation. In contrast, the CP means that the pulse transmitted is slightly longer than the pulse received, maintaining orthogonality in specific frequency streams.

Unfortunately, in the frequency domain, the rectangular pulse is not located, resulting in high out-of-band (OOB) emissions which can be viewed in Fig. 1 and which is the CP-OFDM's largest drawback. Additionally, in frequency-selective channels, the CP simplifies equalization but also reduces spectral efficiency. No CP is required in an AWGN network.

Windowing and filtering, as shown in Fig. 2, could reduce the high CP-OFDM OOB emissions. This, however, leads to the very low and decreased efficiency of spectra as viewed in the T F object and decreasing robustness in selective channels of frequency. In addition, windowing with filtering is again not generate as less OOB emissions with that of FBMC (see next section), that can get the highest T F = 1.

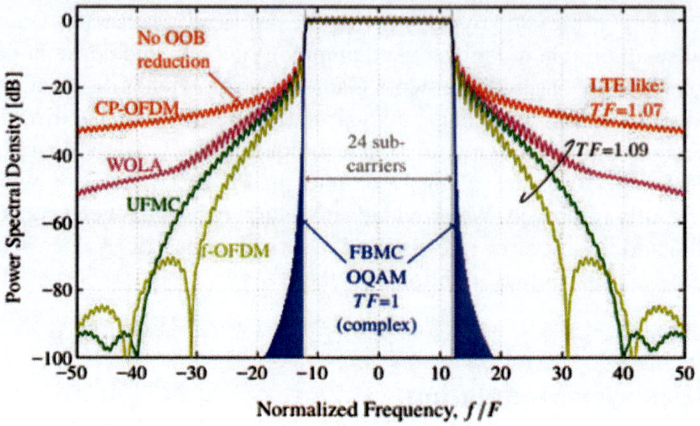

Fig. 1 Similar to CP-OFDM, FBMC has much stronger spectral properties. The PHYDYAS prototype filter with $O = 4$ is considered for FBMC

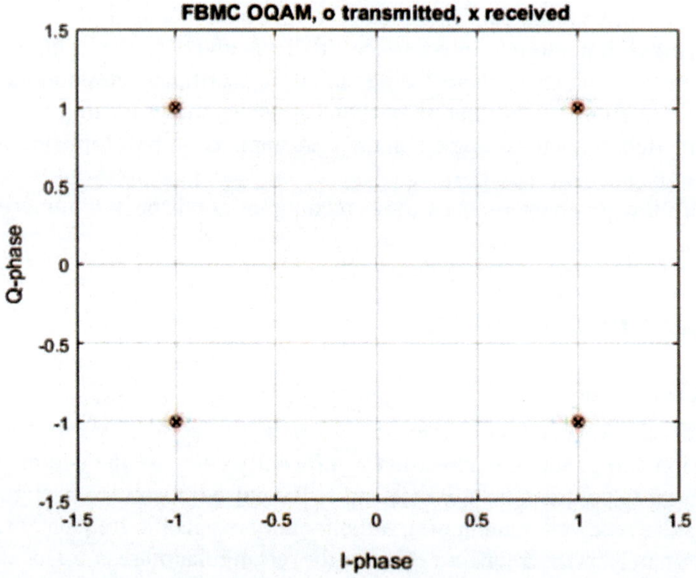

Fig. 2 Q phase and I phase for transmitted and received signals

2.2 *FBMC—QAM*

There is no common FBMC—QAM definition. Few researchers [20] sacrifice frequency localization, in that of dragging the OOB emissions not as good as OFDM. Others [21, 22] sacrifice orthogonality to get the spacing of the time-frequency to that of T F = 1 and localization of time-frequency.

For FBMC—QAM, on the other hand, we find the spacing of the time-frequency to that of T F = 2, thus forgetting about the spectral efficiency to satisfy different desirable properties. Such more time-frequency spacing improves total robustness in a dual-selected stream. Nevertheless, the important decision for taking T F = 2 is the direct application stated in Filter Bank Multicarrier—OQAM.

2.3 Filter Bank Multicarrier—OQAM

Filter Bank Multicarrier—OQAM is Filter Bank Multicarrier-QAM compatible and also has the same number of symbols as OFDM without cyclic prefix. In principle, FBMC-OQAM operates as follows:

- Build a $p(t) = p(-t)$ design filter that is orthogonal for $T = T0$ time spacing and $F = 2/T0$ frequency spacing, leading to T F = 2, see [9].
- Decrease the spacing of time (orthogonal) by a factor of two, i.e. $T = T0/2$ and $F = 1/\mathrm{T0}$.
- The mediated interference is moved to the imaginary domain by the phase shift $\pi l, k = 2\ (l + k)$ in [23].

Even though the time-frequency spacing (density) is comparable to T F = 0.5, we must bear in mind that only real-value information symbols can be transmitted in this way, resulting in a time-frequency spacing equivalent to T F = 1 for complex symbols. Frequently, a complex symbol's real-part is mapped to the first time-slot and the imaginary part to the second time-slot, hence the name offset-QAM. However, there is no need for such self-limitation. FBMC-OQAM's main drawback is the loss of complex orthogonality. This implies particularities for certain Multiple Input Multiple Output techniques, such as space-time block codes [24] or the detection of maximum likelihood symbols [25], as well as channel estimation [26].

2.4 Coded Filter Bank Multicarrier—OQAM: Enabling All Multiple Input Multiple Output Methods

We need to restore complex orthogonality in FBMC—OQAM in order to use all Multiple Input Multiple Output methods and channel estimation techniques known in OFDM straightforwardly. This can be accomplished in time or frequency by spreading symbols. While such spreading is similar to Code Division Multiple Access(CDMA), used in 3 G, coded Filter Bank Multicarrier—OQAM is unique in that there is no need for a rake receiver and no root-raised-cosine filter.

3 Results and Discussions

Specific findings were explored in this paper to explain the principles mentioned in the previous chapters. For different subcarriers, an average BER for OQAM—FBMC is considered and the results are compared to the theoretical values. We can see that the effects of the simulation are much similar to the theoretical predictions. Compared to the simulation results of cyclic prefix-OFDM and coded FBMC—OQAM [27, 28], the proposed FBMC with OQAM gives an improved bit error rate.

The presence of errors in the synchronization of frequencies affects the orthogonality between the subcarriers, resulting in the loss of the output of the device. At this stage, the FBMC technique by proper design of the pulse shaping filter reduces intersymbol interference and intercell interference. From the results of simulation we can understand that, by increasing the number of subcarriers by 64–256, BER is similar to the theoretical performance. The following result shows the proposed method and conventional methods in the BER versus SNR. Compared with traditional methods, BER is improved from the analysis.

Here in Fig. 2 the Q Phase and I Phase for Transmitted and Received signals for the modulation and demodulation techniques are seen. Meanwhile, in Fig. 3 the PAPR of the FBMC has been taken out from our proposed model which can be used in 5G is verified so that it has the best PAPR with different dB levels.

Furthermore, in the next figure which shows the BER which is very less and effective for the use in 5G communication is shown which has a value of 0.000190 only.

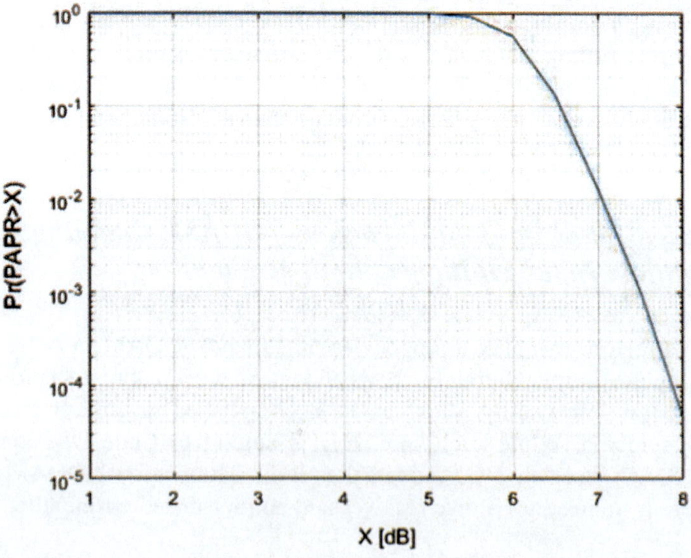

Fig. 3 PAPR plot

4 Conclusion

In this paper, the FBMC principles with OQAM and model filter were used to implement BER analysis for 256 subcarriers. Results obtained indicate an increase in the SNR in BER. In contrast with cyclic prefix—OFDM and FBMC—QAM, we can see from the simulation results that the proposed OQAM-FBMC is best suited for the evolving 5G technology. Spectrum performance can be improved by applying FBMC for MIMO enabled techniques while addressing complexity issues and high PAPR.

References

1. B. Farhang-Boroujeny, OFDM versus filter bank multicarrier. IEEE Signal Process. Mag. **28**(3), 92–112 (2011)
2. D. Chen, D. Qu, T. Jiang, Prototype filter optimization to minimize stop band energy with NPR constraint for filter bank multicarrier modulation systems. IEEE Trans. Signal Process. **61**(1), 159–169 (2013)
3. P. Siohan, C. Siclet, N. Lacaille, Analysis and design of OFDM/OQAM systems based on filterbank theory. IEEE Trans. Signal Process. **50**(5), 1170–1183 (2002)
4. M.U. Rahim, T.H. Stitz, M. Renfors, Analysis of clipping-based PAPR-reduction in multicarrier systems, in *Proceedings of IEEE VTC* (Barcelona, Spain, 2009, April), pp. 1–5
5. Cisco, Visual Networking Index, White paper, Feb. 2015 [Online]. Available: www.Cisco.com
6. T.S. Rappaport, W. Roh, K. Cheun, Wireless engineers long considered high frequencies worthless for cellular systems. They couldn't be more wrong. IEEE Spectr. **51**(9), 34–58 (2014)
7. M. Agiwal, A. Roy, N. Saxena, Next generation 5G wireless networks: a comprehensive survey. IEEE Commun. Surv. Tutorials **18**(3), 1617–1655
8. M. Shafi, A.F. Molisch, P.J. Smith, T. Haustein, P. Zhu, P.D. Silva, F. Tufvesson, A. Benjebbour, G. Wunder, 5G: a tutorial overview of standards, trials, challenges, deployment, and practice. IEEE J. Sel. Areas Commun. **35**(6), 1201–1221 (2017)
9. L. Zhang, P. Xiao, A. Zafar, A. ulQuddus, R. Tafazolli, FBMC system: an insight into doubly dispersive channel impact. IEEE Trans. Veh. Technol. 66(5), 3942–3956 (2017)
10. M. Bellanger, Maurice, D. Le Ruyet, D. Roviras, M. Terré, J. Nossek, L. Baltar, Q. Bai, D. Waldhauser, M. Renfors, T. Ihalainen, FBMC physical layer: a primer. PHYDYAS, **25**(4), 7–10 (2010)
11. J. Zhang, M. Zhao, J. Zhong, P. Xiao, T. Yu, Optimised index modulation for filter bank multicarrier system. IET Commun. 11(4), 459–467 (2017)
12. L. Zhang, A. Ijaz, P. Xiao, M.M. Molu, R. Tafazolli, Filtered OFDM systems, algorithms, and performance analysis for 5G and beyond. IEEE Trans. Commun. **66**(3), 1205–1218 (2018)
13. S. Kaur, L. Kansal, G.S. Gaba, N. Safarov, Survey of filter bank multicarrier (FBMC) as an efficient waveform for 5G. Int. J. Pure Appl. Math. **118**(7), 45–49 (2018)
14. I.A. Shaheen, A. Zekry, F. Newagy, R. Ibrahim, Performance evaluation of PAPR reduction in FBMC system using nonlinear companding transform. ICT Express (2018)
15. S. Patil, S. Patil, U. Kolekar, Implementation Of 5G using OFDM and FBMC (Filter Bank Multicarrier)/OQAM (Offset Quadrature Amplitude Modulation). Int. J. Innovative Sci. Eng. Technol. **5**(1), 11–15 (2018)
16. H. Bouhadda, H. Shaiek, Y. Medjahdi, D. Roviras, R. Zayani, R. Bouallegue, Sensitivity analysis of FBMC signals to non linear phase distortion, in *2014 IEEE International Conference on Communications Workshops (ICC)*. IEEE (2014), pp. 73–78

17. R. Zakaria, D. Le-Ruyet, SER analysis by Gaussian interference approximation for filter bank based multicarrier system in the presence of phase error, in *IEEE International Conference on Communications (ICC)* (2015)
18. H. Bouhadda, H. Shaiek, D. Roviras, R. Zayani, Y. Medjahdi, R. Bouallegue, Theoretical analysis of BER performance of nonlinearly amplified FBMC/OQAM and OFDM signals. EURASIP J. Adv. Signal Process. **2014**(1), 60 (2014)
19. A. Sahin, I. Guvenc, H. Arslan, A survey on multicarrier communications: prototype filters, lattice structures, and implementation aspects. IEEE Commun. Surv. Tutorials **16**(3), 1312–1338 (2012)
20. H. Nam, M. Choi, S. Han, C. Kim, S. Choi, D. Hong, A new filter-bank multicarrier system with two prototype filters for QAM symbols transmission and reception. IEEE Trans. Wireless Commun. **15**(9), 5998–6009 (2016)
21. Y.H. Yun, C. Kim, K. Kim, Z. Ho, B. Lee, J.-Y. Seol, A new waveform enabling enhanced QAM-FBMC systems, in IEEE International Workshop on Signal Processing Advances in Wireless Communications (SPAWC), (2015) pp. 116–120
22. C. Kim, Y.H. Yun, K. Kim, J.-Y. Seol, Introduction to QAM- FBMC: from waveform optimization to system design. IEEE Commun. Mag. **54**(11), 66–73 (2016)
23. S. Schwarz, M. Rupp, Society in motion: challenges for LTE and beyond mobile communications. IEEE Commun. Mag. Feature Topic LTE Evolution **54**(5) (2016)
24. C. Lélé, P. Siohan, R. Legouable, The Alamouti scheme with CDMA-OFDM/OQAM. EURASIP J. Adv. Signal Process **2010**, 1–13 (2010). (Article ID 703513)
25. R. Zakaria, D. Le Ruyet, A novel filter-bank multicarrier scheme to mitigate the intrinsic interference: application to MIMO systems. IEEE Trans. Wireless Commun. **11**(3), 1112–1123 (2012)
26. R. Nissel, M. Rupp, Bit error probability for pilot-symbol aided channel estimation in FBMC-OQAM, in *IEEE International Conference on Communications (ICC)* (Kuala Lumpur, Malaysia, May 2016)
27. L.G. Baltar, J.A. Nossek, Multicarrier systems: a comparison between filter bank based and cyclic prefix based OFDM, in *Proceedings of OFDM 2012, 17th International OFDM Workshop 2012 (InOWo'12)* (VDE, 2012), pp. 1–5
28. R. Nissel, Markus Rupp, OFDM and FBMCOQAM in doubly-selective channels: calculating the bit error probability. IEEE Commun. Lett. **21**(6), 1297–1300 (2017)

Design and Analysis of 2D Extended Reed–Solomon Code for OCDMA

Manisha Bharti

Abstract In this paper, an attempt has been done to design and analyze two-dimensional codes lying on modified extended Reed–Solomon technique. These optical codes possess partition property for optical code division multiple access (OCDMA) applications which require code obscurity, as well as this property, provides a trade-off between code cardinality and code performance. This paper includes a comparative analysis of these codes and multilevel prime codes. Results show that performance of these optical codes is superior to MPC in Multiple Access Interference affected environment. This optical code outperforms MPC due to larger code weight and larger number of available wavelength. This 2-D optical code allows choosing between code cardinality and code performance. E-RS based optical codes are robust to multiple access interference affected operating systems hence are suitable for applications where simultaneous multiple active user access is required.

Keywords Optical code division multiple access · Reed–solomon codes

1 Introduction

This paper focuses on OCDMA technology which is basically code division multiple access in optical domain to enhance the feature of bandwidth provided by optical media. OCDMA deals with the transmission of data by several users at the same time and at the same frequency. OCDMA possesses various features like asynchronous access by multiple users, security and legacy in communication. Due to these features and advantages, there have been steady advancements in OCDMA in the last few decades [1–4]. There are two main schemes of OCDMA, "synchronous" and "asynchronous". Synchronous OCDMA scheme just like its name requires time synchronization for code words transmission whereas asynchronous OCDMA scheme do not need time synchronization and this makes it more useful scheme in providing

M. Bharti (✉)
Department of ECE, National Institute of Technology, Delhi 110040, India
e-mail: manishabharti@nitdelhi.ac.in

P. Bansal et al. (eds.), *Proceedings of International Conference on Artificial Intelligence and Applications*, Advances in Intelligent Systems and Computing 1164,
https://doi.org/10.1007/978-981-15-4992-2_54

multiple simultaneous user access accessing medium, hence more suitable for bursty traffic where a large number of users are using the medium for the transmission of information. As an asynchronous OCDMA scheme does not require time synchronization so here optical pulse intensity is detected using either wavelength/time or both domain. The presence of optical pulse decides the element of unipolar (0, 1) optical code like if the pulse is present or intensity is above preset value then it represents "1" element/symbol in unipolar optical codes hence requires incoherent detection. Transmission of bipolar $(-1, +1)$ code is generally in form of phases hence requires coherent detection. But these types of system suffer from multiple access interference (MAI) which is originated by other users. MAI is the co-channel interference from other users who are accessing the same channel/medium. Like prime codes and optical orthogonal codes (OOC) [3–9] possess good auto- and cross-correlation property. OOC is considered to possess ideal auto- and cross-correlation property with the largest code cardinality (number of code words).

Generally, the family of code is a collection of code words, each of N time slots and L wavelengths that can be represented as

- Autocorrelation function:

$$\sum_{i=0}^{L-1}\sum_{j=0}^{N-1} c_{i,j}c_{i,j\oplus\tau} \leq \lambda_a$$

where $c_{i,j}$ represents ith row and jth column of codeword whereas τ represents a shift.

- Cross-correlation function:

$$\sum_{i=0}^{L-1}\sum_{j=0}^{N-1} c_{i,j}c^{1}_{i,j\oplus\tau} \leq \lambda_a$$

One dimensional codes are called time-spreading where elements (0, 1) are separated in time which is the coding dimension [2–6], whereas, in 2D optical codes along with time, the wavelength is used as another coding dimension and called as 2-D wavelength-time codes [7–15]. In 1D codes code cardinality is proportional to code length whereas for 2-D optical codes it is not true. In optical codes, λ_c value is restricted to 1 or 2. But a larger number of codewords are obtained by compromising the code performance [2, 12]. Larger number of codewords is obtained by increasing value of λ_c to two and three [8–15].

We know about Reed–Solomon codes of primitive length $(p^m - 1)$ with code cardinality p^{mk} and its well-known error-correcting capability of $(p^m - 1 - k)$ symbols whereas extended Reed–solomon codes are algebraically designed over Galois field [15–18], have been designed using modified generator polynomials.

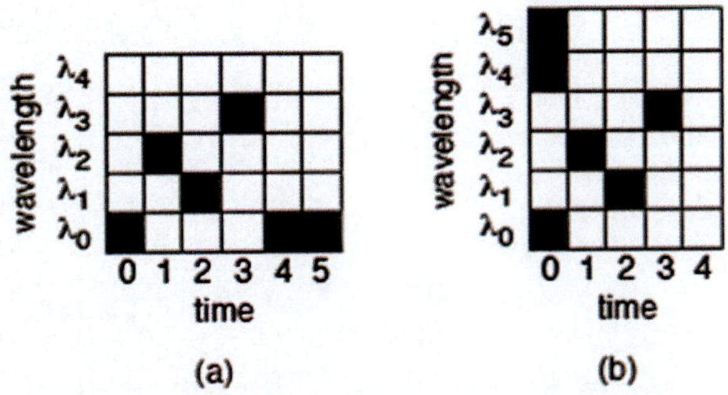

Fig. 1 2D w/t code words, mapped from the coefficient (0, 2, 1, 3, 0, 0) of ERS code

Section 2 demonstrates a family of 2D codes based on E-RS. These codes are algebraically designed over Galois field which gives us p groups of new optical codes with each group has p^{k-1} code words and $L = w$ wavelength which is second coding dimension as explained earlier. Code weight holds an equality $w \leq p + 1$. These 2-D codes have a special partitioning property supporting Multiple Tree structure which allows us to select a number of codewords in applications where we need to focus either on code performance or code cardinality. In Sect. 3, performance analysis of these codes is done and compared with multilevel prime codes (MPC) [5–18] (Fig. 1).

2 Design of E-RS Codes

In 2D wavelength/time codes, each non-zero elements are carried by a single wavelength in one-time slot or instant. To generate a 2-D Wavelength-Time codeword, code coefficient of E-RS code is mapped to time and wavelength. If we take an example where (0, 2, 1, 3, 0, 0) are the code coefficient, then a matrix is generated by representing wavelength number as ith coefficient in ith time slot.

Designing of these codes starts with construction of E-RS code over Galois field GF (p). E-RS code is denoted as RS ($p + 1, k, p - k + 2$) which is a Linear cyclic code and Minimum Distance separable codes (MDS) [16, 18]. E-RS codeword has code cardinality of p^k, length $p + 1$, a minimum hamming distance of $p - k + 2$ where k holds the inequality $k \geq 1$ which is actually coded dimension (Table 1).

(1) As a finite field must be placed on a prime number to ensure that each row and column of its addition and multiplication tables contains a unique value. Hence a prime number p is chosen first. Then for this, p we select an irreducible polynomial (which cannot be factorized further). This root is actually a primitive root of field to derive all values in the field.
(2) Let's say α is the root of $f(x)$ then $f(\alpha) = 0$. A finite field indeed finite but repeats an infinite number of times hence for smallest integer i, $\alpha^i = 0$. With this root α of $f(x)$ we have to calculate all exponential powers of α under all modulo p operations.

Table 1 E-RS Optical code words with $p = 7, k = 2$ and $m = 1$

Generation of E-RS code							
$i_2 = 0$							
i_1	i_0						
	0	1	2	3	4	5	6
0	0000	0145	0213	0351	0426	0564	0632
	0000	5410	3120	1530	6240	4650	2360
1	1455	1523	1661	1036	1104	1242	1310
	4100	2510	0220	5630	3340	1050	6460
2	2133	2201	2346	2414	2552	2620	2065
	1200	6610	4320	2030	0440	5150	3560
3	3511	3656	3024	3162	3230	3305	3443
	5300	3010	1420	6130	4540	2250	0660
4	4266	4334	4402	4540	4615	4053	4121
	2400	0110	5520	3230	1640	6350	4060
5	5644	5012	5150	5225	5363	5431	5506
	6500	4210	2620	0330	5440	3450	1160
6	6322	6460	6535	6603	6041	6116	6254
	3600	1310	6020	4430	2140	0550	5260
$i_2 = 1$							
i_1	i_0						
	0	1	2	3	4	5	6
0	0014	0152	0220	0365	0433	0501	0646
	5541	3251	1661	6301	4011	2421	0131
1	1462	1530	1605	1043	1111	1256	1324
	2641	0315	5061	3401	1111	6521	4231
2	2140	2215	2353	2421	2566	2634	2002
	6041	4451	2161	0501	5211	3621	1331
3	3525	3663	3031	3106	3244	3312	3450
	3141	1551	6261	4601	2311	0021	5431
4	4203	4341	4416	4454	4622	4060	4135
	0241	5651	3361	1001	6411	4121	2531
5	5651	5026	5164	5232	5300	5445	5513
	4341	2051	0461	5101	3511	1221	6631
6	6336	6404	6542	6610	6055	6123	6261
	1441	6151	4561	2201	0611	5321	3031

(continued)

Table 1 (continued)

Generation of E-RS code							
$i_2 = 2$							
i_1	i_0						
	0	1	2	3	4	5	6
0	0021	0166	0234	0302	0440	0515	0653
	3312	1022	6432	4142	2552	0262	5602
1	1406	1544	1612	1050	1125	1263	1331
	0412	5122	3532	1242	6652	4362	2002
2	2154	2222	2360	2435	2503	2641	2016
	4512	2222	0632	5342	3052	1462	6102
3	3532	3600	3045	3113	3251	3326	3464
	1612	6322	4032	2442	0152	5562	3202
4	4210	4355	4423	4561	4636	4004	4142
	5012	3422	1132	6542	4252	2662	0302
5	5665	5033	5101	5246	5314	5452	5520
	2112	0522	5232	3642	1352	6062	4402
6	6343	6411	6556	6624	6062	6130	6205
	6212	4622	2332	0042	5452	3162	1502
$i_2 = 3$							

(3) For each exponential power of the primitive root α we evaluate minimal polynomial. equation to be written are as follows:

$$m_0(x) = x - \alpha^0$$

$$m_j(x) = \left(x - \alpha^j\right)\left(x - \alpha^{jp}\right)$$

$$m_{(p+1)/2}(x) = x - \alpha^{(p+1)/2}$$

(4) Then generator polynomial of E-RS is evaluated as follows:

$$g(x) = m_{\frac{p+1}{2}}(x) \times m_{\frac{p-1}{2}}(x) \times \cdots \times m_{\frac{k+1}{2}}(x)$$

After some manipulations $g(x)$ is minimized in a general form given as:

$$g(x) = g_0 + g_1 x + g_2 x^2 + \cdots + g_{p-k} x^{p-k+1}$$

(5) By multiplying $g(x)$ and $i(x)$ under all modulo p operation codeword polynomial $c(x)$ is generated.

$$c(x) = i(x)g(x)$$

In the matrix forms, codeword polynomial coefficient is obtained by matrix multiplication of message matrix and generator matrix containing coefficient of message polynomial coefficients and generator polynomial coefficients.

(6) E-RS Codeword is the polynomial coefficient of $c(x)$ which is total p^k in number. Resulting code coefficient denotes the time slot location of jth wavelength where $j = \{0, 1, \ldots, p\}$.

3 Analysis of Codes

For analyzing the performance of these E-RS codes, the effect of MAI is taken into account and effects of physical noises are ignored by assuming an MAI limited environment for the OCDMA system. Combinatorial analysis is used for evaluating the hard error-limiting error probability that is denoted by $(w \times p, w, 0, \lambda_c)$. Hard error limiting probability is given by:

$$P_e = \frac{1}{2}(-1)^i \binom{w}{i} \left[\sum_{j=0}^{\lambda_c} q_{\lambda c, j} \frac{\binom{w-i}{j}}{\binom{w}{j}} \right] \quad (1)$$

Here q depicts getting a number of j hits in a time slot where $j = \{0, 1 \ldots, \lambda_c\}$. To calculate P_e we find out q for all values of j as

$$q_{\lambda c, j} = \frac{1}{2} \times \frac{h_{\lambda c, j}}{p(p^{\lambda c} - 1)} \quad (2)$$

where $\frac{1}{2}$ depicts for equal probability of "1" and "0" in OOK. In denominator, p denotes of shifts in codeword whereas $p^{\lambda_c} - 1$ represents the possible number of p^{λ_c} code words. Before calculating for q we evaluate equations by $h_{\lambda c, j}$.

For any arbitrary value of w equality holds $w \leq p + 1$, for which $p + 1 \leq w$. Other equations for $h_{2,1}, h_{2,0}, h_{4,3}, h_{4,2}, h_{4,0}$ are derived by rearranging value of q in hit probability equations.

For $\lambda_c = 2$

$$h_{2,2} = \binom{w}{2}(p - 1) \quad (3)$$

$$h_{2,1} = w(p^2 - 1) - h_{2,2} \quad (4)$$

For $\lambda_c = 4$

$$h_{4,4} = \binom{w}{4}(p-1) \tag{5}$$

$$h_{4,3} = \frac{w(w-1)(w-2)(p-1)(p+4-w)}{6} \tag{6}$$

$$h_{4,2} = \frac{w(w-1)(p-1)}{2} \times \frac{2p^2 - 2pw + w^2 + 6p - 7w + 12}{2} \tag{7}$$

$$h_{4,1} = w\left(p^4 - 1\right) - 4h_{4,4} - 3\mathrm{h}_{4,3} - 2h_{4,2} \tag{8}$$

$$h_{4,0} = 2p\left(p^4 - 1\right) - 4h_{4,4} - \mathrm{h}_{4,3} - h_{4,2} - h_{4,1} \tag{9}$$

Using $p = 7, w = 8, \lambda_c = 4$ and cardinality $p^{\lambda_c} = 2401$ the codes can be divided into 49 groups with $\lambda_c = 4$ within each group.

In Fig. 2, the hard error probability, P_e of $(L \times N, W, 0, \lambda_c)$ optical code is drawn with respect to simultaneous users K for $p = \{7, 11, 13\}$, $L = W = p + 1 = \{8, 12, 14\}$ and $\lambda_c = \{2, 4\}$. Generally, BER improve with p as it indicates the number of available wavelengths to be used are increased ($L = p + 1$), with increased code length ($N = P$). This results in enhanced performance because hit probabilities are reduced and autocorrelation peak is increased but BER gets worse as K increases because with increase in number of users, effect of MAI becomes stronger and prominent.

Apart from this, the performance of system degrades due to increase in λ_c because of increase in number of hits. From Fig. 2a, it is clear that curves of E-RS codes for $\lambda_c = 2$ are always below than curves of other codes for same number of simultaneous users, code weight and λ_c. It is observed from the graph that E-RS codes with smaller

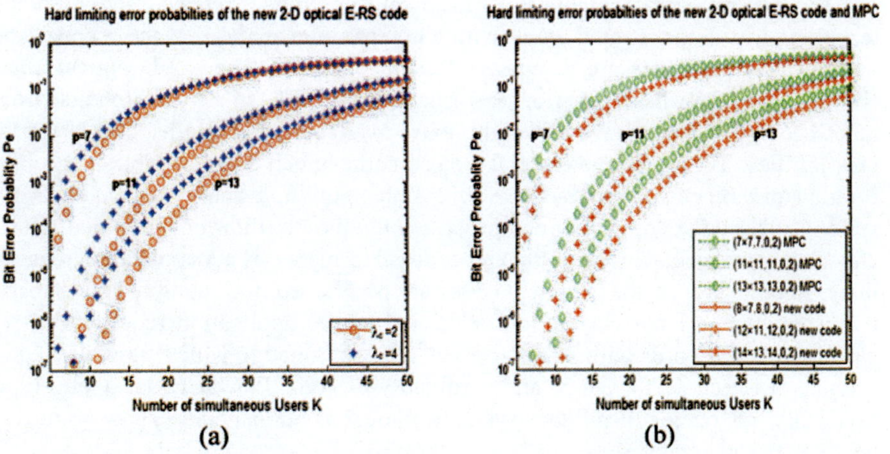

Fig. 2 **a** BER of $((p+1) \times p, p+1, 0, \lambda_c)$ code for $p = \{7, 11, 13\}$ and $\lambda_c = \{2, 4\}$. **b** BER of $((p+1) \times p, p+1, 0, \lambda_c)$ code and the $(p \times p, p, 0, 2)$ MPC [14] for $p = \{7, 11, 13\}$

λ_c performs better than codes with larger λ_c. As larger p value represents more time slots, then with reference to results with increase in prime number, hit probabilities reduce and performance improves.

For MPC hit probabilities are derived by general equations as follows [14]:

$$h_{n,0} = 2p\left(p^n - 1\right) - h_{n,n} - h_{n,n-1} - \cdots - h_{n,1} \tag{10}$$

$$h_{n,1} = w\left(p^n - 1\right) - nh_{n,n} - \cdots - 2h_{n,2} \tag{11}$$

$$h_{n,n} = \binom{w}{n}(p - 1) \tag{12}$$

From these equation hit probabilities for MPC are derived that are similar to E-RS, but the difference is performance of two is because wavelength number, code weight, are equal to p ($L = w = p$) for MPC whereas for E-RS codes, number of wavelength = code weight = p i.e ($L = w = p$). From Fig. 2b, it is clear that curve for E-RS is always lower than the curve for MPC, hence E-RS code performs better than MPC though both codes carry their respective maximum wavelength. Also, E-RS has more weight (=$p + 1$) than MPC (=p). In other words, transmission bandwidth (i.e. LN product) is sacrificed to enhance performance of code. Therefore, with an increase in K, more code matrices exist hence E-RS curves slowly approach MPC curves later due to increased MAI.

4 Conclusion

In this paper, two-dimensional (2D) optical code based on extended Reed–Solomon technique was demonstrated along with a comparative analysis of these codes and multilevel prime codes are discussed. Results show that these 2-D optical asynchronous OCDMA have superior performance than MPC in a MAI prone environment because of heavier code weight, extra available wavelengths, and lesser hit probabilities. These parameters make these codes much more suitable and robust for transmission channels or systems accessible to multiple active users in OCDMA systems. These 2-D optical codes possess a unique partition property that allows choosing between code cardinality and code performance as per application requirement. In other words, these optical codes are partitioned into multiple code subsets corresponding to cross-correlation value and as per application requirement, we can choose codewords with an appropriate cross-correlation which signifies a compromise between performance and cardinality of code. For OCDMA applications especially in strategic or military systems, these 2-D optical codes show promising performance characteristics.

References

1. J.A. Salehi, Code division multiple-access techniques in optical fiber networks—part I: fundamental principles. IEEE Trans. Commun. **37**(8), 824–833 (1989)
2. G.-C. Yang, W.C. Kwong, *Prime Codes With Applications to CDMA Optical and Wireless Networks* (Artech House, 2002)
3. P.R. Prucnal (ed.), *Optical Code Division Multiple Access: Fundamental and Applications* (Taylor & Francis, 2006)
4. W.C. Kwong, G.-C. Yang, *Optical Coding Theory with Prime* (CRC Press, 2013)
5. F.R.K. Chung, J.A. Salehi, V.K. Wei, Optical orthogonal code: design, analysis, and applications. IEEE Trans. Inf. Theory **35**(3), 595–604 (2009)
6. G.-C. Yang, T. Fuja, Optical orthogonal codes with unequal auto-and cross-correlation constraints. IEEE Trans. Inf. Theory **41**(1), 96–106 (1995)
7. L. Tancevsk, I. Andonovic, Hybrid wavelength hopping/time spreading schemes for use in massive optical networks with increased security. J. Lightw. Technol. **14**(12), 2636–2647 (1996)
8. E.L. Titlebaum, L.H. Sibul, Time-frequency hop signals—part II: coding based upon quadratic congruences. IEEE Trans. Aerosp. Electron. Syst. **17**(4), 494–500 (2015)
9. S.V. Maric, E.L. Titlebaum, Frequency hop multiple access codes based upon the theory of cubic congruences. IEEE Trans. Aerosp. Electron. Syst. **26**(6), 1035–1039 (1990)
10. R.M.H. Yim, L.R. Chen, J. Bajcsy, Design and performance of 2-D codes for wavelength-time optical CDMA. IEEE Photon. Technol. Lett. **14**(5), 714–716 (2002)
11. W.C. Kwong, G.-C. Yang, V. Baby, C.-S. Bres, P.R. Prucnal, Multiple-wavelength optical orthogonal codes under prime-sequence permutations for optical CDMA. IEEE Trans. Commun. **53**(1), 117–123 (2005)
12. C.-Y. Chang, G.-C. Yang, W.C. Kwong, Wavelength-time codes with maximum cross-correlation function of two for multicode-keying optical CDMA. J. Lightw. Technol. **24**(3), 1093–1100 (2006)
13. J.-H. Tien, G.-C. Yang, C.-Y. Chang, W.C. Kwong, Design and analysis of 2-D codes with the maximum cross-correlation value of two for optical CDMA. J. Lightw. Technol. **26**(22), 3632–3639 (2008)
14. C.-H. Hsieh, G.-C. Yang, C.-Y. Chang, W.C. Kwong, Multilevel prime codes for optical CDMA systems. J. Opt. Commun. Netw. **1**(7), 600–607 (2009)
15. C.-C. Sun, G.-C. Yang, C.-P. Tu, C.-Y. Chang, W.C. Kwong, Extended multilevel prime codes for optical CDMA. IEEE Trans. Commun. **58**(5), 1344–1350 (2010)
16. V.C. Rocha Jr., Maximum distance separable multilevel codes. IEEE Trans. Inf. Theory **30**(3), 547–548 (2006)
17. S. Lin, D.J. Costello, *Error control coding: fundamentals and Applications*, 2nd edn (Pearson Education, 2017)
18. L. Gyorfi, J.L. Massey, Constructions of binary constant-weight cyclic codes and cyclically permutable codes. IEEE Trans. Inf. Theory **38**(3), 940–949 (2016)

A Computationally Efficient Real-Time Vehicle and Speed Detection System for Video Traffic Surveillance

Ritika Bhardwaj, Anuradha Dhull, and Meghna Sharma

Abstract This research article presents a computationally efficient real-time vehicle and speed detection method for video traffic surveillance. It applies training and testing phases in parallel over pre-trained video surveillance dataset by employing YOLO algorithm. In order to enhance the computation time, dynamic background subtraction method has been applied in conjunction with YOLO algorithm. The aim is to reduce the total number of frames required to be considered for further processing. DBS has strong adaptability to changed background and can detect the movement in frame. In addition to vehicle detection, the proposed algorithm can detect the speed of detected vehicles using a centroid feature extraction method. Our proposed method is a dual combination of speed and accuracy. The performance of the proposed algorithm is analyzed over 2018 NVIDIA AI City challenge dataset and a comparative analysis is done with various state-of-the-art vehicle detection methods (CNN, R-CNN, Fast R-CNN, and FastYOLO).

Keywords Traffic video surveillance · Vehicle detection · Vehicle classification · Speed estimation · YOLO

1 Introduction

Automatic video surveillance means real-time monitoring of people and vehicles leading to understand their actions and interactions. The video traffic surveillance is used to monitor the traffic on road. The main goal of the video-based traffic monitoring system is to detect the vehicles that appear on video image and estimate their position and other important information in real time. In January 2019, more than a million and half new vehicles were bought and registered in India [1]. This shows that the number of vehicles on road is increasing rapidly, so it is difficult to monitor them manually. There is need for automatic traffic monitoring system

R. Bhardwaj (✉) · A. Dhull · M. Sharma
The NorthCap University, Sector-23A, Gurugram, Haryana, India
e-mail: ritika18csp002@ncuindia.edu

P. Bansal et al. (eds.), *Proceedings of International Conference on Artificial Intelligence and Applications*, Advances in Intelligent Systems and Computing 1164,
https://doi.org/10.1007/978-981-15-4992-2_55

which can detect the vehicles, estimate their speed, tracking vehicles, recognizing their number plates, detecting accidents, etc. The real-time traffic information can be collected from various sources like inductive loop detectors, infrared detectors, radar detectors, and video-based systems [2]. The video-based monitoring is cheap and easy to maintain as compared to other methods like radar detectors, infrared detectors, etc.

Since late twentieth-century, the researchers are working in video-based monitoring system [3–5]. The researchers have proposed many computer vision and machine learning algorithms for vehicle and speed detection in real time. The popular methods for vehicle detection and tracking are CNN [6], R-CNN [7], Faster R-CNN [8–10], SSD [11], etc. But these methods have some drawbacks which make them unsuitable to implement in real world. The challenges in video-based traffic surveillance are the high computational power, low-visibility, real-time analysis, environmental disturbance, bad weather, etc.

Keeping in mind the associated problems, a new method has been proposed in this paper for vehicle and speed detection in real time. The overall objective of the proposed work is to identify vehicles and their speed in real time with improved accuracy and reduced computational time. From the detailed literature, it can be seen that it is difficult to maintain speed and accuracy in parallel. However, our proposed algorithm is able to provide dual benefits of vehicle detection as well as speed with improved accuracy with the help of preprocessing of the frames.

The rest of the paper is organized as follows: Sect. 2 gives an overview of the related background work done in the area of image preprocessing, vehicle detection and classification, and speed estimation. Section 3 gives the details of the proposed method followed by result analysis in Sect. 4. At last, Sect. 5 presents the concluding remarks.

2 Related Work

In this section, we briefly review the recent work on vehicle detection and speed detection. The preprocessing is done for separating the background and foreground part of the image. This can help in improving the detection accuracy and can also help in detecting speed of the vehicles. The popular techniques for are background subtraction method [12], frame differencing method [13], and optical flow method [14]. Among these, background subtraction method is quite effective to work in real time. In this method, image frame is subtracted from the original background image. It can obtain complete and accurate information but it is very sensitive to changes in light, weather, and other environmental conditions. Unlike background subtraction method, the current real-time image frame is subtracted from previous image frame in frame differencing method. The advantages of this method are that it has a strong adaptability for changes of light and it can be easily implemented for target detection but it is unable to get complete boundary information of the object. The optical flow method is also popular method for motion detection. It detects motion by constraint conditions of gray gradient method and the brightness preservation of moving target. The advantage of this method is that it can detect moving target without knowing any scene information in advance. But this method cannot be used for real-time detection because of its complex and time-consuming calculations [9].

Object detection is the most crucial part of the video surveillance system as speed, number plate detection, or accident detection largely depend on it for improved accuracy and speed. In 2007, David Acunzo [15] proposed a method for vehicle detection under varying lightning conditions. For daylight and low light conditions, the classifiers trained with Adaboost was used and a tail-light detector was used for night time. In 2016, Joseph Redmon [14] proposed a method you only look once (YOLO) which is a unified and real-time detection algorithm. It is far different from the other CNN approaches. It can process images at 45 fps. Mahalla et al. [9] proposed an SMC faster R-CNN method for multi-object detection. In this, SMC filter is used for improving the efficiency of the faster R-CNN detector. In 2018, Li Suhao [8] used faster R-CNN and RPN for detection and faster R-CNN for classification. This technique has been applied on various datasets and it performed really well. But in this paper, the author has considered only three type of vehicle categories for detection.

There are many techniques to detect the motion and speed of an object. With the help of speed calculation, we can detect the over-speeding vehicles. This will help government in law enforcement. In 2018, intrusion line technique was proposed by Saleh Javadi et al. [16] for speed estimation. Many other techniques have been utilized for speed detection such as motion vector technique [17], ray constraint optical flow algorithm [14], optical flow and Lucas–Kanade method [18].

3 Proposed Method

The prime objective is to detect the vehicle and calculate its speed in real time. Figure 1 shows the pseudo code of the proposed method. The inputs given to the algorithm are threshold (θ), weight assigned to the grid containing center of the object ($\alpha = 5$), weight assigned to the grid containing no object ($\beta = 0.5$), and frame per second (fps). Every frame passes through pre-processing step. After that, in all frames, go through YOLO algorithm, in which probability of bounding box, center of object, and class is calculated. If object is detected in pre-processing step, then this frame is processed further for speed estimation.

Figure 2 shows the general layout of the proposed method. The video dataset is input in DBS for preprocessing. After that, YOLO algorithm is applied for detection and classification. If any movement is detected during preprocessing then the frame is sent for speed estimation.

The step by step working of the various steps is explained in the subsections given below:

Fig. 1 Algorithm of the proposed method

```
Algorithm
Input:
            Video parameters: Θ, α, β, fps

Procedure:
        Obtain each frame of the video
         Preprocess DBS:
              Calculate the Revised background image.
               Compare the current frame with Revised
                background image.
               Detect the object in frame.
         For each frame in the video:
               Divide the image into 13*13 grids.
               Build 5 different sizes of boxes for the
centre
of each grid
               Compare the P(bounding box), P(centre)
and                        P(class)
          End for
          If Object detected in preprocess step:
                Calculate centroids of object in different
time frame
                Calculate pixel distance.
                Estimate the speed of the vehicle.
          End if
          else
            Next frame.
         End else
Output:
          bounding box (x, y, z, h, c), probability of
classes and estimated speed.
```

3.1 Image Preprocessing (Dynamic Background Subtraction)

The image preprocessing is an important step required to improve the accuracy and speed of the detection. Dynamic background subtraction technique has been used for image preprocessing. As it is a combination of frame differencing technique and background subtraction technique which consists of the advantages of both of these techniques and also helps to overcome the disadvantages of both. The background reference image is dynamic in this technique [19]. That is why, it is known as dynamic background subtraction method.

Figure 3 shows the flowchart of the dynamic background subtraction method in detail. Three frames f_t, $f_{t-5,}$ and f_{t+5} are taken and D_{t-5} and D_{t+5} are calculated as shown below. With the help of D_{t-5} and D_{t+5}, we can get the new reference

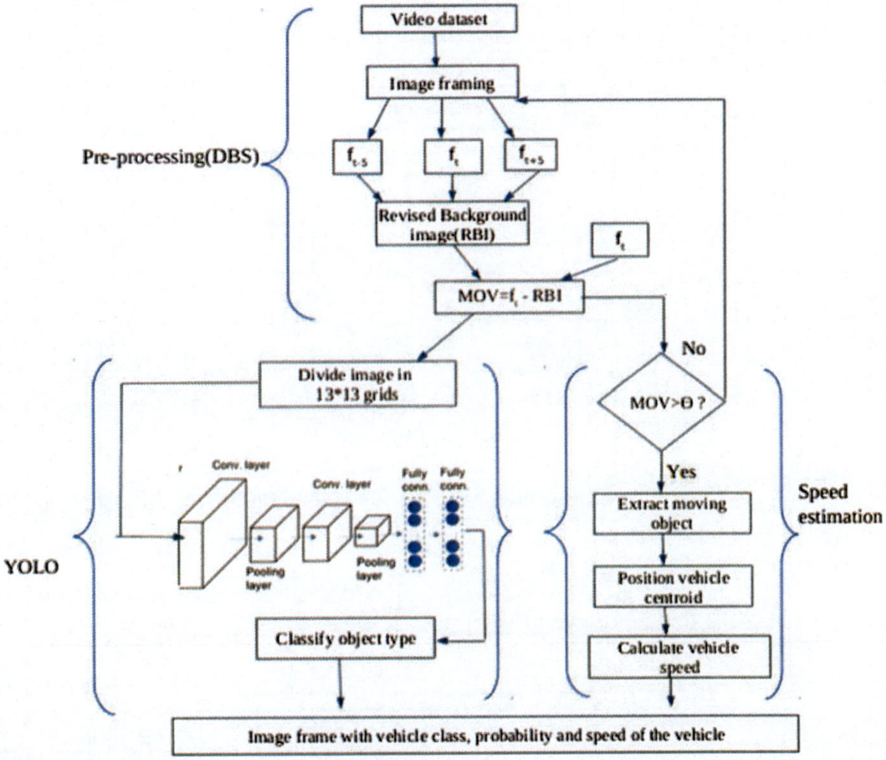

Fig. 2 General layout of the proposed methodology

background image, RBI (*x*, *y*). Finally, subtraction can be made between RBI and f_t will give MOV (*x*, *y*) which is compared with the threshold value (Θ).

$$D_{t-5} = |f_t - f_{t-5}| \tag{1}$$

$$D_{t+5} = |f_t - f_{t+5}| \tag{2}$$

$$\text{MOV}(x, y) = |\text{RBI}(x, y) - f_t(x, y)| \tag{3}$$

Figure 4, the images are shown from the dataset and the image after implementation of dynamic background subtraction technique in proposed method.

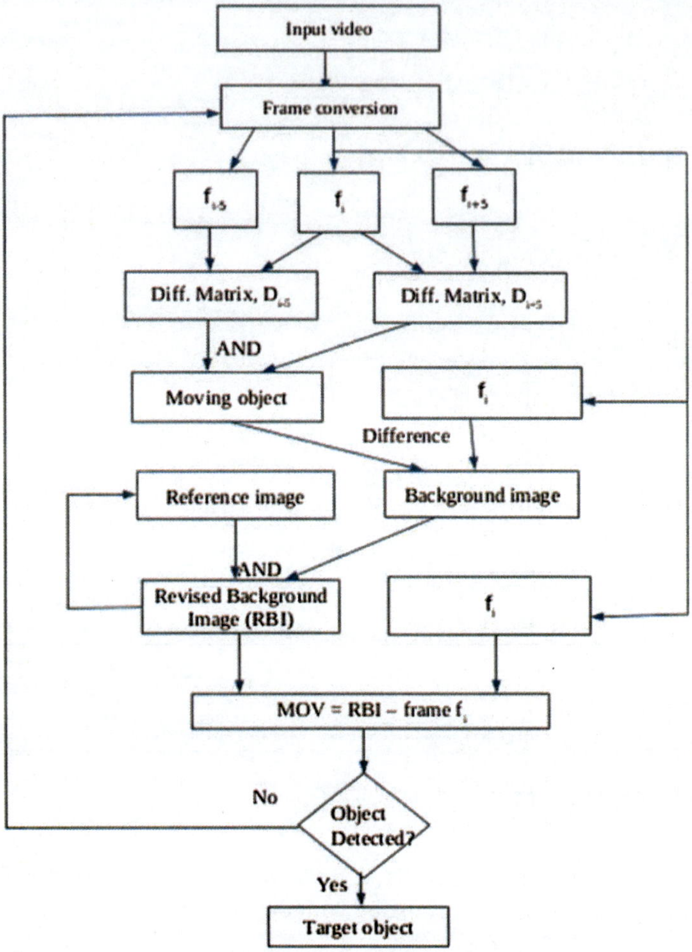

Fig. 3 Dynamic background subtraction method flow chart

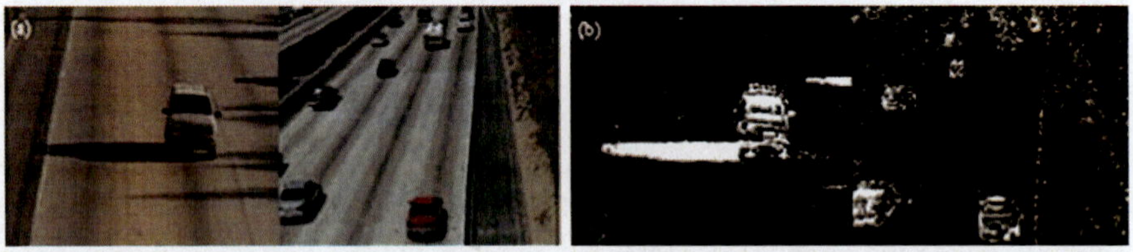

Fig. 4 **a** image from dataset, **b** image after applying DBS in proposed method

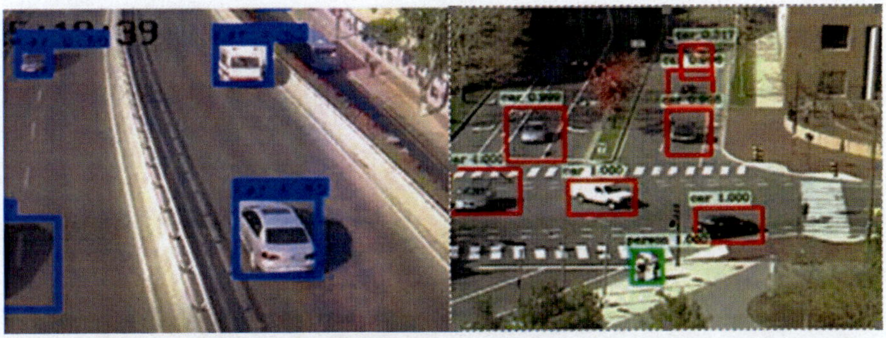

Fig. 5 Vehicle classification with class probability using YOLO algorithm

3.2 *Object Detection and Classification (YOLO Algorithm)*

Many algorithms and techniques have been studied in past for the object detection. Among these, YOLO algorithm is one of the most popular techniques for object detection. You look only once (YOLO) algorithm is one of the fastest objects detecting algorithms which works good in real time. It uses the pre-trained model of YOLOv3. This algorithm classifies the vehicle present in frame and also create the bounding box around the vehicle. It also calculates the class probability of the vehicle and shows in output. The YOLOv3 algorithm has been trained over ImageNet dataset. This dataset consists of more than 14 million images which belong to 20,000 categories of image objects.

The pre-trained model of the YOLOv3 is used. YOLO is basically improved CNN; it uses the convolutional neural network to extract features and uses the fully connected layer to predict the object information.

This algorithm consists of 24 convolutional layers and two fully connected layers. Given below in Fig. 5 shows the output image from the YOLO algorithm, after vehicle classification with class probability.

3.3 *Speed Estimation*

Speed detection is also very crucial part of traffic surveillance system. There are many techniques to detect the motion and speed of the vehicles in traffic monitoring. With the help of the centroids, the speed of the object is calculated. The approach used for speed detection is centroid feature extraction method [20]. In this, the speed is calculated with the help of the centroids. The distance between the centroids is calculated as pixel distance. The advantage of this technique is that it is easy to implement. So, it can be implemented in real time.

The pixel distance is multiplied with the fps of the input video for converting it into real-world distance. Let the video frame rate is fps, and the $(x1, y1)$ and $(x2, y2)$

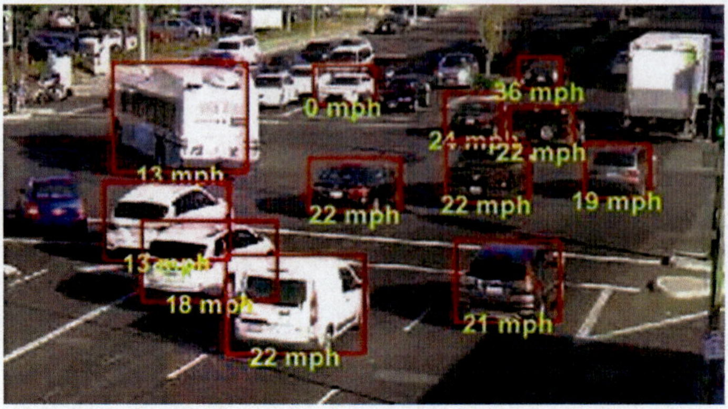

Fig. 6 Vehicle speed estimation using proposed method

are two centroid points in t frame in video and $t1$ frame in video, respectively. So, the speed of the moving object in video can be calculated as:

$$V = \sqrt{(y1 - y2)2 + (x1 - x2)2} * \text{fps} \tag{4}$$

Here, V is the estimated speed of the vehicle. Figure 6 shows the output of the estimated speed.

4 Experimental Analysis and Discussion

The dataset used for implementing the video-based traffic monitoring system is 2018 NVIDIA AI City challenge dataset. This dataset is available online [21]. Every year, AI City challenge comes up with new set of videos and images with different tasks to implement. Table 1 shows the detail about used dataset. The dataset available have been captured by stationary cameras located at different intersections and highways. The dataset used in this paper just to detect the moving vehicle and estimate its speed. This dataset consists of the metadata with it, which helps in checking the accuracy of the estimated speed of the vehicles.

Table 1 AI City challenge dataset description

	Instances	fps	Resolution	Task
Track 1	27 videos (1 min each)	30 fps	1920 × 1080	Traffic flow analysis
Track 2	100 videos (15 min each)	30 fps	800 × 410	Anomaly detection
Track 3	15 videos (0.5–1.5 h each)	30 fps	1920 × 1080	Multisensor vehicle detection

Table 2 Direction and speed limit data for Track 1 dataset

Location	Latitude	Longitude	Direction	Speed
1	37.316788	−121.950242	E → ΘW	65 MPG
2	37.330574	−121.014273	NW → ΘSE	65 MPG
3	37.326776	−121.965343	NW → ΘSE	45 MPG
3	37.326776	−121.965343	NE → ΘSW	35 MPG
4	37.323140	−121.950852	N → ΘS	35 MPG

The maximum speed information can be collected from the meta-data files associated with each video. These videos are captured at four different locations. In locations 1 and 2, highway traffic is considered and in locations 3 and 4, the road intersection traffic is considered. Table 2 shows the location, direction and speed limit of vehicle in Track 1 dataset.

All the experiments were executed on a system with fifth-generation core i5 16 GB RAM. The 2018 NVIDIA AI City challenge was assessed based on composite score, $S1 = \text{DR} * (1 - \text{NRMSE})$, where DR is the detection rate of vehicle and the NRMSE is the RMSE score across all detections of ground-truth vehicle. We have used the methodology and tool of Hoiem et al. [22]. This analysis shows us that our proposed algorithm is better as compared to the other detection techniques. It is more accurate and faster.

Due to preprocessing of the images, detection of vehicle is more accurate and faster. The vehicles can also be detected in bad weather case. As we can see in the below image, Fig. 7, our proposed method is faster as compared to others. After this, the proposed method is compared with other algorithms on the bases of precision and recall. These properties are used to check the accuracy of the method. Table 3 shows the experimental result of comparison of proposed method with other algorithms. This shows that proposed method is better in accuracy than other methods. The comparison is done with the help of the values given by Lu et al. [22]. The formula used for precision and recall is as follows:

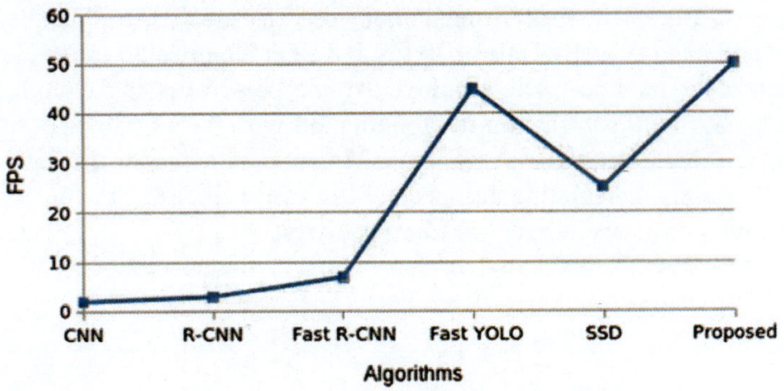

Fig. 7 Comparative analysis with respect to frames per second (FPS)

Table 3 Precision and recall value comparative analysis

Algorithm	Precision (%)	Recall (%)
CNN	80.82	78.38
R-CNN	84.19	83.69
Fast R-CNN	83.96	82.65
Fast YOLO	88.45	86.64
SSD	86.82	84.10
Proposed method	89.54	87.46

Table 4 Average error (%) of proposed method with respect to YOLO, fast R-CNN and PDF

	PDF [23]	YOLO	Fast R-CNN	Proposed method
Average error (%)	1.77	1.85	1.56	1.39

$$\text{Precision} = \frac{\text{TP}}{\text{TP} + \text{FP}} \tag{5}$$

$$\text{Recall} = \frac{\text{TP}}{\text{TP} + \text{FN}} \tag{6}$$

The error rate of the proposed method is 1.39% at 30 fps which is better than other methods. Table 4 shows the average error obtained by different methods and the proposed methodology.

5 Conclusion

In this paper, we have presented a computationally efficient vehicle and speed detection method for real-time traffic surveillance. The performance of the proposed method in terms of speed and accuracy proves the worthiness of proposed method over different state-of-art techniques used for traffic surveillance. The proposed method has 1.5% error rate at 30 fps and can recognize 50 frames per second, better than other methods. The employed frame pre-processing technique named dynamic background subtraction in combination with YOLO algorithm is able to speed up the vehicle detection in real time. Moreover, it achieves the desired performance by parallelly calculating the speed of detected vehicles. The future work is to detect the number plates in over-speeding vehicles.

References

1. https://www.news18.com/news/india/the-single-statistic-that-shows-why-indian-roads-are-getting-morecongested-each-passing-month-2031835.html
2. Engel, J. I., Martin, J., & Barco, R. . A low-complexity vision-based system for real-time traffic monitoring. IEEE Transactions on Intelligent Transportation Systems, **18**(5), 1279–1288 (2016)
3. A. Soto, A. Cipriano, Image processing applied to real time measurement of traffic flow. IFAC Proc. Vol. **28**(19), 43–48 (1995)
4. B. Coifman, D. Beymer, P. McLauchlan, J. Malik, Areal-time computer vision system for vehicle tracking and traffic surveillance. Transp. Res. Part C Emerg. Technol. **6**(4), 271–288 (1998)
5. Sullivan, G. D., Baker, K. D., Worrall, A. D., Attwood, C. I., & Remagnino, P. M. . Model-based vehicle detection and classification using orthographic approximations. Image and vision computing, **15**(8), 649–654 (1997)
6. J. Kurniawan, S.G. Syahra, C.K. Dewa, Traffic congestion detection: learning from CCTV monitoring images using convolutional neural network. Procedia Comput. Sci. **144**, 291–297 (2018)
7. Z. Huo, Y. Xia, B. Zhang, Vehicle type classification and attribute prediction using multi-task RCNN, in *2016 9th International Congress on Image and Signal Processing, BioMedical Engineering and Informatics (CISP-BMEI)*. IEEE (2016), pp. 564–569
8. A. Arinaldi, J.A. Pradana, A.A. Gurusinga, Detection and classification of vehicles for traffic video analytics. Procedia Comput. Sci. **144**, 259–268 (2018)
9. A. Mhalla, T. Chateau, H. Maamatou, S. Gazzah, N.E.B. Amara, SMC faster R-CNN: toward a scene-specialized multi-object detector. Comput. Vis. Image Underst. **164**, 3–15 (2017)
10. L. Suhao, L. Jinzhao, L. Guoquan, B. Tong, W. Huiqian, P. Yu, Vehicle type detection based on deep learning in traffic scene. Procedia Comput. Sci. **131**, 564–572 (2018)
11. C.H. Hilario, J.M. Collado, J.M. Armingol, A. De La Escalera, Pyramidal image analysis for vehicle detection, in *IEEE Proceedings of Intelligent Vehicles Symposium, 2005*. IEEE (2005), pp. 88–93
12. Seki, M., Fujiwara, H., & Sumi, K.. A robust background subtraction method for changing background. In Proceedings Fifth IEEE Workshop on Applications of Computer Vision (pp. 207–213). IEEE.(2000, December)
13. N. Singla, Motion detection based on frame difference method. Int. J. Inf. Comput. Technol. **4**(15), 1559–1565 (2014)
14. Duncan, J. H., & Chou, T. C. . On the detection of motion and the computation of optical flow. IEEE Transactions on Pattern Analysis & Machine Intelligence, (3), 346–352 (1992)
15. Acunzo, D., Zhu, Y., Xie, B., & Baratoff, G. Context-adaptive approach for vehicle detection under varying lighting conditions. In 2007 IEEE Intelligent Transportation Systems Conference (pp. 654-660). IEEE. (2007, September)
16. S.S.S. Ranjit, S.A. Anas, S.K. Subramaniam, K.C. Lim, A.F.I. Fayeez, A.R. Amirah, *RealTime Vehicle Speed Detection Algorithm using Motion Vector Technique* (2012)
17. J. Lan, J. Li, G. Hu, B. Ran, L. Wang, Vehicle speed measurement based on gray constraint optical flow algorithm. Optik **125**(1), 289–295 (2014)
18. J. Gerát, D. Sopiak, M. Oravec, J. Pavlovicová, Vehicle speed detection from camera stream using image processing methods, in *2017 International Symposium ELMAR*. IEEE (2017), pp. 201–204
19. D.S. Alex, A. Wahi: BSFD: background subtraction frame difference algorithm for moving object detection and extraction. J. Theor. Appl. Inf. Technol **60**(3) (2014)
20. J.X. Wang, Research of vehicle speed detection algorithm in video surveillance, in *2016 International Conference on Audio, Language and Image Processing (ICALIP)*. IEEE (2016), pp. 349–352

21. M. Naphade, M.C. Chang, A. Sharma, D.C. Anastasiu, V. Jagarlamudi, P. Chakraborty, J.N. Hwang, The 2018 nvidia ai city challenge, in *Proceedings of the IEEE Conference on Computer Vision and Pattern Recognition Workshops* (2018), pp. 53–60
22. S. Lu, B. Wang, H. Wang, L. Chen, M. Linjian, X. Zhang, Areal-time object detection algorithm for video. Comput. Electr. Eng. **77**, 398–408 (2019)
23. S. Javadi, M. Dahl, M.I. Pettersson, Vehicle speed measurement model for video-based systems. Comput. Electr. Eng. **76**, 238–248 (2019)

A Novel Data Prediction Technique Based on Correlation for Data Reduction in Sensor Networks

Khushboo Jain, Arun Agarwal, and Anoop Kumar

Abstract Environmental monitoring is among the most significant applications of wireless sensor networks (WSNs), which results in sensing, communicating, aggregating and transmitting large volumes of data over a very short period. Thus, a lot of energy is consumed in transmitting this redundant and correlated data to the basestation (BS) making it enormously challenging to achieve an acceptable network lifetime, which has become a bottleneck in scaling such applications. In order to proficiently deal with the energy utilization in successive data aggregation cycles, we propose a data prediction-based aggregation model, which will reduce data transmission by establishing relationship between sensor readings. The purpose of the proposed model is to exempt the sensor nodes (SN) from sending huge volumes of data for a specific duration during which the BS will predict the future data values and thus minimize the energy utilization of WSN. The study suggested an extended linear regression model, which determines resemblance in shape of data curve of contiguous data periods. We have used real sensor dataset of 54 SN that was deployed in the Intel Berkeley Research laboratory. We tested and compared our work with the recent prediction-based data reduction method. Results reveal that the proposed ELR model works better when compared with the other techniques in many assessment indicators.

Keywords Wireless sensor network · Data prediction · Regression analysis · Energy efficiency · Data reduction

---

K. Jain (✉) · A. Kumar
Department of CSE, Banasthali Vidyapith, Tonk, India
e-mail: khushboojain2806@gmail.com

A. Kumar
e-mail: anupbhola@banasthali.in

A. Agarwal
Department of CSE, GGSIPU, Delhi, India
e-mail: arun.261986@gmail.com

P. Bansal et al. (eds.), *Proceedings of International Conference on Artificial Intelligence and Applications*, Advances in Intelligent Systems and Computing 1164,
https://doi.org/10.1007/978-981-15-4992-2_56

1 Introduction

Due to recent changes in climate and increasing natural disasters globally, the significance of monitoring environmental parameters continuously has also increased which has proliferating developed as a major application of WSN [1–3]. WSNs can be periodically used in any environment to sense hydrological and meteorological parameters in the vicinity, like temperature, light intensity, humidity, traffic, wind speed and direction, voltage, air and water quality and many others.

WSNs provide several benefits for monitoring environmental parameters and in all application areas, energy conservation is taken as a significant factor of consideration [4]. The main reason of this conservation is due to the fact that the nodes of a sensor network in many environments are infeasible to be replaced or recharged, so these applications are energy bound. Continuous sensing of data also results in a rapid drainage of energy, which decreases overall network lifetime. Therefore, energy conservation has become a major design issues in all applications on WSNs. Various energy conversation approaches are proposed to conserve energy as discussed in [5–9].

Moreover, in continuous monitoring applications, it has been observed that most of the data changes in long duration at a very slow rate. This results in huge volume of data duplication in spacial and temporal domains and too recurrent communications between the sensor nodes will dissipate fast the scarce energy resource. Therefore, if we reduce the number of transmitted packets, we can the increase of network lifetime by conserving energy. Data aggregation [10] is also considered to be crucial criterion to reduce duplication in the highly variable data of time-series. It helps to minimize the number of repetitive transmissions and leads to energy conservation which results in enhanced network lifetime.

In real-world scenarios, sometimes, it is unnecessary and expensive to find the precise and accurate measurements for all time-periods as it will result in redundancy. Data prediction is a most effective and efficient way to reduce data in WSNs that makes use of the predicted data instead of the real sensor's readings, hence restricting the data transmission. Data prediction technique focuses on reducing the number of transmission from SNs to the BS during the continuous monitoring of an application. To guarantee high prediction accuracy, one of the key concerns is to assign a user-given error bound. The main aim of applying prediction technique is to develop a regression model, which can depict the relationship in datasets due to the resemblance in shape of data curve of contiguous data periods. The proposed prediction model is based on a linear regression model termed as extended linear regression technique (ELR) which is developed by improving the cosine distance method to measure the relatedness present in data and the degree of fitness in data curve between current time-period and earlier time-period. The ELR establishes a relation between total number of data values received by a given SNs and its collected data value.

In order to obtain energy-efficient design for continuous monitoring WSN applications, this paper presents a prediction model for monitoring environmental data in

WSN using extended linear regression model. Deploying ELR reduces data transmission, mean square error and energy consumption while maintaining accuracy in WSNs, consequently prolongs network's lifetime. The rest of the paper is structured as follows: Sect. 2 discusses some related research. Section 3 presents the proposed approach and system model used. Section 4 presents the analysis of the proposed model and highlights graphically the results obtained and discussion and Sect. 5 sum-up the conclusion and future work of this research work.

2 Literature Survey

WSNs can be deployed to collect data remotely and eliminate human involvement from the system. WSN is considered as the most significant solutions for organized, automated and continuous collection of data [11]. To improve performance of WSNs, several protocols, models and algorithms have been proposed, few of them are presented below.

An asynchronous in-network data prediction-based framework has been proposed by the authors of [12] to ensure better quality aggregation results which can increase the network lifetime. The strategy in the framework knows the required aggregation coherency, residual energy of a SN, delay in communication and processing. Experimental results in better quality aggregate for long duration of time when compared with the synchronous scheme. An energy-efficient framework based on adaptively enabling/disabling prediction scheme was proposed in [13] for clustering-based data gathering in WSNs. Each cluster in the framework consists of CHs, which represents all SNs in that cluster, and gathers data values from them. The authors presented an adaptive scheme to control prediction and analyse the performance trade-offs between reducing communication and transmission cost which limiting prediction cost.

A data centre monitoring (DCM) solution describes the system architecture [14] which includes the system hardware and an enterprise software application has been deployed in several location at seven different cities. All the locations of different cities were then monitored from a central enterprise software application control panel. A periodical data prediction algorithm is proposed in [15] called P-DPA in WSNs. In this approach to adjust the data prediction values in a WSN, the P-DPA algorithm takes the potential law hidden in periodicity with a reference in order to improve the accuracy of prediction algorithm.

A data prediction model is presented in [16] that was deployed within the SNs. The system architecture consists of the cloud system to generate data, which is predicted by the model. The model was formulated as a line equation through two n-dimensional vectors in n-dimensional space. Experimental outcome highlights that the model achieves improved error rates as compared to traditional prediction techniques, in addition to extended network lifetime. In [17], to predict the temperature and humidity measures in a greenhouse environment in Nanjing, China, the authors have proposed a novel prediction method based on an extreme learning machine (ELM) algorithm and kernel-based ELM (KELM) algorithm was presented.

In order to predict wireless sensing of data based on historical values, a multi-step and improved HMM [18] model was proposed. This clustering-based sensing HMM model uses multistep prediction by varying patterns whose parameters are optimized by particle swarm optimization (PSO). The model was then evaluated on two real-world datasets and the comparison analysis was performed between traditional HMM, grey system, Naive Bayesian and BP neural networks. Experimental outcomes showed that the improved HMM model has provided high accuracy in sensing data predicting. A brief survey about available prediction techniques is presented in [19]. In this, analysis and the classification of prediction techniques have been listed. The authors have suggested a systematic procedure for selecting a scheme to make predictions in WSNs. The contribution was based on constraints of WSN's, characteristics of prediction methods and data under monitoring.

The authors of [20] aim at providing a data prediction model, which reduces load on SNs to enhance network lifetime. The proposed approach computes the degree of relatedness between the data values to establish a relation, which can predict future value. Experimental results with the proposed model proved to achieve better accuracy and lesser energy consumption when compared to traditional data prediction methods like linear regression model, dual prediction model and least mean square model.

3 Proposed Approach

Temporal correlation can occur in two ways, one due to the consecutive data samples which are broadly used in data prediction models. The other occurs due to the resemblance in shape of data curve of contiguous data periods (for instance temperature, light, wind, fog and humidity are periodically changes in a day). This can be taken as a reference in data prediction algorithm to improve the quality of predicted data. This is of huge significance and supports the developments of sensor networks, which are involved with periodic environmental phenomenon.

In this work, a prediction model is deployed in all the SNs and the same prediction model is also deployed in the BS. The SN will only transmit the data reading to the BS when the prediction fails that means the error threshold is higher than required. Both the prediction models will update actual data reading for synchronization in case of a failed prediction. However, when the prediction is successful, the SN will not transmit data reading to the BS and the BS will save its own predicted value, which is also equal to the SN's predicted value.

Figure 1 illustrates the diagram of the proposed method; the SN predicts a new data value in each period according to the previous available data readings. Our proposed model makes first prediction based on first-order autoregression AR (1) algorithm, and then evaluates the data change direction of current time-period and previous time-period. When SN get the data sample, it compares the predicted value with

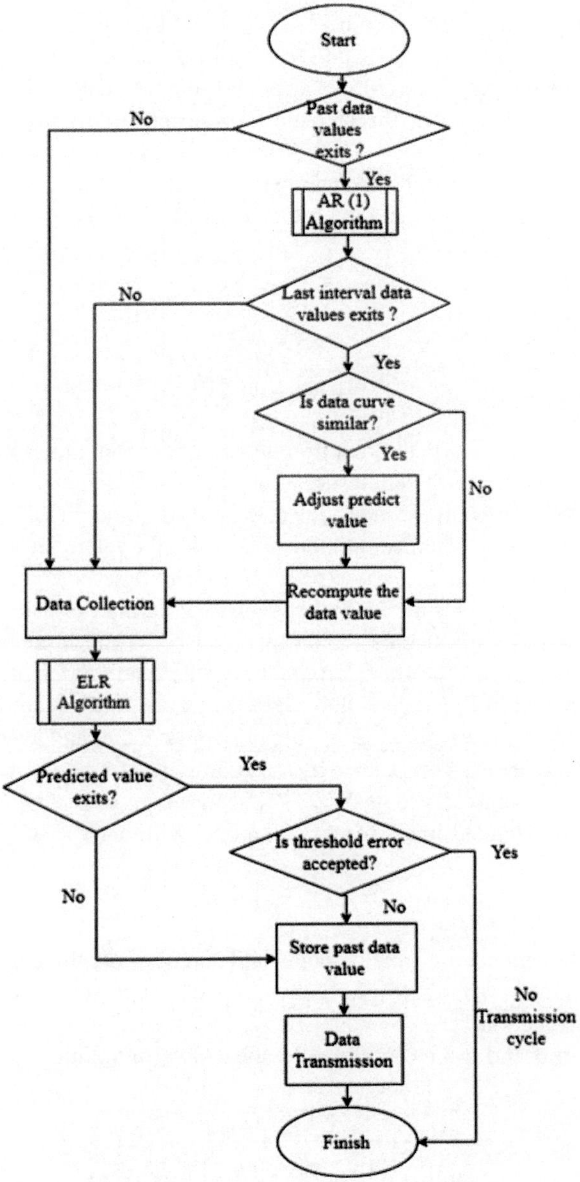

Fig. 1 Flowchart of proposed ELR technique

the actual readings, and if the prediction error is within the user-defined threshold, there is no need to send data to BS, and just wait for the next data sampling period; otherwise, update the prediction model and send sampled data to BS.

The complete technique of the algorithm proposed in this paper is as follows. First, we define factors of this model: t is the current time, $t + 1$ is the next sample time. The sampled data sequence of t is normal $[x_1, x_2, \ldots, x_t]$, we can use a "first order autoregressive process", so the predicted value of time $t + 1$ is:

$$x_{t+1} = \varphi x_t + \alpha_{t+1} \tag{1}$$

$$\varphi = \rho = \frac{D \sum_{t=2}^{D} x_{t-1} x_t}{(D - 1) \sum_{t=1}^{D} x_t^2} \tag{2}$$

where ρ denotes sample AR (1) function and α_{t+1} denotes the stochastic noise of $t + 1$ sampling time while D denotes the number of historic data.

Then, the model uses an extended linear regression model (ELR) which is built by improving the cosine distance method to measure the relatedness present in data and the degree of fitness in data curve between current time-period and earlier time-period. The ELR establishes a relation between total number of data values received by a given SNs and its collected data value. Let N be the number of SN in the network, T be the fixed time duration and D be the data values for each SN. The data values collected from N SNs will be aggregated and three quantitative attributes have been determined, i.e. minimum value, maximum value and average value. The data aggregation will be accomplished by these simple aggregations functions. These three values will be represented as $X_{\min}$, $X_{\max}$, and X_{avg}, respectively.

The proposed extended linear regression model will take the following form:

$$x_{t+1} = \alpha x_t + \beta \tag{3}$$

Here, α is the extended cosine distance and it represents the current slope coefficient of data curve. β is the fitting constant and it represents the previous slope coefficient of data curve.

The values of α and β are calculated by the following relation.

$$\alpha = \frac{\prod_{i=1}^{N} X_{\min} + \prod_{i=1}^{N} X_{\max} + \prod_{i=1}^{N} X_{\text{avg}}}{\prod_{i=1}^{N} \sqrt[2]{(X_{\min})^2 + (X_{\max})^2 + (X_{\text{avg}})^2}} \tag{4}$$

$$\beta = \sum_{i=1}^{N} \sqrt[2]{\left(\frac{X_{\max} - X_{\min}}{X_{\text{avg}}}\right)^2} \tag{5}$$

4 Simulation and Result Analysis

The proposed prediction-based ELR technique is implemented on "The Intel Berkeley Research Laboratory Data" [21] where the data readings were collected with Mica2Dot sensors once in each 31 s using the TinyDB in-network query processing system. The data values are continuously sensed by 54 SNs which are randomly deployed in the sensing area as illustrated in Fig. 2. The data values comprise of four different parameters namely temperature, humidity, light intensity and voltage. We have only simulated the prediction technique for the dataset of temperature and humidity as they are the environmental monitoring attribute but not for light intensity and voltage. We conducted several experiments to evaluate and compare the performances of ELR with P-DPA [15].

We have defined ε as an error prediction threshold parameter whose value lies between $0 < \varepsilon < 1$. As the value of ε increases, less data will be transmitted to the BS owing to lack of data accuracy. Thus, to determine the value of ε, there is a tradeoff between the number of data transmissions and the quality of predicted data. Therefore, we have taken value of ε ranging between 0.1 and 0.9 to experiment with ELR and P-DPA techniques.

Comparison of rate of successful prediction at various parameters at varying threshold

To prove the validity of our proposed algorithm, we have compared the rate of successful prediction of ELR with P-DPA at varying thresholds. We took ten data gathering cycles with 48 data values of both parameters in each data cycle. Clearly, ELR has a great improvement compared with P-DPA in temperature and humidity as demonstrated in Figs. 3 and 4, respectively, which clearly demonstrates that the rate of successful prediction is higher for both parameters of around 15–30%, respectively, in each case.

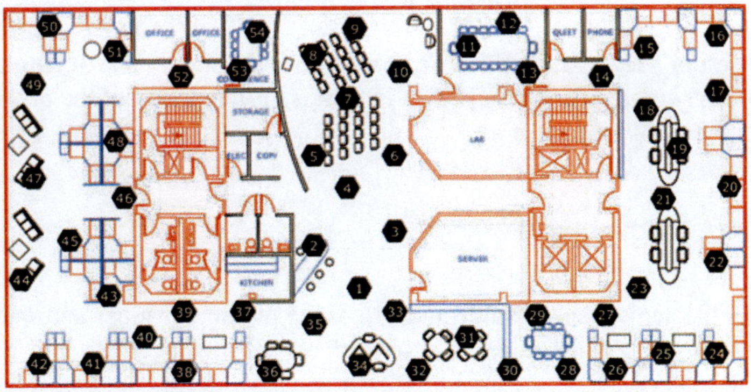

Fig. 2 Network architecture [21]

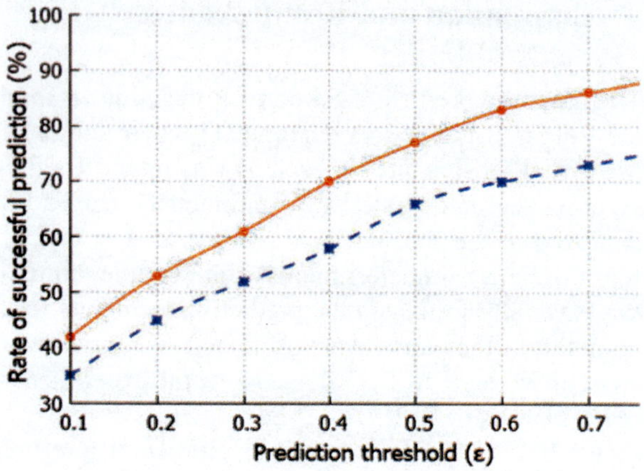

Fig. 3 Rate of successful prediction versus prediction threshold for temperature

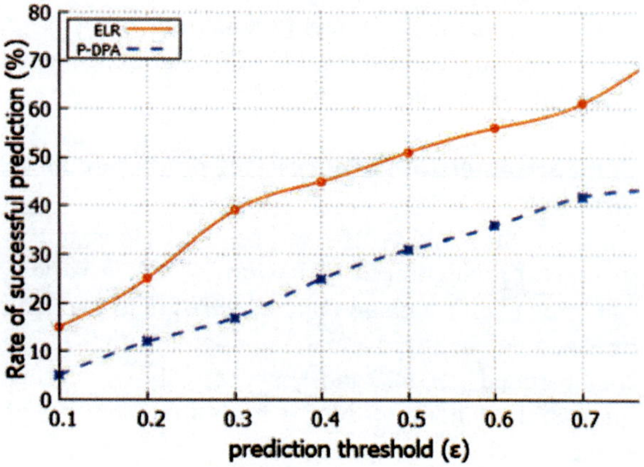

Fig. 4 Rate of successful prediction versus prediction threshold for humidity

Comparison of Mean Square Error at various parameters at varying threshold We use mean square error (MSE) as an indicator of the quality of the predicted data values at the BS. We define it as

$$\text{MSE} = \frac{1}{D} * \sum_{i=1}^{D} (x_{i+1}^2 - x_i^2) \tag{6}$$

where x_i is the actual sensed value and x_{i+1} is the predicted values and D is the total data value sensed. Figures 5 and 6 demonstrates the data quality achieved by ELR and P-PDP techniques. Corresponding to the high MSE attained by P-DPA of each parameter, ELR technique delivers much less MSE with change in prediction error threshold in each case.

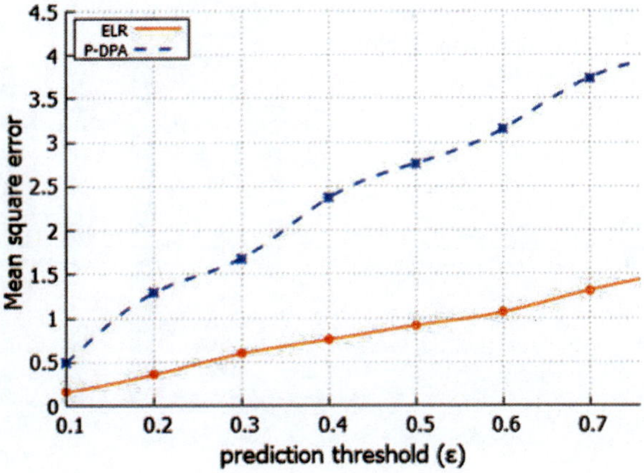

Fig. 5 Mean square error versus prediction threshold for temperature

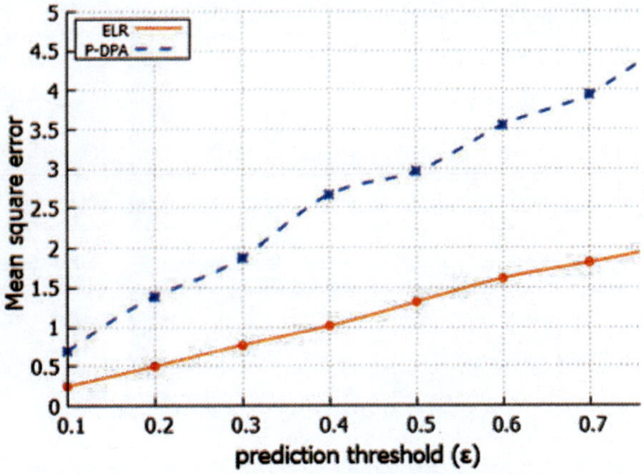

Fig. 6 Mean square error versus prediction threshold for humidity

Comparison of transmission reduction of various parameters over varying threshold

We compare the effectiveness of both algorithms by comparing the number of transmitted data at varying threshold. For simplicity and better visualization of the results, Figs. 7 and 8 will be illustrating the percentage of the aggregated sum of the data transmitted by the 54SNs for each attribute. The rate of transmitted data is less for both the parameters of around 10–20% in ELR technique as compared with the P-DPA method.

Comparison of Energy Consumption of various parameters over varying threshold

As demonstrated in Figs. 9 and 10, the energy overhead decreases with ELR algorithm as compared with P-DPA method while generating successful predictions.

The analysis and experiments are performed on the data sequence D. The transmission overhead is computed as follows:

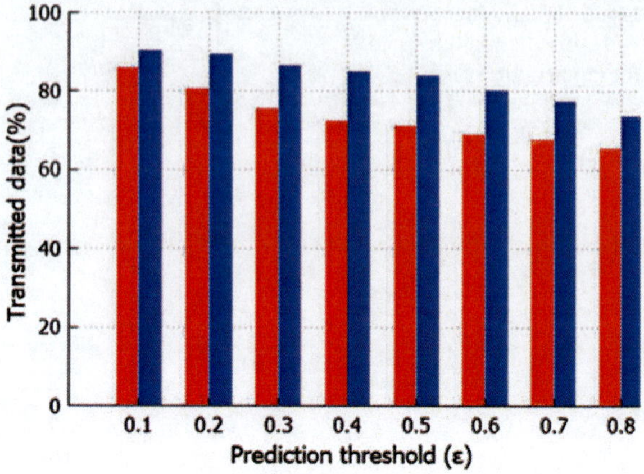

Fig. 7 Transmitted data versus prediction threshold for temperature

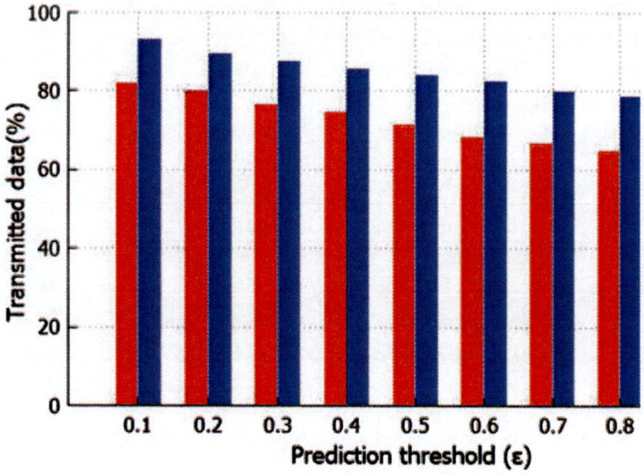

Fig. 8 Transmitted data versus prediction threshold for humidity

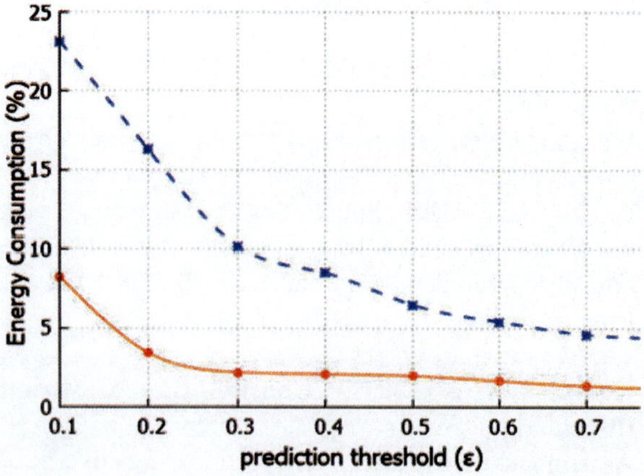

Fig. 9 Energy consumption versus prediction threshold for temperature

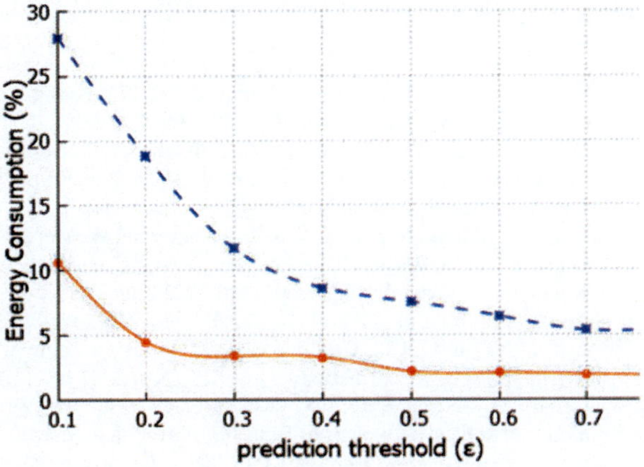

Fig. 10 Energy consumption versus prediction threshold for humidity

$$\text{Transmission Overhead} = 1 - \frac{\left(\sum_{i=0}^{t} x_{n+1}(t_{n+1})(i)\right)}{D} * 100 \quad (7)$$

where $\sum_{i=0}^{t} x_{n+1}(t_{n+1})(i)$ signifies the total number of successful predictions computed in time t over ten data collection cycles. It is experientially shown that the energy consumption at each time interval has been reduced. The results depict an improved performance of ELR technique with better energy savage than the P-DPA.

5 Conclusion and Future Work

In this paper, we proposed a novel prediction technique for reducing redundant data in a sensor network. In our prediction model, an extended version of linear regression is used. The ELR model makes first prediction based on first-order autoregression AR (1) algorithm, and then estimates the data change direction of current time-period and previous time-period owing to the resemblance in the space of curve of continuous data periods to adjust the prediction values. This framework will provide a simple yet effective solution to continuously monitor applications in WSNs as it will significantly reduce energy utilization by providing successful predictions while reducing data transmissions. The experimentation results based on real dataset highlights that the proposed algorithm delivers better performance. In the future, we will explore the cluster-based topology for reducing data in sensor network. Also will determine how such scalable architectures can decrease the energy utilization.

References

1. E. Osterweil, J. Polastre, M. Hamilton et al., Habitat monitoring with sensor networks. Commun. ACM Wirel. Sens. Netw. **47**, 34–40 (2004)
2. O.I.S. Ingelrest, G. Barrenetxea, G. Schaefer et al., SensorScope: application-specific sensor network for environmental monitoring. ACM Trans. Sens. Netw. **6**, 1–32 (2010). https://doi.org/10.1145/1689239.1689247
3. H. Liu, Z. Meng, S. Cui, A Wireless sensor network prototype for environmental monitoring in greenhouses. Wirel. Commun. Netw. Mob. Comput. 2344–2347
4. K. Gupta, V. Sikka, Design issues and challenges in wireless sensor networks. Int. J. Comput. Appl. **0975–8887**(112), 26–32 (2015). https://doi.org/10.5120/19656-1293
5. A. Liu, X. Jin, G. Cui, Deployment guidelines for achieving maximum lifetime and avoiding energy holes in sensor network. Inf. Sci. (Ny) **230**, 197–226 (2013)
6. K. Jain, A. Kumar, C.K. Jha, Probabilistic-based energy-efficient single-hop clustering technique for sensor networks, in Bansal J., Gupta M., Sharma H., Agarwal B. (eds) *Communication and Intelligent Systems*. ICCIS 2019. Lecture Notes in Networks and Systems, vol 120 (Springer, Singapore, 2020)
7. K. Bicakci, I.E. Bagci, B. Tavli, Communication/computation tradeoffs for prolonging network lifetime in wireless sensor networks: the case of digital signatures. Inf. Sci. Int. J. **188**, 44–63 (2012)
8. Y.-H. Zhu, W.-D. Wu, J. Pan, Y.-P. Tang, An energy-efficient data gathering algorithm to prolong lifetime of wireless sensor networks. Comput. Commun. **33**(5), 639–647 (2010)
9. A. Agarwal, K. Gupta, K. Yadav, A novel energy efficiency protocol for WSN based on optimal chain routing, in *IEEE Xplore 2016 International Conference on Computing for Sustainable Global Development (INDIACom)* (2016), pp. 488–493
10. K. Jain, A. Bhola, Data aggregation design goals for monitoring data in wireless sensor networks. J. Netw. Security Comput. Netw. **4**(3), 1–9 (2018)
11. K. Gupta, K. Yadav, Data collection method to improve energy efficiency in wireless sensor network, in *International Conference of Advance Research and Innovation (ICARI—2015)* (2015)
12. P. Edara, A. Limaye, K. Ramamritham, Asynchronous in-network prediction: efficient aggregation in sensor networks. ACM Trans. Sens. Netw. **4**, 25–34 (2008)
13. H. Jiang, S. Jin, C. Wang, Prediction or not? An energy-efficient framework for clustering-based data collection in wireless sensor networks. IEEE Trans. Parallel Distrib. Syst. **22**(6), 1064–1071 (2011)
14. S.K. Vuppala, A. Ghosh, K.A. Patil, A scalable WSN based data center monitoring solution with probabilistic event prediction, in *2012 IEEE 26th International Conference on Advanced Information Networking and Applications* (2012), pp. 446–453
15. J. Zhao, H. Liu, Z. Li, W. Li, Periodic data prediction algorithm in wireless sensor networks. Adv. Wirel. Sens. Netw. **334**, 695–701 (2013)
16. S. Samarah, A Data predication model for integrating wireless sensor networks and cloud computing. Procedia Comput. Sci. **52**, 1141–1146 (2015). https://doi.org/10.1016/j.procs.2015.05.148
17. Q. Liu, Y.Y. Zhang, J. Shen, B. Xiao, N. Linge, A WSN-based prediction model of microclimate in a greenhouse using an extreme learning approach. Adv. Commun. Technol. 133–137 (2015)
18. Z. Zhang, B. Deng, S. Chen L Li, An improved HMM model for sensing data predicting in WSN. Web-Age Inf. Manag. WAIM 2016 Lect. Notes Comput. Sci. **9658**:31–42 (2016)
19. G.M. Dias, B. Bellalta, S. Oechsner, A survey about prediction-based data reduction in wireless sensor networks. ACM Comput. Surv. **49** (2016)
20. A. Agarwal, A. Dev, A data prediction model based on extended cosine distance for maximizing network lifetime of WSN. WSEAS Trans. Comput. Res. **7**, 23–28 (2019)
21. S. Madden, Intel lab data (2004)

Image Enhancement Using Exposure and Standard Deviation-Based Sub-image Histogram Equalization for Night-time Images

Upendra Kumar Acharya and Sandeep Kumar

Abstract In this paper, a novel exposure and standard deviation-based sub-image histogram equalization technique is proposed for the enhancement of low-contrast nighttime images. Initially, the histogram of the input image is clipped to avoid the over-enhancement. The clipped histogram is partitioned into three sub-histograms depending on the exposure threshold and standard deviation values. After that, the individual sub-histogram is equalized independently. At last, a new enhanced image is produced after combining each equalized sub-images. The simulation results reveal that our proposed method outperforms over other histogram equalized techniques by providing a good visual quality image. The proposed method minimizes the entropy loss and preserves the brightness of the enhanced image efficiently by reducing the absolute mean brightness error (AMBE). It also maintains the structural similarity with the input image and controls the over-enhancement rate effectively.

Keywords Image enhancement · Exposure threshold · Standard deviation · Histogram equalization

1 Introduction

The method of improving the interpretability in images or improving the perception of information in images is known as image enhancement. So, to improve the quality of low-contrast images, many enhancement techniques have been developed. Histogram equalization (HE) [1] is one of the very popular enhancement techniques in spatial domain. But histogram equalization changes the brightness of an image to a large extent due to its flattening property. So, histogram equalization-based image subdivision methods like brightness preserving bi-histogram equalization (BBHE)

U. K. Acharya
Galgotias College of Engineering and Technology, Greater Noida, Uttar Pradesh, India

U. K. Acharya · S. Kumar (✉)
National Institute of Technology, New Delhi, Delhi, India
e-mail: sandeep@nitdelhi.ac.in

P. Bansal et al. (eds.), *Proceedings of International Conference on Artificial Intelligence and Applications*, Advances in Intelligent Systems and Computing 1164,
https://doi.org/10.1007/978-981-15-4992-2_57

[2] and equal area dualistic sub-image histogram equalization (DSIHE) [3] have been developed to preserve the brightness and enhance the contrast. To enhance the image as well as to preserve more brightness for the application of consumer electronics, recursive mean separate histogram equalization (RMSHE) [4], recursive sub-image histogram equalization (RSIHE) [5], recursively separated and weighted histogram equalization (RSWHE) [6] have been designed.

To enhance the low-exposure images, Singh et al. presented two new enhancement techniques named as recursive exposure-based sub-image histogram equalization (R-ESIHE) and recursively separated exposure-based sub-image histogram equalization (RSESIHE) [7]. M. Kanmani proposed a new technique [8] for improving the visual quality and entropy of the grayscale image using particle swarm optimization algorithm and adaptive gamma correction technique. In 2018, Abhisek Paul et al. suggested a new method known as plateau limit-based tri-histogram equalization for image enhancement [9]. This approach partitioned the clipped histogram into three sub-histograms based on standard deviation, and then, sub-histogram equalization is performed. To improve the contrast and maximize the entropy of satellite images, another enhancement technique has been designed [10], which is based on gamma correction and histogram equalization. In 2019, a new contrast enhancement technique [11] is introduced which is named as triple clipped dynamic histogram equalization based on standard deviation (TCDHE-SD). In TCDHE-SD, each histogram is clipped separately, and then, corresponding sub-image followed the method of histogram equalization. For illumination boosting of nighttime images, a new approach [12] is designed by Zohair, which preserves the brightness in bright area and boosts the brightness in low-exposure area.

In this paper, our proposed method gives more focus to meet the challenges like reducing the information loss, preserving the brightness and controlling the enhancement rate. So, a new exposure and standard deviation-based image subdivision framework is proposed for harvesting more information content and preserving brightness. The problem of over-enhancement is eradicated by using clipping technique. The comparative study with other conventional algorithms is also reported using quality metrics like entropy, AMBE and SSIM.

Other parts of this paper are arranged as follows. Details about the proposed method are described in Sect. 2. Results and discussions are illustrated in Sect. 3. The paper is concluded in Sect. 4.

2 Proposed Method

This section describes the flow process of our proposed method. It is based on image subdivision and histogram equalization (HE). By the method of histogram equalization, information loss is more, and it causes shifting of mean brightness. So, the AMBE between input and enhanced image is more in histogram equalization technique. It produces over-enhanced image. So, the proposed method is illustrated in the following steps to overcome the problems like over-enhancement, artifacts, information loss, shifting the mean brightness. Initially, the exposure threshold and standard deviation of the image are calculated. Then, histogram is clipped using clipping threshold. The clipped histogram is divided into three sub-histograms based on threshold parameters. Then, each sub-histogram is mapped to a new dynamic range and equalized independently.

2.1 Exposure Threshold and Standard Deviation Calculation

In this sub-section, exposure of the image is calculated to find whether it is a low-exposure or high-exposure image. Then, exposure threshold [7] is computed to find the boundary between overexposed and underexposed sub-images. Image exposure (ex) value and exposure threshold (T_x) are calculated in (1–2).

$$\text{ex} = \frac{\sum_{k=0}^{L-1} H(k) \times k}{L \sum_{k=0}^{L-1} H(k)} \tag{1}$$

$$T_x = L(1 - \text{ex}) \tag{2}$$

In (2), the variable L represents the maximum gray level, and $H(k)$ represents the histogram of the input image. Then, the standard deviation [9] of the image is calculated using (3)

$$s = \left(\frac{\sum_{k=0}^{L-1} (k - \mu)^2 \times H(k)}{\sum_{k=0}^{L-1} H(k)} \right)^{1/2} \tag{3}$$

where μ represents the mean [9] of the image, and it is calculated in (4)

$$\mu = \frac{\sum_{k=0}^{L-1} k \times H(k)}{\sum_{k=0}^{L-1} H(k)} \tag{4}$$

2.2 Histogram Clipping

In this sub-section, histogram of the original image is clipped using clipped threshold. The main purpose of clipping the histogram is to avoid over-enhancement. The clipped threshold (C_t) is computed by taking the average of median and mean intensity value of the images.

$$C_t = \frac{I_{\text{median}} + I_{\text{mean}}}{2} \tag{5}$$

The clipped histogram ($H_{ch}(k)$) is represented as,

$$H_{ch}(k) = \begin{cases} C_t, & \text{if } H(k) \geq C_t \\ H(k), & \text{if } H(k) < C_t \end{cases} \quad (6)$$

2.3 Image Subdivision and Equalization

The clipped histogram is divided into three sub-histograms such as lower histogram (H_l), middle histogram (H_m) and upper histogram (H_u). The division of histogram is based on threshold parameters X_{ls} and X_{us}.The threshold parameters X_{ls} and X_{us} are calculated using (7) and (8).

$$X_{ls} = T_x - s \quad (7)$$

$$X_{us} = T_x + s \quad (8)$$

The range of intensity level (k) of these three histograms H_l, H_m and H_u is 0 to $X_{ls} - 1$, X_{ls} to $X_{us} - 1$ and X_{us} to $L - 1$, respectively. Then, histogram equalization is applied to each sub-histogram using cumulative density function (CDF). Probability density function of each sub-histogram is calculated, using (9)

$$P_i(k) = \frac{H_{ch}(k)}{N_i} \quad (9)$$

Cumulative density function of each sub-histogram is calculated, using (10)

$$C_i(k) = \sum_k P_i(k) \quad (10)$$

Mapping function for each sub-histogram (11)

$$f_i = X_l + (X_u - X_l)C_i(k) \quad (11)$$

Here, X_l and X_u represent the minimum and maximum intensity level for corresponding sub-histograms. Then, the mapping function for the enhanced image is represented in (12)

$$f = \begin{cases} f_1, \text{ for } 0 \leq k \leq X_{ls} - 1 \\ f_2, \text{ for } X_{ls} \leq k \leq X_{us} - 1 \\ f_3, \text{ for } X_{us} \leq k \leq L - 1 \end{cases} \quad (12)$$

In (12), the functions f_1, f_2 and f_3 represent the mapping functions of lower histogram, middle histogram and upper histogram, respectively.

3 Results and Discussion

In this section, the performance of the proposed method is evaluated and demonstrated. The comparative analyses are carried out between the proposed method and the existing image enhancement techniques such as HE [1], DSIHE [3], RSESIHE and R-ESIHE [7] by taking some of the nighttime images from database [13]. MATLAB 2018 on Windows 7 operating system has been used for the experimental work. The performance of the proposed method is evaluated in two ways, qualitative and quantitative approach.

3.1 Qualitative Evaluation

The qualitative evaluation includes visual quality, over-enhancement, brightness preservation and natural appearance of the images. Here, Figs. 1, 2 and 3 represents the input images and enhanced images produced by different image enhancement techniques. From Figs. 1b, 2b and 3b, it has been observed that the method of HE enhances the visual quality of the image but produces an over-enhanced image. More artifacts and noise amplification are also noticed in histogram equalized images. Figure 1c, 2c and 3c represents the DSIHE-based enhanced image. From these figures, it has been observed that DSIHE technique preserves more brightness as compared to HE but produces some artifacts. Information loss is more in HE-,

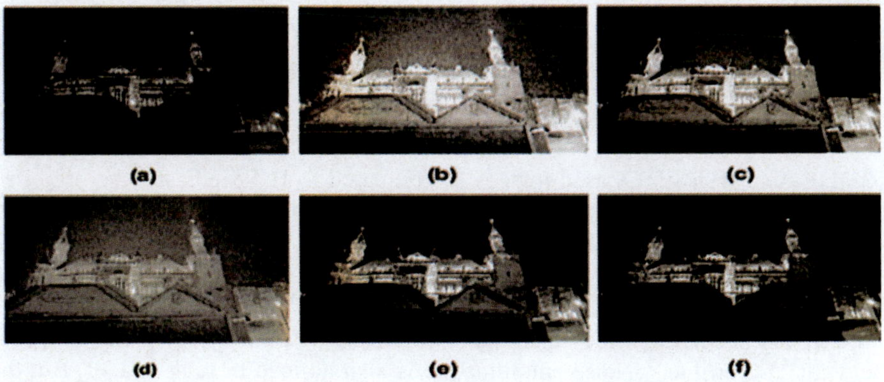

Fig. 1 Simulation result of image-1, **a** input image, **b** HE, **c** DSIHE, **d** RSESIHE, **e** R-ESIHE, **f** proposed method

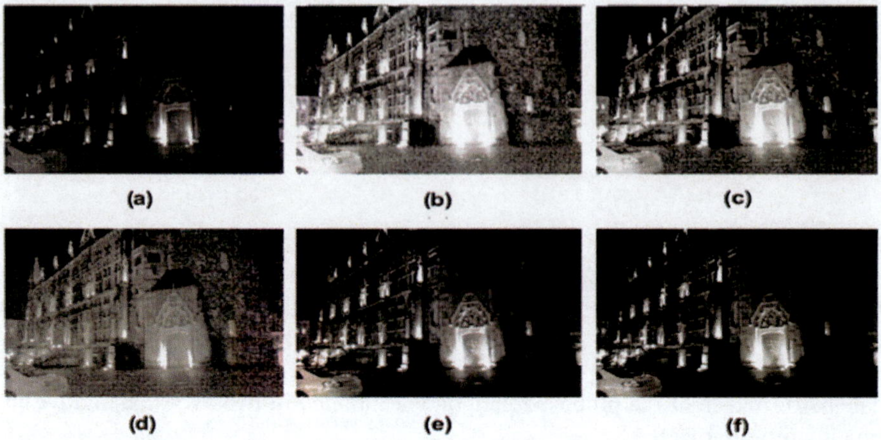

Fig. 2 Simulation result of image-2, a input image, b HE, c DSIHE, d RSESIHE, e R-ESIHE, f proposed method

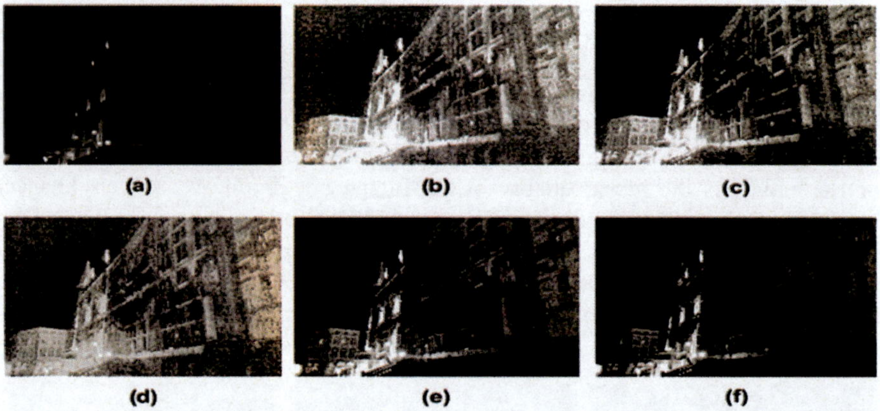

Fig. 3 Simulation result of image-3, a input image, b HE, c DSIHE, d RSESIHE, e R-ESIHE, f proposed method

DSIHE- and RSESIHE-based images as compared to R-ESIHE and the proposed method images. There is no controlling element used in HE and DSIHE to control the over-enhancement rate. But in RSESIHE, R-ESIHE and proposed method, over-enhancement is controlled by clipping the histogram. Figures 1d, 2d and 3d represents the enhanced image produced by RSESIHE technique. From this figure, it has been noticed that the visual quality of RSESIHE-based image is very poor and contains artifacts. Noise amplification is also noticed in such images. But the information loss is less in RSESIHE-based images as compared to HE-based images but produces blurred images. Figures 1e, 2e and 3e represents the R-ESIHE-based images. The R-ESIHE-based enhanced images are free from over-enhancement but contain some artifacts. Figures 1f, 2f and 3f represents the enhanced image produced by our proposed method. From these figures, it has been noticed that the proposed method is completely free from such over-enhancement and artifacts. It enhances the visual quality of the image without noise amplification and gives a natural look to the enhanced images. Mean brightness is also preserved in the proposed method. The natural look and maximization of entropy are only possible due to the novel way of subdivision of the histogram.

3.2 Quantitative Evaluation

Besides visual analyses, quantitative evaluation is also very important to measure the performance of the different image enhancement techniques. The quantitative evaluation includes entropy, AMBE and structural similarity index measure (SSIM).

3.2.1 Entropy

The average information content of the image is known as entropy. A contrast enhancement is said to be the appropriate one, if it enhances the natural contrast without any information loss. The larger value of entropy represents the more information content of the image. The entropy [7] is measured using the following Eq. (13).

$$E = -\sum_{k=0}^{L-1} \mathrm{PDF}(k) \times \log_2 \mathrm{PDF}(k) \tag{13}$$

In (13), PDF(k) represents the probability density function of the image at intensity level 0 *to* $L - 1$. As per Table 1, the entropy obtained by our proposed method is higher than the other existing methods for all images, and this entropy value is very close to the entropy of the input image.

Table 1 Performance evaluation of different enhancement techniques based on entropy

Image/methods	Input image	HE	DSIHE	RSESIHE	R-ESIHE	Proposed method
Image-1	5.9206	5.0331	5.6431	5.564	5.633	5.7935
Image-2	5.9175	5.1198	5.7297	5.6137	5.794	5.8005
Image-3	4.9419	4.5322	4.7965	4.7723	4.7718	4.8772
Average	5.5933	4.8950	5.3897	5.316,667	5.3996	5.4904

Table 2 Performance evaluation of enhancement techniques based on AMBE

Image/methods	HE	DSIHE	RSESIHE	R-ESIHE	Proposed method
Image-1	99.034	48.256	76.688	27.822	14.932
Image-2	98.427	52.593	75.154	29.101	14.709
Image-3	113.04	66.048	105.16	48.828	16.072
Average	103.5003	55.632	85.667	35.2503	15.237

3.3 AMBE

AMBE is used to measure the change in brightness between input and enhanced image. The AMBE value is calculated using (14)

$$\text{AMBE} = \left|X_\mu - Y_\mu\right| \tag{14}$$

In (14), X_μ, Y_μ are the mean brightness of the input and enhanced image, respectively. The lower value of AMBE indicates that the mean brightness of the enhanced image is closed to mean brightness of input image. From Table 2, it is observed that the proposed method preserved more brightness as compared to other methods.

3.3.1 SSIM

SSIM is used to compare the structural similarity between input and enhanced image. The SSIM is measured using (15)

$$\text{SSIM} = \frac{(2\mu_x\mu_y + k_1)(2\sigma_{xy} + k_2)}{(\mu_x^2 + \mu_y^2 + C_1)(\sigma_x^2 + \sigma_y^2 + C_2)} \tag{15}$$

In (15), $\mu_x, \mu_y, \sigma_x, \sigma_y$ represent the local mean and standard deviation of input and enhanced image. From Table 3, it has been observed that our proposed method produces the enhanced image with better structural similarity with input image as compared to other methods.

Table 3 Performance evaluation of different enhancement techniques based on SSIM

Image/methods	HE	DSIHE	RSESIHE	R-ESIHE	Proposed method
Image-1	0.15669	0.47962	0.19739	0.58823	0.8819
Image-2	0.18109	0.40351	0.23728	0.64569	0.8632
Image-3	0.065189	0.25741	0.0752	0.22989	0.7050
Average	0.1343	0.3801	0.1699	0.4879	0.8167

4 Conclusions

In this paper, a robust image enhancement technique is proposed to achieve multiple objectives. The exposure threshold and standard deviation-based histogram partition took the important role to maximize the entropy and brightness preservation. The enhancement rate of the proposed method is controlled by clipping technique. The experimental results show that our proposed method performs superior over the other existing methods in terms of visual quality, entropy maximization, SSIM, brightness preservation and controlling the enhancement rate.

References

1. R.C. Gonzalez, E.W. Richard, *Digital Image Processing*, 3rd edn. (Prentice Hall Press, Upper Saddle River, NJ, USA, 2002)
2. Y.T. Kim, Contrast enhancement using brightness preserving bi-histogram equalization. IEEE Trans. Consum. Electron. 1–8 (1997)
3. Y. Wang, Q. Chen, B. Zhang, Image enhancement based on equal area dualistic sub-image histogram equalization method. IEEE Trans. Consum. Electron. 68–75 (1999)
4. S.D. Chen, A.R. Ramli, Contrast enhancement using recursive mean-separate histogram equalization for scalable brightness preservation. IEEE Trans. Consum. Electron. 1301–1309 (2003)
5. K.S. Sim, C.P. Tso, Y.Y. Tan, Recursive sub-image histogram equalization applied to gray scale images. Pattern Recog. Lett. 1209–1221 (2007)
6. M. Kim, M.G. Chung, Recursively separated and weighted histogram equalization for brightness preservation and contrast enhancement. IEEE Trans. Consum. Electron. 1389–1397 (2008)
7. K. Singh, R. Kapoor, S.K. Sinha, Enhancement of low exposure images via recursive histogram equalization algorithms. Optik 2619–2625 (2015)
8. M. Kanmani, N. Venkateswaran, An image contrast enhancement algorithm for grayscale images using particle swarm optimization. Multimedia Tools Appl. 23371–23387 (2018)
9. A. Paul, P. Bhattacharya, S.P. Maity, B.K. Bhattacharyya, Plateau limit-based tri-histogram equalization for image enhancement. IET Image Process. 1617–1625 (2018)
10. H. Singh, A. Kumar, L.K. Balyan, G.K. Singh, Swarm intelligence optimized piecewise gamma corrected histogram equalization for dark image enhancement. Comput. Electr. Eng. 462–475
11. M. Zarie, A. Pourmohammad, H. Hajghassem, Image contrast enhancement using triple clipped dynamic histogram equalization based on standard deviation. IET Image Process. 1081–1089 (2019)
12. Z. Al-Ameen, Nighttime image enhancement using a new illumination boost algorithm. IET Image Process. (2019)
13. Image database: www.visuallocalization.net

Author Index

P. Bansal et al. (eds.), *Proceedings of International Conference on Artificial Intelligence and Applications*, Advances in Intelligent Systems and Computing 1164,
https://doi.org/10.1007/978-981-15-4992-2